Oceanography and Marine Science

Oceanography and Marine Science

Edited by **Theodore Roa**

SYRAWOOD
PUBLISHING HOUSE
New York

Published by Syrawood Publishing House,
750 Third Avenue, 9th Floor,
New York, NY 10017, USA
www.syrawoodpublishinghouse.com

Oceanography and Marine Science
Edited by Theodore Roa

International Standard Book Number: 978-1-68286-334-3 (Hardback)

Printed in China.

Contents

Preface

This book covers various theories, concepts and applications of oceanography and marine science. It aims to shed light on some of the unexplored aspects of marine life and the recent researches in the fields of ocean dynamics, marine geology and plate tectonics, aquaculture and ocean engineering, etc. The various sub-fields of oceanography along with technological advancements that have future implications for further research and exploration are glanced at in this book. This text aims to provide a broad overview of the oceanic surface structures and diverse processes that occur on it. It includes contributions of experts and scientists which will provide innovative insights into this field.

The information contained in this book is the result of intensive hard work done by researchers in this field. All due efforts have been made to make this book serve as a complete guiding source for students and researchers. The topics in this book have been comprehensively explained to help readers understand the growing trends in the field.

I would like to thank the entire group of writers who made sincere efforts in this book and my family who supported me in my efforts of working on this book. I take this opportunity to thank all those who have been a guiding force throughout my life.

Editor

Isolation and identification of *Streptomyces* ST1 and ST2 strains from Tsunami affected soils: Morphological and biochemical studies

Suneetha V.[1]*, Karthick Raj[2] and Prathusha .K.[2]

[1]School of Bio Sciences and Technology (SBST), Vellore, Tamil Nadu, India
[2]Vellore Institute of Technology University (VIT), Vellore-632014, Tamil Nadu, India.

Tsunamis are one of the natural disasters that are a series of long oceanic waves. A large volume of water gets displaced from the ocean because of undersea earthquakes or land slides or volcanic eruptions or extra terrestrial collisions. Our current research focuses on screening and characterization of microbes in water, soil and other extraneous material from the Tsunami affected areas along with substrates. Our present research work rests with *Actinomycetes* spp. possessing wide diversity of Genera. The morphological and biochemical characteristics of two *Streptomyces* spp. were identified from collected samples at different places of Pondicherry and Chennai. The screening methods, biochemical characterization and tentative identification methods will be discussed.

Key words: Tsunami, *Actinomycetes* spp., substrates, *Streptomyces*.

INTRODUCTION

The genus *Streptomyces* is represented in nature by the largest number of species and varieties among the family Actinomycetaceae. They differ greatly in their morphology, physiology and biochemical activities, producing the majority of known enzymes (Suneetha and Zaved, 2011). Among the genera of Actinomycetes, the genus streptomyces is represented in nature by the largest number of species and varieties, which differ greatly in their morphology, physiology and biochemical activities. Interestingly, the majority of the antibiotic-producing Actinomycetes are found among those species, which led to a growth in economic importance for this group of organisms. Enrichment procedures always gives wonderful organisms which have potential applications in present world. In the 2004 Tsunami crisis blown off the south east coast of India. States of Tamil Nadu and the Union Territory of Puducherry were predominantly affected. Sea water entered into the cultivable lands increasing the soil pH notably from 6.5 to 7.5 to 8.5 inducing huge marked damage to the Indian agriculture (Dietz, 1986; Bentley et al., 2002) (Figure 1). In this study we screened various soil samples for screening of Actinomycetes from various Tsunamis affected areas.

The *Actinomycetes* are a morphologically, physiologically and ecologically diverse group of bacteria. Absolute majority of antibiotic-producers are encountered among these species making them economically potent. Keeping in view the significant contribution of *Actinomycetes* in the areas of soil ecology and industrial exploitation, an attempt has been made to isolate and identify potential *Actinomycetes* from Chennai and Kalapet (Puducherry). Two most dominant *Streptomyces* strains from Tsunami soil samples were screened and further characterized for micro morphology, cultural characterization followed by physiological and biochemical characterization. Based on these characteristics, the two dominant *Actinomycetes* strains were identified as *Streptomyces* ST1 and ST2 which could be further utilized for Biodegradation of Industrial effluents and microbial biodiversity studies. *Streptomyces*, the wonderful pigmented organisms bearing fungal characteristics but still are deliberated to be prokaryotic by their cellular status (Gulyas, 1978). Few species are pathogenic to both humans and plants. This group is an excellent reservoir for maximum number of antibiotics with commercial application (Bibb, 1996). The classification of the Prokaryotes is a complex issue and takes into consideration different characters.

*Corresponding author. E-mail: vsuneetha@vit.ac.in.

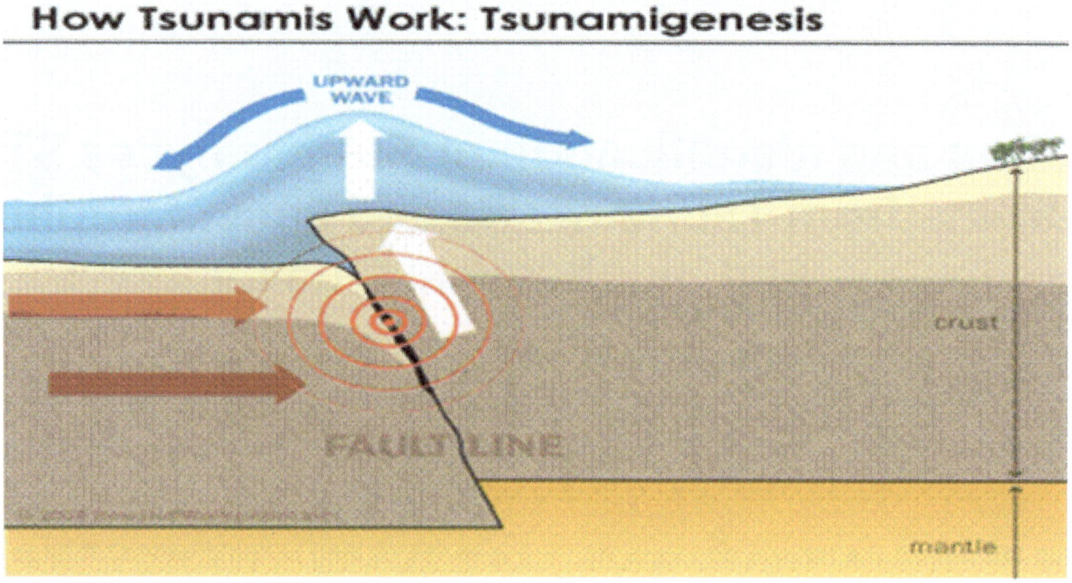

Figure 1. Tsunamigenesis.

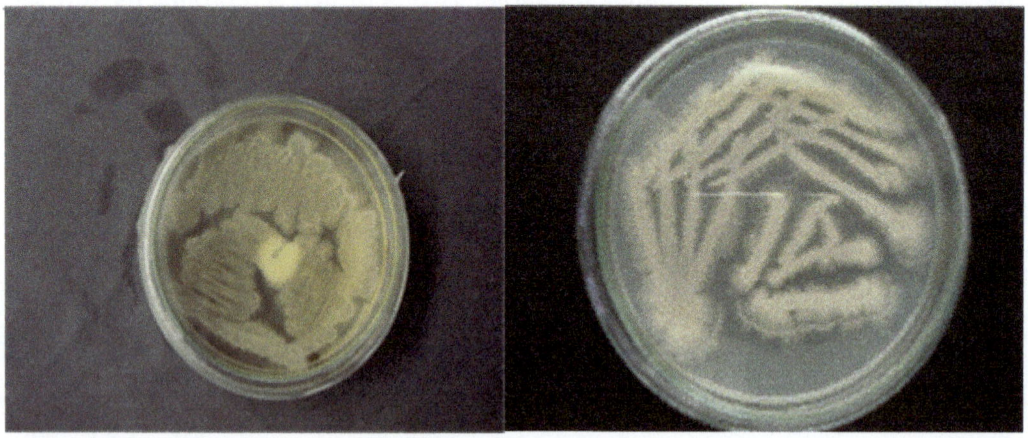

Figure 2. Screened *Actinomycetes* on Petri plates.

Major subdivisions account to bacteria (Schizophyta), the blue green algae (Cyanophyta) and the Actino-mycetales (Teddi et al., 2006, 1999; Suneetha and Zaved, 2011).

MATERIALS AND METHODS

Media/Buffers and Reagents Employed

Basal medium

KH_2PO_4	1.5
$ZnSO_4$	0.5
$MgSO_4 .7 H_2O$	0.5
Glycerol	10 mM
$CaCl_2$	0.25
$(NH_4)_2SO_4$	10.0
$FeSO_4$	0.015
Agar	20.0

pH of the medium was adjusted to 8.3 before sterilization. Cyclohexamide (50 µg/ml) was added to the media to facilitate selective isolation of Actinomycetes (Malviya et al., 1992) (g/L).

Basal media without glycerol

Same as aforementioned without glycerol.

Starch casein agar medium (g/L)

Starch	10.0
Casein (vitamin free)	0.3
KNO_3	2.0

Table 1. Identification of the cultures.

Characters	Characteristics of ST1	Characteristics of ST2
Spore chain morphology	Rectiflexibilis	Straight to flexuous
Substrate myclia	Yellow to Green in colour	White in colour
Colour of spore mass	Dark green colour	White in colour
Diffusible pigment	No pigments	No pigments
Melanin pigment formation	+	+

NaCl	2.0
K_2HPO_4	2.0
$MgSO_4$ 7H $_2O$	0.05
$CaSO_4$	0.02
$FeSO_4$	0.01
Agar	20.0

pH of the medium was adjusted to 8.2 before sterilization.

Production medium

Galactose	4.0
K_2HPO_4	1.5
$FeSO_4$ $7H_2O$	0.015
$MgSO_47H_2O$	0.05
$(NH_4)_2SO_4$	5.0
$ZnSO_4.7H_2O$	0.005
$CaC1_2$	0.25

pH was adjusted to 7.8 before sterilization (Mukhopadhyay and Chandra, 1990) (g/L).

CSPY-ME medium (g/L)

K_2HPO_4	0.5
Casein	3.0
Maize starch	10.0
Peptone	1.0
Yeast extract	1.0
Malt extract	10.0
Agar	20.0

pH of the media was adjusted to 7.5.

***Yeast extract Malt extract medium* (g/100 ml)**

Yeast extract	0.5
Malt extract	1.0
Glucose	1.0
Agar	2.0

pH was adjusted to 7.8 before sterilization. This media was used to maintain Actinomycetes cultures.

***Ringers solution* (g/litre)**

NaCl	2.15
$CaCl_2$ (anhy)	0.12
$Na_2S_2O_3$ 5 H_2O	0.5
KCl	0.075

pH was adjusted to 6.6 before sterilization.

Screening of Actinomycetes from Tsunami soil samples

Soil is a natural source for the isolation of microbes. For the isolation of Actinomycetes, soil samples were collected from different places of Tsunami affected areas. The superficial loose soil layers were removed and the underlying soil sample was dug. The samples were collected in polythene bags, labeled and the collected soil samples were transported to the laboratory and were screened for Actinomycetes (Suneetha and Lakshmi, 2006).

Cultural and biochemical screening

Cultural characterization followed for Actinomycetes isolates were identified up to the generic level (tentatively) by studying morphological, physiological and ecological characters. Morphological characters were studied by observing the growth on cover slip under (10X and 45X lens). Bio-chemical characterization was done by using different tests like spore chain morphology, substrate mycelia, glucose, mannitol, xylose, lactose, starch, saccharose and l-Ramose (Mahr et al., 2000) (Tables 1 - 2 and Figure 5).

Genus identification and morphological characteristics

Morphological studies were carried out by microscopic observation and studies on growth characteristics in Petri dishes. Genus identification and morphological characteristics were observed. Visual observation of both morphological and microscopic characteristics using light microscopy, acid fastness and gram-staining properties were performed. All morphological characters were observed on CSPY-ME and were used for classification and differentiation as follows:

Aerial mass color: The mass color of the mature sporulating aerial mycelium was observed following growth on the CSPY-ME plates. The aerial mass was classified according to the Bergey's manual of systemic bacteriology (Locci, 1989) in the following color series: grey (Gy), white (W).

Substrate mycelium: Distinctive colors of the substrate mycelium were recorded. The observed colors were: yellow (Y), green (G) and white (W).

Melanoid pigments: The production of melanoid pigments was also considered. It was reported as a positive result and the production of melanoid pigments was clear.

Spore chain morphology: According to the shape of the spore chains observed under light microscopy, the isolates were grouped as follows: Rectus-Flexibilis (RF) and spores in straight flexous chains (Table 2 and Figure 5).

Table 2. Physiological characteristics of growth temperature.

Characters	Characteristics of ST1	Characteristics of ST2
4°C	-	-
37°C	+	+
45°C	+	-
Use of carbon source (1% w/v)	+	+
Glucose	+	+
Mannitol	+	+
Xylose	+	+
Lactose	+	+
Starch	+	+
Saccharose	-	-
L-Rhamnose	-	-

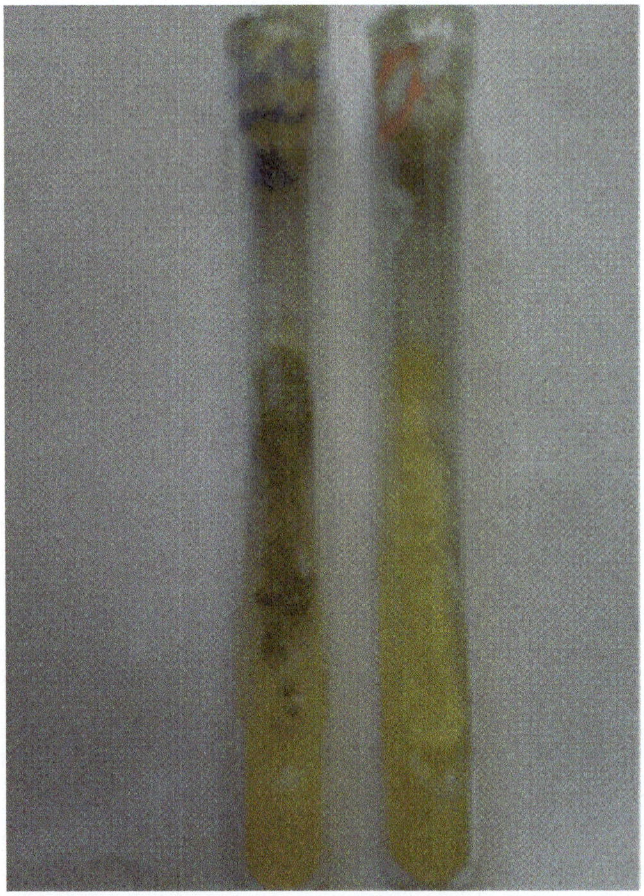

Figure 3. *Actinomycetes* on slants.

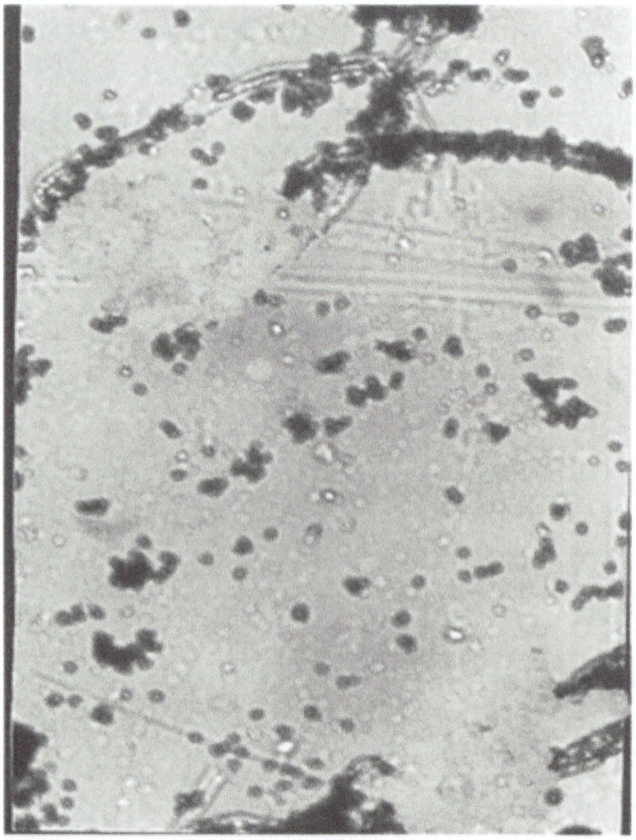

Figure 4. Micromorphology of *Actinomycetes*.

Biochemical screening: Physiological criteria such as ability to degrade casein, tyrosine and xanthine as substrates by the various *Streptomyces* strains were used for genus confirmation. The utilization of different carbon sources and degradation of amino acids as well as the production of melanine and utilization of urea were studied in order to reach a possible classification to species. (Shirling and Gottileb, 1966; Lechevalier, 1968; Korn-Wendisch and kutzner, 1991). Figure 4 and 5.

RESULTS AND DISCUSSION

Most of the colonies that grew on CSPY-ME plates belong to the genus S*treptomyces* since the colonies were slow growing, aerobic, glabrous (Figure 1-3) or chalky, heaped, folded and with aerial and substrate mycelia of different colors. In addition, all colonies possessed an earthy

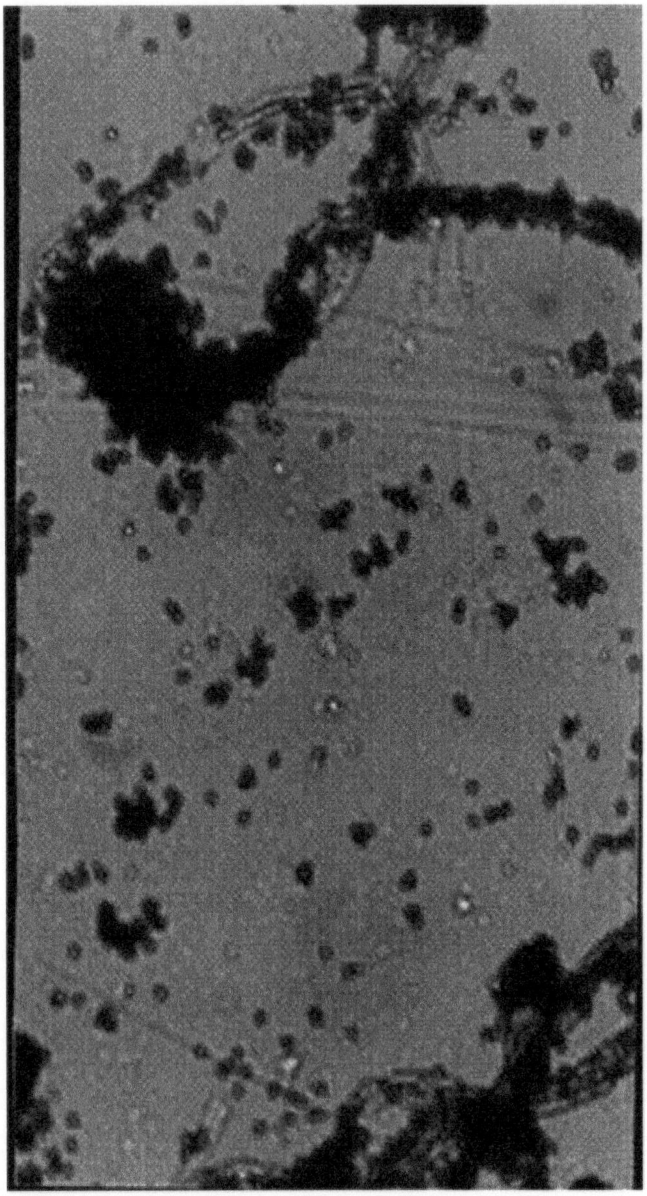

Figure 5. Micromorphology of *Actinomycetes*.

odour. Some of them also produced antibiotics and enzymes (Suneetha, 2011) as reflected by zones of growth inhibition among other inhabitants of soil samples. A confirmatory identification to genus was based on acid-fastness, gram-stain and degradation of casein, tyrosine and xanthine. All *streptomyces* strains were acid-fast negative and gram-stain positive. The degradation of substrates casein, tyrosine and xanthine was variable according to each isolate. Microscopically, it was observed that the morphology of the sore chains varied depending on the species, showing the exact straight and flexuous forms, hooks, open loops and coils, which were used, among other features, to establish differences between them (Table 1 and Figure 5). It was also observed that some of the strains produced diffusible melanoid pigments in the surrounding medium. In accordance with the aerial mycelium color series established in the Bergey's manual of determinatives bacteriology (Buchanan and Gibbons, 1974) and in the category 4 of the Bergeys manual of systemic bacteriology, the isolates could be grouped in the following series: grey and white.

Various colors of mycelia were also observed, the shades beige, yellow and ivory being the most pre dominant. Few of them had substrate mycelia violet-purple or red-violet. As per the utilization of the several carbohydrates, it was noted that strain ST1 were able to use all of the tested carbon sources, while the ST2 were able to use cellulose as carbon source.

REFERENCES

Bentley SD, Chater KF, Cerdeno-Tarraga AM, Challis GL, Thomson NR, James KD, Harris DE, Quail MA, Kieser H, Harper D, Bateman A, Brown S, Chandra G, Chen CW, Collins M, Cronin A, Fraser A, Goble A, Hidalgo J, Hornsby T, Howarth S, Huang CH, Kieser T, Larke L, Murphy L, Oliver K, O'Neil S, Rabbinowitsch E, Rajandream MA, Rutherford K, Rutter S, Seeger K, Saunders D, Sharp S, Squares R, Squares S, Taylor K, Warren T, Wietzorrek A, Woodward J, Barrell BG, Parkhill J, Hopwood DA (2002). Complete genome sequence of the model actinomycete *Streptomyces coelicolor A3 (2).* Nature 417: 141-147.

Bibb MJ (1996). The regulation of antibiotic production in *Streptomycetes.* Microbiology 142: 1335-1344.

Buchanan RE, Gibbons NE (1974).Bergey's Manual of Determinative bacteriology, eight ed. Williams and Wilkins, Baltimore,pp.747-829.

Dietz A (1986). Structure and taxonomy of *Streptomyces.* In: Queener an, st., Day. L. E. (Eds.), The Bacteria, vol. 9: Antibiotic-producing *Streptomyces.* Academic press, Orlando, pp. 1-25.

Gulyas F (1978). "Studies of pigment formation by *Actinomycetes,*" Chem. Abstracts 89 (15): 294 abstract 125900n & Soil Biol. Conserv. Biosphere (Proc. Meet.) 7th 1975 (Pub. 1977), 265-270.

Lechevalier MP (1968). Identification of aerobic actinomycetes of clinical importance. J. Lab. Clin. Med., pp. 71, 934-944.

Locci R (1989). *Steptomyces* and related genera. In: Williams, s.t., sharpe, m.e., Holt, J.G (Eds.), Bergey's Manual of systemic Bacteriology. Williams and Wilkins, Baltimore, pp. 2451-2493.

Mahr K, van Wezel GP, Svensson C, Krengel U, Bibb MJ, Titgemeyer F (2000). Glucose kinase of *Streptomyces coelicolor A3* (2): Large-scale purification and biochemical analysis. Antonie van Leeuwenhoek 78: 253-261.

Shirling EB, Gottlieb D (1966). Methods for characterization of *Streptomyces* species. Int. J. Syst. Bacteriol., 16, 313-340.

Suneetha (2011). Actinomycetes: Sources for Soil Enzymes Soil Enzymology, Soil Biology-22, G. Shukla and A. Varma (eds) Springer-Verlag BerlinHeidelberg 2011, pp. 259-269.

Suneetha V, Lakshmi VV (2002). Fermentative production of Kerazyme with isolates from Tirumala hills, Biotechnology in Agriculture, Industry & environment. pp. 198-203.

Suneetha V, Lakshmi VV (2004). Identification of collagenase producing Actinomycetes. Asian J. Microbio. Biotech. Environ. Sci., 2005, 7, 53-58.

Suneetha V, Lakshmi VV (2006). Isolation and characterization of keratinase producing organisms from Tirumala and Tirupati, Thesis awarded. pp. 45-49.

Suneetha V, Zaved AK (2011). Screening, Characterisation and Optimization of Microbial Pectinase, Soil Enzymology, Soil Biology-22, G. Shukla and A. Varma (eds) Springer-Verlag Berlin Heidelberg 2011. pp. 329-337.

Taddei A, Jose RM, Marquez-Vilchez E, Castelli C (2006). "Isolation and identification of *Streptomyces* spp. From Venezuelan soils:

Morphological and biochemical studies. Microbial Research. 161: 222-231.
Taddei A, Tremarias MM, Hartung de capriles C (1999). Viability studies on actinomycetes. Mycopathologia 143, 161-164.

2

Mortality and survival of eggs and chicks of Western Reef Heron *Egretta gularis* on Nakhilo Island, Persian Gulf, 2008

Behrouz Behrouzi-Rad[1]*, Rahman Rasekhi[2] and Sima Momeni[1]

[1]Islamic Azad University, Science and Research Branch, Kuzestan, Iran.
[2]Shaid Chamran University, Ahwaz, Iran.

In a study on breeding processes of Western Reef Heron *Egretta gularis* carried out from March to September 2008 on Nakhiloo Island, we estimated the breeding population, mortality, natality and the breeding success of this species. The breeding success during the incubation, nestling, post-nestling and overall stages were 62.1, 46.4, 39.2 and 49.2%, respectively. Eating of the eggs by Black Rat *Rattus rattus*, high temperature and leaving of nests without cause were more important reasons of their mortality. The average weight of pulli at birth was 22.2±0.61 and at the flying stage was 590±10 g.

Key words: Breeding, *Egretta gularis*, mortality, survival, Nakhiloo Island, Persian Gulf, Iran.

INTRODUCTION

The Western Reef Heron *Egretta gularis* occurs mainly on the coasts in tropical West Africa, the Red Sea, the Persian Gulf and east to India (Birdlife International, 2009). The medium-sized heron inhabits coastal areas, mainly on rocky or sandy shores and reefs, but is also found around estuaries, mudflats, salt marshes, mangroves, tidal creeks and lagoons (del Hoyo et al., 1996). It feeds diurnally but also at night, depending on the tides, and roosts at night in large numbers of between 500 and 1,000 individuals, in mangroves or on rocky cliffs and islets. The species shows a preference for rocky or sandy shores and reefs (IUCN, 2009; Grewal et al., 2002; del Hoyo et al., 1996). The diet of this species consists mainly of fish, crustaceans and molluscs, but crickets, grubs and earthworms are also taken (Grewal et al., 2002).

During the breeding season, both forms develop red lores and distinctive plumes on their head, chest and back. The legs are dark while the feet are bright yellow, except during their courtship when they turn pinkish red. Sexes are similar. Generally silent, but when disturbed gives a guttural "kawww" (Cramp and Simmons, 1977; Voisin 1991). It is a monogamous and colonial breeder. Nesting is done in colonies. The clutch is two to four greenish blue eggs. Incubation takes about 25 days and it is done by both parents. The young leaves the nest when they are about 30 days old. Out of breeding season is seen alone or in pairs (Milon, 1946). Two to five pale green to blue-white eggs are laid on an untidy platform of sticks built by both sexes, located in a tree, on a rock ledge, under shrubs, or on the ground. Eggs are incubated for 20–25 days by both sexes. Semi-altricial young stay in nest about 30 days (Grewal et al., 2002).

In the Persian Gulf, Western Reef Heron's breeding habitat is coastal wetlands and islands (Behrouzi-Rad, 2009). Western Reef Heron is one of the resident birds of Nakhiloo Island selects there for the breeding activities. For determining the reproduction success, and as confidence coefficient (47 nests), by studying the colonies increases and more nests can be studied (Scarton et al., 1994). The percentage of mortality and natality reflects habitat quality and environment conditions. This study is also important for conservation measures.

MATERIALS AND METHODS

From beginning of nest-building activity (late March) to leaving time (September) the field study was done once a week in one

*Corresponding author. E-mail: behrouzirad@yahoo.com.

colony of 47 nests (Figure 1). Selected nests were determined by numerical tags and some parameters such as length and width of nest, diameter and distance between nests were measured by measuring tape (Holloway, 1993). In order to estimate the breeding population of the Western Reef Heron on Nakhiloo Island, we used total count method (Symens, 1996). After beginning of egg-lying in the marked nests, the length and diameter (breadth) of eggs was measured using callipers with accuracy of 0.1 mm and were weighed by digital scale with accuracy of 0.1 g. For determining the volume and the index of egg shape equation (Symens, 1996; Schjorring and Bregnball, 1999) the equation was:

V = kL (cm). B^2 (cm), where, L is the length of eggs, B is diameter or breadth of eggs and K is coefficient which is 0.4866.

The index of egg shape = B÷L.100. If the index of eggs shape is more, it means that the eggs are more spherical.

Chicks of the marked nests were weighed by the digital scale at birth and every week (for distinguishing them, a red color was used for them). The chicks of *E. gularis* do not leave the nest after birth till short time. The success in egg maintenance represents the number of eggs hatched and foster success of chicks means how many percent of chicks leave the nest and are able to fly (Scarton, 1994; Symens, 1996). By regular checking of the 47 marked nests, the number of eggs, successful nestling and the number of chicken which are able to leave the nest, were registered. The loss was studied for: a) before the coming out: some of the eggs were eaten by *Rattus rattus*, or damaging their nests by other birds like gulls and also left nests by parents due to unknown reason. b) At the beginning of nestling: the nests in which 3-day-old chicks were, was lost, c) loss as a result of human disturbances (fishermen and other visitors to the island). Data analysis has been done using SPSS and Excel softwares.

Study area

Mond protected area in the south of Iran and in a distance of about 180 km from Busheher City situated 27°15'N, 51°35'E was the study area used (Figure 2). The mond protected area includes Nakhiloo, Om-Al-Goram, Tahmadon and Khan Islands. Nakhiloo Island, 110 km of the southeast to Busheher and 20 km to Dayer City, stretches from 27°49'19"N to 27°49'47"N and 51°27'54"E to 51°28'16"E. It has an area size of about 35 ha which its area at total ebb reaches c. 36.2 ha. The highest point of this island is 3 m a.s.l. Nakhiloo Island is one of the most important areas for breeding *E. gularis* in Iran. The land is covered mainly by *Atriplex* and *Salsola* ssp. Nakhiloo Island is one of the 105 Iranian Important Bird Areas (IBAs) (Evans, 1994) which sometimes holds 20,000 to 30,000 migratory water birds (Scott, 1995; Behrouzi-Rad, 2009).

RESULTS

According to the field observations, arrival to the island takes place mid-April and departure in early September. The Western Reef Heron bred at three colonies on Nakhiloo Island (16 nests at 27°49'17.5"N, 51°28'13.3"E, 19 nests at 27°49'15.5"N, 51°28'11.5"E, and 57 nests at 27°49'19.5"N, 51°28'13.8"E), and a total of 92 nests were counted by Total Count Method. The breeding population was concluded as 184 individuals. While total population was 202 individuals, 18 were non-breeders.

Nest

They nest in colonies, often with other wading birds, usually on platforms of sticks in trees or shrubs. Western Reef Herons build their nests on higher *Atriplex* bushes averaging 62.8 cm and located in the southeast of the Nakhiloo Island. They also build their nest, on the average, with c. 600 wood pieces of *Suaeda fruticosa*, *Atriplex leucoclada* and *Arthrocnemum macrostachyum*. The average length of nest is 38.42±0.53 cm and the width of nest 27.59±0.38 and the big perimeter is 107.03±2.7 and small one is 45.63±0.86. The maximum distance between nests was 8.50 m while their minimum distance was only 80 cm (Table 1).

Egg

Egg-laying took place between mid-May to mid-June and parents leave their nests between late June and late August. In the 47 studied nests on Nakhiloo Island, in total 153 eggs were calculated. Average clutch size was 2.3. Clutch size of 3 eggs had the highest proportion (51.06%). The clutch size ranged between 1 and 5 (Table 2). The weight of eggs was 27.38±0.09 g. The average egg length and breadth were measured as 47.02±0.16 mm and 34.12±0.08 mm, respectively. The volume of eggs was calculated as 26.63±0.17 and the index of egg shape as 72.35±0.32.

Breeding success

The proportion of hatched eggs in 47 nests were 62.1%, success in nestling 46.4%, success in post-nestling 39.21%, overall success in incubation 49.23% (Figure 3). The average weight of the chicken at birth was 22.19±0.61 g, seven-day olds 140.91±12.3 g, and 14-day olds as 251.63±14.9 g (Figure 4). Mortality of eggs in 47 nests and 153 eggs was 37.9%, nestling 15.68% and post-nestling 7.18%.

DISCUSSION

Population

Western Reef Heron is widespread along the Persian Gulf and Oman sea coasts. Its wintering population in Bushehr Province has been counted to be about 500 individuals and breeding population about 20 pairs on each of Shidvar and Gabre-Nakhoda Islands, 18 pairs on Omm-Al-Gorm, 10 pairs in Hormoz Island, 70 pairs on Khan and 92 pairs on Nakhiloo Island (Behrouzi-Rad, 2008) (Table 3). However, the breeding population of Western Reef Heron increased from 10 to 44 nests during the 2005 to 2007periods. The breeding population

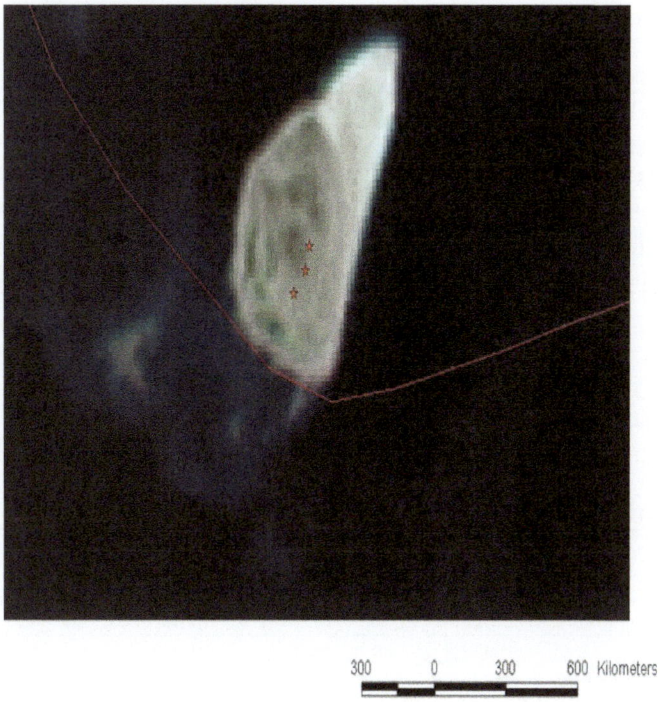

Figure 1. Breeding place of *Egretta gularis* on Nakhilo Island.

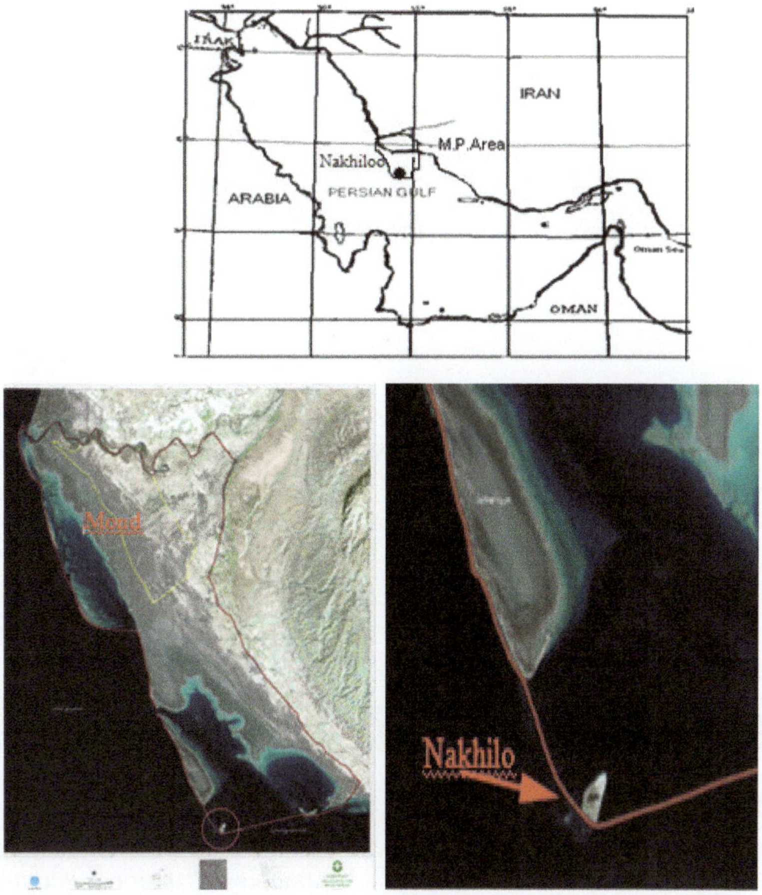

Figure 2. Geographical situation of Mond Protected Area and Nakhilo Island.

Table 1. Average of nest characteristics of *Egretta gularis* on Nakhilo Island.

Dimension of nests (cm)	N=47
Length or big diameter	38.42±0.53
Width or small diameter	27.59±0.38
Thickness	13.14±0.31
Big perimeter	107.03±2.07
Cup perimeter	45.63±0.88
Average length	62.75±1.81

Table 2. Clutch sizes of *Egretta gularis* on Nakhilo Island in 2008.

	Five egg	Four egg	Three egg	Two egg	One egg	Total
Number nest	3	14	24	4	2	47
Frequency (%)	6.38	29.7	51.06	8.5	4.2	100

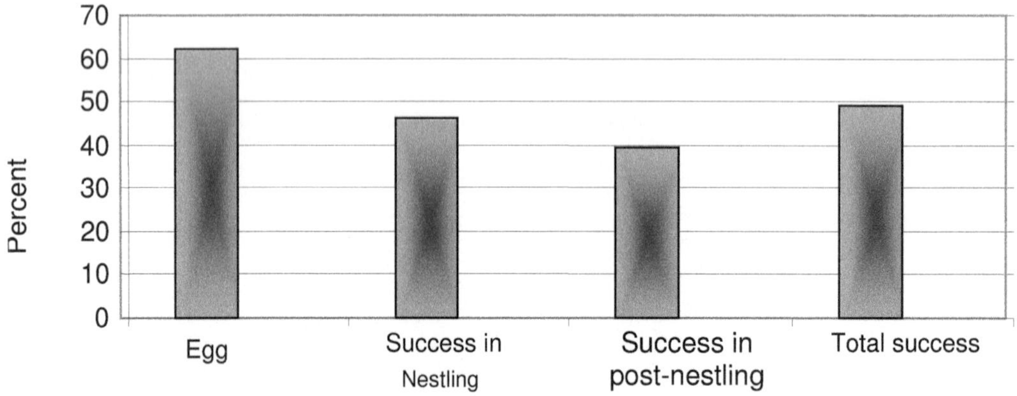

Figure 3. Successful of breeding of *Egretta gularis* on Nakhiloo Island in 2008.

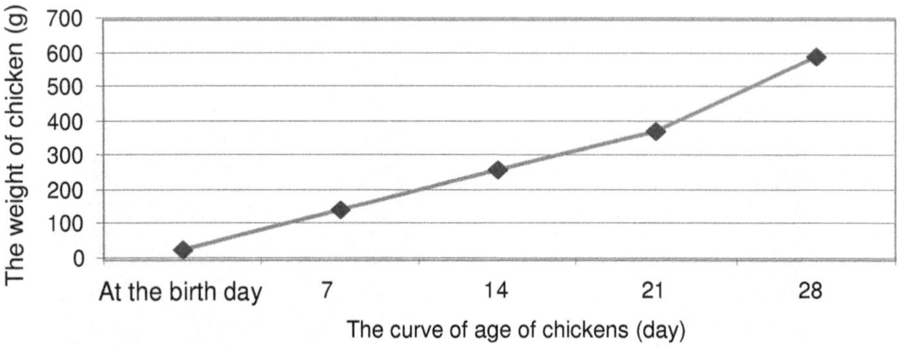

Figure 4. Relationship between age and weight of chicken of *Egretta gularis* in Nakhiloo in Bushehr -2008.

of the Western Reef Heron on Nakhiloo Island was 10 pairs in 2005, 34 in 2006, 44 in 2007 and 92 in 2008 (Behrouzi-Rad and Tayfeh, 2008). On the southern coast of the Persian Gulf, over 350 pairs seasonally nest around Hawar close to dense nesting areas where large numbers of herons congregate to catch fish in the shallows (Gallagher and Rogers, 1978; Gallagher et al., 1984).

Table 3. Average success rate of reproduction stages on Nakhilo Island in 2008.

Average number of eggs in every nest	Average number of Chicken in every nest	The number of successful nestling in every nest	The number of successful post-nestling	The number of losses eggs in every nest	The number of nestling losses in every nest	The number of post-nestling in every nest
3.2	2.02	1.18	1.27	1.23	0.5	0.2

Breeding season period

The nesting season of Western Reef Heron is between April and August in all parts of their range; except for one January record in eastern Nigeria (Cramp and Simmons, 1977). Western Reef Heron nests colonially, but no more than a dozen pairs or so, and seldom in association with other species (Hancock and Elliott, 1978). On Nakhiloo Island no other breeding heron occurs. On this island, the species is associated with *Sterna bengalensis*, *sterna bergii* and *Sterna anaethetus* (Behrouzi-Rad and Tayfeh, 2008). On Nakhiloo Island breeding activities begins in mid-April and they leave the island in early September. Nesting commences from early March through until August. Breeding occurs between April and July and by October in Africa (Cramp and Simmons, 1977; Kushlan and Hancock, 2005), with some individuals forming solitary pairs, whilst others gather in small colonies of up to 100 pairs (Kushlan and Hancock, 2005). The Reef Heron is classed in the UAE as a resident breeder.

Nest

Breed on coastal bays, coastal mudflats, islands and estuaries, nest materials are sticks and branches (del Hoyo et al., 1996). The species shows a preference for rocky or sandy shores and reefs, but will also frequent other coastal habitats such as estuaries, mudflats, salt marshes, mangroves, tidal creeks, islands, and lagoons (Grewal et al., 2002). The nest is a platform of twigs and seaweed positioned on the ground, in reed beds and mangroves, and on ledges or rocks (Kushlan and Hancock, 2005). Similarly, on Nakhiloo Island, the nest was rough platform of twigs about 50 cm in length, constructed by the female. During the 2005-2007 period, this species bred on tall bushes in the southwest of the island. Nests are built on the top of dense vegetation and are scattered randomly across the vegetation sometimes almost being touched by human, the capability of the vegetation to support a nest being the only apparent criteria (Kushlan and Hancock, 2005). It was difficult to obtain exact dimensions of the nests due to their untidy construction. If, as in one case, the extreme dimensions were taken the general impression given would be incorrect, because a single twig extended the size to almost 30" across. The nests are usually round or oval in shape, measuring an average of between 15" and 22" across, the overall thickness being about 7" with a shallow depression in the centre, depending on the stoutness of the construction, of between 2 and 4 inches. It was difficult to see the sitting birds but in the one case observed, it was overhanging the edge of the nest. Western Reef Herons generally nest in colonies but are also known to do so in isolation, as evidenced by these observations. It was interesting to note that the three isolated nests were built on the leeward side of the trees away from the prevailing westerly to north westerly winds (del Hoyo et al., 1996).

Clutch

The clutch of pale blue eggs varies from 2 to 3 in *dimorpha*, 3 to 4 in the other subspecies (Cramp and Simmons, 1977). On Nakhiloo Island, totally 93.4% of eggs were light blue and 6.5% were white or light green. Two to five pale green to blue-white eggs are laid on an untidy platform of sticks built by both sexes, located in a tree, on a rock ledge, under shrubs, or on the ground (Cramp and Simmons, 1977). The maximum clutch size in the present study was 3 (51.60%) and the minimum was nests with one egg (4.2%).

Egg

Recorded eggs measurements are 44×32 mm in nominate *gularis*, 44.9×34.3 mm in *asha* and 46.5×33.6 mm in *dimorpha* (Hancock and Elliott, 1978). These are oval, smooth, pale blue/green, with mean average measurements of 45×34 mm (Grewal et al., 2002). The average length and breadth of Western Reef Heron eggs on Nakhiloo Island (length 47.02±16 mm and breadth 34.22±.08 mm) had no significant difference with Africa (47.02×34 mm; Cramp and Simmons, 1977) had no significant difference with samples in Africa. The average weight of eggs at the beginning of egg-lying was 27.38±0.09 g. The weight of egg on Nakhiloo with the samples in Africa had significant difference ($p<0.01$) (Table 4; Cramp and Simmons, 1977). Eggs are laid at approximately 48-h intervals (Behrouzi-Rad and Momeni, 2008).

Incubation

They lay their eggs at intervals of one or more days and

Table 4. Comparison of egg parameters on Nakhiloo Island with Africa.

Studied parameters of eggs	Nakhiloo Island (mm)	Africa (mm)	P-value
Length	47.02	47	0.86
Breadth	34.22	34	0.15
Weight	27.38	29	0.000
Incubation period	26.36	23-26	0.000

incubation begin only when the clutch size completes. Incubation is shared by parents. In the Ardaidae family, incubation period takes at least 16 days (in some Little Bitterns) to a maximum of 30 days (Goliath Heron) and hatching is asynchronous (Hancock et al., 1978). Western Reef Heron eggs hatched earlier than any other species on Nakhiloo Island (Behrouzi-Rad and Tayfeh, 2008). Eggs are incubated for 20 - 25 days by both sexes (Grewal et al., 2002). The incubation period was observed in only one nest in Abu Dhabi and would appear to be about 26–28 days (Gallagher and Rogers, 1978). While on Nakhiloo Island it takes at least 25 days by both sexes. In 2006, at the time of the first egg-laying, there were five nests with clutches of 2 - 4 eggs (mean clutch size of 2.4). After this brood had left, three pairs laid further eggs in these nests; the clutches being 3 or 4 eggs (mean clutch size 3.3).

Chick

Semi-altricial young stay at nest about 30 days (Grewal et al., 2002). Milon (1946) found that *dimorpha* chicks begin clambering out of the nest after about 15 days but are incapable of flight for another month. On Nakhiloo Island, chicks begin clambering out of the nest after 10 to 15 days, returning only when a parent bird approaches with food. They young could flight after 45–55 days (Behrouzi-Rad and Momeni, 2008). The rapid growth continues and in the fourth week the young are very active. Feathers cover the main body and tops of the wings. During the fifth and sixth weeks the feathers turn dark grey and white feathers develop under the throat. The beak becomes paler and the slight down on the head stands upright like a crest. The ring around the iris appears white, the underbelly is pale and the bottom half of the legs are green, the top half grey with greenish rings (Behrouzi-Rad and Momeni, 2008). Chicks were cared and fed by both sexes (Hancock, 1978). On Nakhiloo, young were cared and fed by both sexes.

Survival

Having suitable herbaceous coverage for reproduction, the variety of aquatics around the island, makes the Nakhiloo Island as one the best area for breeding of Western Reef Heron in the south of Iran. Nakhiloo Island is a safe place and has suitable covering for bird to make shelter and shadow. The most preferred conditions for the egg and chicken survival is desirable climate of the region, the second is food. Fish is the best food for Western Reef Heron and is available around the island so the parents provide required fish during the incubation period. The chicken's growth shows that the average weight at the beginning of bird is 22.19±0.61 g and the average weight of a mature chicken is 590±10 g (Behrouzi-Rad and Momeni, 2008). The success of egg was 62.09% and for nestlings as 46.4% while survival in the post-nestling stage reduced to 39.21%, giving an overall survival of 49.23% for the total 153 eggs (Behrouzi-Rad and Momeni, 2008). Various factors contributed to this rate of attrition. Certainly during the egg-laying stage the clumsiness of the parent birds must be of considerable importance. In another study conducted on Ardeidae in northern Iran, 81 of 98 eggs of the Little Egret *E. garzetta* hatched and 79 reached their post-nestling stage, calculating breeding successes as 80.6% (Ashoori, 2010). Except in areas or time of food abundance, chicks seldom survives (Hancock et al., 1978). Eggs are laid at regular interval, normally a few days apart, many will not hatch, and large numbers of inert eggs litter the breeding sites once the season is over. When first is born, one parent normally stays close to the nest and can often be seen shading the chicks with its body. Heat is the greatest threat to young chicks, but both parents will often leave chicks for extended periods as they will go off to fish (Kushlan and Hancock, 2005). Breeding success in term of a high percentage of young reared to the flying stage is needed very dependent on favorable weather-warmth in higher latitudes, rainfall in the tropics-which in turn governs food availability and the timing of the breeding season itself (Hancock et al., 1978). The possibility of limited disturbance by humans cannot be ruled out but it is thought that this would be, at most, of a minor nature.

Conservation

This species has an extremely large range, and hence does not approach the thresholds of the vulnerable under the range size criterion (extent of occurrence <20,000 km^2 combined with a declining or fluctuating range size, habitat extent/quality, or population size and a small number of locations or severe fragmentation). The population trend appears to be stable, and hence the species does not approach the thresholds for vulnerability under the population trend criterion (>30% decline over ten years or three generations). The population size may be moderately small to large, but it is not believed to approach the thresholds for vulnerability under the population size criterion (<10,000 mature individuals with a continuing decline estimated to be >10% in ten years or three generations, or with a specified population structure). For these reasons the species is evaluated as least concern (IUCN Red List, 2009).

Western Reef Heron is not suitable for hunting in Iran, so humans do not hunt this species. The most amount of losses are related to the period in which the birds do not come out of its egg so it can not move any more. Natural events happen for it more than other period of nestling and post-nestling. The losses in nestling period are more at the beginning of birth, because in these days they are not able to move and leave the nest. It is suggested that the three colonies on Nakhiloo Island be studied regularly for long time. Ringing of chicken should be done as well as training the fishermen in order to protect the safe place for Western Reef Heron.

ACKNOWLEDGEMENTS

This work was financially supported by Bushehr Provincial Office of the Department of the Environment. We are very grateful to Mr Mahdaviani, former Director General and Mr Delshab, Director General, DOE, Bushehr and in particular to Mr Mehdi Jamadi, Head of Bord Khon Environment Office, Mr Foladi, boat driver (personnel of the Bord-e Khon Environment Office) and Mr Hoseini Tayefeh, expert of the Bushehr DOE.

REFERENCES

Ashoori A (2010). Breeding biology of Little Egret *Egretta garzetta* in Karfestan Ab-Bandan, Roudsar, Gilan Province, Northern Iran, 4(2): 12-18.

Behrouzi-Rad B (2009). Dictionary of Waterbirds of Persian Gulf and Oman Sea. Deportment of the Environment, p.156

Behrouzi-Rad B, Tayfeh FH (2008). Nest Counts for Western Reef Heron *Egretta gularis* and Four *Sterna* species in Nakhiloo Island in the Persian Gulf from 2005 to 2007. Podoces, 3(1/2): 45-52.

Birdlife International (2009). *Egretta gularis*. In: IUCN 2009. IUCN Red List of Threatened Species. Version 2009. <www.iucnredlist.org>. Downloaded on 17 January 2010.

BirdLife International (2009).. Species Data Zone on Western Reef Heron *Egretta gularis*. BirdLife International. http://www.birdlife.org/datazone/species/index.html?action=SpcHTM Details.asp&andsid=3709&andm=0.

Cramp S, Simmons KEL (1977). Handbook of the birds of Europe the Middle East and North Africa, the birds of the Western Palaearctic, Terns to Woodpeckers, Oxford university press, 5: 287-289.

Evans M (1994). Important Birds Area in the Middle East. Published by IUCN.

Gallagher MD, Rogers TD (1978). On the Breeding Birds of Bahrain. Bonn. Zool. Beitr., 29: 5-17.

Gallagher MD, Scott DA, Ormond RFG, Connor RJ, Jennings MC (1984). The Distribution and Conservation of Seabirds Breeding on the Coasts and Islands of Iran and Arabia. *In:* Croxall P.J., Evans P.G.H. and Schreiber R.W. (eds), Status and Conservation of the World's Seabirds. ICBP Technical Publication No.2. ICBP, Cambridge, U.K., pp. 421-456.

Grewal B, Harvey B, Pfister O (2002). A Photographic Guide to the Birds of India and the Indian Subcontinent, Including Pakistan, Nepal, Bhutan, Bangladesh, Sri Lanka, and the Maldives. Princeton University Press, Princeton, pp.123-127

Hancock J, Elliott H, Gillmor R (1978). The Herons of the World, published in 1978 by London Edition Limited, 30 Uxbridge Road, London W12 8ND, pp. 208-210

Holloway M (1993). The variable breeding success of the Little Tern *(Sterna albifrons)* in South-East India and protective measures needed for its conservation. Biol. Conserv, 65: 1-8.

IUCN (2009). IUCN Red List. http://www.iucnredlist.org accessed March 2009.

Kushlan JA, Hancock JA (2005). Bird Families of the World: Herons. Oxford University Press, Oxford, pp.120-126.

Milon P (1946). Observation sur quelques Oisaux de Madagascar. *L' Oiseau* et al R.F.O., 16: 2-28

Scarton F, Valle R, Borella S (1994). Some comparative aspects of the breeding biology in – Black-headed Gull (*Larus ridibundus*), Common Tern (*Sterna hirundo*) and Little Tern *(Sterna albifrons)* in the lagoon of Venice, NE Italy. Avocetta, 18: 119-123.

Schjorring S, Bregnball T (1999). Prespecting enhance breeding success of first time breeder in the Great Cormorant. Animal Behav., 57(3): 664-674.

Symens H (1996). Status of the breeding population's of (Sternidae) along the eastern of Saudi Arabic following the 1991 Gulf war. NCWCD, Riyadh and Senckenberg Research Institute, Frankfurt, pp. 404-420.

Voisin C (1991). The Herons of Europe. T and AD Poyser. London, p. 364.

Temporal and spatial variations of heavy metals (Cu, Pb and Ni) concentration in the sediment from intertidal zone along the Iranian coasts of the Oman Sea

Fatemeh Einollahi Peer[1]* and Alireza Safahieh[2]

[1]Institute Research of Hamoon International Wetland, Zabol University, Iran.
[2]Department of Marine Biology, Faculty of Oceanic and Marine Science, Khoramshar University of Marine Science and Technology, Iran.

In order to determine distribution and seasonal patterns of heavy metals concentration in the intertidal sediment of the north Oman sea, samples of sediment were collected from five different station in August 2007 and February 2008. The samples were oven dried and acid digested. Heavy metals content of acid digested samples were analyzed by atomic absorption spectrophotometery method. Results showed that mean Cu, Pb and Ni concentration during summer were 38.28 ± 0.37, 50.37 ± 2.41 and 26.90 ± 2.72 $\mu g.g^{-1}$ respectively. The average heavy metals concentration in winter was sharply dropped to 16.45 ± 1.77, 13.56 ± 1.24 and 14.68 ± 1.93 $\mu g.g^{-1}$ for Cu, Pb and Ni respectively. Though Cu and Ni concentration did not significantly change over various station, seasonal difference between heavy metals concentration in sediment was significant ($P < 0.05$). It is suggested that Cu and Ni originate from non point sources. The heavy metals concentration in the sediment may hardly be affected by monsoon.

Key words: Heavy metals, intertidal zone, sediment.

INTRODUCTION

Marine environment particularly intertidal zone is very complicated natural system, since it is affected by many processes including human activities (Qiao et al., 2007; Pote et al., 2008), physicochemical and biological processes (Fulton et al., 2004; Gonza´lez-Mac´ias et al., 2006) and even atmospheric depositions (Witt et al., 2010). Human activities in the coastal area always result in production and discharge of various kinds of pollutants such as heavy metals into the marine ecosystems (Censi et al., 2006).

Heavy metals are normally present in marine environment at low concentration (Karuppiah and Obbard, 2008; Garc´ıa et al., 2001; Kim et al., 2003; Santos et al., 2005; Catsiki and Florou, 2006; Nobi et al., 2010), however, considerable amount of them are released to the marine ecosystems from anthropogenic sources (Kamala-Kannan et al., 2008). These heavy metals may settle or accumulate in the coastal area apparently due to their affinity to be bounded to particulate organic matter and deposition in marine sediment (Tam and Wong, 2000; Santos et al., 2005; Mil-Homens et al., 2007; El Nemr et al., 2007). Heavy metals are considered as persistent contaminants and due to their toxic properties, the elevated levels of them in aquatic systems could create several problems for marine organisms. Moreover their accumulations in marine organisms and biomagnification throughout the food chain may affect human health (Kowalski, 1994; Valls and Lorenzo, 2002; Gochfeld, 2003; Yi et al., 2008).

The accumulation, biomagnification and toxic behaviors of heavy metals in the marine environment is greatly influenced by their bioavailability, which in turn depends on many factors such as nature of sediment, environmental conditions including pH, temperature, redox potential, water turbulence and properties of the absorbed compounds (Tam and Wong, 2000; Santos et

*Corresponding author. E-mail: fateme.eynollahi@yahoo.com.

al., 2005; Mil-Homens et al., 2007; El Nemr et al., 2007).

Normally sediment act as a marine sink for discharged heavy metals into the marine environment (Di Tao et al., 1992; Morillo et al., 2004; Wang et al., 2004; IP et al., 2007). Meanwhile deposited metals in the sediment may return to the water column again as a result of bottom turbulence created by strong waves (Simpson and Batley, 2007; Alagarsamy, 2006; Anu et al., 2010). Infact coastal sediments have been widely used to evaluate sea water quality because, it has more stability and less variability of heavy metals concentration compared to the sea water (RuiźFernández et al., 2003; Cuong and Obbard, 2006). In some circumstances, information about the concentration of heavy metals in sediment can help to detect the source of pollution in the aquatic ecosystems (Birch et al., 2001; Liu et al., 2003; Alagarsamy, 2006; Silva et al., 2006).

Located at Iranian coasts of the Oman Sea, Chabahar is a developing area. Traditionally this area is used for fisheries and aquaculture activities of the local people. However it has developed as an important free zone for country import and export, economy, urbanization and industries have also grown in the area parallel to economic development (Amini and Miraki, 2006). This area has a subtropical weather with two distinct seasons, dry season in summer and rainy season in winter. It also experience monsoon (seasonal western winds) during summer and winter. The sea water is completely disturbed by huge waves during monsoon period (Zareii, 1995). Therefore considerable amount of sediment is displaced by the power of waves. This condition could result in redistribution of heavy metals in the coastal environment. Although the area is still growing and especial weather seems to greatly influence on distribution and redistribution of heavy metals in coastal ecosystems, there is a lack of information concerning heavy metals contamination in Chabahar. Such information could be used to establish baselines in the area. The Study of De Mora et al (2004) in southern coasts of the Oman Sea showed, some metals like the Ni have natural source in this area. Mainwhile, the human activity including shipping, fishing activity and cannery industry can increase the Pb concentration in coastal area. These activity are conducted in Iranian coasts of the Oman Sea. These metals also can be accumulated in tissues of organisms (Gochfeld, 2003; Yi et al., 2008). Therefore, the objective of this study was to investigate Cu, Pb and Ni concentration in sediment of northern coasts of the Oman Sea and determine the seasonal variation of these metals in this area.

MATERIALS AND METHODS

Study area

Sediment samples were taken from five different locations (including Guatr bay, Beriss, Ramin, Shahid Beheshti and Chabahar bay) along intertidal zone of the north coasts of the Oman Sea in February 2007 and August 2008 (Figure1). Geographical position of the sampling station and current sources of metals input are shown in Table 1.

Sample collection and preparation

From each location, three sediment samples were taken from 3 to 5 cm surface with a stainless steel grab sampler during low tide. Samples were collected in acid washed polyethylene bottles and transferred to the laboratory with icebox. Each sample was mixed thoroughly and dried at 105°C in drying oven for at least 16 h until a constant weight was obtained (Delman et al., 2006). Dry samples were sieved with 63 µm stainless steel seive to remove shells and other debrises, were powdered in a glass mortar, stored in plastic container which were acid washed in 10% solution of nitric acid, rinsed with double distillated water and preserved for the determination of heavy metal concentration (Delman et al., 2006; Orescanin et al., 2006).

Sample analysis

For analysis of heavy metals in sediment of each station, 1 g of each dry sample was digested in a mixture of 10 cc concentrated perchloric acid (60% Merck) and Nitric acid (65%Merck) in the 1:4 ratio with hotplate digester first at 40°C for 1h and then in 140^{0}C for 3h. After digestion, the samples were cooled in laboratory temperature, diluted to certain volume using double distilled water and then filtered by Whattman 42 µ filter paper (Delman et al., 2006; Yap et al., 2002). Heavy metal concentration including Cu, Pb and Ni were measured using Unicam 919 air-acetylene atomic absorption spectrophotometer and detailed measurements of each metal is listed in the Table 2. The metals concentration were measured and expressed as microgram per gram in dry weight (µg/g dw).

Data analysis

All data were tested for normal distribution first. The normality test revealed all data were normally distributed. In order to compare differences between seasonal variation of heavy metal concentration t-test analysis was used. One-way analysis of variance (ANOVA) was used to find any significant difference in metal concentration among different station. If significant difference was observed, Tukey post hoc test was employed to determine different station.

RESULTS

The overall pattern of heavy metals concentration in the sediment was Pb > Cu > Ni. Copper concentration in the sediment of various sampling station were not significantly different ($P > 0.05$) either in Summer or Winter. The concentration of Cu in Summer was 38.47 ± 1.00, 38.69 ± 0.62, 37.88 ± 0.70, 38.07 ± 0.10 and 38.29 ± 0.15 µg/g dw for Guatr, Beriss, Ramin, Shahid Beheshti and Tiss respectively. A sharp decrease in Cu concentration was appeared in Winter. During this season the level of Cu was 16.23 ± 1.88, 16.96 ± 2.26, 16.96 ± 2.26, 15.87 ± 1.25 and 16.25 ± 1.21 µg/g dw for

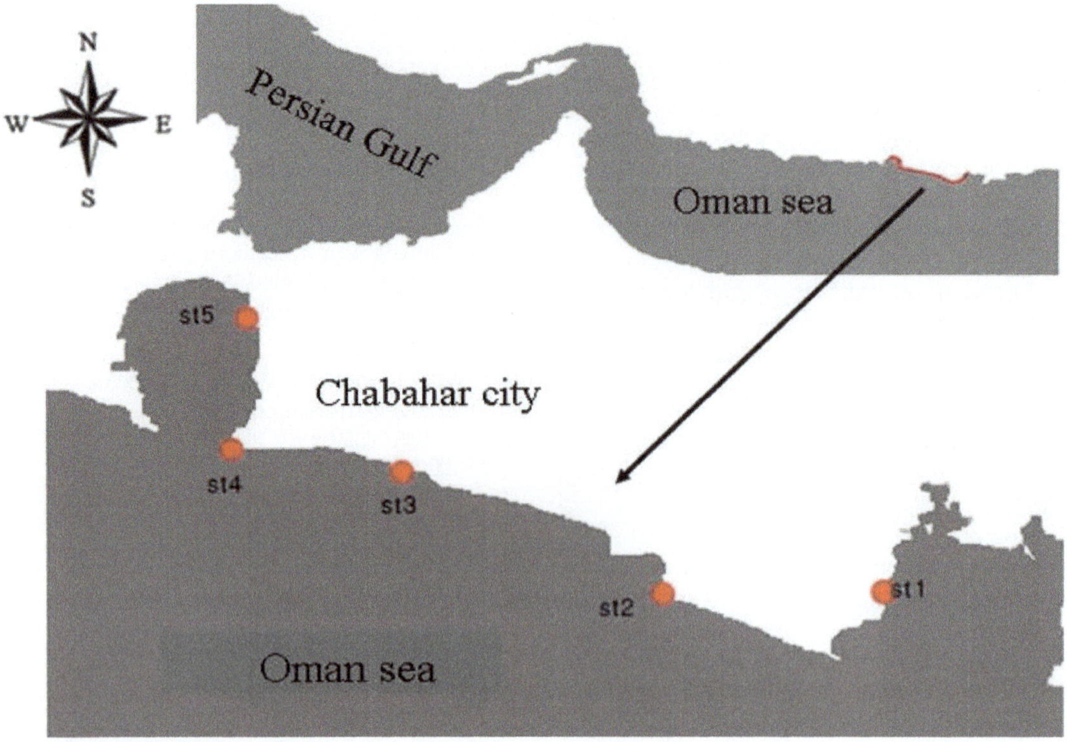

Figure 1. Map showing study area.

Table 1. Position of stations along Chabahar coasts.

Sources of heavy metal input	Latitudes	Station	
Aquaculture and Fishing boats	" 3/9 ' 30 ◌ْ61 E " 1/53 ' 9 25 N	Guatr	St1
Port, transportation, food industries, aquaculture, domestic discharges	" 6/5 ' 10.3 ◌ْ61 E " 55/7 ' 8 ◌ْ25 N	Beriss	St2
Port, transportation, food industries, aquaculture, domestic discharges	" 50 ' 44 ◌ْ60 E " 3/9 ' 16 ◌ْ25 N	Ramin	St3
Domestic discharges, Port and transportation	" 7/12 ' 37 ◌ْ60 E " 92/9 ' 21 ◌ْ25 N	Shahid Beheshti	St4
Agriculture, aquaculture, domestic discharges	" 5/21 ' 37 ◌ْ60 E " 2/39 ' 17 ◌ْ25 N	Tiss	St5

Guatr, Beriss, Ramin, Shahid Beheshti and Tiss respectively. The Cu content of the sediment in August and February were significantly different (Figure 2).

Comparison between Pb concentration in the sediment showed significant difference among different station. The maximum concentration of Pb was recorded in Ramin (56.45 µg/g dw) during summer. On the other hand, Beriss was characterized by significantly higher Pb concentration among other studied station in winter ($P < 0.05$). While, during summer Pb concentration was measured as 50.25 ± 1.53, 44.93 ± 2.65, 56.45 ± 4.06, 48.48 ± 4.60 and 51.78 ± 3.27 µg/g dw for Guatr, Beriss, Ramin, Shahid Beheshti and Tiss respectively. A significant decrease was observed in Pb level during Winter ($P < 0.05$). It was measured as 13.33 ± 1.29, 15.79 ± 1.52, 15.61 ± 1.90, 12.09 ± 0.35, and 10.98 ± 1.14 µg/g dw for Guatr, Berris, Ramin, Shahid Beheshti and Tiss in February (Figure 3).

Table 2. The details of atomic absorptaion spectrophotometer Unicam 919 for each metal.

Metal	Wavelength (MA)	Lump current (Cathodic lump) (NM)	Band width (NM)	Fuel
Cu	324	3.5	0.5	Acetylene gas
Pb	216.2	5	1	Acetylene gas
Ni	224	3.5	0.2	Acetylene gas

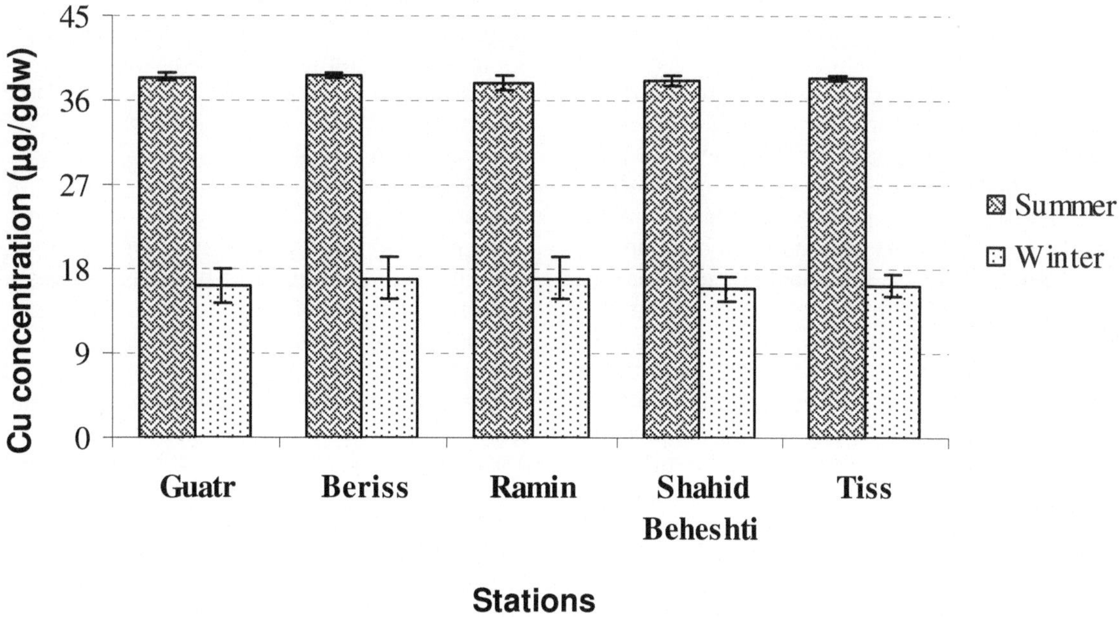

Figure 2. Cu concentrations (µg/g dw) in sediment in summer and winter.

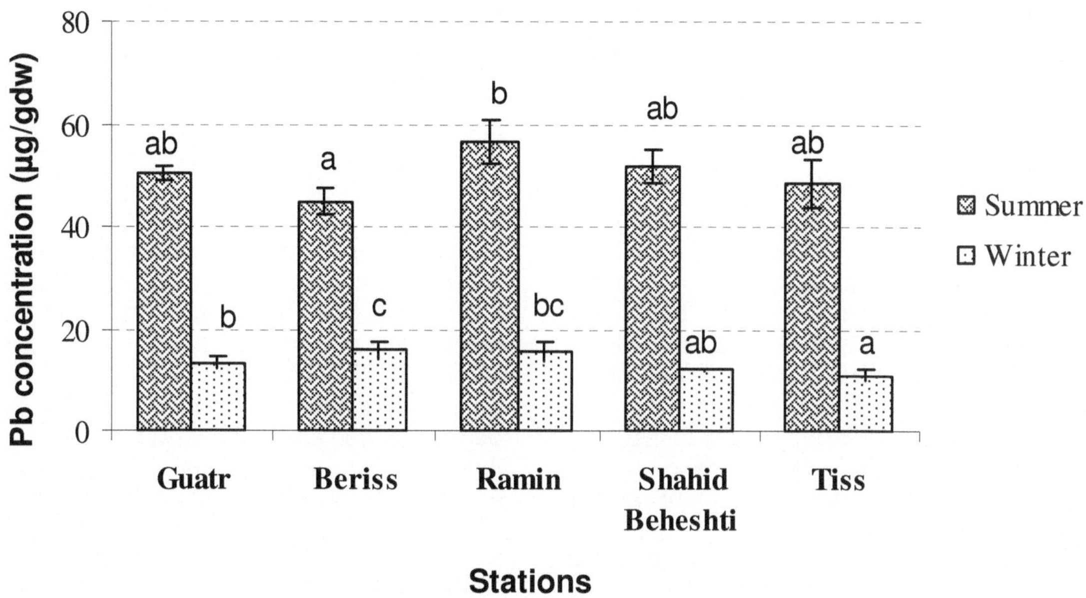

Figure 3. Pb concentrations (µg/g dw) in sediment in summer and winter.

The level of Ni in the sediment did not vary significantly among different station either in August or February. During Summer Ni concentration in the sediment was 25.69 ± 2.22, 30.28 ± 2.40, 28.42 ± 2.54, 26.41 ± 5.42

and 23.74 ± 1.06 µg/g dw for Guatr, Beriss, Ramin, Shahid Beheshti and Tiss respectively. As for other metal again significant decrease was found in Ni concentration in February. During Winter Ni level dropped to 12.39 ± 1.86, 17.10 ± 1.78, 14.46 ± 2.33, 13.55 ± 0.98 and 15.91 ± 2.71 µg/g dw in station Guatr, Beriss, Ramin, Shahid Beheshti and Tiss respectively (Figure 4).

DISCUSSION

The maximum concentration of Pb was observed in Ramin during summer and in Beriss during winter. Elevated Pb concentration in mentioned station, could be related to human activities in these locations. Intense marine transportation, ship painting, fishing boats activities, aquaculture, polluted wastes and discharges from residential area could be the major factors for Pb input in these station. Another factor which could be involved in Pb input in Ramin and Beriss station is adjacent mountains. The distance between mountains and sea is minimum in these station compared to others. Therefore weathering of the lead containing rocks may increase the level of Pb in coastal sediment. Kilemade et al. (2004) suggested that short distance between mountains and coasts can be an important factor to increase the concentration of heavy metals in the marine environment.

Zhou and Hao (2007) suggested that anthropogenic sources including antifouling paint, domestic sewage, surface runoff, wastewater, vehicle emissions, marine transportation, ship repainting, dental clinics, electronic/ chemical industries and rock weathering sea could be important factors affecting metals concentration in the coastal sediments of Hong Kong.

Despite the existence of the wastes in station Beriss, the concentration of Pb measured in this station during summer was low. This could be related to high disturbance of the sediment created by huge waves during monsoon. In fact during monsoon season, sea water in Beriss is so disturbed and wavy that most local activities including shipping and fishing stop in some locations (Zareii, 1995). Subsequently, after this reduction of human activities, the level of Pb input by the local vessels might be decreased leading to occurrence of low Pb concentration in sediment. After monsoon, by the starting of local activities, Pb input starts to increase due to the beginning of local shipping and human activities in the area. The sediment is more stable, leading Pb level to rise up again.

The concentration of Cu and Ni did not changed significantly in studied station ($P > 0.05$). Both in summer and winter it is suggested that Cu and Ni mainly originate from natural sources. Geological studies of the Oman Sea have revealed that the sea bed is reach in Ophiolits. This mineral contains considerable amounts of Nickel sulphide (Leblanc and Ceuleneer, 1991). Therefore, it seems that the more or less constant concentration of Ni in different station in much influenced by local mineralogy derived from Oman ophiolites than anthropogenic input. De Mora et al. (2004) found high concentration of Ni and Cr in sediment from Akkah in southern coasts of the Oman Sea where located is far from sever human activities and anthropogenic inputs. They also suggested that Nickel sulphide rich Ophiolits are the main sources of Ni found in the sediment.

Comparisons between heavy metals content of sediment in summer and winter indicated that all metals concentration were higher in summer (August). Increasing of human activities such as shipping and transportation together with enhanced growth of phytoplankton in the water column due to upwelling and nutrient availability and favorite environmental factors (temperature) during monsoon, are suggested to be the main factors. Increasing of human activities may result in elevated Pb input, while the occurrence of high phytoplankton growth in the water column promotes the production of higher amounts of suspended organic matter. These materials have affinity to bound with metal ions which are dissolved in seawater. The complex particles finally deposit on the bottom sediment and increase the metals concentration. In addition, seasonal variation of is also depend on the availability of other metals and compounds in the same site which may bind or unbind the presently studied metals, so it is not totally depended on the anthropogenic studies. Seasonal growth of phytoplankton may also influence the concentration of heavy metals.

Similar to present results, Lee et al. (2008) found that heavy metals concentration in the sediment of the South Korean coasts had increased during summer. They suggested that optimization of physicochemical factors of seawater (ie. dissolved oxygen, temperature and pH) during spring may stimulate phytoplankton growth which finally in turn results in appearance of algal blooms. The production of great amounts of organic carbon in sea water due to phytoplankton growth and chelation of dissolved metals and settlement of the chelated metals increase heavy metal concentration in the sediment (Lee et al., 2008). Like phytoplankton macrophytes can also affect heavy metals concentration in the sediment by the same way. Rigollet et al. (2004) concluded that when organic production by Zostrea is low, heavy metals were found at lower concentration in sediment while, after a period of Zostrea growth at the end of summer heavy metals concentration in sediment rose up.

Conclusion

The Level of Cu and Ni did not significantly change among different station suggesting that these two elements originate from natural sources. The concentration of Pb varied among different station with maximum

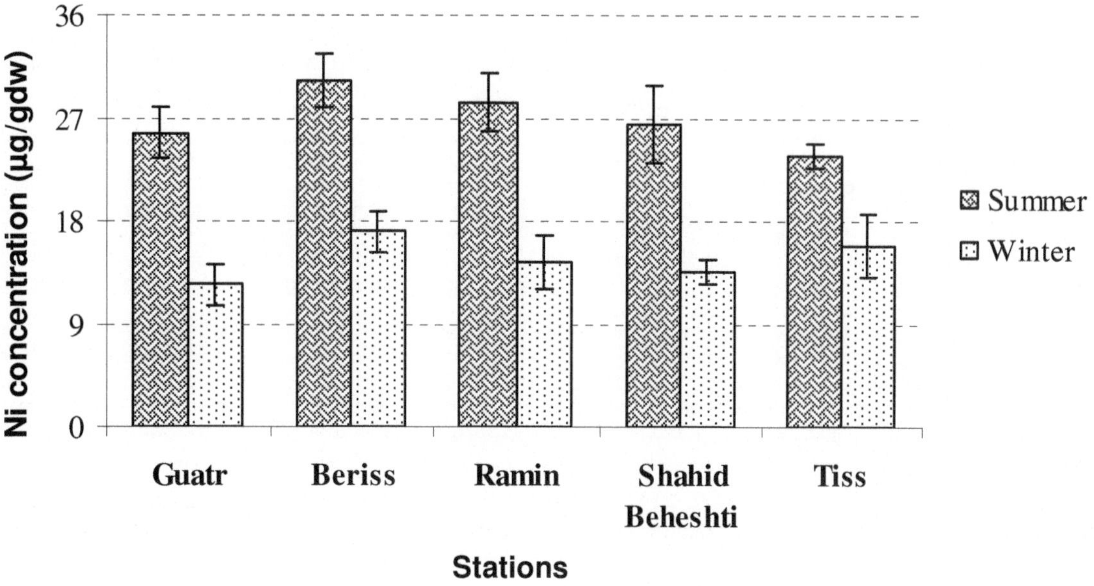

Figure 4. Ni concentrations (µg/g dw) in sediment in summer and winter.

levels in Beriss and Ramin. Human activities such as shipping, marine transportation and fisheries are might associate in heavy metals enrichment in the sediment. During summer heavy metals concentration in Chabahar sediment increased markedly. It is suggested that increasing of nutrient availability due to upwelling during monsoon season (summer) enhanced phytoplankton growth followed by increasing of suspended organic matter which should be involved in heavy metals enrichment of sediment. So can the variation in heavy metals concentration be correlated to the evaporation rate of water during the winter and summer. To understand the exact effect of monsoon on heavy metal concentration in the sediment more detailed study is recommended.

REFERENCES

Alagarsamy R (2006). Distribution and seasonal variation of trace metals in surface sediments of the Mondovi estuary, west coast of India. Estuarine, Coastal Shelf Sci., 67:333-339.

Amini Ranjbar Gh, Miraki Gh (2006). Evaluation of quantity and quality of the Chabahar bay. Sistan and Baluchestan University, pp. 7-9.

Anu G, Nair SM, Kumar NC, Jayalakshmi KV, Pamala D (2009). A baseline study of trace metals in a coral reef sedimentary environment, lakesshadweep , Archipelago, Environmental Earth Scencei., 59: 1245-1266.

Birch FG, Taylor SE, Matthai C (2001). Smal-scale spatial and temporal variance in the concentration of heavy metals in aquatic sediments: a review and some new concepts. Environ. Pollut., 113: 357-372.

Catsiki VA, Florou H (2006). Study on the behavior of the heavy metals Cu,Cr, Ni, Zn, Fe, Mn and 137Cs in an estuarine ecosystem using Mytilus galloprovincialis as a bioindicator species: the case of Thermaikos gulf, Greece. J. Environ. Radioactivity, 86: 31-44.

Censi P, Spoto SE, Saino F, Sprovieri M, Mazzola S, Nardone Di Geronimo G, Punturo SI, Ottonello D (2006). Heavy metals in coastal water systems: A case study from the northewestern Gulf of Tailand. Chemosphere, 64: 1167-1176.

Cuong DT, Obbard JP (2006). Metal speciation in coastal marine sediments from Singapore using a modified BCR-sequential extraction procedure. Appl. Geochem., 21: 1335-1346.

Delman O, Demirak A, Balci A (2006). Determination of heavy metals (Cd, Pb) and trace elements (Cu, Zn) in sediments and fish of the southeastern Aegean sea (Turkey) by atomic absorption spectrometry. Food Chem., 26: 157-162.

De Mora S, Fowler SW, Wyse E, Azemard S (2004). Distribution of heavy metals in marine bivalves, fish and coastal sediments in the Gulf and Gulf of Oman. Mar. Pollut. Bull., 49: 410-424.

Di Tao DM, Mahuny JD, Hansen D, Scott KJ, Carlson AR, Ankly GT (1992). Acid-volatile sulfide predicts the acute toxicity of Cadmium and Nickel in sediments. Environ. Sci. Technol., 26: 96-101.

EL Nemr A, Sikaily EL, Khaled A (2007). Total and Leachable heavy metals in muddy and sandy sediments of Egyptian coasts along Mediterranean Sea. Estuar. Coastal Shelf Sci., 192:151-168.

Fulton EA, Smith ADM, Johnson CR (2004). Biogeochemical marine ecosystem models I: IGBEM—a model of marine bay ecosystems. Ecol. Model., 174: 267-307.

Garc´ıa F, Ortega A, Domingo JL, Corbella J (2001). Accumulation of metals in autopsy tissues of subjects living in Tarragona County, Spain. Toxic/Hazardous Substances and Environmental Engineering. J. Environ. Sci. Health. Part A, 36(9): 1767-1786.

Gonza´lez-Mac´ias C, Schifter I, Lluch-Cota DB, Mendez-Rodriguez L, Hernandez-Vazquez S (2006). Distribution, enrichment and accumulation of heavy metals in coastal sediments of Salina Cruz Bay, Mexico. Environ. Monit. Assess., 118: 211-230.

Gochfeld M (2003). Cases of mercury exposure, bioavailability and absorption. Ecotoxicol. Environ. Saf., 56:174-179.

Ip CCM, Xi XD, Zhang G, Wai OWH, Li YS (2007). Trace metal distribution in sediments of the Pearl River Estuary and thesurrounding coastal area, South China. Environ. Pollut., 147(2): 311-323.

Kamala – Kannan S, Drabho Dass Batravi B, Jae Lee K, Kannan N, Krishnamoorthy R, Shanthi K, Jayaprakash M (2008). Assessment of heavy metals (Cd, Cr and Pb) in water, sediment and seaweed (Ulva lactuca) in the Pulicat Lake, south East India. Chemosphere, 71: 1233-1240.

Karuppiah CDT, Obbard SJP (2008). Distribution of heavy metals in the dissolved and suspended phase of the sea-surface microlayer, seawater column and in sediments of Singapore's coastal environment. Environ. Model Assess.,138: 255-272.

Kilemade M, Hartl MGJ, Sheehan D, Mothersill C, Pelt FNAM, Berien

NMO, Halloran JO (2004). An assessment of the pollutant status of surficial sediment in Cork harbour in the south east of Ireland with particular reference to polycyclic aromatic hydrocarbons. Mar. Pollut. Bull., 49: 1084-1096.
Kim SJ, Rodriguez-Lanetty M, Suh JH, Song JI (2003). Emergent effects of heavy metal pollution at a population level: Littorina brevicula a study case. Mar. Pollut. Bull., 46: 74-80.
Kowalski Z (1994). Treatment of Chromic tannery wastes. J. Hazard. Mater., 37: 137-144.
Leblanc M, Ceuleneer G (1991). Chromite crystalization in a multicellular magma flow: evidence from a chromatite dike in the Oman ophiolite. Lithos, 27: 231-257.
Lee M, Bae W, Chung J, Jung H, Shimd H (2008). Seasonal and spatial characteristics of seawater and sediment at Youngil bay, Southeast Coast of Korea. Mar. Pollut. Bull., 225: 467-474.
Liu WX, Li XD, Shen ZG, Wang DC, Wai OWH, Li YS (2003). Multivariate statistical study of heavy metal enrichment in sediments of the pearl river estuary. Environ. Pollut., 121: 377-388.
Mil-Homens M, Stevens RL, Cato I, Abrantes F (2007). Regional geochemical baselines for Portuguese shelf sediments. Environ. Pollut., 148: 418-427.
Morillo J, Usero G, Gracia I (2004). Heavy metal distribution in marine sediments from the southwest of Spain. Chemosphere, 55: 431-442.
Nobi EP, Dilipan E, Thangaradjou T, Sivakumar K, Kannan L (2010). Geochemical and geo-statistical assessment of heavy metal concentration in the sediments of different coastal ecosystems of Andaman Islands, Iindia. Estuar. Coastal Shelf Sci., 87: 253-264.
Orescanin V, Lovrencic I, Mikelic L, Barisic D, Matasin Z, Lulic S, Pezelj D (2006). Biomonitoring of heavy metal and arsenic on the east coast of the middle Aderiatic sea using Mytilus galloprovinciallis. Nuclear Instr. Methods Phis. Res., 56: 495-500.
Pote J, Haller L, Loizeau J, Bravo AG, Sastre V, Wildi W (2008). Effects of a sewage treatment plant outlet pipe extension on the distribution of contaminants in the sediments of the Bay of Vidy, Lake Geneva, Switzerland. Bioresour. Technol., 99: 7122-7131.
Qiao S, Yang Z, Pan Y, Guo Z (2007). Metals in suspended sediments from the Changjiang (Yangtze River) and Huanghe (Yellow River) to the sea, and their comparison. Estuar. Coastal Shelf Sci., 74: 539-548.
Rigollet V, Sfriso A, Marcomini A, Casabianca MLD (2004). Seasonal evolution of heavy metal concentration in the surface sediments of two Mediterranean Zostera marina L. beds at Thau lagoon (France) and Venice lagoon (Italy). Bioresour. Technol., 95: 159-167.
RuiźFernández AC, Hillaire-Marcel C, P_aez-Osuna F, Ghaleb B, Soto-Jim_enez M (2003). Historical trends of metal pollution recorded in sediments of the Culiacan River Estuary, Northwestern Mexico. Appl. Geochem., 18: 577-588.
Santos IR, Silva EV, Schaefer CEGR, Albuquerque MR, Campos LS (2005). Heavy metal contamination in coastal sediments and soils near the Brazilian Antarctic station, King George Island. Mar. Pollut. Bull., 50: 185-194.
Silva CAR, Smith BD, Rainbow PS (2006). Comparative biomonitors of coastal trace metal contamination tropical south America (N. Brazil). Mar. Environ. Res., 61: 439-455.
Simpson SL, Batley GE (2007). Predicting metal toxicity in sediments: a critique of current approaches. Integrated Environmental Assessment And Management, 3:18–31.
Tam NFY, Wong YS (2000). Spatial variation of heavy metals in surface sediments of Hong Kong mangrove swamps. Environ. Pollut., 110: 195-205.
Valls M, Lorenzo VD (2002). Exploiting the genetic and biochemical capacities of bacteria for the remediation of heavy metal pollution. FEMS Microbiol., 26: 327-338.
Wang F, Richard RG, Chapman PM (2004). Testing sediment biological effects with the freshwater amphipod Hyalella azteca: the gap between laboratory and nature, Chemosphere, 57: 1713-1724.
Witt MLI, Mather TA, Baker AR, de Hoog CJ, Pyle DM (2010). Atmospheric trace metals over the south-west Indian Ocean: Total gaseous mercury, aerosol trace metal concentrations and lead isotope ratios. Marine Chemistry., 121, 2-16.
Yap CK, Ismail A, Tan SG, Omar H (2002). Correlations between speciation of Cd, Cu, Pb and Zn in sediment and their concentration in total soft tissue of green-lipped mussel Perna viridis from the west coast of Peninsular Malaysia. Environ. Int., 28: 117-128.
Yi Y, Wang Z, Zhang K, YU G, Duan X (2008). Sediment pollution and it's effect on fish through food chain in the Yangtze river. Int. J. Sediment Res., 23: 338-347.
Zareii A (1995). Survey of hydrobiology of main Estuary of Sistan and Baloochestan, Iran, the Central Research of distant water fisheries.
Zhou F, Guo H, Hao Z (2007). Spatial distribution of heavy metals in Hong kongs marine sediments and their human impacts: A GIS based chemometric approach. Mar. Pollut. Bull., 54: 1372- 1384.

Breeding biology of South American Tern from Cardos Island, Santa Catarina State, Brazil

Hélio Augusto Alves Fracasso[1]*, Joaquim Olinto Branco[2], Joanna Burger[3], Luís Fabio Silveira[4] and José Roberto Verani[1]

[1]Department of Hydrobiology, Federal University of Sao Carlos, Sao Paulo, Brazil.
[2]Centro de Ensino em Ciências Tecnológicas da Terra e do Mar, Santa Catarina, Brazil.
[3]Division of Life Sciences, Rutgers University, New Jersey, USA.
[4]Museu de Zoologia da USP, CEP 04263-000 - São Paulo - SP - Brazil.

The South American Tern *Sterna hirundinacea* Lesson, 1831 occurs in the Atlantic Ocean, from Tierra del Fuego (Argentina) to Bahia (Brazil) and in the Pacific Ocean all around the southern part of the continent, up the Peruvian coast. Daily observations were conducted from April to November in 2003, 2005 and 2006 to gather information on the reproductive biology of the South American Tern in the Cardos Island (SC, Brazil). Although, adults arrived earlier in 2005 and 2006, their breeding chronology followed the same pattern in all three years, with well-characterized reproductive stages. In 2006, the average time of copulation was about three minutes and the number of inseminations (sexual intercourses) stayed constant throughout the day. The period of egg-laying and total number of nests was higher in 2005 (2417 nests), than in 2003 (2124) and 2006 (1852). The smallest average egg sizes (length, width, weight) occurred at the end of egg-laying in 2003, while the largest occurred in the beginning of 2006. Eggs hatched earlier in 2005 and 2006 compared to 2003, with the earliest hatching at the end of May.

Key words: Behavior ecology, seabird breeding, *Sterna hirundinacea*.

INTRODUCTION

The South American Tern *S. hirundinacea* Lesson, 1831 is a small seabird (38-41 cm) with red beak and legs, distributed along the Atlantic coast, from Tierra del Fuego (Argentina) to Bahia (Brazil) (Sick, 1997) and along the Pacific coast from the southern part of the South American continent up to the Peruvian coast (Higgins and Davies, 1996). Although there is little information about the population that breeds in the Atlantic coast, it is known that breeding in Argentina occurs from December to April followed by migration to wintering areas of Uruguay and southern Brazil (Olrog, 1979; Yorio et al., 1994; Scolaro et al., 1996; Quintana and Yorio, 1997). Along the Brazilian coast, the breeding occurs from April to October (Sick, 1997; Efe et al., 2000; Branco, 2003a, b).

Mixed colonies of South American Tern and Cayenne Terns *Thalasseus sandvicensis* (Latham, 1787), were recorded on the coasts of Rio de Janeiro (Sick and Leão, 1965; Alves et al., 2004), Espirito Santo (Efe et al., 2000), São Paulo (Campos et al., 2004), Paraná (Krul, 2004), Santa Catarina (Soares and Schiefler, 1995; Branco, 2003a, b) and Patagônia, Argentina (Yorio et al., 1994; Scolaro et al., 1996; Quintana and Yorio, 1997).

The South American Tern nests close together (nearly 50 cm) in merely depressions in the ground lined with grasses. Colonies can be as large as 10,000 pairs in the Falklands Islands. In general, clutch size varied in one to three eggs that are, predominantly brown-yellow with

*Corresponding author. E-mail: h_fracasso@yahoo.com.br.

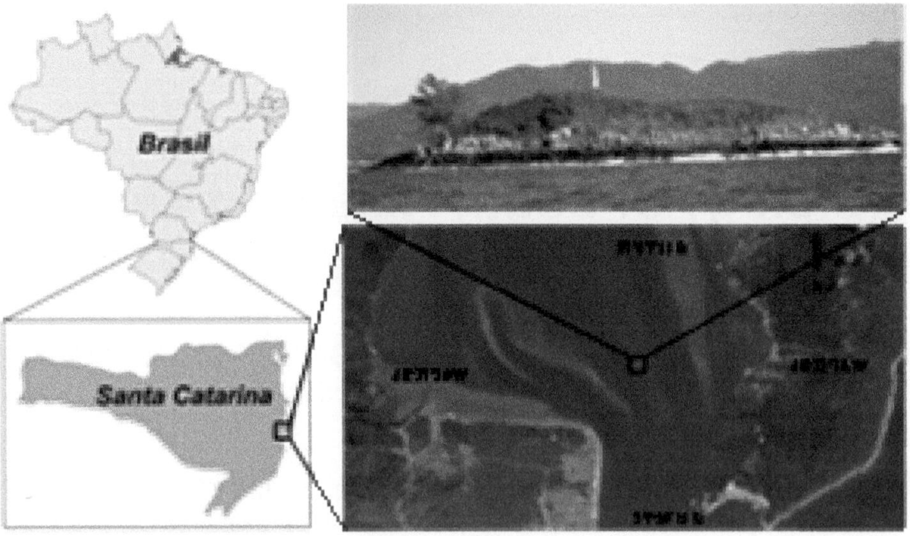

Figure 1. Map and overview of the study area, showing the location of Cardos Island in the coastal portion of Santa Catarina, Brazil.

black spots colours cryptic in appearance and they confuse predators (Scolaro et al., 1996, Branco, 2003b). The pressure from predators (hawks, vultures, lizards and kelp gulls), lack of food, and humans (eggs collection, fisheries, pollution, lack of habitat) are some of the factors that induced shifts in breeding sites from closer islands (Yorio et al., 1994; Scolaro et al., 1996; Efe et al., 2000), although along the Santa Catarina coast on the Itacolomis and Cardos Islands they have bred since 2001 (Branco, 2003b).

There is information available on the distribution and abundance of the species in Argentina (Duffy et al., 1988) and Brazil (Sick, 1997; Soares and Schiefler, 1995; Efe et al., 2000; Branco, 2003a; Branco, 2004). However, reproductive ecology was studied only in Punta Loma (Scolaro et al., 1996) and Santa Catarina (Branco, 2003b). This paper presents data on the reproductive behavior of the South American Tern during three different breeding seasons on The Cardos Island (Santa Catarina, Brazil). More specifically the reproductive and copulation period, number of inseminations, egg size and hatching timing of the species.

METHODOLOGY

Field work

Daily observations were made on terns on the Cardos Island (27°48'54"S, 48°34'52"W; area of 1.5 km^2), Florianopolis, Santa Catarina, Brazil (Figure 1), during the reproductive seasons of 2003, 2005 and 2006 (totaling 63, 121 and 144 days and 315, 847 and 1440 h of observation, respectively). Data on number of adults, reproductive behavior, mate choice, courtship feeding (delivery of fish), time and frequency of copulations and nest-building activities were obtained through observation with binoculars (10x50 Bushnell) and photography from a fixed point in the colony. Using the number of adults and nests and the egg-laying time, the reproductive season was categorized into three stages: beginning (12/04 to 04/07), middle (05/07 to 11/08) and late (after 12/08).

The duration of copulations (total time of male on the female's back) and the number of inseminations were recorded, except when the couples were disturbed (sound of helicopters, aircraft or boats, predators, or even attack from other individuals). The settlement and copulation period in 2005 was not observed because the adults arrived at the island earlier than expected (compared to 2003). In 2006, a sailor and a boat were hired to enable a greater period of observation on the island, extending the observation from 6:00AM to 06:00PM. The observation time was divided into 59 minute-intervals: that is, first observation interval started at 6:00AM and ended at 6:59AM.

We recorded: the number of nests built, number and size of eggs by laying order (marked with a Hidrocor pen "A", "B" and "C"), distance between nests (to 0.1 cm), and the number of newborn chicks. The length of the major axis of eggs (Lt) and width (Wid) in cm were determined with a caliper (0.1 cm) and weight (Wt) in grams was determined with a dynamometer type PESOLA (60 g and accuracy of 0.5 g). The volume of eggs was determined using Hoyt's (1979) equation: Vol (cm^3) = Kv.Lt.Wid^2 once: Kv = V/Lt.Wid^2, where volume estimated Kv = coefficient of 0.5205; Lt = total length; Wid = greater width of the axis (Branco, 2003a).

Statistical analyses

The distance between nests, biometrics and average volume of eggs, incubation period, order of laying in the seasons of 2003 and 2005 were examined through analysis of variance (ANOVA) and tested for homogeneity of variance (Bartlett test) and normality of

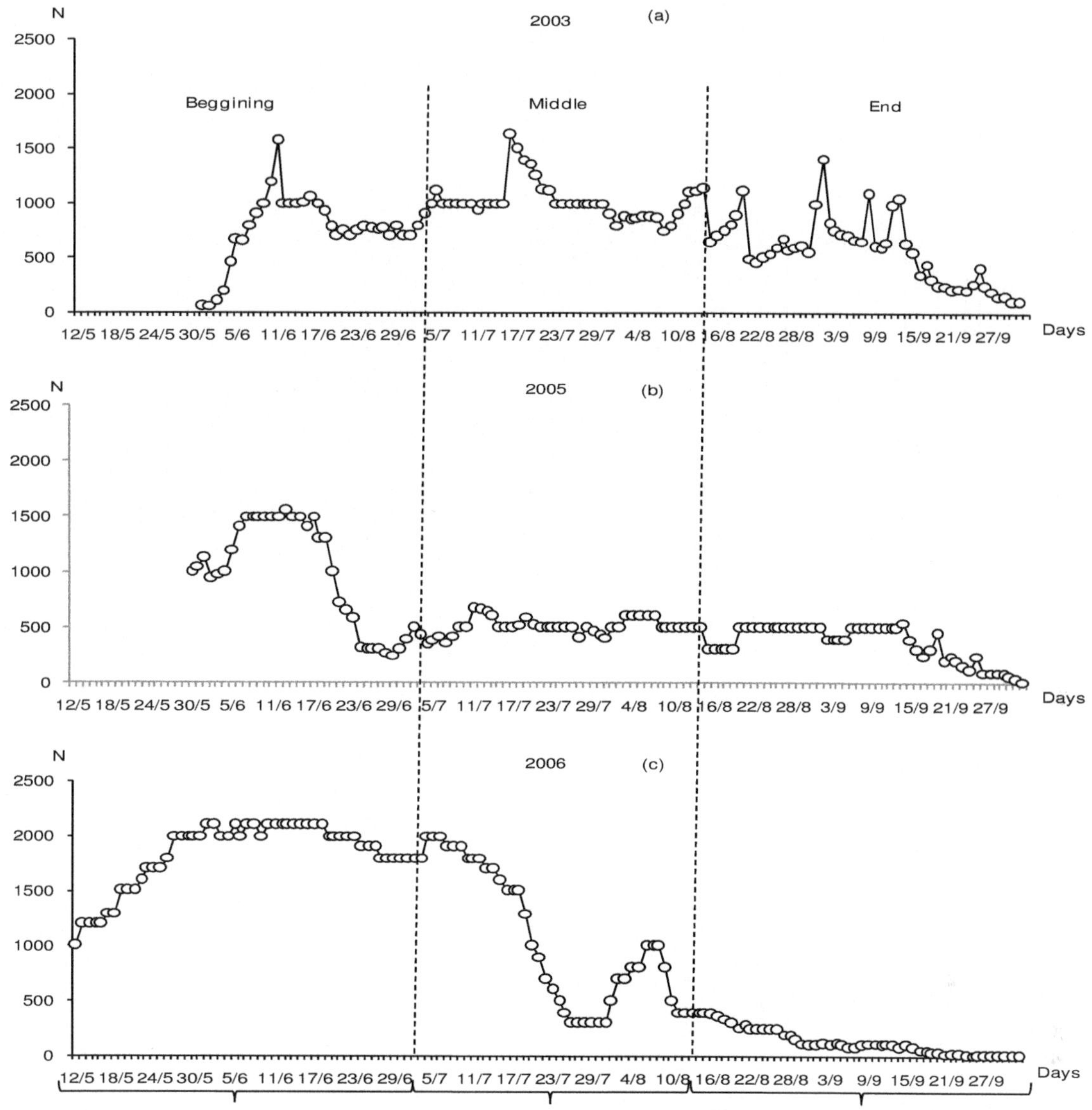

Figure 2. Daily variations in the absolute abundance of adults of *Sterna hirundinacea* during the period of (a) 2003, (b) 2005 and (c) 2006, at Cardos Island, Santa Catarina state, Brazil.

the distribution (Kolmorov-Smirnov) (Zar, 1999). The contrast of means (Tukey-Kramer test) was used to indicate which means were significantly different. The Student *t*-test ($p < 0.05$) was used to verify the existence of significant differences in the nests with two eggs between the beginning, middle and late reproductive periods of 2003 and 2005, and the center and edge of colony in 2006.

RESULTS

Establishment of the colony and Reproductive behavior

The reproductive season of 2003 began with the arrival of the first adults in mid-May, followed by a gradual increase and oscillations in the number of breeders until June, and greatest abundance in July. The island was abandoned in October (Figure 2). Colony formation in 2005 and 2006 started at the end of April, with the periods of beginning (late June), middle (July) and end (August) with the departure of adults in beginning October (Figures 2b and c).

In 2003, 138 copulations were observed with an average duration of 85.8 ± 11.8 s at 1:00 PM and 161.0 ± 85.8 s at 8:00 AM, and inseminations were repeated for up to seven times, ranging from 1.6 ± 0.4 (number of sexual intercourses) at 1:00 PM to 3.5 ± 0.5 at 11:00 AM

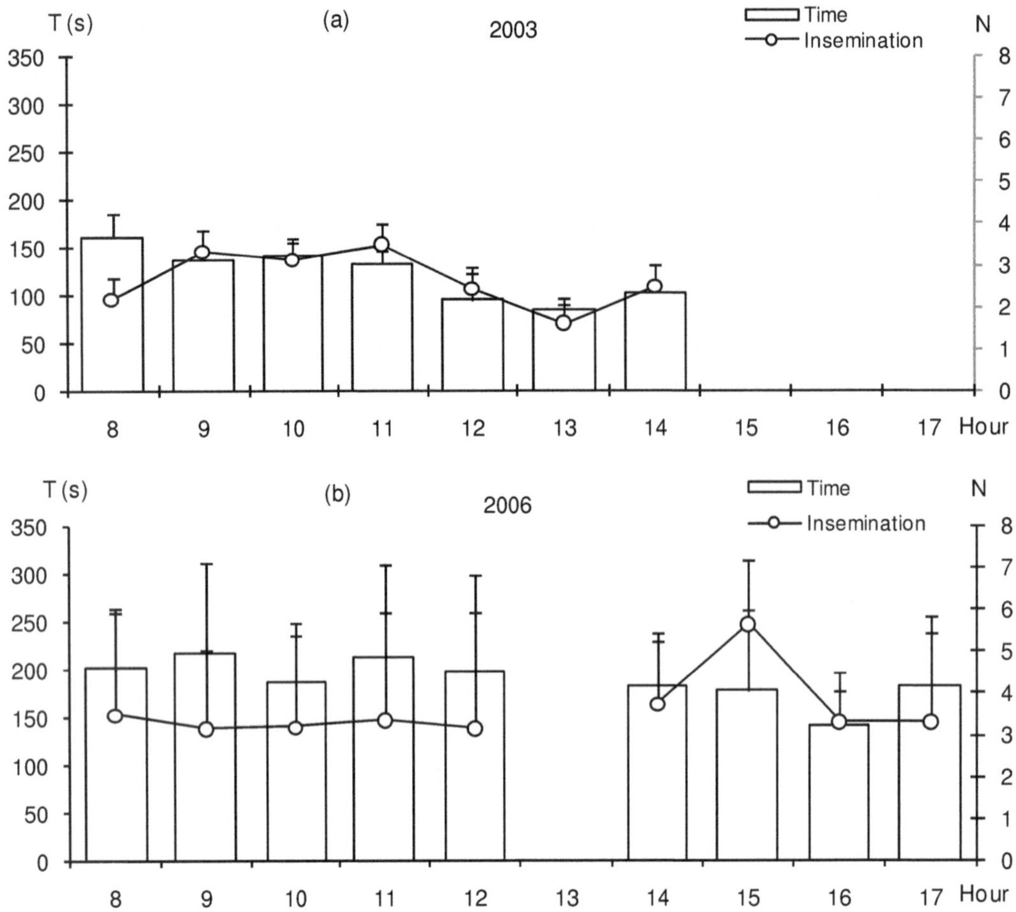

Figure 3. Number of inseminations (represented by continuous line) and duration of each insemination (represented by bars) in seconds measured during the mating of *Sterna hirundinacea* throughout the day, during the reproductive seasons of (a) 2003 and (b) 2006, at Cardos Island, Santa Catarina state, Brazil. (Vertical bars indicate the average standard error).

(Figure 3a). In 2006, the average time of copulation was 200 s, with the highest records at 9:00 AM and 11:00 AM and the lowest at 3:00 PM and 4:00 PM; the number of inseminations remained constant throughout the day, except at 3:00 AM when they were highest (Figure 3b).

Construction of nests and eggs' incubation period

After mating, the males and females started to build the nest, usually found on the ground (rarely in shrubs) using materials available on the island, such as fragments of shells and rocks, grass, dried grasses, grasses and small kindling.

The terns nested on most area of Cardos Island, excepting the Eastern side which had exposed rocks and bushes. Terns built an average of six nests/m^2, with significant differences among years ($F_{3\text{-}16} = 3.99$, $p < 0.01$ in 2003; $F_{3\text{-}236} = 23.42$, $p < 0.001$ in 2005 and $F_{3\text{-}266} = 14.51$, $p < 0.01$ in 2006) (Figure 4). The larger distances between nests were found in the highest area of the colony, covered by shrubs and big rocks, and in the lowest area, surrounded by rocks and grasses (2003) (Figure 4a) and other locations in 2005 and 2006 (Figure 4b and c; Tukey test).

In all the three years, the largest number of new nests were found at the beginning of the reproductive period (June of 2003 and 2005 and May of 2006), followed by peaks in July, with fewer new nests at its end (09/09, 25/08 and 01/09, respectively) (Figure 5).

Egg size

The smallest and largest eggs (length, width and weight) were recorded at the end and at the beginning of 2003

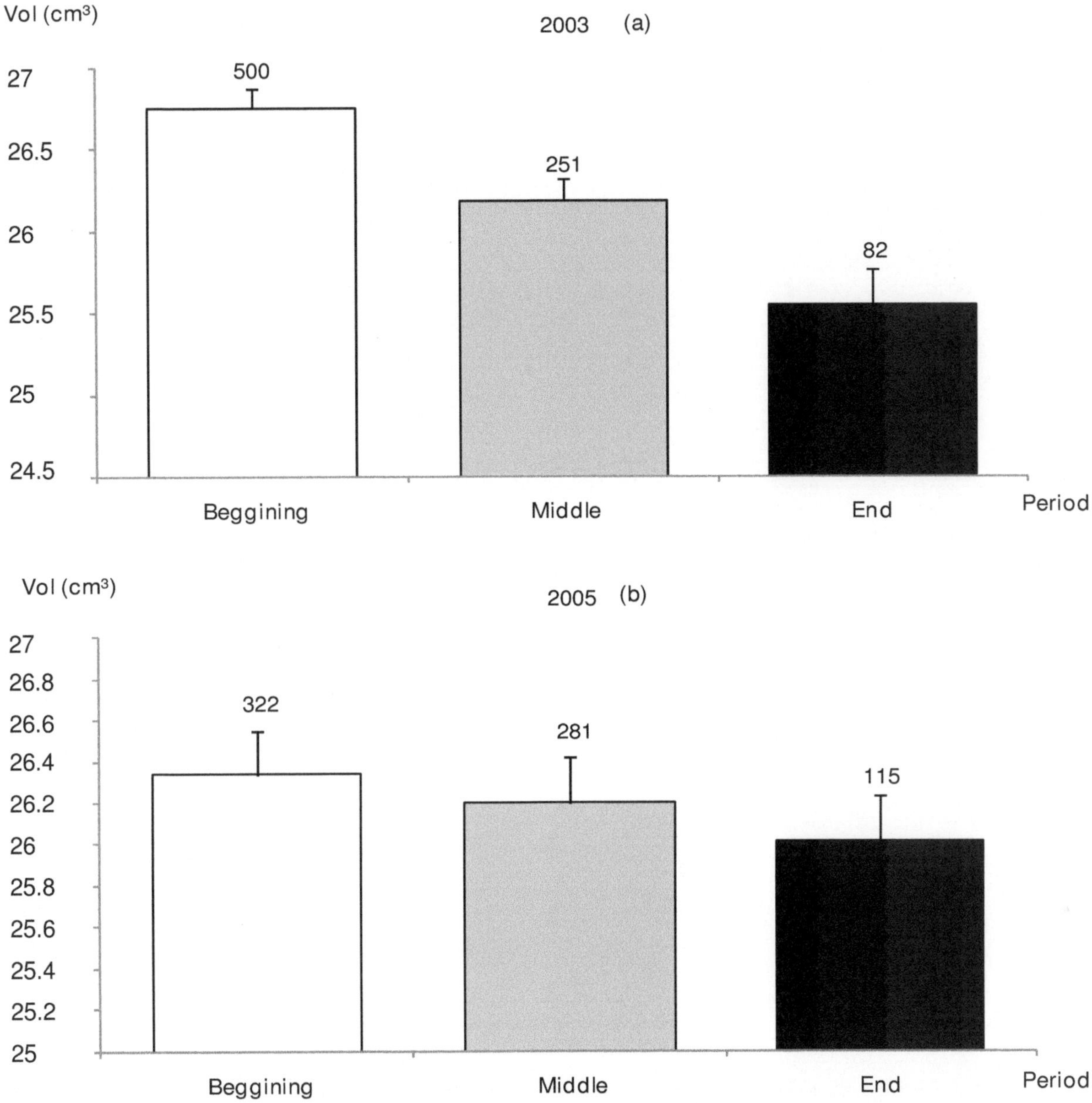

Figure 4. Distance between nests of *Sterna hirundinacea* constructed in different substrates during (a) 2003, (b) 2005 and(c) 2006 at Cardos Island, Santa Catarina state, Brazil. (The number above bars indicates quantify of measured distances, and vertical bars indicate the average standard error).

season, respectively, although weight was highest at the end of 2005 breeding season (Table 1). Eggs from 2003, regardless of laying order, did not differ significantly in length ($F_{2\text{-}830} = 1.506$, $p > 0.05$). However, they did differ in width ($F_{2\text{-}830} = 5.413$, $p < 0.05$) and weight ($F_{2\text{-}830} = 5.421$, $p < 0.05$) between periods. Differences in egg's size occur mainly at the end of the season (Table 1). In 2005, there were no differences in length ($F_{2\text{-}715} = 0.179$, $p > 0.05$) and width ($F_{2\text{-}715} = 1.186$, $p > 0.05$), although they differ significantly in weight ($F_{2\text{-}715} = 11.573$; $p < 0.001$) between the beginning and late periods due to lower values found at the beginning of the season (Table 1).

Eggs incubated in different periods in 2003 differed significantly in volume ($F_{2\text{-}830} = 10.907$, $p < 0.001$), attributed to lower value (smaller eggs) at the end of the season (Figure 6a). Eggs from 2005 did not differ

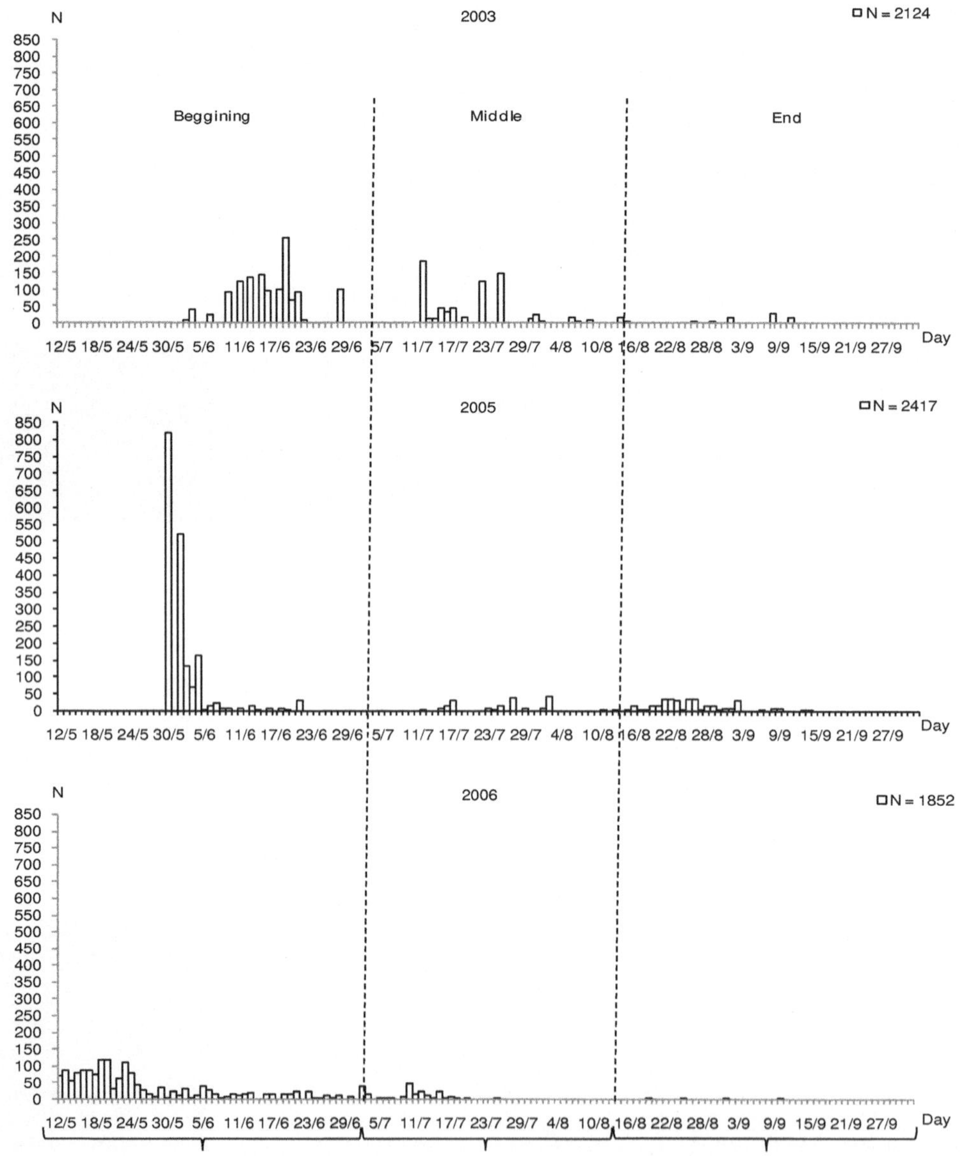

Figure 5. Daily abundance of new nests of *Sterna hirundinacea* during the beginning, middle and end of (a) 2003, (b) 2005 and (c) 2006 breeding seasons at Cardos island, Santa Catarina state, Brazil.

Table 1. Number of eggs of *Sterna hirundinacea*, and its biometry (length, width, and weight) recorded in the Cardos Island, Santa Catarina State, Brazil. (< = lower limit, > = higher limit values).

Year	Period	Length (cm)			Width (cm)			Weight (g)			N
		<	>	Mean ± error	<	>	Mean ± error	<	>	Mean ± error	
	Beggining	4.10	5.60	4.59 ± 0.01	3.00	3.50	3.29 ± 0.01	18.00	40.00	26.00 ± 0.12	500
2003	Middle	3.90	5.20	4.57 ± 0.01	3.00	3.20	3.26 ± 0.01	20.50	31.00	25.83 ± 0.13	251
	End	3.90	5.40	4.55 ± 0.02	3.00	3.40	3.23 ± 0.01	20.00	29.50	25.07 ± 0.23	82
	Beggining	3.90	5.30	4.57 ± 0.02	3.00	3.5	3.27 ± 0.01	18.00	32.00	25.46 ± 0.23	322
2005	Middle	4.00	5.10	4.57 ± 0.02	3.00	3.5	3.26 ± 0.01	20.00	30.50	26.06 ± 0.23	281
	End	4.10	5.10	4.56 ± 0.02	3.00	3.5	3.26 ± 0.01	22.00	31.00	26.41 ± 0.22	115

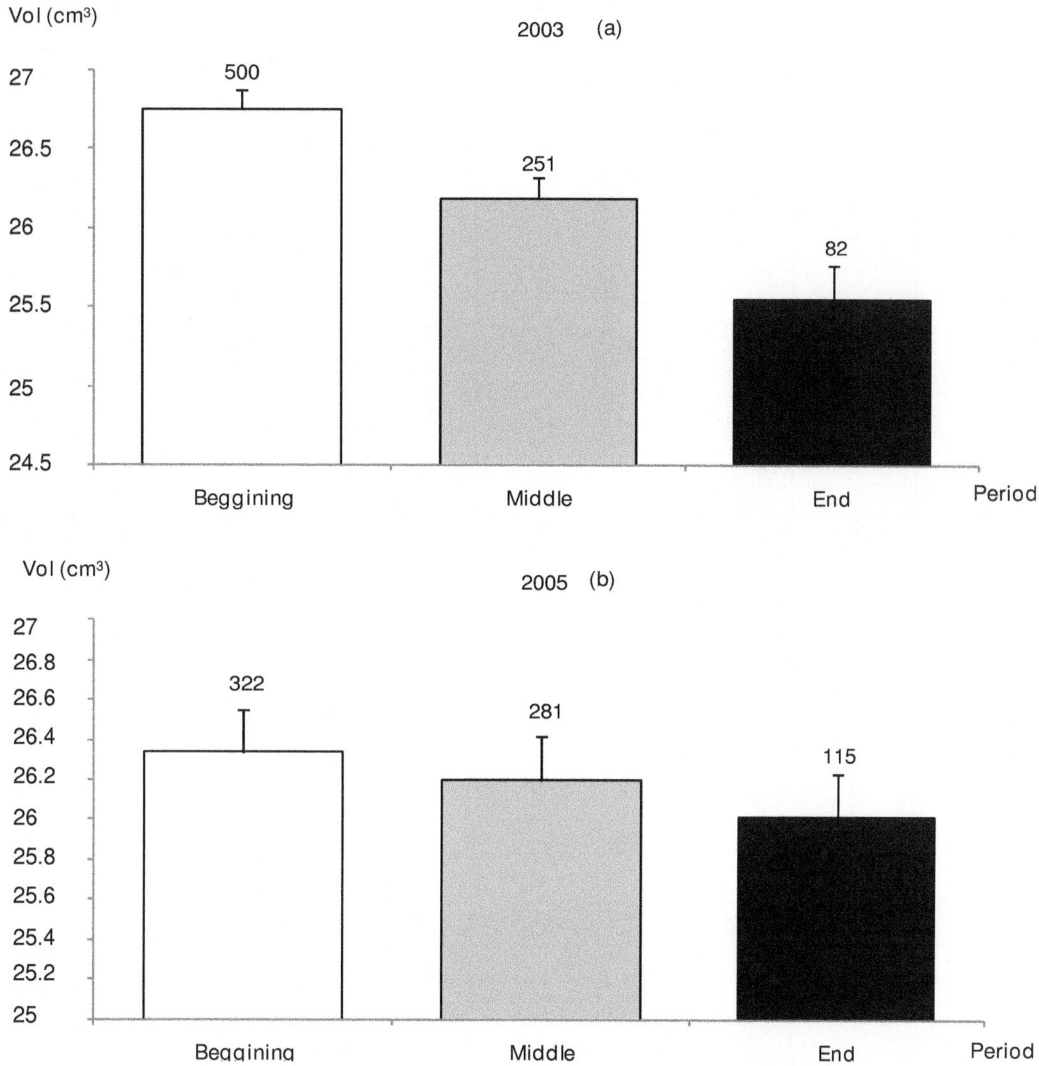

Figure 6. Average volume of *Sterna hirundinacea* eggs in the beginning, middle and end of the breeding season in (a) 2003 and (b) 2005 at Cardos Island, Santa Catarina state, Brazil. (The number above bars indicates quantify of measured eggs, and vertical bars indicate the average standard error).

significantly in volume between periods ($F_{2\text{-}715} = 1.214$, $p > 0.05$) (Figure 6b).

In nests with only one egg there were no differences in size in the beginning, middle and late periods of 2003 ($F_{2\text{-}87} = 2.424$, $p > 0.05$) and 2005 ($F_{2\text{-}48} = 1.622$, $p > 0.05$). Similarly there were no differences in nests with two eggs ($t = 0.188$, $p > 0.05$) and ($F_{2\text{-}160} = 0213$, $p > 0.05$), respectively (Figure 7). However, there were significant differences between the volume of "A" and "B" eggs in nests with two eggs between beginning ($t = 4.564$, $p < 0.001$), ($t = 7.681$, $p < 0.001$) and middle nesting periods ($t = 2.370$; $p < 0.05$), ($t = 3.207$; $p < 0.05$) in 2003 and 2005, respectively, but there was no difference between eggs in late nests in 2005 ($t = 1.800$, $p > 0.05$) (Figure 7b).

The 236 eggs found in 2006 had an average length of 4.61 ± 0.01 cm, average width of 3.30 ± 0.01 cm and average weight of 27.19 ± 0.12 g. Despite having recorded lower volume values at the edge than in center of the colony, there were no significant differences in the size of eggs "A" ($t = 1566$, $p > 0.05$) and "B" ($t = 1016$; $p > 0.05$) (Figure 8).

Abundance of newborn chicks

In 2003 the first eggs hatched in late June, with a gradual increase in the number of chicks until the peak in July 7 (N = 525), followed by decrease until August 4 (N = 31)

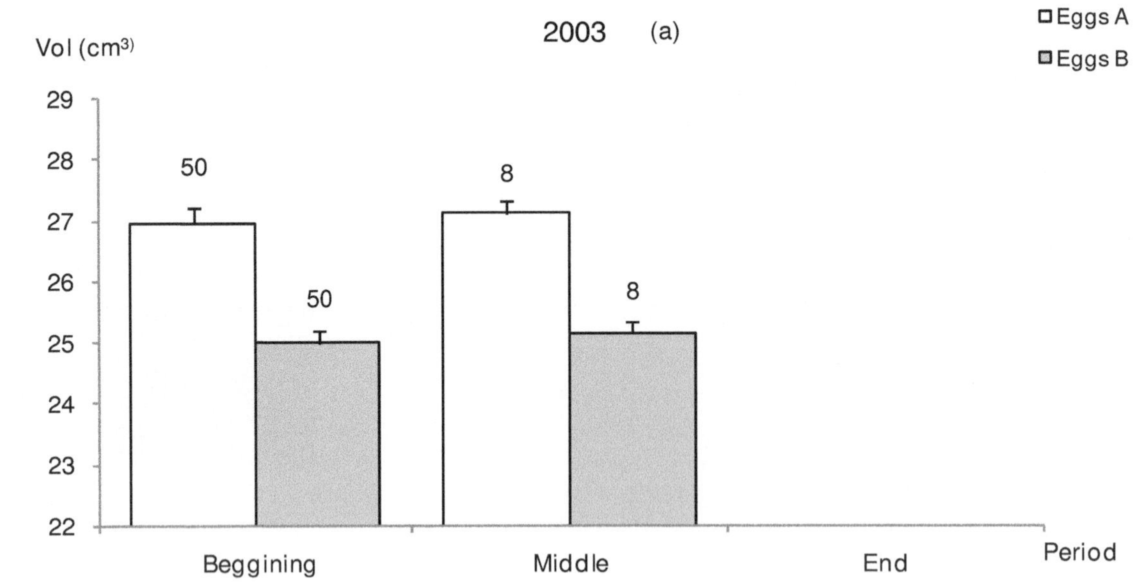

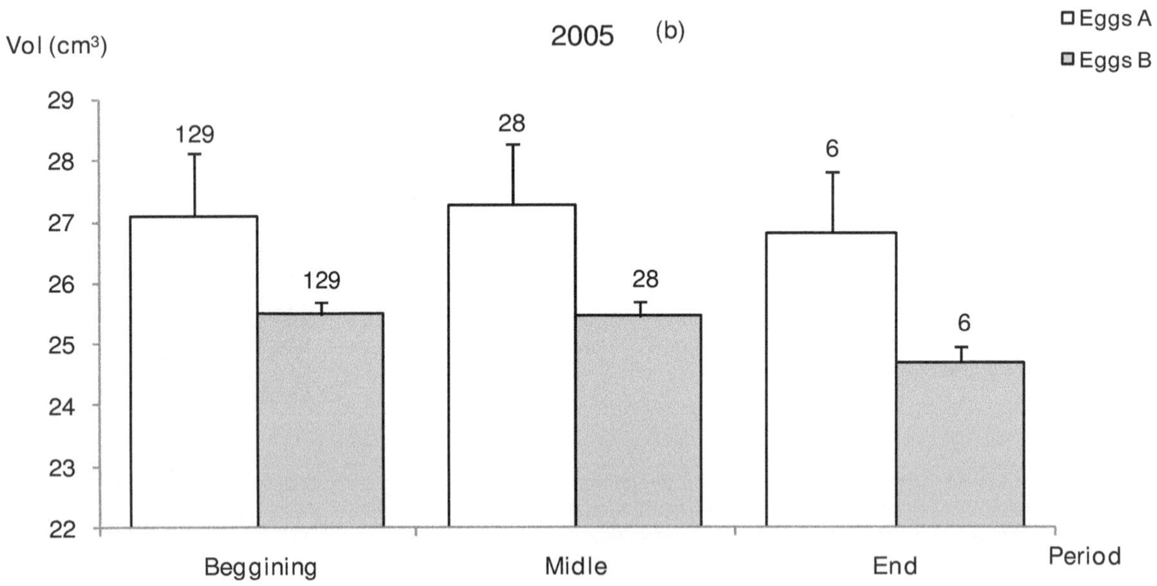

Figure 7. Average volume of *Sterna hirundinacea* A and B eggs tern during different periods of (a) 2003 and (b) 2005 breeding seasons at Cardos Island, Santa Catarina State, Brazil. (The number above bars indicates quantify of measured eggs, and vertical bars indicate the average standard error).

and oscillation from the middle towards the end of the season (September 15) (N = 14) (Figure 9a). In 2005, the first chicks appeared at the end of May, and followed the same trend of past seasons, with three peaks of abundance: June (N = 590), August (N = 54) and September (N = 26) (Figure 9b). The hatching of first egg in 2006 was in May 31st, with the highest values at the beginning of the season (N = 485), decreased in the middle and ceased at the end of the season (Figure 9c).

DISCUSSION

Many species of terns breed in large colonies like *Sterna antillarum* (Lesson, 1847), *Sterna hirundo* (Linnaeus, 1758) and *Sterna dougalli* (Montagu, 1813) that usually nest on the coast of the United States in beginning April or May (Jackson and Jackson, 1985; Spendelow, 1982; Nisbet et al., 1984; Hébert, 1985). Most seabirds are philopatric. However, some Sternidae that nest in

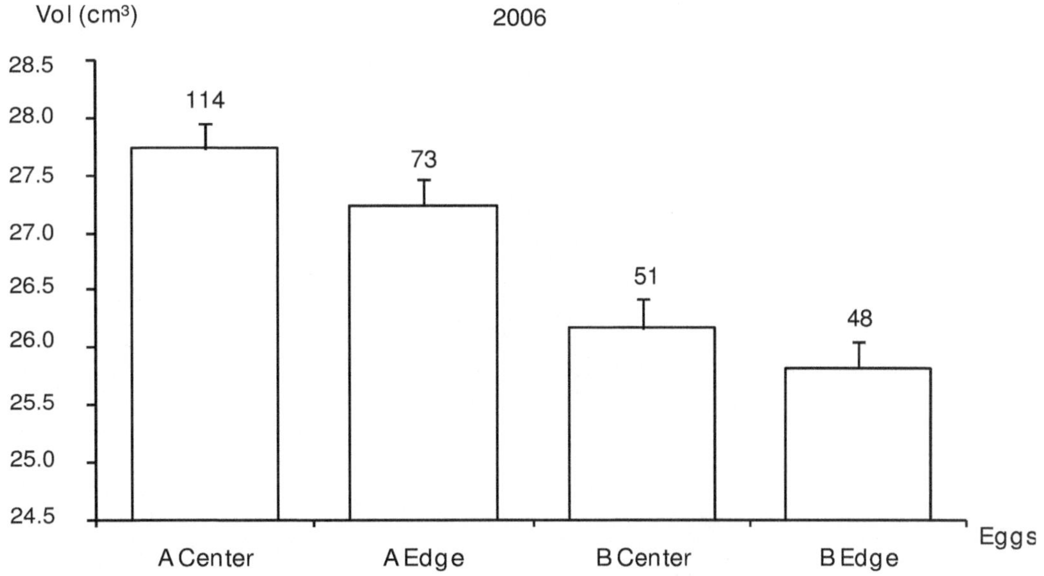

Figure 8. Average volume of *Sterna hirundinacea* A and B eggs tern in the center and edge of the colony, during the 2006 breeding season at Cardos Island, Santa Catarina State, Brazil (The number above bars indicates quantify of measured eggs, and vertical bars indicate the average standard error).

unstable environments are prone to change locations from year to year (Baird, 1990). *S. hirundinacea* and *S. eurygnatha* frequently and massively abandon their nesting areas due to disturbances (Yorio et al., 1994; Soares and Schiefler, 1995; Scolaro et al., 1996; Quintana and Yorio, 1997; Efe et al., 2000; Alves et al., 2004; Campos et al., 2004; Krull, 2004; Branco, 2003a, b).

The reproductive period of South American Tern in Santa Catarina coast, from May to October (Soares and Schiefler, 1995; Branco, 2003a) is similar to the one in the islands of Espirito Santo (May) (Efe et al., 2000), to the season in the coast of Rio de Janeiro (from March to October) (Sick and Leão, 1965; Sick, 1997), Paraná (from June to September) (Krull, 2004) and in São Paulo (from May to August) (Campos et al., 2004). According to Branco (2003b) the first adult of *S. hirundinacea* were observed at Cardos island in mid-April (2002), with a gradual population increase until July, which corroborates our observations for 2005 and 2006. In 2003 the occupation of our site was late (end of May), which can be attributed to variation in average temperature and food availability near the colony. In Argentina the breeding season is from November to December (Scolaro et al., 1996) and genetic homogeneity was found among South American tern breeding colonies from Brazil (Faria et al., 2009). Low, but significant, genetic differentiation was detected between the Patagonian and some Brazilian colonies (providing evidence that the Atlantic populations of these terns are not completely panmictic. These findings are supported by two different types of molecular markers (799 bp of mtDNA and five microsatellites), which despite their dissimilar inheritance models and evolutionary rates, provided concordant results (Faria et al., 2009).

The reproductive behavior of observed for *S. hirundinacea* at Cardos island is consistent with that described for Terns in general (Nelson, 1980; Krebs and Davies, 1996) and specifically reminds that of *S. eurygnatha* on the island Itatiaia (ES) (Efe, 2004). In Itatiaia the season began in mid-May and displays were more intense in the beginning hours of the day (Efe, 2004).

The pattern and spread of egg-laying in colonial species is viewed as an adaptive function to reduce losses by predation (Houde, 1983). That is, highly synchronous breeding reduces predation through swamping (predators can eat only one egg so many each day). The population of South American Terns that nest in Punta Loma (Argentina) follows the pattern established for other species of Terns in temperate zones (Langham, 1983), with dense colonies (Buckley and Buckley, 1980; Burger, 1981; Scolaro et al., 1996). At Cardos Island we observed a similar trend of colony occupation in various types of substrates, with disputes with Cayenne Terns occurring frequently in higher areas above the tide.

South American Terns occupy small areas with high concentrations of nests; on Cardos Island the average

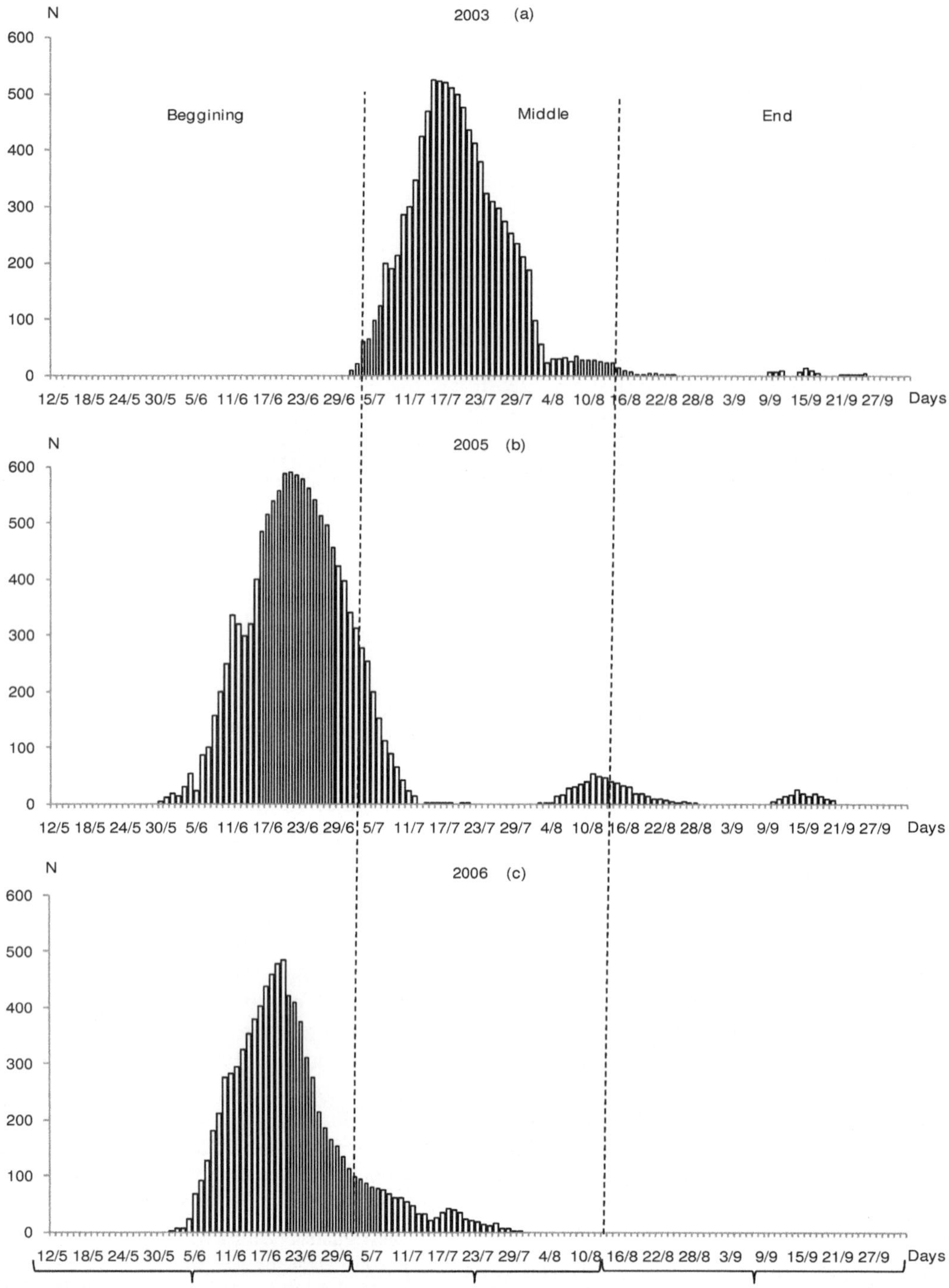

Figure 9. Daily variations in the number of newborn chicks of *Sterna hirundinacea* along the reproductive seasons of (a) 2003, (b) 2005 and (c) 2006 at Cardos Island, Santa Catarina state, Brazil.

Distance between nests was 0.7 m in an area of approximately 0.5 ha (six nests/m^2), while on Itaçuce Island (coast of São Paulo) there were 400 nests in 0.6 ha, (four nests/m^2) (Campos et al., 2004). The distance between the nests of Common Terns in New Jersey (USA) marshes was lower (between 1.50 to 2.12 m) than in areas of beach (2.69 to 2.71 m) (Erwin and Smith, 1985).

Data on egg-laying dates for South American Terns are not available for enough years to establish a pattern for egg-laying activities, but Common Terns have become earlier egg-layers over the last years (Nisbet et al., 1984), and they also have higher average number of eggs per clutch (Spendelow, 1982). The egg-laying period for South American Terns in the coast of Argentina began in December and showed two peaks, mid-December and beginning January, with late nests at the end of January (Scolaro et al., 1996), while in Espirito Santo and Santa Catarina it began in May with the largest number of eggs in June (Efe et al., 2000; Branco, 2003b), indicating two distinct populations, corroborated by Faria et al. (2009).

The eggs of South American Terns on Cardos Island were smaller, narrower, lighter and of less volume than eggs from Deserta Island (27.35 ± 5.23 cm^3) (Branco, 2003b). This difference may be related to climate variations, plus the availability of food near the colony, once that Deserta is part of a Biological Marine Reserve, and is forbidden to fish and in Florianopolis there are few fisheries of by-catch discards.

On the coast of Espirito Santo the first chicks of Cayenne Terns appeared in beginning of July (Efe, 2004) and the South American Terns of Cardos Island between May 20 to 25, 2002 (Branco, 2003b). In the following season (2003), chicks appeared in June, and 2005 and 2006 their presence was brought forward to May, which may have been favored by a better suited to the climate and growth due to lower harassment from predators.

S. hirundinacea breeds uninterruptedly at Cardos Island since 2001. Some characteristics of this island, such as the physical conditions (low probability of flooding), lower number and action of predators, protection by vegetation or rocks, little distance from the area of foraging, low exposure to the prevailing winds, and the presence of *Sterna eurygnatha*, which often helps in defense against predators defense, despite the competition for food and space, allowed uninterrupted breeding of *S. hirundinacea* on the island of Cardos since 2001, making this island an important reference for the conservation of the species in the Brazilian coast (Buckley and Buckley, 1980; Burger, 1981; Thompson and Slack, 1982).

ACKNOWLEDGMENTS

Our gratitude goes to the Brazilian Research Council for Scientific and Technological Development and the Projeto Aves Marinhas who funded this project. The Foundation on the Environment of Santa Catarina - FATMA, who granted the authorizations to work on the Island of Cardos and to IBAMA (CEMAVE) for all licenses to work on birds. At colleges and trainees: Cristiano L. Evangelista, Fernando K. Decker, Juliano C. Hillesheim, Felipe Freitas Jr., among others, the valuable field assistance and sorting of the material.

REFERENCES

Alves VS, Soares ABA, Couto GS (2004). Aves marinhas e aquáticas das ilhas do litoral do Estado do Rio de Janeiro, p. 83-100. *In*: J.O. Branco [ed.], Aves marinhas insulares brasileiras: bioecologia e conservação. Editora da UNIVALI, Itajaí, SC.

Baird PH (1990). Influence of abiotic factors and prey distribution on diet and reproductive success of three seabird species in Alaska. Ornis Scandinavia, 21: 224-235.

Branco JO (2003a). Reprodução das aves marinhas nas ilhas costeiras de Santa Catarina, Brasil. Revista Brasileira de Zoologia, 20(4): 619-623.

Branco JO (2003b). Reprodução de *Sterna hirundinacea* Lesson e *S. eurygnatha* Saunders (Aves, Laridae), no litoral de Santa Catarina, Brasil. Revista Brasileira de Zoologia, 20(4): 655-659.

Branco JO (2004). Aves marinhas das Ilhas de Santa Catarina, p.15-36. *In*: J. O. Branco [ed.], Aves marinhas e insulares brasileiras: bioecologia e conservação. Editora da UNIVALI, Itajaí, SC.

Buckley FG, Buckley AP (1980). Habitat selection and marine birds, 4: 69-112. *In:* J. Burger, B. Olla and H. E., Winn [eds.], Behavior of marine animals, New York. Columbia University Press. New York

Burger J (1981). A model for the evolution of mixed-species colonies of Ciconiiformes. Quaternary Rev. Biol., 56: 143-167.

Burger J, Gochfeld M (1991). The common Tern. Columbia University Press, New York, New York.

Campos FP, Paludo D, Faria PJ, Martuscelli P (2004). Aves insulares marinhas, residentes e migratórias, do litoral do Estado de São Paulo, p. 57-82. *In*: J.O. Branco [ed.], Aves marinhas insulares brasileiras: bioecologia e conservação. Editora da Univali, Itajaí, SC.

Duffy DC, Ryan PG, Wilson RP, Wilson MP (1988). Spring seabird distribution in the Strait of Magellan. Cormorant, 16: 98-102.

Efe MA (2004). Aves marinhas das ilhas do Espírito Santo, pp. 101-118. *In:* J. O. Branco [ed.], Aves marinhas insulares brasileiras: bioecologia e conservação. Editora da UNIVALI, Itajaí, SC.

Efe MA, Nascimento JLX, Nascimento ILS, Musso C (2000). Distribuição e ecologia reprodutiva de *Sterna sandivicensis eurygnatha* no Brasil. Melopsittacus, 3(3): 110-121.

Erwin RM, Smith DC (1985). Habitat comparisons and productivity in nesting common terns on the mid-atlantic coast. Colonial Waterbirds, 8(2): 155-165.

Faria PJ, Campos FP, Branco JO, Musso CM, Morgante JS, Bruford MW (2009). Population structure in the South American tern Sterna hirundinacea in the South Atlantic: two populations with distinct breeding phonologies. J. Avian Biol., 41: 378-387.

Hébert PN (1985). Breeding failure and decline of a common tern colony in Southern Manitoba. Colonial Waterbirds, 8(2): 183-185.

Higgins PJ, Davies SJJF (1996). Handbook of Australian, New Zealand and Antarctic Birds. Vol. 3. Snipe to Pigeons. Oxford University Press, Melbourne.

Houde AE (1983). Nest density, habitat choice and predation in a common tern colony. Colonial Waterbirds, 6: 178-184.

Hoyt DF (1979). Practical methods of estimating volume and fresh weight of bird eggs. Auk, 82: 507-508.

Jackson JA, Jackson BJS (1985). Status, dispersion and population changes of the least tern in coastal Mississippi. Colonial Waterbirds, 8(1): 54-62.

Krebs JR, Davies NB (1996). Introdução à Ecologia Comportamental. Atheneu, Editora São Paulo, SP.

Krull R (2004). Aves Marinas costeiras do Paraná, p. 37-56. *In:* J.O. Branco [ed.], Aves marinhas insulares brasileiras: bioecologia e conservação. Editora da Univali, Itajaí, SC.

Langham NP (1983). Growth strategies in marine terns. Studies in Avian Biology, 8: 73-83.
Nelson B (1980). Seabirds: their biology and ecology. The Hamlyn Publishing Group Limited, Italy, p. 224.
Nisbet ICT, Winchell JM, Heise AE (1984). Influence of age on the breeding biology of common terns. Colonial Waterbirds, 7: 117-126.
Olrog CC (1979). Nueva lista de la avifauna argentina. Opera Lilloana. 27: 1-324.
Quintana F, Yorio P (1997). Breeding biology of royal and cayenne terns at a mixed-species colony in Patagonia. Wilson bull., 109(4): 650-662.
Scolaro JA, Laurenti S, Gallelli H (1996). The nesting and breeding biology of the south american tern in northern Patagônia. J. Field Ornithol., 67(1): 17-24.
Sick H, Leão AP (1965). Breeding sites of *Sterna eurygnatha* and other sea birds of the Brazilian coast. Auk, 82: 507-508.
Sick H (1997). Ornitologia Brasileira. Rio de Janeiro, Nova Fronteira, p. 862.
Soares M, Schiefler AF (1995). Aves da Ilhota da Galheta, Laguna, SC, Brasil. Arquivos de Biologia e Tecnologia, 38(4): 1101-1107.
Spendelow JA (1982). An analysis of temporal variation in, and the effects of habitat modification on, the reproductive success of roseate terns. Colonial Waterbirds, 5: 19-31.
Thompson BC, Slack RD (1982). Physical aspects of colony selection by terns on the Texas Coast. Colonial Waterbirds 5: 161-168.
Yorio P, Quintana F, Campagna C, Harris G (1994). Diversidad, abundancia y dinamica espacio-temporal de la colonia mixta de aves marinas en Punta Leon, Patagonia. Ornitologia Neotropical 6(2): 69-77.
Zar JH (1999). Biostatistical Analysis, 4a ed. Prentice-Hall Inc., New Jersey, p. 663.

5

Morphotectonics and formation of the Taiwan Bank Canyon, Southwest offshore Taiwan Island

Weiwei Ding[1,2]*, Jiabiao Li[1,2], Xiqiu Han[1,2], Erwin Suess[3], Yongyang Huang[4], Xuelin Qiu[5] and Mingbi Li[1,2]

[1]Key Laboratory of Submarine Geoscience of the State Oceanic Administration, Hangzhou 310012, China.
[2]Second Institute of Oceanography, State Oceanic Administration, Hangzhou 310012, China.
[3]Leibniz-Institute for Marine Sciences, 24148 Kiel, Germany.
[4]Guangzhou Marine Geological Survey, Guangzhou, 510075, China.
[5]Key Laboratory of Marginal Sea Geology, South China Sea Institute of Oceanology, Chinese Academy of – Sciences, Guangzhou, 510301, China.

The South Taiwan Bank Canyon (STBC) is a large canyon on the southwest offshore Taiwan Island, bearing significant implications in fossil energy exploration. In this study, we analyze the morphology and sedimentary structure of the STBC, as well as those in nearby continental margin, using high-resolution swath bathymetry and seismic data. The the STBC is multiple-headed, steep in the upper part as a V-shaped incised trough and gentle in the lower as a U-shaped one; the direction changes from NW-SE to E-W when it joins the Manila Trench at around 2200 m. The regional tectonic framework shows that the cannon was born in the Neogene under the influence of onshore rivers, followed by rapid thermal subsidence after the mid-Miocene, enhancing the sediment transport with frequent slides, slumps, and other sediment flows, and resulting in an incised slope with multiple gullies and heads. The NW tenso shear faulting triggered by the collision of the Luzon Arc against the Eurasian Plate after mid-Miocene probably pre-determined the position of the canyon and caused heavy down-cutting erosion. Strong magmatism in the underlying oceanic plate segment after the cessation of the South China Sea's spreading could also play roles in reducing slope stability and generating seamounts that blocked the southeastward flow in the canyon and coerced the direction change into the east.

Key words: South Taiwan Bank Canyon, morphology, tectonic setting, controlling factors.

INTRODUCTION

Submarine canyons are common features on continental margins wordwide, either active or passive (Farre et al., 1983; Hagen et al., 1996; Klaus and Taylor, 1991) Most of them are located entirely on the upper slope, while there are also some large canyons with head that deeply indent the shelfbreak. They are generally in "V" shape (erosional situation) or "U" shape (depositional situation). What makes investigation of a submarine canyon interesting is that submarine canyons are prominent topographic features on the sea floors of continental margins.

They not only shape the present morphology of continental margins, but also acted as the main conduits for transferring terrestrial sediments toward the deep sea and fed the fan system (Kolla and Perlmutter, 1993). Apart from constituting an important archive for documenting the sedimentation process history of an area, submarine canyons are also studied as modern analogues for deepwater hydrocarbon reservoirs because of their association with sand-rich turbidites.

The South Taiwan Bank Canyon (STBC) is a large shelf-indenting canyon developed on the northeast continental margin of the South China Sea, off Southwest Taiwan Island. The canyon begins south of the Taiwan Bank high, featuring shelf-indenting heads, crosses the continental slope and deep sea basin, and finally merges

*Corresponding author. E-mail: wwdingsio@yahoo.com.cn

Table 1. Acquistition parameters for seismic profiles used in this study.

Profile names	973-01	973-04	E53
Acquired by	GMGS*	SCSIO**	CNOOC***
Acquisition date	2001	2001	1989
Streamer channel	240	48	240
Record length (s)	10.024	12.0	5.0
Sampling rate (ms)	2	2	2
Receiver interval (m)	12.5	25.0	12.5
Shot interval (m)	50	50	25
Airgun volumn (inch3)	3000	1520	3946

* GMGS: Guangzhou Marine Geological Survey, Chinese Ministry of Land and Resource; ** SCSIO: South China Sea Institute of Oceanology, Chinese Academy of Sciences; *** CNOOC: China National Offshore Oil Corperation;

into the Northern Manila Trench for a distance of about 315 km. Many efforts had been made to understand the morphology, sedimentary and forming mechanism. Jin (1989) described the origin, extending direction, length and changes of the shape with the depth. Wu and Bao (2000) considered that the STBC was strongly related to a NW-direction fault, named Hanjiang fault formed in the late Cenozoic.

Yu and Zhou (2001) discussed the geomorphic characteristics offshore southwest Taiwan and argued that the morphology was mostly related to structural and sedimentary processes. Based on geophysical data, Sibuet et al. (2002) and Hsu et al. (1998) proved the existence of a transverse plate boundary in the Northeast South China Sea. The STBC seemed to have been controlled since its formation by this boundary. However, the lack of high-resolution multi-beam data on this area hindered, until now, the study of detailed morphology. The age, origin and its relationship with the structures, sea level changes, onland-rivers and ocean currents are almost unclear. Using the multi-beam bathymetric mapping, high-resolution seismic profiles and gravity piston columns, we discussed the geomorphic characteristics and define sedimentary processes that marked its evolution. The purpose of the paper is to evaluate the development of this large shelf-indenting canyon and to identify the main canyon-forming forces.

GEOPHYSICAL DATA AND METHODS

The bathymetric data set over the region of 118°15´- 119° 25´E, 19°55´- 22°10´N, northeast South China Sea (Figure 1), was collected during the joint Chinese-German RV SONNE Cruise 177, in June 2004, using a SIMRAD EM120 multi-beam echo-sounder system. The EM120 system can provide accurate bathymetric mapping up to depths greater than 11000 m. It is composed of two transducer arrays fixed on the hull of the ship, which send successive frequency coded acoustic signals (11.25 to 12.6 kHz). Data acquisition is based on successive emission-reception cycles of this signal. The emission beam is 150° wide across track, and 2° along track direction.

The reception is obtained from 191 overlapping beams, with widths of 2° across track and 20° along it. During deep tow profiling dense line, offset was used to reduce the opening angles of the beams in order to receive higher resolution images while still covering the complete area. Bathymetric data were processed routinely onboard during the survey, using the NEPTUNE software from Simrad, available on board and the academic software MB-System from Lamont – Doherty Earth Observatory. In order to show morpho-structural features more clearly, the data were processed to generate shaded relief images after being edited (Figure 2).
The seismic data come from other cruises. Acquisition parameters for these seismic profiles are listed in Table 1. Locations are shown in Figure 1.

Geological background

The northeastern part of the South China Sea continental margin is a special region in many aspects of its tectonics. It lies in the junction area of several major tectonic regimes of distinct evolutionary histories. To its immediate west lies the Pearl River Basin which formed through continental stretching and rifting since 50 Ma (Zhou et al., 1995). To the south is cooling oceanic crust of the South China Sea which formed during the subsequent oceanic spreading during the Oligo-Miocene (~30-15 Ma) (Briais et al., 1993). The northwest movement of the Philippine Sea Plate and the following arc-trench collision against the Eurasian margin in 6.5 Ma made a subduction zone along the Manila Trench and Taiwan Orogen, which lies in the east and northeast of the study area, respectively (Huang et al., 1997, 2001).

Most of the researchers agreed that the northeast South China Sea continental margin had the same origin as the north continental margin of the South China Sea, which experienced Mesozoic Andean type convergent margin and Cenozoic continental stretching and rifting (Taylor et al., 1983; Lüdmann and Wong, 1999; Li et al., 2007). But there are others who argued that a sector of the northeast South China Sea is Paleo-oceanic crust that differs from the central South China Sea crust, but which may have some genetic relationships to a portion of the present-day Huatung Basin (Sibuet et al., 2002, 2004). After the cessation of South China Sea's sea-floor spreading, the convergence between the Eurasian and the Philippine Sea Plate changed the stress regime of the South China Sea. The arc-trench collision between the Luzon Arc and Eurasian, made the northeast South China Sea continental margin be in extrusion regime. Many inverse structures such as thrusts, faulted bend folds were formed, together with some renewed NW-trending tenso-shear faults

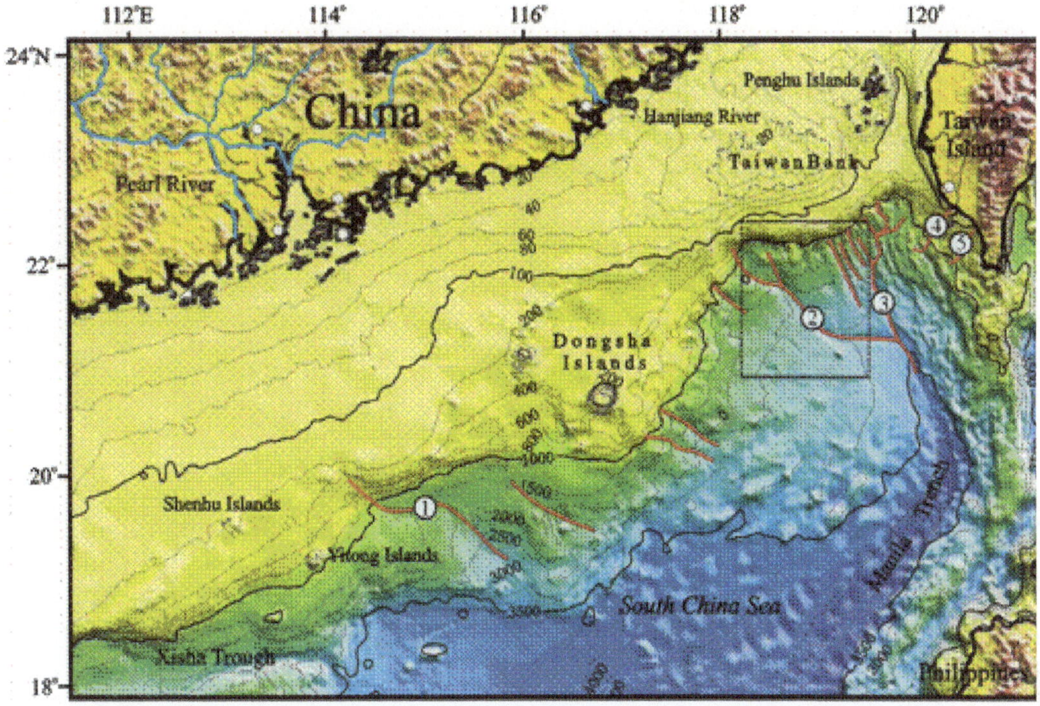

Figure 1. Morphologic map of the north continental margin, South China Sea, showing the study area (The red lines show the location of the gullies and canyons; 1: Pearl River Mouth Canyon; 2: STBC; 3: Penghu Canyon; 4: Kaoping Canyon; 5: Fangliao Canyon.

(Lüdmann and Wong, 1999; Ding et al., 2006). The continental basins in study areas begun their fore-land basin development, terrestrial sedi-ments originated from the Taiwan Oregon accumulated in the west front (Chou and Yu, 2002; Lin et al., 2003; Chiang et al., 2004).

MORPHOLOGICAL CHARACTERISTICS

General morphology

The research area is located towards the east of the northern continental shelf of the South China Sea. Tectonically, it is a passive margin setting grading into the South China Sea Basin and which abuts the accretionary wedge formed off-shore south-western Taiwan Island. This area covers about 2.97×10^4 km^2. Generally, the topography is very complicated and shows highly-changeable vertical gradients. In the northern part the topography alternates between steep slopes and flat-topped ridges extending from the west to east and descend gradually from northwest to southeast into the deeper part of the South China Sea Basin. The average slope is about 7° and the water depth ranges from 150 m to over 3200 m.

According to the topographic features, the slope can be divided into an upper and a lower portion. The upper portion slopes between 1 to 3°, the water depth is between 150 to 2200 m and it is about 30 to 70 km wide, whereas the lower portion slopes between 2-4° and is between 2200 to 3200 m deep. The morphology is especially complicated in the northern portion in which steep scarps and canyons have developed with erosional channels, valleys and ridges in between (Figures 2 and 3).

In the southern part of the research area, the topography is relatively simple with water depths gradually increasing southward from 2200 to 3200 m. On the northern flank of the STBC, there are several vertical cliffs with displacements of up to 30 m. They appear to result from down-sloping moving sediment waves which break off as they are intersected by the channel. A coniform sea mount can be clearly identified at 119°05'E, 21°17'created by the magmatic activities after the cessation of the South China Sea's spreading (Suess, 2005).

Morphology and structure of the STBC

The STBC is the most remarkable morphologic unit in this region. It crosses the entire area from the NW to the SE and then turns into EW direction at around 21°17'N, 119°05'E before finally joining the Manila Trench. The morphological features along the canyon and the orientation vary significantly along the canyon, as well as gradient of the channel floor along the axial thalweg.

From its head region at the shelf edge and upper slope areas, the canyon initially runs NW-SE with multi-heads. Figure 2 shows the canyon separates into two channels north of 21°45' with different orientation. Due to the lack of multi-beam data, only a part of the west channel is shown in our figures. Previous works on the STBC showed that the head of the west channel may reach to the South Taiwan Bank of ~30 m water-depth [6]. The east channel begin at <200 m water-depth with multi-heads. At water depth shallower than 1500 m the course is characterized by a steep thalweg gradient (Figure 4) with the maximum which can reach to 12°8'. Bathymetric cross-sections across the canyon heads show deeply incised, narrow and V-shaped troughs with steep walls (1 and 2 cross-sections in Figure 5). The maximum incision of the seabed by this segment reaches to 1200 m, as determined by relief between canyon bottom and edges (2 in Figure 5). At the depth greater than 2000 m where gradient reduces, the width of the canyon increase steadily, and the two channels gradually merge, showing U-shaped cross-sectional morphology (3 and 4 in Figure 5) with a low ridge

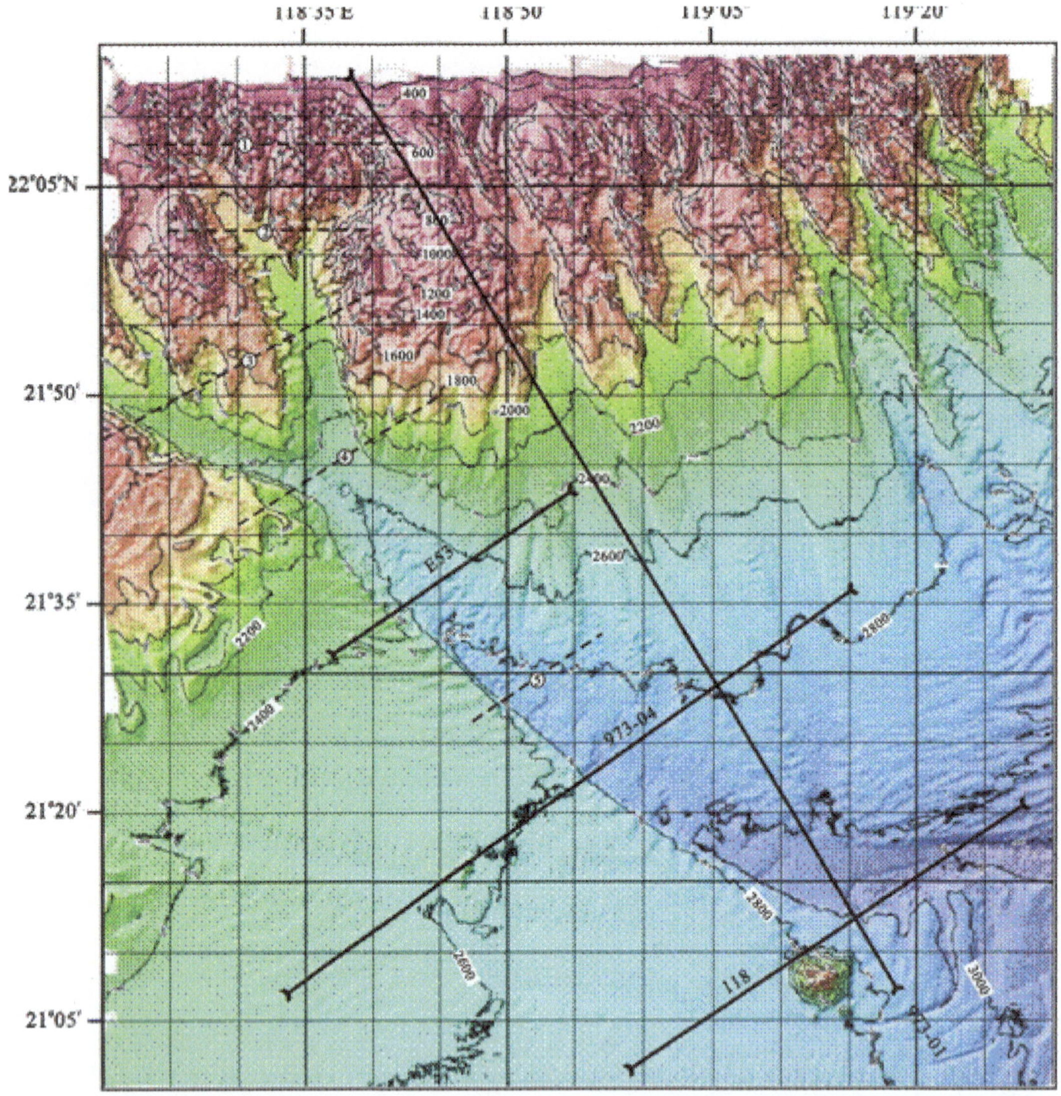

Figure 2. Bathymetric 3D map of the STBC, obtained from EM120 data. The black real lines are the multi-channel seismic reflection sections, and the broken lines are the bathymetric profiles.

separating the east channel from the west one (visible in cross-section 3 but in cross-section 4 of Figure 5).

Seismic profile E53 can offer more details of the sedimentary structure (Figure 6) in the above-mentioned area. It shows that truncations of parallel reflectors against the east and west walls, indicating down-cutting of the canyon. Stepped and curved surfaces in the middle ridge indicate that sliding or slumping of the strata. The flat reflector at the canyon bottom is considered to be a cut-and-fill feature. Sediments transported from eroded canyon walls nearby or from up-canyon have been deposited and have partially filled the canyon bottom, resulting in a flat surface. During canyon development, multiple canyon cut-and-fill features are common. Further to south the course is asymmetrical (cross-section 5 in Figures 5 and 7). The west wall is still steep, while the east wall has been buried by sufficient sediments transported through the channels northeast of the canyon and becomes flat.

Further downstream to at around 3000 m, the canyon course turns suddenly to E-W direction in overall before joining the Manila Trench, showing a broad, flat, and U-shaped in cross-section (Figure 8), from 3200 to 3400 m at the center of canyon and 2900-3200 m on two flanks.

DISCUSSION

Controlling factors on the development of the STBC

Analysis on the controlling factors was conducted based on the facts of onshore river, margin physiography, tectonics, sea level change, and ocean current; these factors co-determined the patterns of sedimentation and growth of the canyon.

Onshore rivers

It is obvious that many submarine canyons are related

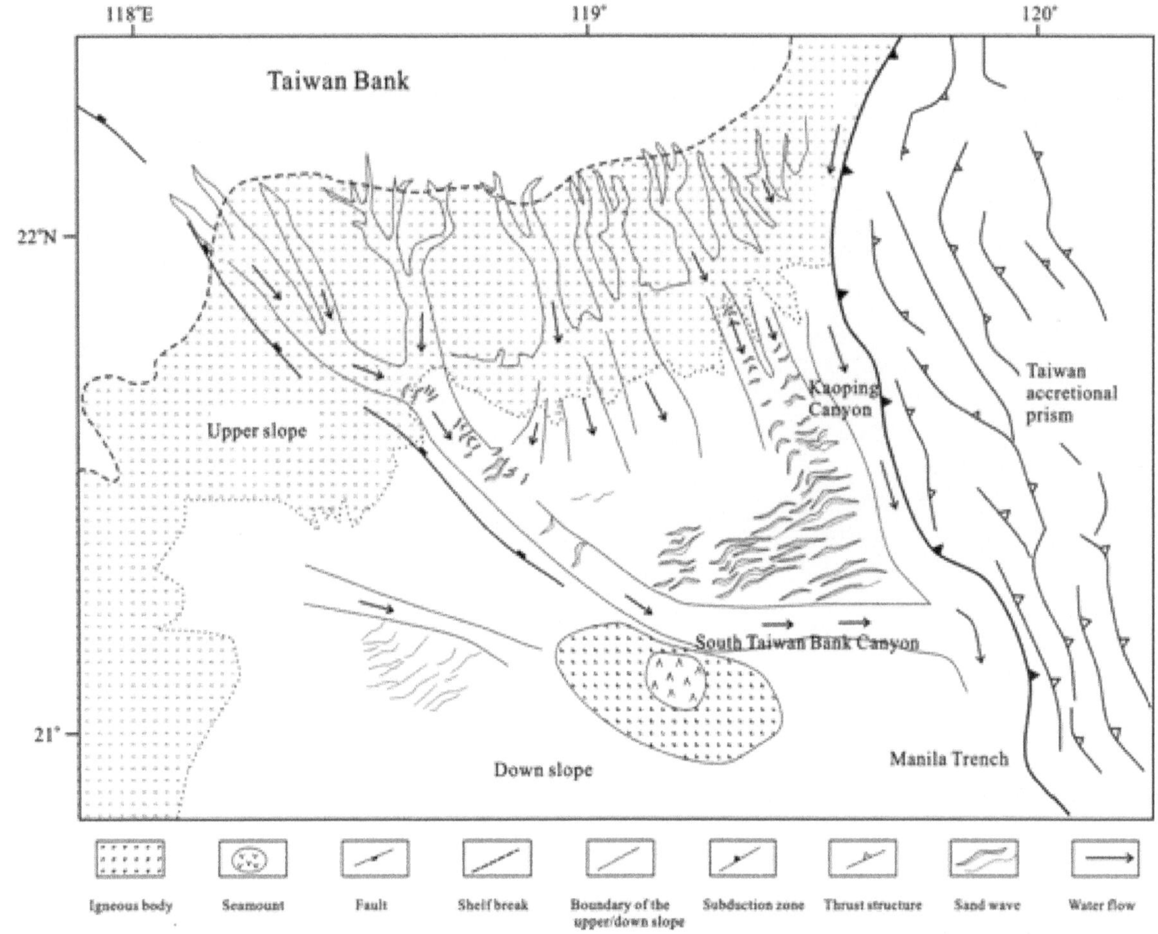

Figure 3. Interpretation of the bathymetric map (The part east of 119°30′ is according to Yu and Zhou (2001). Break gray line is the location of relic transform fault).

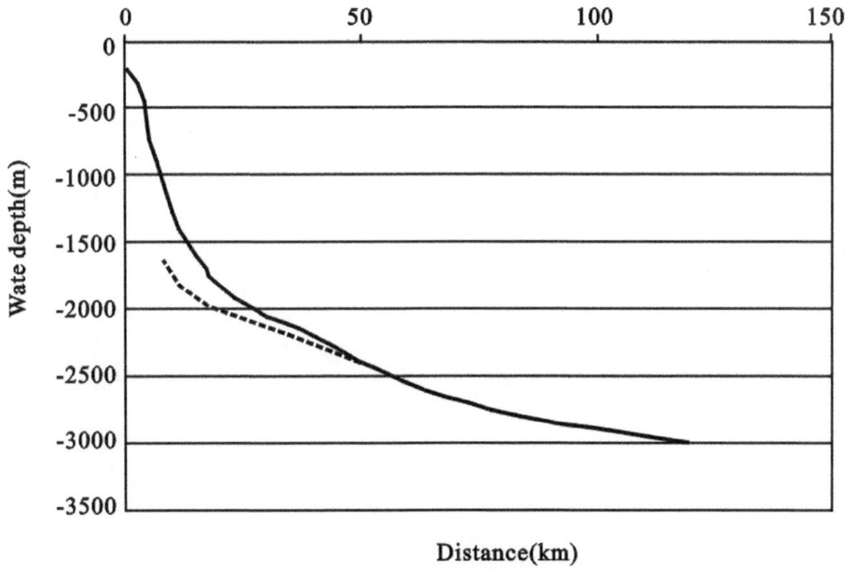

Figure 4. Longitudinal depth profile along the axial thalweg of the STBC obtained from EM120 data. The gradient reduces sharply around 1700 m water depth. The left broken line represents the west channel partly shown in Figure 2. The right real line is the ECS.

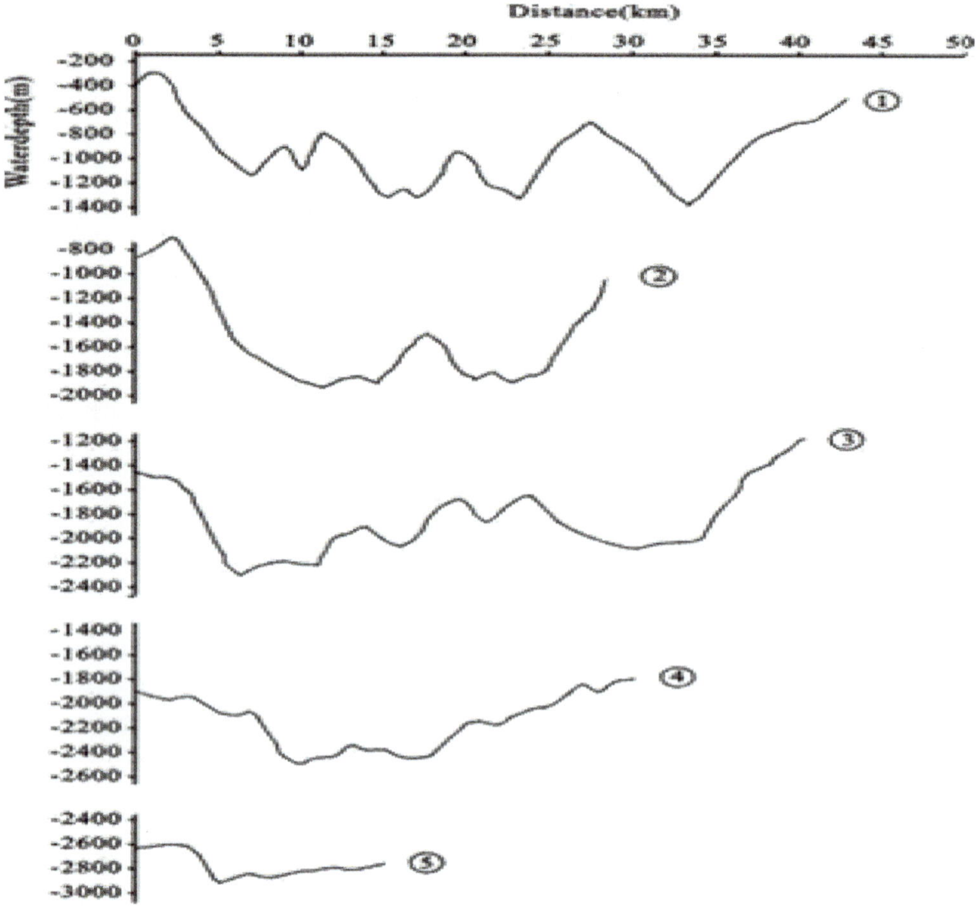

Figure 5. Cross-sectional morphology of the STBC revealed by bathymetric transects. Locations of bathymetric profiles are shown in Figure 2.

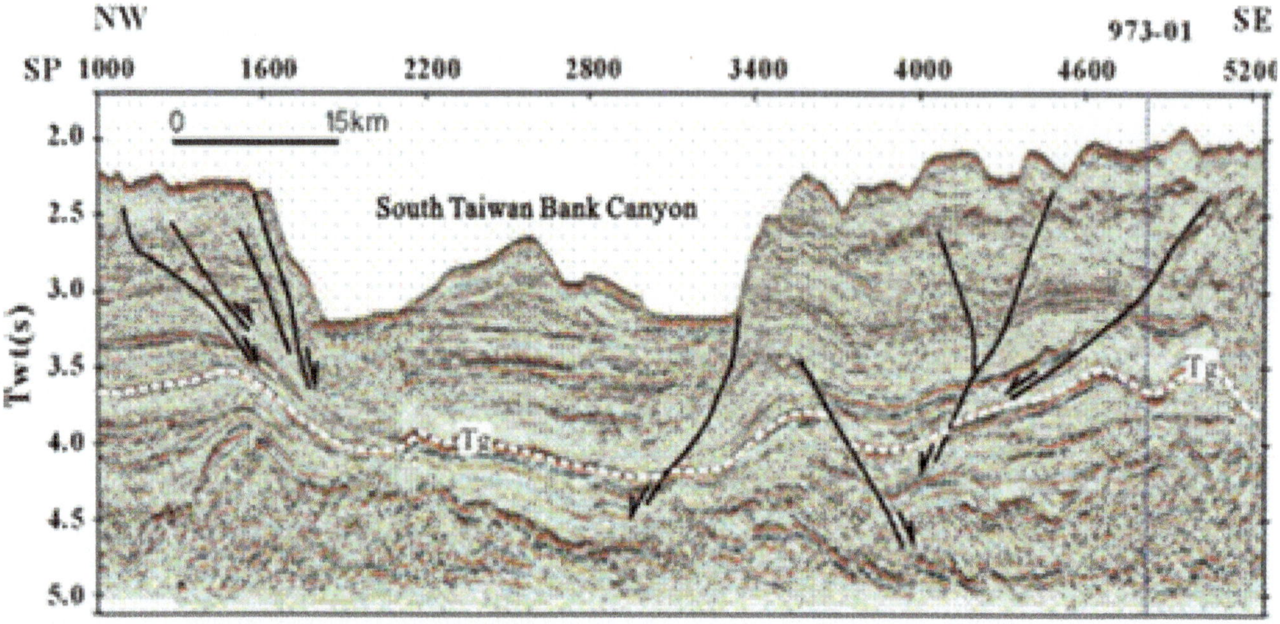

Figure 6. Seismic profile E53 shows truncations of parallel reflectors against the east and west flanks, indicating down-cutting along the fault planes (Location in Figure 2; Twt: two way time).

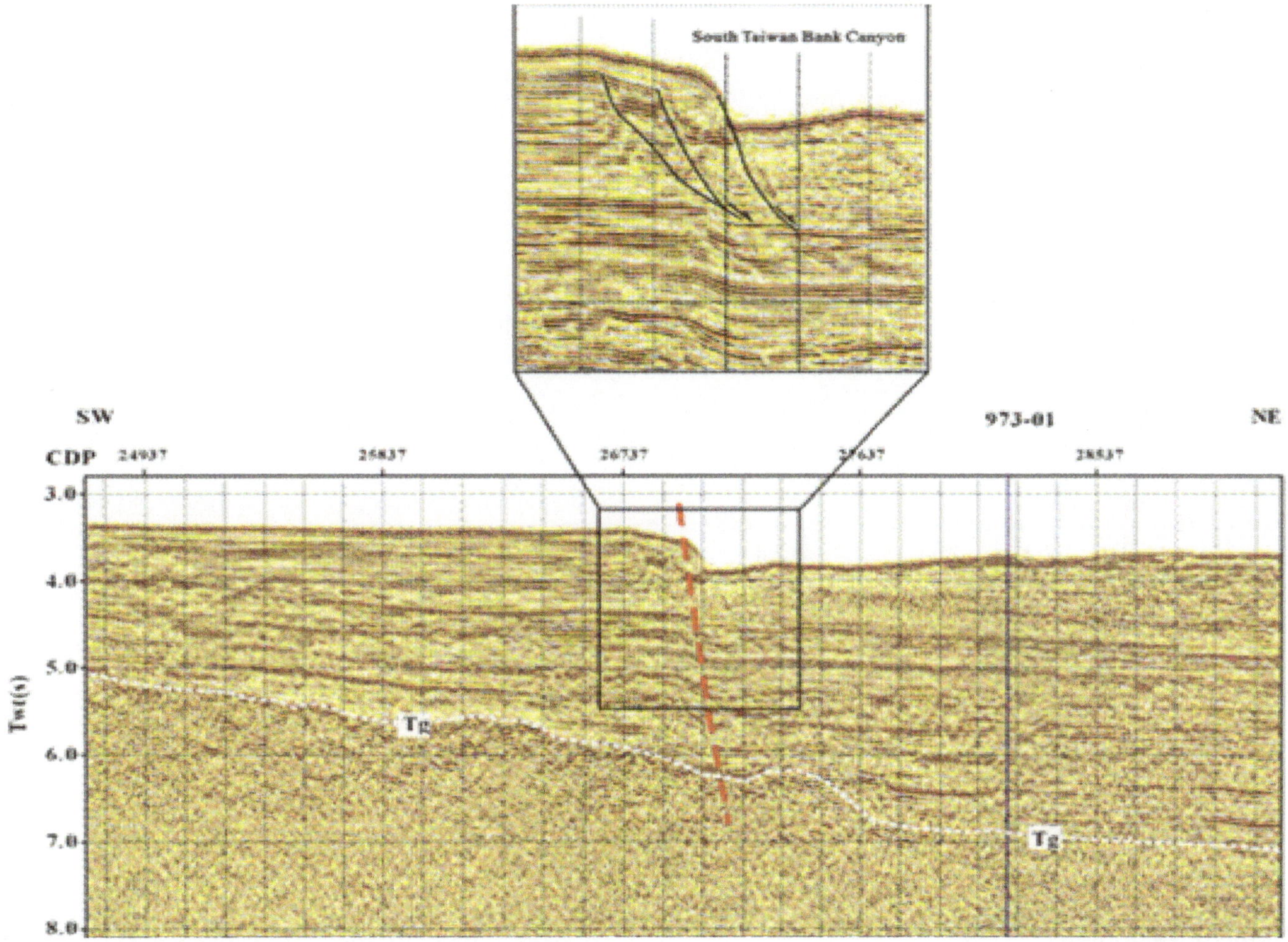

Figure 7. Seismic profile 974-04 shows the canyon structure in the down slope. The Cenozoic sediments are offset consistently, which may indicate the existence of a relict transform fault. Furthermore, the Cenozoic strata west of the relic transform fault drop down to northeast made a seabed terrace along the fault (Location in Figure 2).

closely to onshore rivers. Onshore rivers may connect canyons directly into the river mouth (Yu and Liu, 1995; Liu et al., 2002; Babonneau et al., 2002), or erode their way on a continental shelf during major glaciation event at a lowstand sea level (Damuth, 1994; Popescu et al., 2004; Anderson, 2005). Hyperpycnal currents generated in river push sediment transport in the sea bottom, which favors the formation of valleys. In our study, in the north of the STBC is the Hanjiang River, not a big river, whose annual sediment discharge is about 7×10^6 t/a (Wang, 1995) (only 10% of the Pearl River's, the biggest river in the South China). Most of the terrestrial sediments from the river deposited right at the north of the Taiwan Bank and can hardly go farther across the broad continental shelf (~180 km) (Liu et al., 2005). Therefore, the direct impact on the STBC is small. However, during the lowstand of the South China Sea, the continental shelf exposed largely, ancient Hanjiang River could run through the shelf and reach the shelf break, allowing fluvial sediments transporting directly to the shelf edge and then the STBC. Wu and Bao (2000) and Yu and Zhou (2001) pointed out that the forming of the Kaoping Canyon (see Figure 1 for location) were related to the paleo-Minjiang River in Fujian, South China, making us to believe the spatial relationship between the paleo-Hanjiang River and the canyon.

We can also find many onshore rivers in the southwest Taiwan Island close to the study area, such as the Kaoping River and Zengwen River, having close relationship with the canyons developed on the southwest slope off Taiwan Island. Take Kaoping River for instance, the great amount of precipitation makes the present annual sediment discharge reach 6.7×10^7 t despite the drainage area is 1% of the Pearl River (Liu et al., 2005). The Kaoping Canyon is just located in an area of high sediment supply close to the Kaoping River mouth. Sufficient sediment input formed NE-SW-oriented Kaoping Canyon. Although the sediments carried by these rivers may reach the study area, no direct connection can be inferred between rivers from Taiwan Island and the STBC

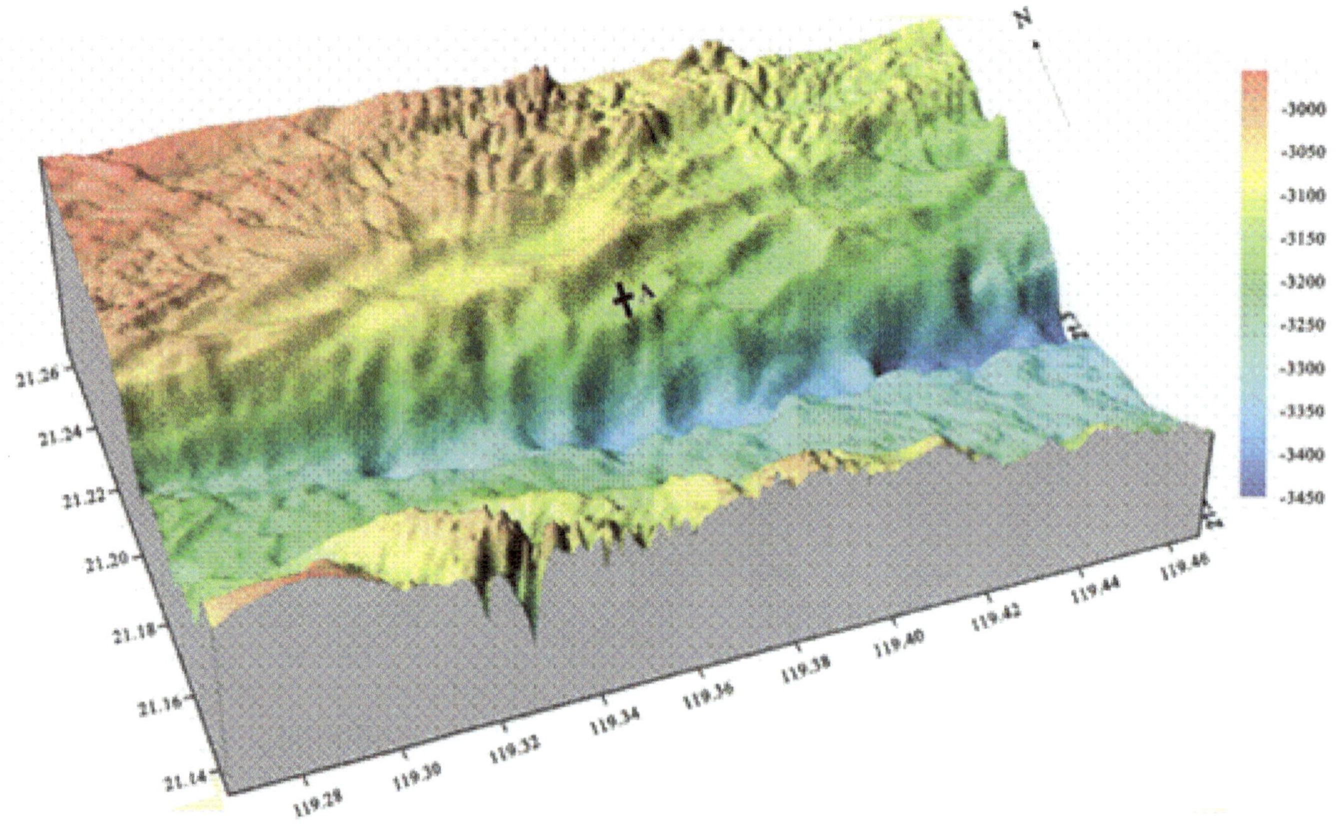

Figure 8. Bathymetric 3D map of the EW-oriented course of the STBC.

Margin physiography

Margin physiography is important in the evolution of STBC; the gradient of the continental slope is a key factor. High gradients up to 6°0′ in the study area is responsible for causing instability events, such as slumping, sliding, and mass flows, transporting a great amount of sediment from continental shelf to abyssal plain, and shaping geomorphology of continental margins. Large-scale sediment failures are considered one of the key factors in canyon formation and evolution for canyon widening and shelf incision (Milliman and Syvitski, 1992; Pratson and Coakley, 1996).

In seismic profile 973-01, a slump structure is obvious just beneath the shelf break (Figure 9). Truncation of the parallel and continuous cenozoic reflectors indicates sediment slumping in the south with chaotic reflectors. Slope failure (Figure 2) covers the slope between 500-2000 m. Bottom currents swirl around it and form two channels on both sides.

As mentioned above, canyon heads could undergo retrogressive growth as instability proceeded, by over-excavation of the heads along shelf edge, characterizing the high gradient. The east head of the STBC may change its course as the slump interruption before it joins the west head to form the STBC. Sediments in the canyon head and upper canyon are removed and flushed into the lower canyon and distal abyssal plain as a result of large slope failures evolving in high-energy sediment flows capable of eroding the canyon floor and producing pervasive grooves and other bedforms.

Where the gradient is reduced, the flow suffers a hydraulic change with subsequent flow thickening, flow velocity decrease, and sediment waves at the end of the channel-levee system (Figure 2).

Tectonics

A fault-controlled weak zone would pre-determine the initial location of a canyon (Pratson and Coakely, 1996; Laberg et al., 2000). Many canyons of a passive continental margin would be in a sinuous course when gradient reduces, as a river does in a continent. While the straight course and different orientations of the STBC shows structural background controlling under regimes of faulting and magmatism. On seismic profile 973-04 at the STBC, the Cenozoic strata are consistently offset, to which Hsu et al (1998) explained as a relic transform fault roughly along the STBC, indicative of the southwestern end of the Ryukyu trench-arc system. Sedimentation and structural geology shown on profile 973-04 (Figure 7) suggested also the existence of the relic transform fault. First, the Cenozoic reflections to the southwest of relic

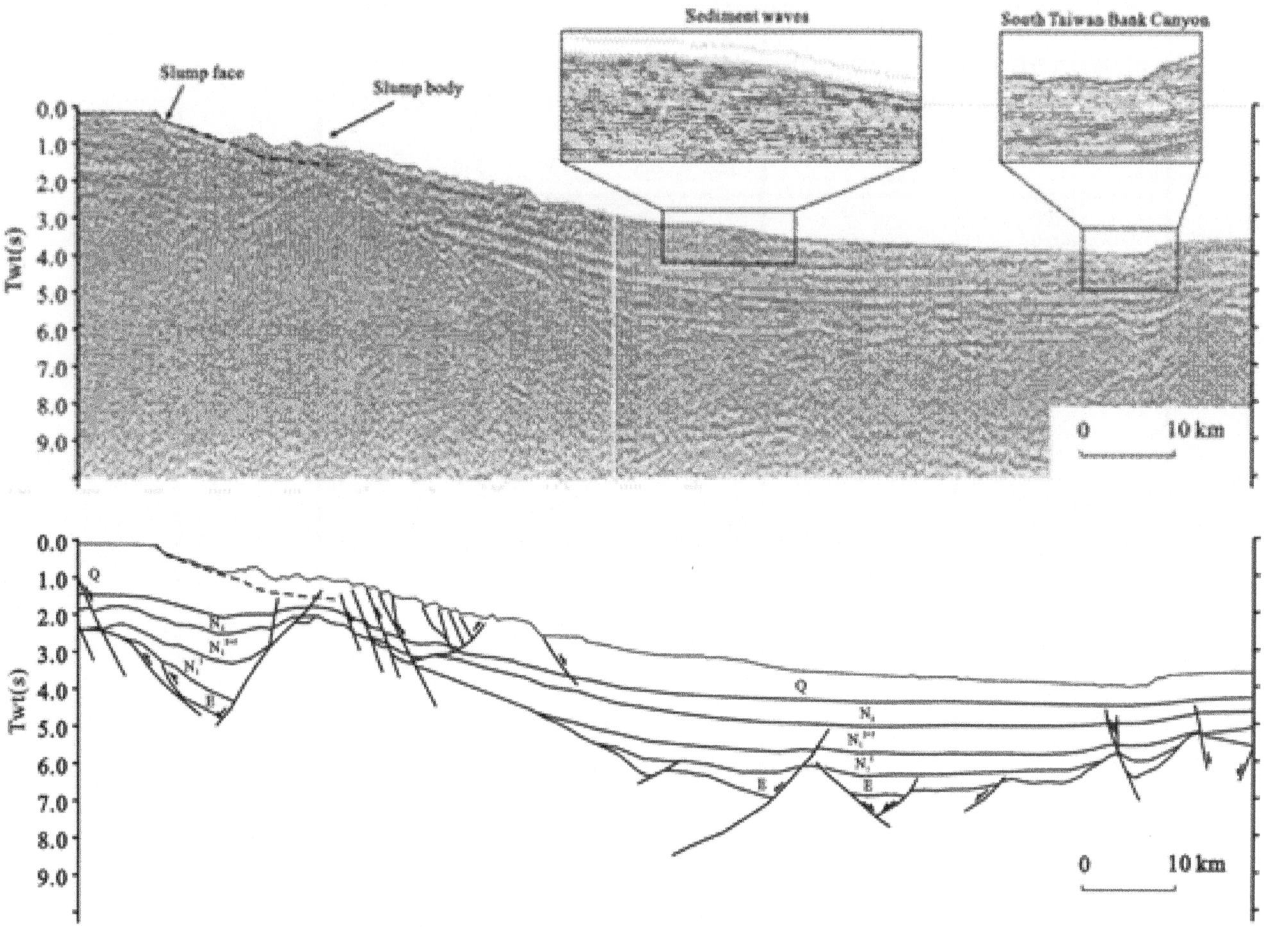

Figure 9. Slump and sediment waves on the slope revealed by the 973-01 seismic profile.

transform fault are high frequent, wavy and continuous, while those to the northeast of the relic transform fault are relatively in lower frequency and lateral continuity. Second, the Cenozoic basement (*T*g) has a breakpoint near the relic transform fault. To the southwest, *T*g is easily identifiable, undulating locally, in almost linear SW-NE trend. In contrast, this basement trend makes a sharp turn at the breakpoint in a smaller dipping gradient to the northeast. These differences prove the existence of the proposed relic transform fault. The Cenozoic strata west of the relic transform fault drop down to northeast, making a terraced seabed along the fault, which determined the course direction of the STBC. In seismic profiles, we also find some normal faults near the seabed (Figures 6 and 7). As the canyon walls were continuously oversteepened by downcutting, the instability in both walls increased on a fault.

Magmatism is another geological factor affecting the STBC; it would trigger an instability event in continental slope, and form a seamount, altering the margin physiography. In the study area, a seamount stands right in the NW-SE trending course of the STBC (Figure 2); it turns the southwastward flow into eastwards in the canyon suddenly. After that the seamount formation, tectonic influence on the canyon tended to discrease.

Sea level change

Sea level change played a role in the evolution and development of the STBC. Relative sea level change is responsible for the formation of sequence boundaries along continental margins, especially passive ones (Farre et al., 1983)

A previous study shows that after the Middle Miocene (15.5 Ma), the down slope section first subsided and the upper slope steepened (Ding et al., 2008), which resulted from the post-rifting thermal cooling of the South China Sea. Terrestrial materials were transported in the canyon over the continental shelf and into the abyssal basin. A sea level change curve of the South China Sea was reconstructed biostratigraphically from data of 50 wells on the north continental margin. The curve reveals 13 eustasies after 15.5 Ma (Pang et al., 2007), of which five times are the sea level down to the shelf break.

In general, during a regressive event, sea level is low in an early transgressive stage, the coastline is located near

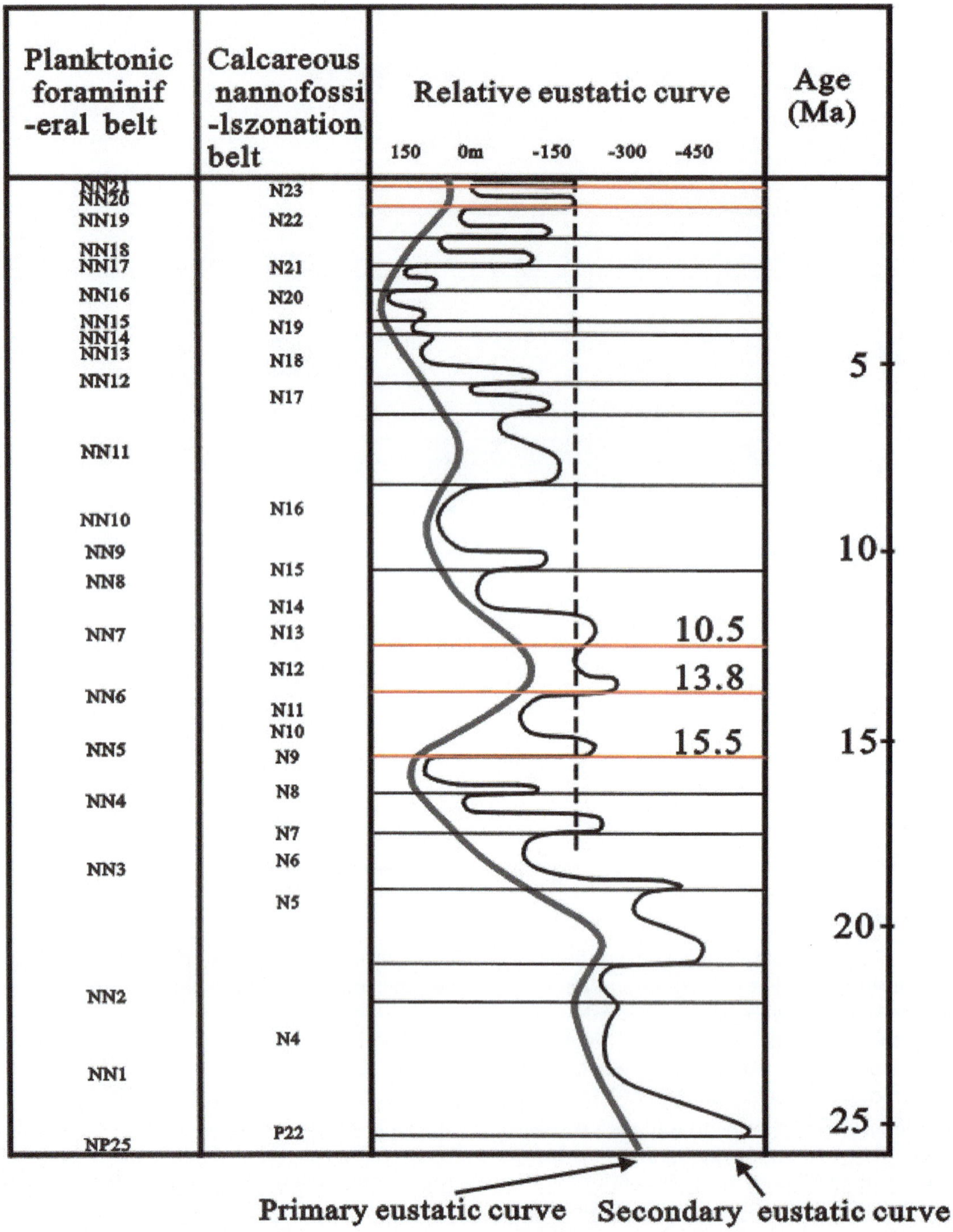

Figure 10. Sea level change curve of the South China Sea (after Taylor B, Hayes D E. 1983).

the shelf-edge, sediments would be moved easily in high-energy processes such as storm waves and tidal currents. In this case, instability processes (mass-flows or channelized turbidity flows) is common alongslope, favoring the shaping of a valley. Leveed channels would grow by turbidity currents, causing sedimentation in levees and channels in a valley, while gullies would grow by differential aggradation. During a major sea level falling, continental shelf would be exposed for erosion, resulting in sequential incision of submarine canyons and a regional unconformity or sequence boundaries in other words. During a sea level low, canyon channels are generally active causing significant erosion and deep down-cutting into the sediment underlain, or vice versa, rising sea level would reverse such depositional phenol-mena, causing active landward down-cutting in canyon channel.

Sea level changes cause variations in amount and type

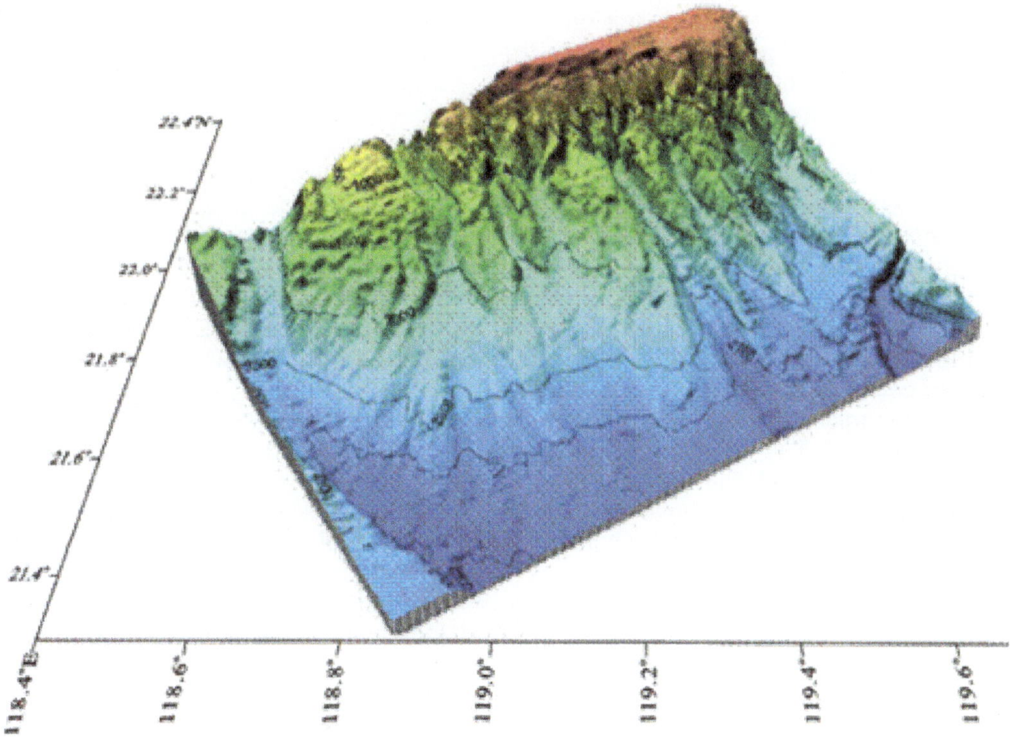

Figure 11. Bathymetric 3D map of the ECS.

of terrigenous materials into the basin. For spatial relationship between river and canyon, recognizing the location of the paleo-Hanjiang River is a key in this study. While unfortunately, no direct proofs of buried fluvial channels, as a diagnostic feature of ancient drainage systems, have been obtained for lacking the seismic data on the continental shelf north of the study area.

Ocean currents

High-energy marine processes, such as ocean currents, waves, and tides, can erode seabed and trigger instability events (Liu et al., 2002; Shanmugam, 2003). Especially, as a force at the bottom, ocean currents would channel the course. Ocean currents in the northeast South China Sea include monsoon-driven surface currents that could hardly work in deep water (mostly >500 m), and so the Kuroshio current (Shaw and Chao, 1994). After entering the South China Sea through Bashi Channel, the Kuroshio separates into two branches. One enters the East China Sea via Taiwan Strait; the other forms an anti-clockwise circulation in the South China Sea. Shao et al (2007) pointed out that the bottom currents forced by the Kuroshio flowed southwestwards at virtually all isobath, and formed NE-SW trending channel on the slope in flow direction nearly perpendicular to the NW-SE trending STBC. We believe that the ocean currents have minor effects in the evolution of the STBC.

Difference between the STBC and the eastern channels

Forming time and mechanism of the eastern channels

At least 9 channels developed on the slope between the STBC and the Penghu Canyon. These NNW-oriented channels erode the seabeds of the continental slopes with relatively straight courses (Figure 11). The sections of both channels over 2200 m water-depth are characterized by a deeply incised, narrow, V-shaped thalweg with multi-heads. Incised valleys can indent the shelf-break at 200 m water-depth. With the decrease of the gradient after entering the lower continental slope, sediment flows accumulate and form NE-oriented sediment waves at the end of these channels.

These channel-sediment waves progressed with the uplifting of Taiwan Island. Analyses on surface sediments in 188 stations east of the South China Sea show that surface sediments north of 21° are mainly terrestrial clastic ones originated from the Taiwan Strait and Taiwan Island (Yang et al., 2004). Collision between the Luzon Arc and Eurasian Plate since 6.5 Ma triggered the Taiwan Orogeny (Sibuet and Hsu, 2004). Uplift of Taiwan Island resulted in erosion, and the physical denudation rate was 1300 $mg/cm^2 \cdot a$ (Li, 1976), generating abundant river-borne terrestrial clastics, and deposited and later filled up the West Taiwan Foreland Basin rapidly. As the collision propagated southward, focuses of the erosion and material supply shifted also southward, thus enhance the scale and frequency of sediment flows generated in the rivers in the southwest of Taiwan Island.

A high sedimentation rate would result in the indentation of continental slope and formation of straight NWW-oriented channels east of the STBC and the sediment waves. Although it is difficult to determine the formation time of the eastern channels as we lack high-resolution seismic data and well-loggings in the study area, we deduced the age by refereeing to the forming time of the sediment waves for having a close relationship. The average sediment rate of the gravity column sampled from the wave region is about 16.65 cm/ka (See Figure 8 for sample location). The wave's depth is about 188 m. Therefore, we determined that the waves were initiated about 1.2 Ma ago. From the above discussion, we conclude that the forming time of these channels is after the Taiwan Orogeny.

Difference between the STBC and the eastern channels

STBC and the eastern channels are very different from each other in morphology and sedimentology. First, the orientations are different, which is from structural backgrounds of them. The STBC is in NW-SE trending north of the submarine mount, while the eastern channels are in NNW-SSE trending in general. Second, the part of the STBC in the lower continental slope still indents into seabed with an obviously course, while the eastern channels are fully filled by sediments and covered by well-developed sediment waves.

Third, the STBC is the westmost canyon in the study area. If the STBC was also resulted from the uplifting and erosion of the Taiwan Island, the terrestrial sediment input should be much less than and thus the grain size should be much less than the east ones that of the eastern channels for the farther distance from the provenance, that is, the size should be smaller than the eastern channels. On the contrary, the STBC is the biggest submarine canyon in the region. Therefore, we believe that the STBC and the eastern channels are different in formation mechanism and time, although the both are affected by the Taiwan Island at present.

Forming process and time

Ding et al. (2008) studied the Cenozoic tectonic-sedimentation of the study area with the 973-01 seismic profiles. The results show that from the Oligocene to the Early Miocene, the study area was under the tectonic regime of rifting. Terrestrial materials are deposited mainly in rifted-sags on the continental shelf. Structural highs on continental shelf break were still above the sea level and blocked material from being transported to the abyssal basin (Figure 12a). After the Middle Miocene, the study area was in the stage of regional subsidence resulted from the post-rifting thermal cooling, during which faulted-sags in continental shelf started to be filled up; terrestrial materials were able to reach the lower continental slope over the structural highs, resulting in an overall change in morphology similar to the modern shape (Figure 12b).

The whole world experienced a global cooling events and a low-stand sea level in recent geological history (Shevendell et al., 2004; Holbourn et al., 2005, 2007), and five ecstasies' events when sea level fell down to the shelf break have been shown in the sea level change curve of the South China Sea. The first event took place in 13.8 Ma (Figure 10). Thus, we believed that the event resulted in the exposure of continental shelf. The paleo-Hanjiang River in the South China ran through the continental shelf and reached the shelf break, allowing direct fluvial sediment delivery to the shelf edge. Rapid thermal subsidence occurring after Middle Miocene increased the slope gradient, triggering instability events, such as slumping or sliding. Sediments in forms of slides, slumps, and sediment flows, incised the upper continental slope and formed gullies and multiple canyon-heads. The pre-existed NW-SE-direction relic transformation fault in the South China Sea defined the canyon system in a liner seabed terrace, and trigged normal faulting near seabed.

We believed that the sediment flows would run along the fault and finally form the NW-SE oriented straight course of the STBC (Figure 12c).

In the lower continental slope, a seamount located in the south blocked the southwestward flow in the canyon and forces the course to turn sharply to the east before finally joining the Manila Trench, then the tectonic impact on the canyon diminished gradually. Collision between the Luzon Arc and Eurasian Plate since 6.5 Ma triggered the Taiwan Orogeny. Continuous uplifting and eroding of the Taiwan Island generated sufficient terrestrial sediments to the areas off southwestern Taiwan, creating straight NWW-oriented ECS and the sediment waves. However, these sediments were hard to reach the STBC course as it was far from the provenance, the Taiwan Island. That is why the STBC is still in a negative landform in the lower continental slope.

Conclusion

1. The STBC is a large shelf-indenting canyon on the northeast slope of the South China Sea. The course displays two distinct trends. From its head region at the shelf edge and upper slope areas, the canyon initially runs NW-SE and indents the shelf and upper slope with multiple heads. At water depth shallower than 1500 m, the course is characterized by a steep thalweg gradient and V-shaped incised trough. At depths deeper than 1500 m and with reduction in gradient, the cross-sectional mor-

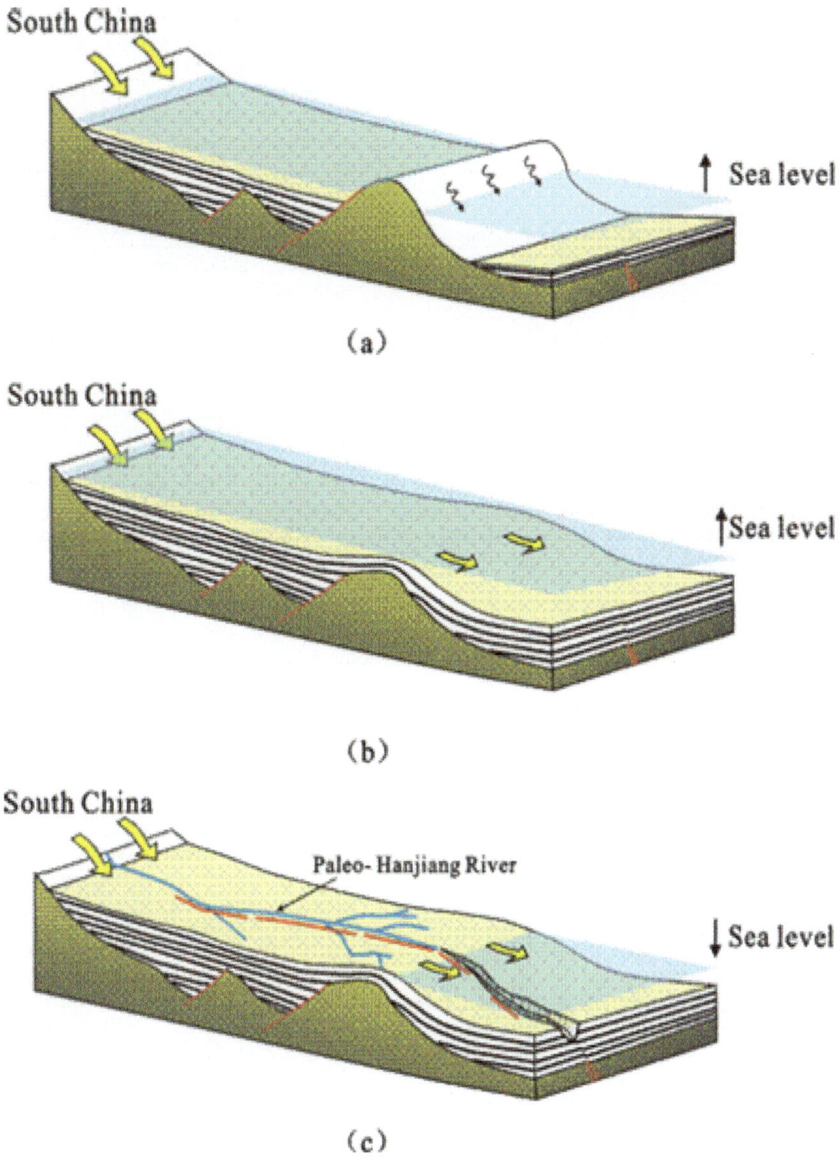

Figure 12. Evolution of the STBC.

phology changes into U-shape. At around 2200 m water depth, the course is sharply diverted to a predominantly E-W direction. This abrupt shift in course divides the canyon into two main morphological domains.

2. The beginning of the STBC may be in middle Miocene. The lowstand sea level related global cooling event in 13.8 Ma, the high slope gradient caused by the quick thermal subsidence of the lower continental slope, the pale-onshore river and the NW-SE oriented relic transform fault together controlled the development of the canyon. A sea mount made the shift of the course orientation possible.

3. The STBC has different forming time and mechanism with respect to the channel-levee system in the east. The latter is directly related with the uplifting and erosion of the Taiwan Island.

Acknowledgments

This work is financially supported by the National Basic Research Program (973) of China (No. 2007CB41170403), the National Natural Science Foundation of China (No.40806023), and the Scientific Research Fund of the Second Institute of Oceanography, SOA (No. JT0704)

REFERENCES

Anderson DE (2005). Holocene fluvial geomorphology of the Amargosa River through Amargosa Canyon, California. Earth-Sci. Rev., 73: 291-307.

Babonneau N, Savoye B, Cremer M, Klein B (2002). Morphology and architecture of the present canyon and channel system of the Zaire deep-sea fan. Marine Petroleum Geol., 19: 445-467.

Briais A, Patriat P, Tapponnier P (1993). Updated interpretation of magnetic anomalies and seafloor spreading in the South China Sea: Implications for the Tertiary tectonics of Southeast Asia. J. Geophysical Res., 98(B4): 6299-6328.

Chiang CS, Yu HS, Chou YW (2004). Characteristics of the wedge-top depozone of the southern Taiwan foreland basin system. Basin Research 16: 65-78

Chou YW, Yu HS (2002). Structural expressions of flexural extension in the arc-continental collisional foredeep of western Taiwan. Geological Society Am. (Special Paper) 358: 1-12.

Damuth JE (1994). Neogene gravity tectonics and depositional processes on the deep Niger Delta continental margin. Marine Petroleum Geol., 11: 320-346.

Ding WW, Li JB, Li MB, et al (2008). A Cenozoic tectono-sedimentary model of the Tainan Basin, the South China Sea: evidence from a multi-channel seismic profile. J. Zhejiang Univer. (Science A), 9(5): 702-713.

Ding WW, Yang SF, Chen HL, Cheng XG, Wu NY (2006). Arc-continent collision orogeny in offshore Taiwan during Neogene. Chinese J. Geol., 41(2): 195-201.

Farre JA, Mcgregor BA, Ryan WBF, Robb JM (1983). Breaching the shelfbreak; passage from youthful to mature phase in submarine canyon evolution. In: The Shelfbreak, Critical Interface on Continental Margins. Special Publication. Society of Economic Paleontologists Mineralogists, 33: 25-39.

Hagen RA, Vergara H, Naar DF (1996). Morphology of San Antonio submarine canyon on the central Chile forearc. Marine Geol., 129(3-4): 197-205.

Holbourn A, Kuhnt W, Schulz M, Erlenkeuser H (2005). Impacts of orbital forcing and atmospheric carbon dioxide on Miocene ice-sheet expansion. Nature, 438: 483-487.

Holbourn A, Kuhnt W, Schulz M, Flores JA, Andersen N (2007). Orbitally-paced climate evolution during the middle Miocene "Monterey" carbon-isotope excursion. Earth. Planet. Sci. Lett., 261: 534-550.

Hsu SK, Liu CS, Shyu CT, Liu JC, Sibuet S (1998). New gravity and magnetic anomaly maps in the Taiwan-Luzon region and their preliminary interpretation. T.A.O. 9: 509-532.

Huang CH, Wu WY, Chang CP (1997) Tectonic evolution of accretionary prism in the arc-continent collision terrene of Taiwan. Tectonophysics, 281: 31-51.

Huang CH, Xia KY, Perter B (2001). Structural evolution from Paleogene extention to Latest Miocene-Recent arc-continent collision offshore Taiwan: comparison with on land geology. J.Asian. Earth Sci. 19: 619-639.

Jin QH (1989). Geology and Hydrocarbon Resource of the South China Sea Geological Publishing House, Beijing, China. pp. 84-111.

Klaus A, Taylor B (1991). Submarine canyon development in the Izu-Bonin forearc: A SeaMARC II and seismic survey of Aoga Shima Canyon. Marg. Geophy. Res., 13: 131-152.

Kolla V, Perlmutter MA (1993). Timing of turbidite sedimentation on the Mississippi Fan. AAPG Bulletin 77: 1129-1141.

Laberg JS, Vorren TO, Dowdeswell JA, Kenyon NH, Taylor J (2000). The Andøya Slide and the Andya Canyon, north-eastern Norwegian-Greenland Sea. Mar. Geol., 162: 259-275.

Li CF, Zhou ZY, Li JB, Hao H, Geng JH (2007). Structure of the northeasternmost South China Sea continental margin and ocean basin: geophysical constraints and tectonic implications. Marg. Geophy. Res., 28: 59-79.

Li YH (1976). Denudation of Taiwan island since the Pliocene Epoch. Geology, 4: 105-107.

Lin AT, Watts AB, Hesselbo SP (2003). Cenozoic stratigraphy and subsidence history of the South China Sea margin in the Taiwan region. Basin Res. 15: 453-478.

Liu JT, Liu KJ, Huang JC (2002). The effect of a submarine canyon on the river sediment dispersal and inner shelf sediment movements in southern Taiwan. Mar. Geol., 181: 357-386.

Liu ZC, Liu BH, Huang ZZ, (2005). Geomorphology in Chinese offshore waters and adjacent sea areas. Ocean Press, Beijing, China. pp. 181-183.

Lüdmann T, Wong HK (1999). Neotectonic regime on the passive continental margin of the northern South China Sea. Tectonophysics, 311: 113-138.

Milliman JD, Syvitski JP (1992). Geomorphic/tectonic control of sediment discharge to the ocean: the importance of small mountainous rivers. J. Geol., 100: 525-544.

Pang X, Chen CM, Peng DJ, Zhou D,Sun Z (2007). The Pearl River Deep-water Fan System and Petroleum in South China Sea. Science Press, Beijing, China. p. 147.

Popescu I, Lericolais G, Panin N, Normand A, Dinu C (2004). The Danube submarine canyon (Black Sea): morphology and sedimentary process. Mar. Geol., 206: 249-265.

Pratson LF, Coakley BJ (1996). A model for the headward erosion of submarine canyons induced by downslope-eroding sediment flows. GSA Bull., 108(3): 225-234.

Shanmugam G (2003). Deep-marine tidal bottom currents and their reworked sands in modern and ancient submarine canyons. Mar. Petrol. Geol., 20: 471-491.

Shao L, Li XJ, Geng JH, Pang X,Lei YC, Qiao PJ, Wang JL (2007). Deepwater bottom currents sedimentation in the north South China Sea. Sci. in China (Ser. D), 37(6): 771-777.

Shaw PT, Chao SY (1994). Surface circulation in the South China Sea. Deep-Sea Res., I 41: 1663-1683.

Shevendell AE, Kennett JP, Lea DW (2004). Middle Miocene Southern Ocean cooling and Antarctic cryosphere expansion. Sci., 305: 1766-1770.

Sibuet CS Hsu SK (2004). How was Taiwan created. Tectonophysics, 379: 159-181.

Sibuet JC, Hsu SK, Pichon XL, Formal JP, Reed D (2002). East Asia plate tectonics since 15 Ma: constraints from the Taiwan region. Tectonophysics, 344: 103-134.

Suess E (2005). RV SONNE Cruise Report SO 177, Sino-German Cooperative Project, South China Sea Continental Margin: Geological Methane Budget and Environmental Effects of Methane Emissions and Gas Hydrates. IFM-GEOMARReports, http://store.pangaea.de/documentation/Reports/SO177.pdf.

Taylor B, Hayes DE (1983). Origin and history of the South China Sea Basin. In: Hayes D E ed. The Tectonic and Geologic Evolution of Southeast Asian Seas and Islands: Part 2, Geophysical Monograph. AGU, Washington, 27: 23-56.

Wang PX (1995). South China Sea since 15Ma. Tongji University Press, Shanghai, China. p. 12. (in Chinese).

Wu LS, Bao CW (2000). Types and characteristics of potential subbottom geological hazards in the eastnorthern South China Sea. Geol. South China Sea., 12: 87-101.

Yang QH, Zhang FY, Lin ZH, Zhou HY (2004). On mineralogical and geochemical records of paleosedimentary environmental variation in the northeastern South China Sea since the late Pleistocene. Acta Oceanologica Sinica, 26(2): 72-80

Yu HS, Liu JC (1995). Development of the shale diapir-controlled Fangliao Canyon on the continental slope off southwestern Taiwan. J.Southeast Asian Earth Sci., 11: 265-276.

Yu HX, Zhou YW (2001). Physiographic and geological characteristics of shelves in north and west of Taiwan. Science in China (Ser. D), 31(6): 486-493.

Zhou D, Ru K, Chen HZ (1995). Kinematics of Cenozoic extension on the South China Sea continental margin and its implications for the tectonic evolution of the region. Tectonophysics, 251(1): 161-177.

6

Impacts of surface wave-induced Coriolis-Stokes forcing on the upper ocean circulation

Zengan Deng[1,2]*, Guijun Han[1,2], Xuefeng Zhang[1,2] and Kejian Wu[3]

[1]Key Lab of Marine Environmental Information Technology, SOA, Tianjin 300171, China.
[2]National Marine Data and Information Service, Tianjin 300171, China.
[3]Ocean University of China, Qingdao 266100, China.

Using the Hybrid Coordinate Ocean Model (HYCOM), six experiments configured for three different domains: Global Ocean, South China Sea (SCS) and Western North Atlantic Ocean (WNA), are designed to investigate the effects of wave-induced Coriolis-Stokes forcing (CSF) on ocean surface phenomena including circulation, temperature and mixing processes. CSF calculated using wave parameters simulated by the WaveWatch III (WW3) model, which is used as a part of the driving force for HYCOM. The results indicated that: 1) CSF does not fundamentally change the circulation pattern in the upper ocean mixed-layer (ML), 2) over most of the global ocean, the direction of Stokes transports ($\vec{T}_S$) is different from that of the changes in depth-integrated current transports ($\Delta\vec{T}_c$) caused by CSF, 3) the monthly-mean $\Delta\vec{T}_c$ in the ML varies from month to month in both direction and magnitude, and the CSF plays a more significant role in regions of intensive gyre, such as the area near Yucatan Channel, than in regions of weak currents, 4) both Sea Surface Temperature (SST) and Mixed-Layer Depth (MLD) are noticeably influenced by the CSF.

Key words: Coriolis-Stokes forcing, upper ocean mixed layer, currents, stokes transport.

INTRODUCTION

The motion of the ocean surface water has been investigated extensively for its critical role in transportation of momentum, mass and energy as well as in air-sea interaction. Two main types of motion exist in the upper ocean: waves and currents. Ocean surface waves have three main effects on turbulence in the ocean surface mixed-layer (ML), that is wave breaking, Langmuir circulations (McWilliams et al., 1997; Teixeira and Belcher, 2002; Kantha and Clayson, 2004; Carniel et al., 2005; Ardhuin and Jenkins, 2006), and Coriolis-Stokes Forcing (hereinafter referred to as CSF) (Hasselmann, 1970; Huang, 1979; McWilliams et al., 1999; Polton et al., 2005). Interaction of turbulent vorticity with Stokes drift results in Langmuir circulations, while interaction of large-scale planet vorticity with Stokes drift produces CSF which can be understood by a divergence of the wave-induced stress (Hasselmann, 1970). Both wave breaking and Langmuir circulations serve as additional sources of Turbulent Kinetic Energy (TKE) in the upper ocean (Ardhuin and Jenkins, 2006), whereas CSF changes the upper layer dynamics by acting on the momentum equation. Kantha and Clayson (2004) and Carniel et al. (2005) included the Stokes production into turbulence models, and found that the wave-induced source term extracts energy from wave motions and transfers it into turbulence which then consequently results in changes in upper layer properties, such as temperature and velocity. However, the large-scale CSF has never been implemented in a three-dimensional numerical model for real ocean simulating, thus there is a need to do so.

In the absence of surface wind stress, the ocean surface currents are mainly attributed to pure Stokes drift induced by waves (Bye, 1967; Kenyon, 1970). The ocean surface Stokes drift velocity increases linearly with increasing wind speed, and contributes significantly to the

*Corresponding author. E-mail: dengzengan@gmail.com.

total mean surface currents in the ocean (Kenyon, 1970). Korvin-Kroukovsky (1972) and Ianniello and Garvine (1975) calculated the Stokes and Ekman drift separately, and both found that the wave-induced drift is dominant in their uncoupled model. Hasselmann (1970) found that the interaction between the planetary vorticity and the Stokes drift produces a forcing on Eulerian momentum balance known as CSF. Huang (1979) demonstrated that, the CSF changes the usual Ekman balance in the wind-driven mixed-layer (ML) and the current profiles in combination with Coriolis forcing and the divergence of vertical momentum transfer by turbulent stress. Based on field observations and theoretical analysis, Huang (1979) concluded that any model of the surface drift, currently neglecting the contribution of wave motion would be incomplete. The study of McWilliams and Restrepo (1999) showed that, the wave-driven current effects sometimes could be significant in the basin-scale circulation compared with the wind-driven counterpart. Polton et al. (2005) suggested that, according to the observations of the Ekman current profile, Stokes drift might dynamically influence the current profile. By comparing his research results with observations, he concluded that the changes of the current profile due to CSF are consistent with the differences between observations and the classical Ekman profile. He also showed that the current profile could be influenced by CSF throughout the Mixed-Layer Depth (MLD) by changing the effective boundary condition on the standard wind-driven Ekman solution. Recently, Deng et al. (2007, 2009) estimated the wave-induced Stokes transport over the Pacific by using ECMWF data and the wave parameters simulated by the WaveWatch III (WW3), and noticeable swell transports are found over the Eastern Pacific. These studies also suggested that, the Stokes drift over the Eastern Pacific originates from the Westerlies and contributes to the basin-wide ocean circulation. Over all, CSF induces the momentum transfer between currents and waves, thus changes current field and temperature gradient in the upper ocean.

In order to estimate the effect of CSF, numerical method is used in this work. Sensitivity experiments are conducted using a HYCOM-WW3 one-way coupling model over the global ocean, South China Sea (SCS) and Western North Atlantic Ocean (WNA) region, respectively. Some background knowledge of SCS and WNA will be given in the following. [5]SCS is the largest semi-enclosed marginal sea (3.5×10^6 km^2) in the west Pacific Ocean. It has a shallow shelf in the south, the Vietnamese-Chinese coast to the north, and a deep basin at the center. A chain of islands separates SCS from adjacent oceans. SCS is connected to East China Sea through Taiwan Strait, to Pacific Ocean through Luzon Strait which is the deepest strait with maximum depth of 2400 m, to Sulu Sea via the Mindoro Strait, to Indian Ocean through the Strait of Malacca, and to Java Sea through the Karimata and Gasper Straits. The topography of SCS is complicated, and the shelf that extends from the Gulf of Tonkin to Taiwan Strait is about 70 m in depth and 150 km in width. The Sunda Shelf which links SCS to Malaysia, Southeast Asia, Sumatra, Java and Borneo, is a submerged passage with 100 m depth in the middle. The depth of the center of Thailand Gulf is about 70 m. In the interior, the deep basin is 1100 km southeast-northwest and 1800 km northeast-southwest.

One of the striking characteristics of SCS is the seasonally reversing monsoon winds that are important in driving the upper ocean circulation. The weaker southwesterly summer monsoon winds have a wind stress of over 0.1 Nm^{-2} from April to August, and the stronger northeasterly winter monsoon winds correspond to a maximum wind stress of approximately 0.3 Nm^{-2} from November to March (Chu et al., 1999). Many studies about SCS have been carried out previously by different researchers. Shaw and Chao (1994) demonstrated that the northward jet in summer flows away from the coast of Vietnam at about 12°N and the southward jet in summer moves along the western boundary. A series of numerical experiments conduced by Metzger and Hurlburt (1996) suggested that, the local monsoon has little effect on the circulation of SCS and it is just a function of model geometry and large-scale forcing over the Pacific. Qu et al. (1999) investigated the upper-layer circulation by using all available historical temperature data combined with climatological temperature-salinity relationships but without winds in the SCS, and they revealed that two cyclonic eddies, namely, the West Luzon eddy and the East Vietnam eddy exist at around 18°N, 118°E from late fall to early spring coinciding well with a positive wind stress curl both in location and time and at about 14°N, 110°E from late summer to early fall. Chu et al. (1999) found that during the summer (winter) monsoon period, the SCS surface circulation is generally anti-cyclonic (cyclonic) with a strong western boundary current, the Vietnam coastal jet, which moves northward (southward) during summer (winter) with a mean maximum speed of 0.5 ms^{-1} (0.95 ms^{-1}), and a mean volume transport of 5.5 Sv (10.6 Sv) extending to a depth of around 200 m (500 m).

The second experiment domain (WNA) consists of the Gulf of Mexico, Caribbean Sea and part of the western North Atlantic Ocean. The Gulf of Mexico is a semi-enclosed sea that connects in the south to the Caribbean Sea through the Yucatan Channel, and in the east to the Atlantic Ocean through the Strait of Florida. The Gulf has a central basin (about 3500 m in depth) of relatively limited extent (about 90°~94°W and 23°~25°N), surrounded by continental rise, slope and shelves (Oey et al., 2005). The circulation of WNA region, especially of the Gulf of Mexico region, has been well studied in the past decades for its dominant flow feature of the Caribbean Current in the Caribbean Sea, the Loop Current in the Gulf of Mexico and the Gulf Stream in the North Atlantic Ocean. Currents through the Caribbean

Sea, the Gulf of Mexico and the Florida Straits constitute an important component of the subtropical gyre circulation of the North Atlanta Ocean, which has been pointed out by Fratantoni (2001) and other researchers. The Loop Current originates at the Yucatan Channel and is the formation region of the Florida Current-Gulf Stream system with peak speeds of 1.5~1.8 ms^{-1} near the surface (Carder et al., 1977; Ochoa et al., 2001). It episodically sheds warm-core rings (Cochrane, 1972; Vukovich, 1995; Sturges and Leben, 2000) at intervals of about 3~17 months. These rings' diameters are 200~300 km with swirl speeds in the range of 1.8~2 ms^{-1}, and they propagate westward at approximately 2~5 $km.day^{-1}$ (Cooper et al., 1990).

APPLICATION OF CSF IN HYCOM

Application of CSF in the momentum equation in the ocean surface Ekman layer

For monochromatic deep water waves, there is a mean Lagrangian volume transport produced in their propagation direction, the Stokes drift (Phillips, 1977), which is expressed as

$$\vec{u}_s = \vec{U}_s e^{2kz} \text{ and } \vec{U}_s = c(ak)^2 \cdot \vec{D}\,, \qquad (1)$$

where $\vec{U}_s$ is the Stokes drift at the ocean surface, a is wave amplitude, k is wave-number, c is wave phase speed, z is depth that is zero at the mean sea level and decreasing downward, and $\vec{D}$ is wave propagation direction. The interaction between the planetary vorticity and the Stokes drift yields a force on the Eulerian momentum balance (Hasselmann, 1970), $\vec{f}\times\vec{u}_s$. Hasselmann (1970) illustrated CSF as the divergence of a wave-induced stress $\tau_w = \overline{\tilde{v}\tilde{w}}$, where $\tilde{v}$ and $\tilde{w}$ are along wave crest component and vertical component of the rapidly varying wave orbital velocity, respectively, which is given by

$$-\rho_w \frac{\partial}{\partial z}\overline{\tilde{v}\tilde{w}} = -\rho_w \,|\, \vec{f}\times\vec{u}_s \,|. \qquad (2)$$

In the classical Ekman layer model, the momentum equation describing the non-steady-state, ageostrophic current in the surface layer is (Wu and Liu, 2008)

$$\frac{\partial \vec{U}}{\partial t} + \vec{f}\times\vec{U} = \frac{\partial}{\partial z}\left(A_z \frac{\partial \vec{U}}{\partial z}\right), \qquad (3)$$

where $\vec{U}$ is the horizontal current, and A_z is the vertical momentum diffusivity. The corresponding boundary condition for Equation (3) is

$$\begin{cases} \rho_w A_z \dfrac{\partial \vec{U}}{\partial z} = \vec{\tau}_0 & z = 0 \\ \vec{U} \to 0 & z \to -\infty \end{cases}, \qquad (4)$$

in which $\vec{\tau}_0$ is the sea surface wind stress. Introducing the wave-induced stress in the vertical turbulent diffuse term, Equation (3) can be rewritten as

$$\frac{\partial \vec{U}}{\partial t} + \vec{f}\times\vec{U} = \frac{\partial}{\partial z}(\vec{\tau}_* + \vec{\tau}_w), \qquad (5)$$

where $\vec{\tau}_* = A_z \dfrac{\partial \vec{U}}{\partial z}$ is turbulent stress, and $\vec{\tau}_w = \overline{\tilde{v}\tilde{w}}$ is wave-induced stress. Incorporating Equation (2) into (5) obtains

$$\frac{\partial \vec{U}}{\partial t} + \vec{f}\times\vec{U} = \frac{\partial}{\partial z}\left(A_z \frac{\partial \vec{U}}{\partial z}\right) - \vec{f}\times\vec{u}_s\,. \qquad (6)$$

Equation (6) is consistent with Polton et al.'s (2005) statement, that when the momentum equations are averaged over the wave periods, CSF appears as an extra term in the mean horizontal momentum equations that govern the ageostrophic motions. In Equation (6), the Coriolis force-like CSF is a large-scale wave-induced forcing contributing to the vertical turbulent diffuse term.

Application of CSF in the HYCOM momentum equation

Integration of Equation (2) gives

$$\vec{\tau}_w = -\rho \vec{f}\times\vec{T}_s, \qquad (7)$$

$$\vec{T}_s = \int_{-\infty}^{0} \vec{u}_s dz = c(ak)^2 \delta_s \cdot \vec{D} = \pi \frac{a^2}{T} \cdot \vec{D} \qquad (8)$$

where $\vec{\tau}_w$ is the depth-integrated CSF through the whole Stokes depth δ_s, ρ the water density, $\vec{T}_s$ the depth-integrated wave-induced transport, and T the wave period. The Stokes drift $\vec{u}_s$ and CSF $f\times\vec{u}_s$ penetrate to a certain depth from the ocean surface, named Stokes depth scale ($\delta_s = 1/2k$). Typically, the Stokes depth is about 5 m in the open ocean and is much shallower than wind-driven Ekman layer depth (Polton et al., 2005).

Polton et al. (2005), and Wu and Liu (2008) extended

the analytical solutions to the standard Ekman layer model that include CSF, and found that CSF changes the structure of the current profile within the Ekman layer. So, it is reasonable to assume that CSF is confined to the upper ocean ML. Here, the depth-integrated CSF based on the theory of Polton et al. (2005) is added to the momentum Equation of HYCOM (Bleck, 2002) as a boundary condition to examine the effects of Stokes drift on ocean surface circulation. As a boundary condition, CSF does not always enhance the momentum input, but also decreases the momentum input when the wind and CSF are in opposite direction.

In this study, we mainly focus on the ocean surface ML. The mixing strategy used here is the K-Profile Parameterization (Large et al., 1994, 1997). The depth-integrated CSF is also applied to the KPP mixing sub-model as a surface stress (equivalent wind stress) in addition to the traditional one depends solely on the wind. In order to reflect the effects of CSF clearly, here we use the default values for all the tuneable constants in KPP model. Incorporating CSF into the momentum equation of HYCOM in (*x*, *y*, *s*) coordinates (where *s* is an unspecified vertical coordinate) produces,

$$\frac{\partial \vec{V}}{\partial t_s}+\nabla_s\frac{\vec{V}^2}{2}+(\varsigma+f)\times\vec{V}+(\dot{s}\frac{\partial p}{\partial s})\frac{\partial \vec{V}}{\partial p}+\nabla_s M-p\nabla_s\alpha=-g\frac{\partial \tau_*}{\partial p}+(\frac{\partial p}{\partial s})^{-1}\nabla_s\cdot(\upsilon\frac{\partial p}{\partial s}\nabla_s\vec{V}), \quad (9)$$

where $\vec{V}$ = (*u*, *v*) is the horizontal current velocity vector, *p* pressure, $\alpha = 1/\rho$ the potential specific volume, ς the relative vorticity, $M = gz + p\alpha$ the Montgomety potential, $gz = \phi$ the geopotential, *f* the Coriolis parameter vector, υ the eddy viscosity/diffusivity coefficient, and τ_* the shear stress vector induced by the combination of wind stress, bottom drag and the depth-integrated CSF. Wind stress data in use are all from ECMWF reanalysis. $\dot{s}\partial p/\partial s$ Represents the vertical mass flux across an *s* surface. CSF influences the horizontal currents by changing both the shear stress and the eddy viscosity/diffusivity coefficient calculated in the KPP sub-model. The function of CSF in the KPP sub-model and the momentum equation is reflected by changing the effective boundary conditions.

Effects of CSF on the thickness of ocean surface boundary layer (BL) and ML

In the KPP mixing sub-model of HYCOM, the ocean surface Boundary Layer Depth (BLD) h_b is based on the bulk Richardson number, Ri_b, that in the following form

$$Ri_b = \frac{(B_r - B)d}{(\vec{V}_r - \vec{V})^2 + V_t^2}, \quad (10)$$

$$V_t^2 = \frac{C_s(-\beta_T)^{1/2}}{Ri_c\kappa^2}(c_s\varepsilon)^{-1/2}dN\omega_s, \quad (11)$$

$$\omega_S = \kappa(a_S u_*^3 + c_S\kappa\sigma\omega_*^3)^{1/3} \to \kappa(c_S\kappa\sigma)^{1/3}\omega_* \quad \sigma<\varepsilon, \quad (12a)$$

$$\omega_S = \kappa(a_S u_*^3 + c_S\kappa\varepsilon\omega_*^3)^{1/3} \to \kappa(c_S\kappa\varepsilon)^{1/3}\omega_* \quad \varepsilon\le\sigma<1, \quad (12b)$$

where *B* is buoyancy, *d* is depth, the subscript *r* denotes reference values, and the two terms in the denominator represent the influence of resolved vertical shear and unresolved turbulent velocity shear, respectively. Reference values are averaged over the depth range εd, where $\varepsilon = 0.1$. At depth of $d = h_b$, the reference depth εh_b represents the thickness of the surface layer where the Monin-Obukhov similarity theory applies. The unresolved turbulent velocity shear is expressed as Equation (11), where C_S is a constant between 1 and 2, β_T is the ratio of entrainment buoyancy flux to surface buoyancy flux, $\kappa = 0.4$ is the von Karman constant, and ω_S is the salinity/scalar turbulent velocity scale, which is estimated from Equation (12). In Equation (12), a_S and c_S are constants, $\omega_* = (-B_f/h)^{1/3}$ is the convective velocity scale with B_f representing the surface buoyancy flux , σ equals to d/h_b, and the friction velocity u_* is from the relation $u_*^2 = |\tau|/\rho$, where τ includes surface wind stress (ECMWF Reanalysis) and CSF. Terms on the right side of the arrows in both Equations (12a) and (12b) are the convective limits. BLD is the depth range over which the turbulent boundary layer eddies can penetrate before becoming stable relative to the local buoyancy and velocity, and is often characterized by very small vertical property gradients over large portions of its central region. BLD is estimated as the minimum depth at which Ri_b exceeds the critical Richardson number $Ri_c = 0.3$. As mentioned above, h_b is based on Ri_b, Ri_b is a function of u_*, and u_* is influenced by CSF, thus BLD h_b can be influenced by CSF.

In the KPP mixing strategy, MLD is estimated diagnostically depending on a temperature jump, e.g. 0.3℃. The base of the ML is set to an interpolated depth where the density jump is equivalent to the temperature jump. Usually, BLD and MLD are similar, since KPP mixes strongly down to BLD, so density should be relatively homogeneous throughout the whole layer. BLD

Table 1. The model settings of the 6 experiments.

Experiments		Exp. 1	Exp. 2	Exp.3	Exp. 4	Exp. 5	Exp.6
Domain		**Global**	**Global**	**SCS**	**SCS**	**WNA**	**WNA**
Resolution	Latitude	$2.5°\times cos\varphi$	$2.5°\times cos\varphi$	$1/12°\times cos\varphi$	$1/12°\times cos\varphi$	1/6 °	1/6 °
	Longitude	2.5°	2.5°	1/12 °	1/12 °	1/6 °	1/6 °
Coriolis-Stokes		Yes	No	Yes	No	Yes	No

and MLD are determined by different criteria, the former is by buoyancy and the latter temperature. Actually, buoyancy and temperature can be linked through density. Thus, the effects of CSF on MLD are similar to those on BLD.

Effects of CSF on SST

The temperature conservation equation in HYCOM is

$$\frac{\partial}{\partial t_s}(\frac{\partial p}{\partial s}T)+\nabla_s\cdot(\vec{V}\frac{\partial p}{\partial s}T)+\frac{\partial}{\partial s}(\dot{s}\frac{\partial p}{\partial s}T)=\nabla_s\cdot(\upsilon\frac{\partial p}{\partial s}\nabla_s T)+S_T \quad (13)$$

where T is temperature, υ the eddy viscosity/diffusivity coefficient, S_T the diabatic source terms including diapycnal mixing. As presented under application of CSF in the HYCOM momentum equation, CSF impacts horizontal current velocity $\vec{V}$ through the momentum equation. Since $\vec{V}$ plays its part in the second term of temperature conservation equation Equation (13), CSF also influences SST. Additionally, CSF intensifies/weakens turbulent mixing in the upper ocean by modifying the diffusivity coefficient as an effective top boundary condition for the KPP sub-model. Thus, it also leads to the change of SST. The effects of CSF on salinity are similar to that on temperature, but they are beyond the scope of this study.

MODEL SETTINGS AND EXPERIMENTS

Three sets of experiments (Table 1) with different horizontal resolutions and in different regions are designed for the current study. HYCOM and WW3 are chosen to conduct those numerical experiments. Considering the length limit, introductions, such as numerical technique, about HYCOM and WW3 please refer to the User's Menu.

Global experiments

Two experiments, Experiments 1 and 2 modeled with and without CSF, respectively, are conducted for the global ocean. The calculation domain for HYCOM is from 64.43911°S to 64.43911°N with a resolution of $2.5^\circ\times\cos\varphi$ in latitude, where φ is the corresponding latitude, and 180°W to 180°E with a resolution of 2.5° in longitude, with an array size of 144×69 horizontal grid points. 26 hybrid layers are used in the vertical. HYCOM is initialized by the Polar Science Center Hydrographic Climatology (PHC) 3.0. This monthly climatology is also used for lateral boundary nudging and relaxation of salinity and temperature on the ocean surface. The model is spun up for 5 years forced by ERA15 ECMWF reanalysis monthly climatology. The forcing variables are 10 m winds, ocean surface air temperature, precipitation, radiation heat flux, short-wave radiation, and water vapor mixing ratio.

Then, 24-hourly ECMWF ERA40 wind anomalies are added to ECMWF ERA15 climatology to produce the 2001 run. In addition, the CSF is added in Experiment 1. WW3 is configured on a grid from 90°S to 90°N with a resolution of 2.5° and 180°W to 180°E with a resolution of 2.5°, which gives a total of 144×73 horizontal grid points. The first frequency is 0.0418 Hz, the frequency increment factor is 1.1, the number of frequencies is 30, and the directional resolution is 15°. Minimum source term time step is 600 s. Time interval of wind input is 86400 s. Input winds are from ECMWF ERA40 re-analysis data sets.

SCS experiments

Another two experiments, Experiments 3 and 4 run with and without CSF, are conducted for SCS region. The model domain for HYCOM is from 3.24696°S to 22.705°N with a resolution of $1/12^\circ\times\cos\varphi$ degree in latitude and 102°E to 120.9924°E with a resolution of 1/12° in longitude, and a total grid size of 229×320. The number of the vertical hybrid layers is 26. For WW3, a horizontal grid size of 39×55 is configured for a domain from 4°S to 23°N with a resolution of 0.5° in latitude and 102°E to 121°E in longitude with a resolution of 0.5°. The boundaries of the calculation domain are closed. The Taiwan Strait, Luzon Strait, Gaspar Strait, Karimata Strait and Strait of Malacca are treated as closed boundary condition, which means no water pass through those boundaries and only the salinity and temperature are relaxed in the buffer zones. The rest of the settings are the same as that of the global run.

WNA experiments

Experiments 5 and 6 are simulations with and without the effect of CSF in WNA region. The WNA domain has a grid size of 181×205 with a horizontal resolution of 1/6°, from 263°E to 293°E in longitude and 8°N to 42°N in latitude, and 26 vertical layers. It is nested in the North Atlantic 2.00 (horizontal resolution is 2°) domain which provides boundary conditions to the finer WNA domain. The WW3 is configured for the same domain, but with a different horizontal resolution of 0.5°. The other settings are the same as that of the global case.

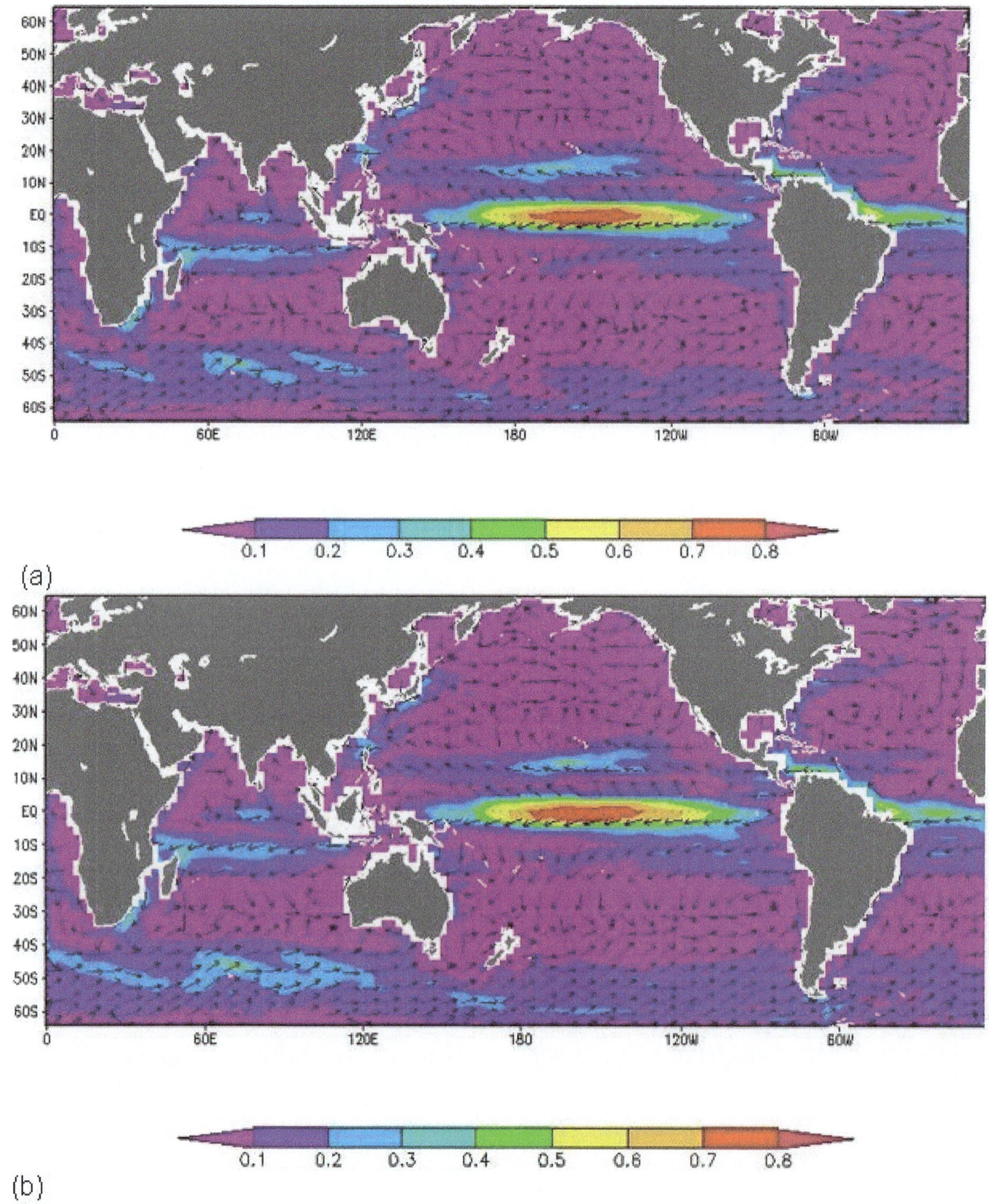

Figure 1. a) Annual-mean mixed-layer current field of 2001 from Exp. 2; b) Annual-mean mixed layer current field of 2001 from Exp. 1 (ms^{-1}).

MODELING RESULTS

Global case

Figures 1a and b show the details of the global ML current fields from Experiments 1 and 2, respectively. The South Equatorial Current (SEC), North Equatorial Current (NEC), Equatorial Current (EC), and Western Boundary Current (WBC) such as the Kuroshio Current and the Gulf Stream, are clearly shown. The resolution used in the global case is 2.5° which is not fine enough to resolve WBC, so WBC is not as noticeable as EC in this situation. By comparing Figure 1a with 1b, we find that the two figures have the same pattern. It indicates that although the wave-induced CSF exerts an influence, as illustrated in Figure 2c, in the upper ocean ML, especially over the Westerly region, it can not fundamentally change the structures of the current field. Figures 1 and 2a show that, both the global ocean surface circulation and waves are basically subject to the wind field (as a common sense, the wind field is not shown).

Figures 2a and b present the annual-mean depth-integrated Stokes transport $\vec{T}_s$ which is calculated by the wave variables from the WW3 simulation and the ECMWF reanalysis, respectively. The general patterns of $\vec{T}_s$ in both cases agree well over the global ocean, and the directions of $\vec{T}_s$ are quite similar over the Westerly regions. This indicates that the wave variables modeled by WW3 are reasonable. The ML depth-integrated current transport is expressed as $\vec{T}_c = \int_{-d}^{0} \vec{V}_z dz$, where $\vec{V}_z$ is the current velocity and *d* refers to the MLD. CSF-induced changes in ML depth-integrated current transport $\Delta\vec{T}_c$ are presented in Figure 2c, with most of the values range from 0 to 3 m^2s^{-1}. Almost all the large values are located over the high-latitude Westerly wind belt with a maximum of 13 m^2s^{-1}, where relatively strong winds force large waves. Note that the values of Stokes transport $\vec{T}_s$, from both WW3 simulation and ECMWF reanalysis, are in the ranges of about 0 to 2.2 m^2s^{-1}. It is safe to assume that most of the values of both $\vec{T}_s$ and $\Delta\vec{T}_c$ are in the similar order. Figure 2d gives the wave-induced percentage changes in $\vec{T}_c$ (defined as $\Delta\vec{T}_c/\vec{T}_c$) which reveals the significant effect of CSF on the upper ocean currents. Changes in the tropical region and the western boundary current regions are relatively small, about 10%. Over the mid-latitude region, especially those under the Westerlies, the changes are more than 30%; with largest value reaches 90%, due to the strong wind waves acting on the relatively weak currents there. In the ocean model, the ocean surface Ekman layer is usually characterized as an ocean surface BL. Since BL and ML have the same depth scale, the wave-induced percentage changes in $\vec{T}_c$ can be considered as a gauge of how important the wave-induced $\Delta\vec{T}_c$ is, relative to Ekman transport. From Figures 2a, b and c, it is interesting to note that, there are large differences between the directions of $\Delta\vec{T}_c$ and $\vec{T}_s$, except for the Southern Hemisphere Westerlies (especially in the Pacific) where comparatively good agreement can be found.

According to Equation (3), the direction of CSF is perpendicular to that of the waves, so the expected direction of $\Delta\vec{T}_c$ will also be normal to $\vec{T}_s$. As shown in Figures 2a and c, the direction of $\Delta\vec{T}_c$ is similar to that of $\vec{T}_s$, where strong winds directly force large waves; in the swell-dominant tropical region, the direction of $\Delta\vec{T}_c$ appears to be normal to that of the waves or $\vec{T}_s$ since swell propagates from remote areas (Deng et al., 2007,2009), while in some other places, the direction of $\Delta\vec{T}_c$ and $\vec{T}_s$ are almost opposite, which can be explained as, that the wave-induced Stokes transport $\vec{T}_s$ is compensated by the CSF.

Theoretically, the direction of $\Delta\vec{T}_c$ over the southern and northern Westerly regions should be symmetrical since the CSF is equator-ward for both of the two regions where wind direction is eastward. However, the actual situation of the northern Westerly region is out of our expectation. Possible reason for that could be the different topography and different wind wave intensity or other unclear influence over these two regions. This phenomenon is worthy of further investigation. In the global ocean surface ML, the CSF-caused changes in MLD (Δthi) are in the range of -15 to 80 m (Figure 3a). There is an obvious enhancement of MLD in the southern Westerlies where the value of Δthi is 24 m on average. It is due to the fact that the strong wind waves prevail in this region and CSF intensifies the mixing in the upper ocean by interacting with the currents. In most regions of the northern hemisphere, the values of Δthi are decreased, because the wave-induced CSF weakens the net forcing (a combination of wind stress, CSF and so on) in these places. For example, if the wave-induced CSF and wind stress are in the same (different) direction, the net forcing will be increased (decreased).

As for the influences of CSF on ocean surface temperature over the global ocean (Figure 3b), the Δtem value varies from -0.8 to 0.8℃; the maximum value, as large as 0.8℃, appears along the very close eastern boundary of the southern Pacific where both $\Delta\vec{T}_c$ and $\vec{T}_s$ originating from the cold southern Westerly region (carrying cold water). Deng et al. (2007) presented that, the total wave-induced equator ward Stokes transport from Westerlies is 3.672 *Sv* on average over the swell-prevailed Eastern Pacific region, and $\vec{T}_s$ over the eastern boundary of the Southern Pacific area is noticeable. The majority of the positive (negative) Δthi values correspond to the negative (positive) Δtem values, since mixing intensified (weakened) by CSF decreases (increases) ocean surface temperature by entraining

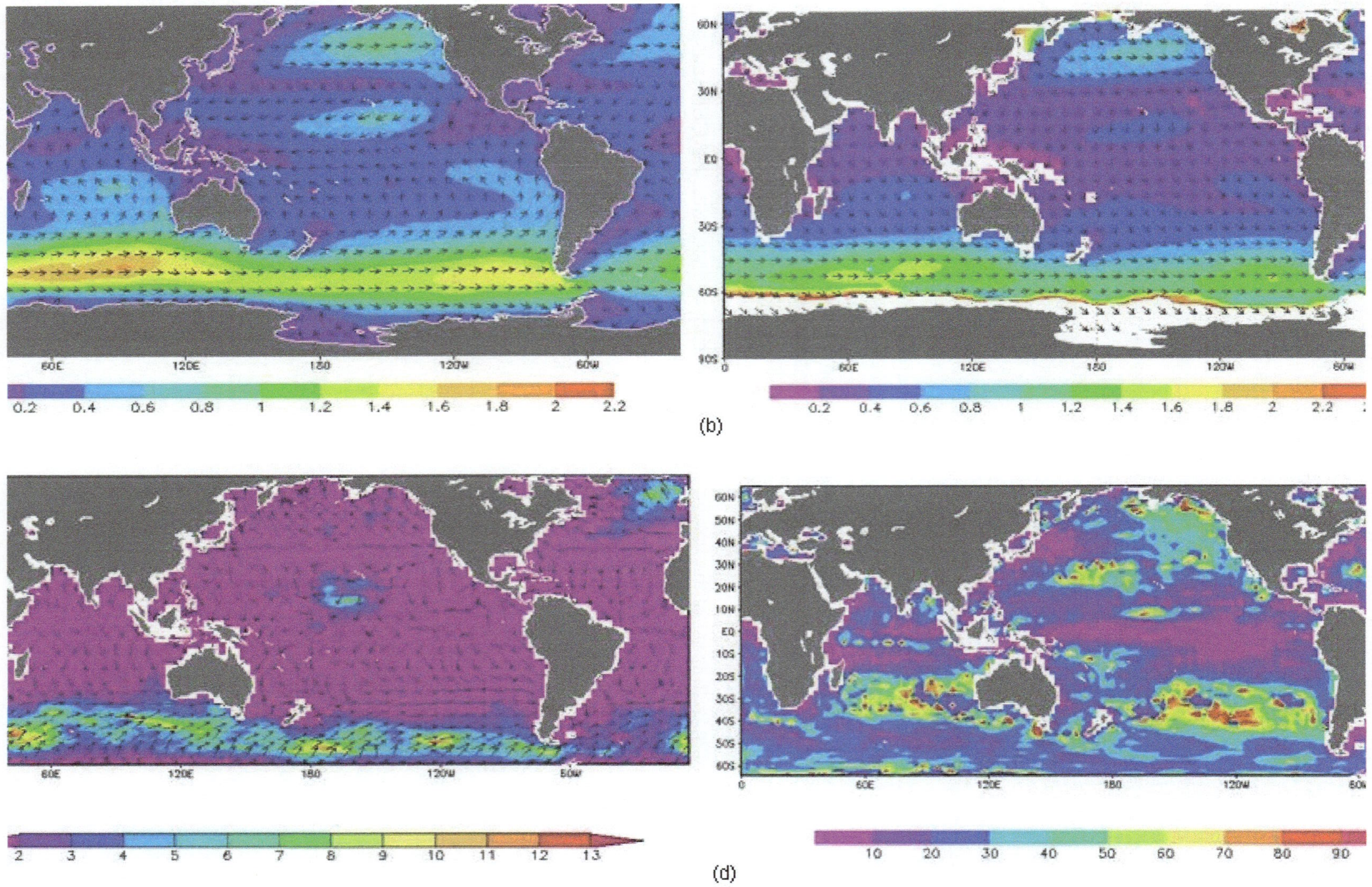

Figure 2. a) Annual-mean depth-integrated wave-induced Stokes transport, $\vec{T}_s$, of 2001 (simulated by WW3 model, m^2s^{-1}), b) Annual-mean depth-integrated wave-induced Stokes transport, $\vec{T}_s$, of 2001 (calculated by ECMWF reanalysis wave variables, m^2s^{-1}), c) Annual-mean changes (results of Exp. 1 minus that of Exp. 2) in depth-integrated current transport through the mixed layer, $\Delta\vec{T}_c$, of 2001 (m^2s^{-1}), and d) percentage change of $\vec{T}_c$ ($\Delta\vec{T}_c/\vec{T}_c$). (Color indicates the magnitude and vector presents the direction).

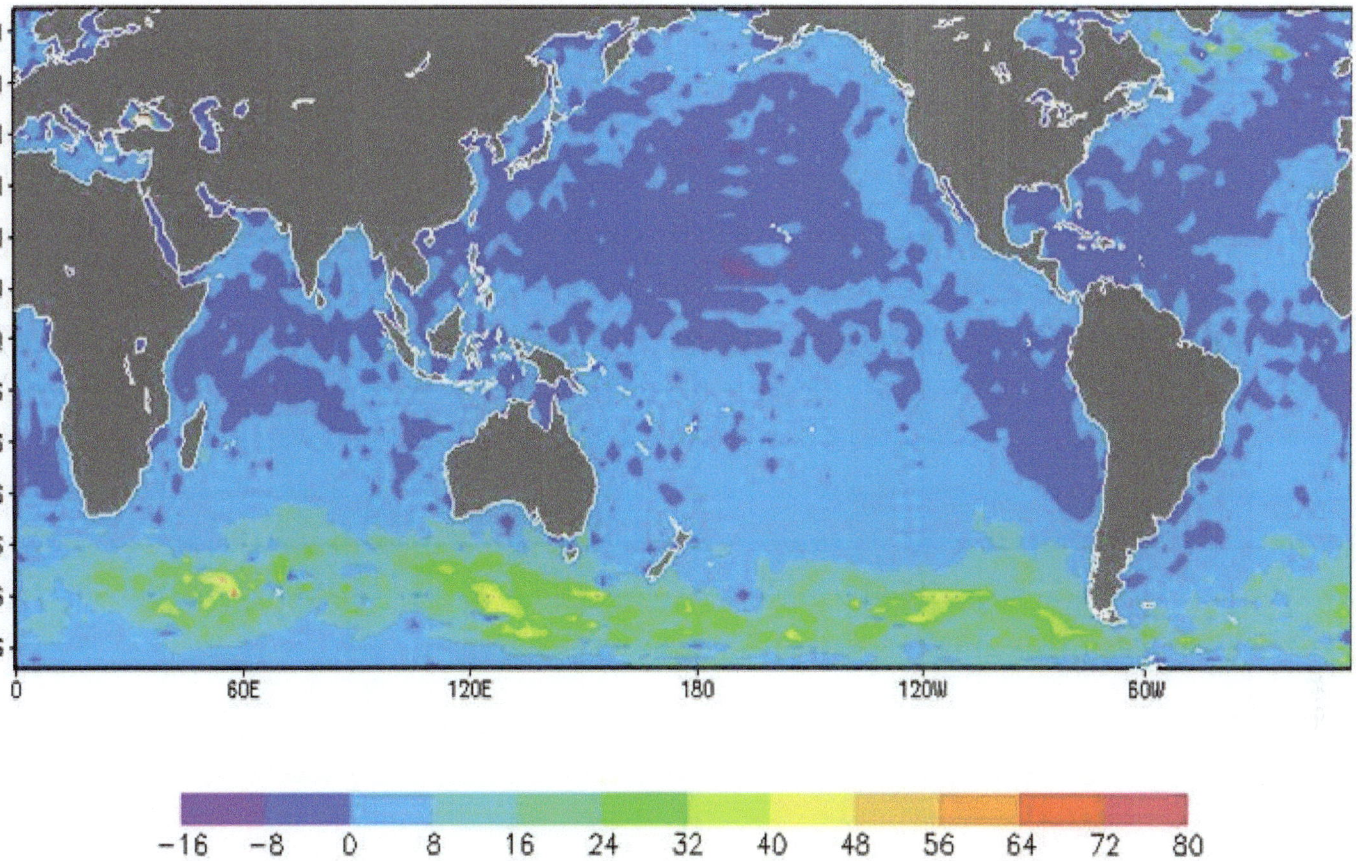
60N
50N
40N
30N
20N
10N
EQ
10S
20S
30S
40S
50S
60S
0
60E
120E
180
120W
60W
−16 −8 0 8 16 24 32 40 48 56 64 72 80

(a)

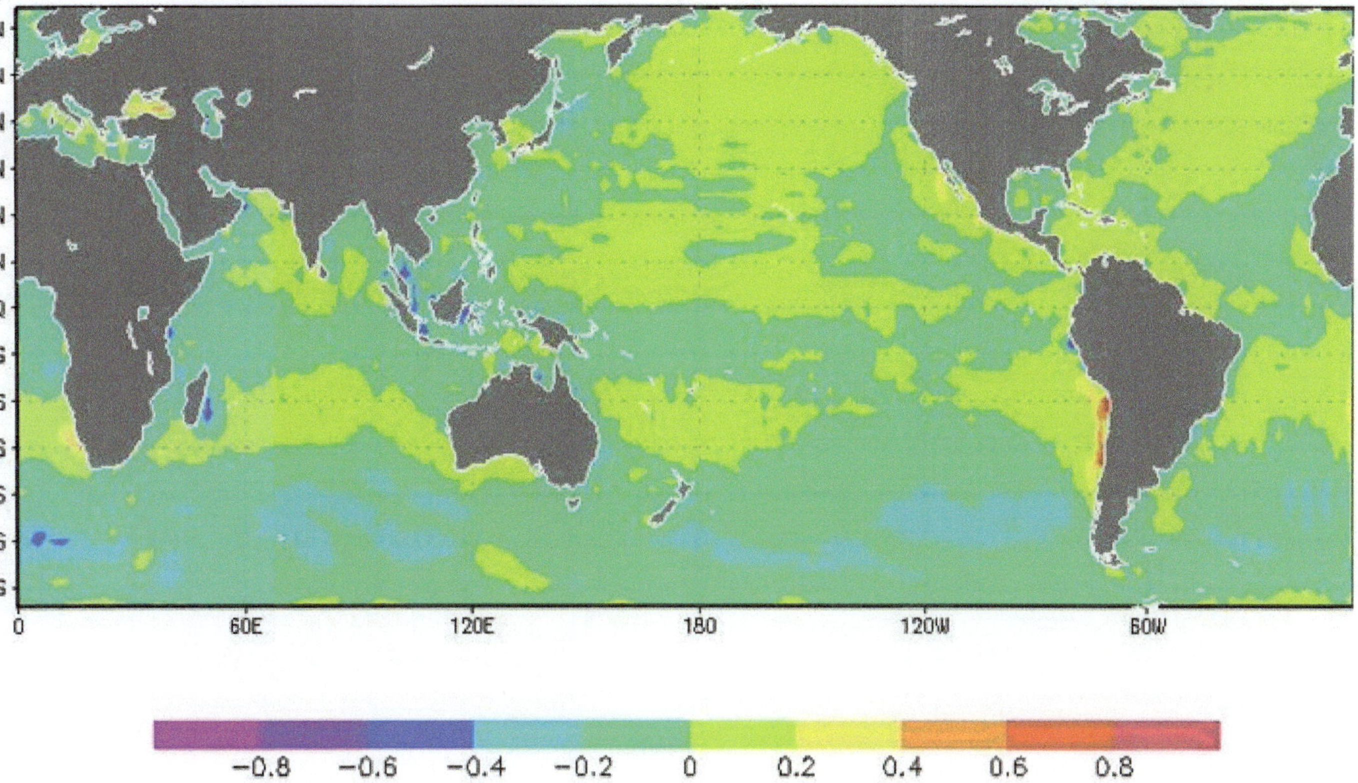
60N
50N
40N
30N
20N
10N
EQ
10S
20S
30S
40S
50S
60S
0
60E
120E
180
120W
60W
−0.8 −0.6 −0.4 −0.2 0 0.2 0.4 0.6 0.8

(b)

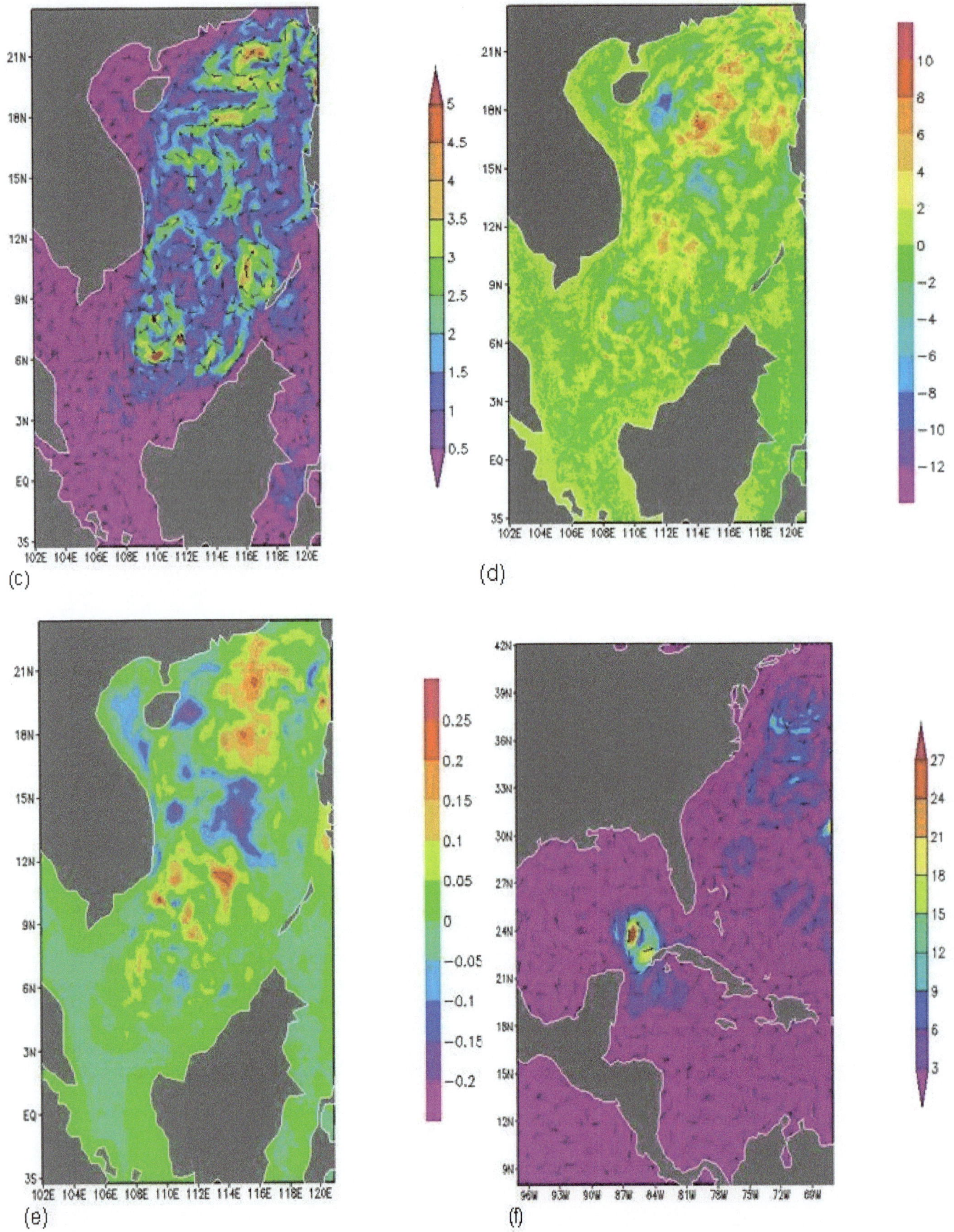
(c)
(d)
(e)
(f)
102E 104E 106E 108E 110E 112E 114E 116E 118E 120E
3S EQ 3N 6N 9N 12N 15N 18N 21N
96W 93W 90W 87W 84W 81W 78W 75W 72W 69W
9N 12N 15N 18N 21N 24N 27N 30N 33N 36N 39N 42N

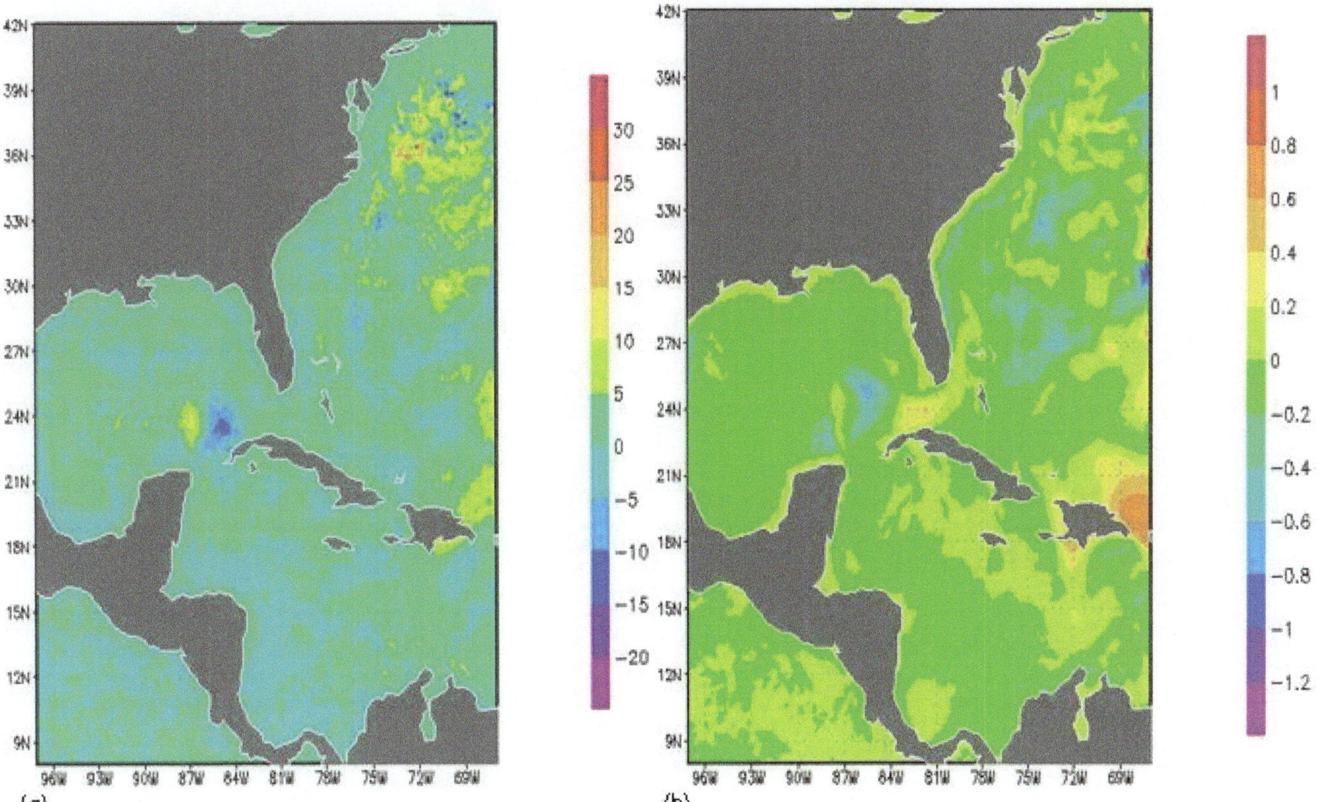

(g) (h)

Figure 3. a) Annual-mean changes in mixed layer depth , Δthi, of 2001in the global ocean (results of Exp. 1 minus that of Exp. 2, *m*); b) Annual-mean changes in mixed layer temperature, Δtem, of 2001 in global ocean (results of Exp. 1 minus that of Exp. 2, m); c) Annual-mean changes in mixed layer depth-integrated current transport, $\Delta\vec{T}_c$, of 2001 in SCS (results of Exp. 3 minus that of Exp. 4, m^2s^{-1}); d) Annual-mean changes in mixed layer depth, Δthi, of 2001 in SCS (results of Exp. 3 minus that of Exp. 4, *m*); e) Annual-mean changes in mixed layer temperature, Δtem, of 2001 in SCS (results of Exp. 3 minus that of Exp. 4, °C); f) Annual -mean changes in mixed layer depth-integrated current transport, $\Delta\vec{T}_c$, of 2001 in WNA (results of Exp. 5 minus that of Exp. 6, m^2s^{-1}); g) Annual-mean changes in mixed layer depth, Δthi, of 2001 in WNA (results of Exp. 5 minus that of Exp. 6, *m*); and h) Annual-mean changes in mixed layer temperature, Δtem, of 2001 in WNA (results of Exp. 5 minus that of Exp.6, °C). (Color indicates the magnitude and vector presents the direction).

more (less) cold water into the ML.

SCS Case

Figures 3c to e present the distribution of annual-mean $\Delta\vec{T}_c$, Δthi, and Δtem in SCS, respectively. The magnitude of annual-mean $\Delta\vec{T}_c$ varies from 0 to 5 m^2s^{-1} with large values located in regions, where the two main gyres emerge in both winter and summer. The influences of wave-induced CSF on $\vec{T}_c$ are apparently more significant in these gyre regions. In other regions, the values of $\Delta\vec{T}_c$ are less than 1 m^2s^{-1}. The direction of $\Delta\vec{T}_c$ seems random as annual-mean vectors cannot capture the variations of the impacts of CSF that varies from month to month. The values of annual-mean Δthi are in the range of -10 to 10 m, with large positive values existing to the west of the Luzon Strait where cyclonic gyres exist. The relatively small ML temperature change varying from -0.2 to 0.25°C compared with the global case, indicates that the horizontal resolution influences the modeling results.

The annual evolution of the ML current field in SCS (Figure 4, only odd months are shown) characterized by two main basin-scale eddies can be found in each month. Figure 5 shows the variability of $\Delta\vec{T}_c$ resulted from CSF for each odd month. Since all the connections that link SCS to the surrounding seas are closed, in the case of

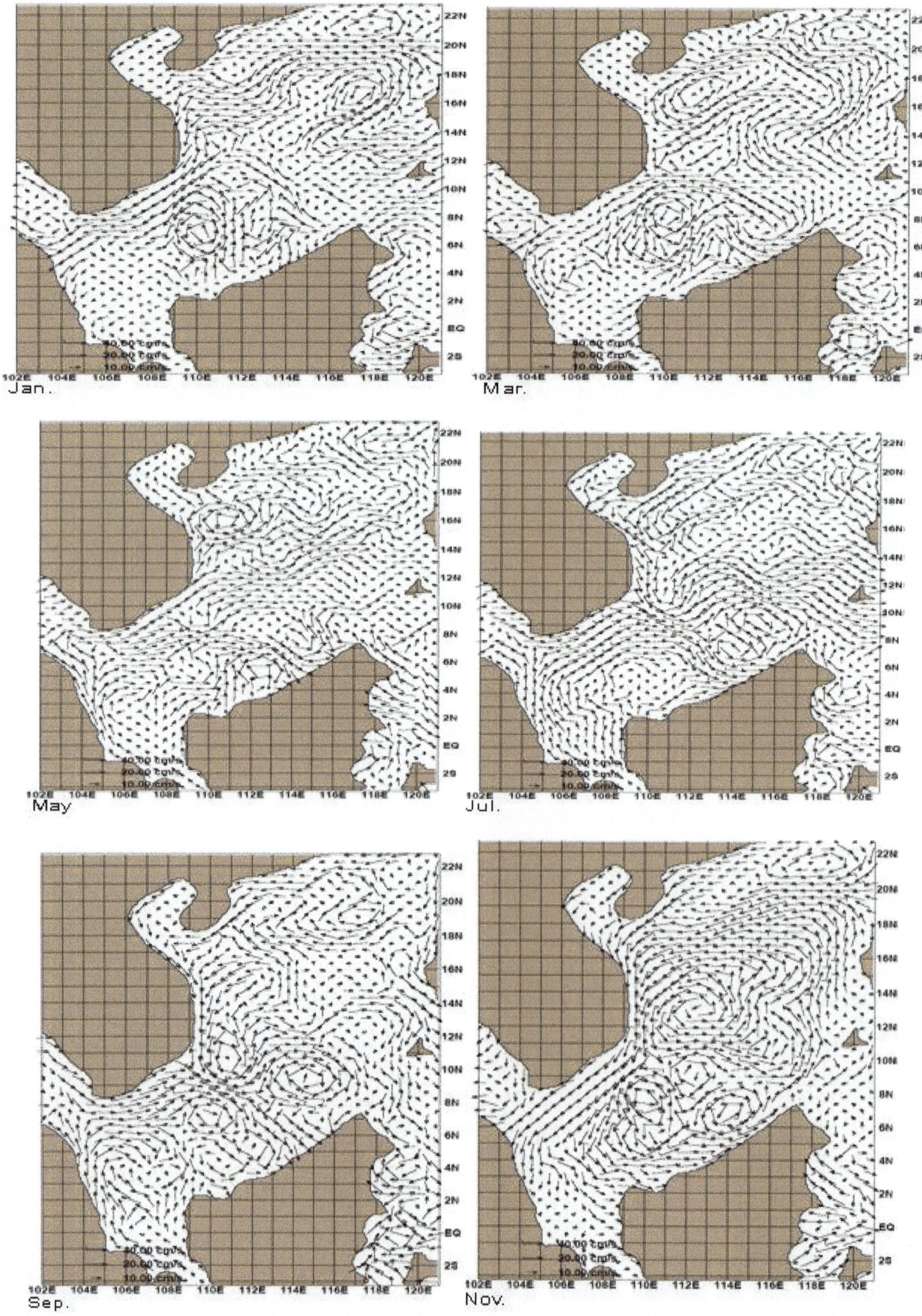

Figure 4. Monthly-mean mixed layer current field in 2001 over SCS (from Exp. 4, $cm \cdot s^{-1}$).

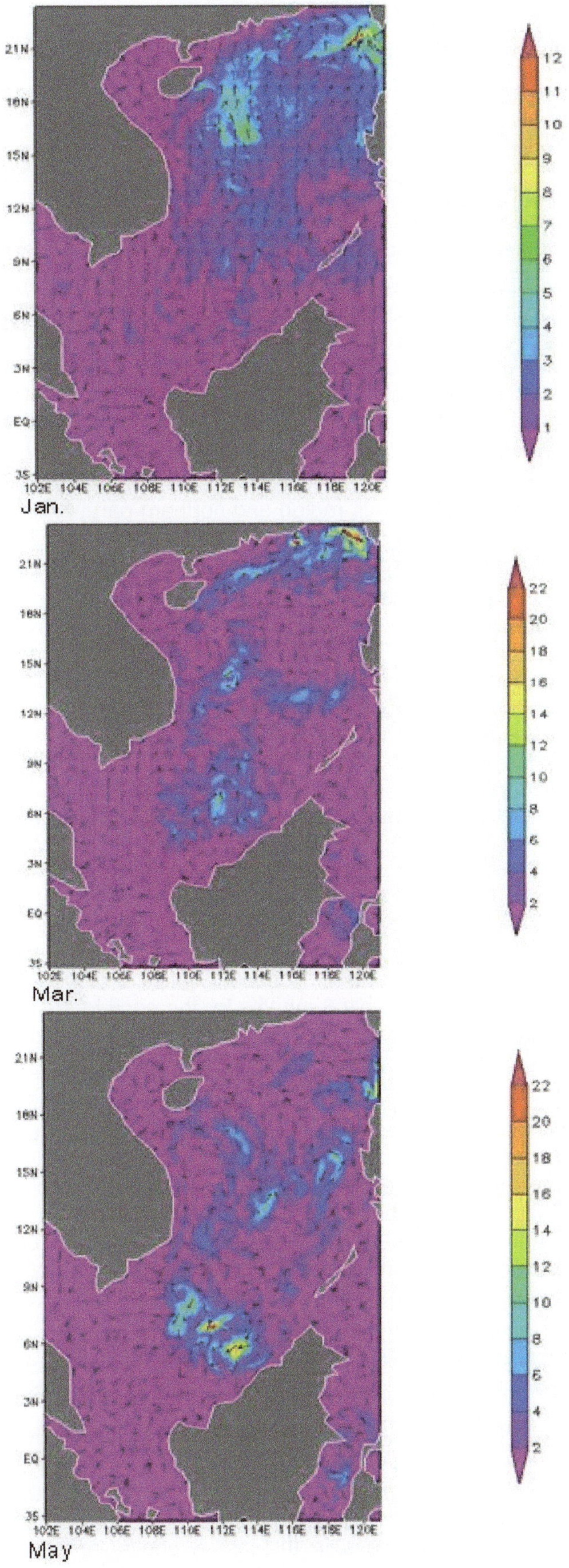

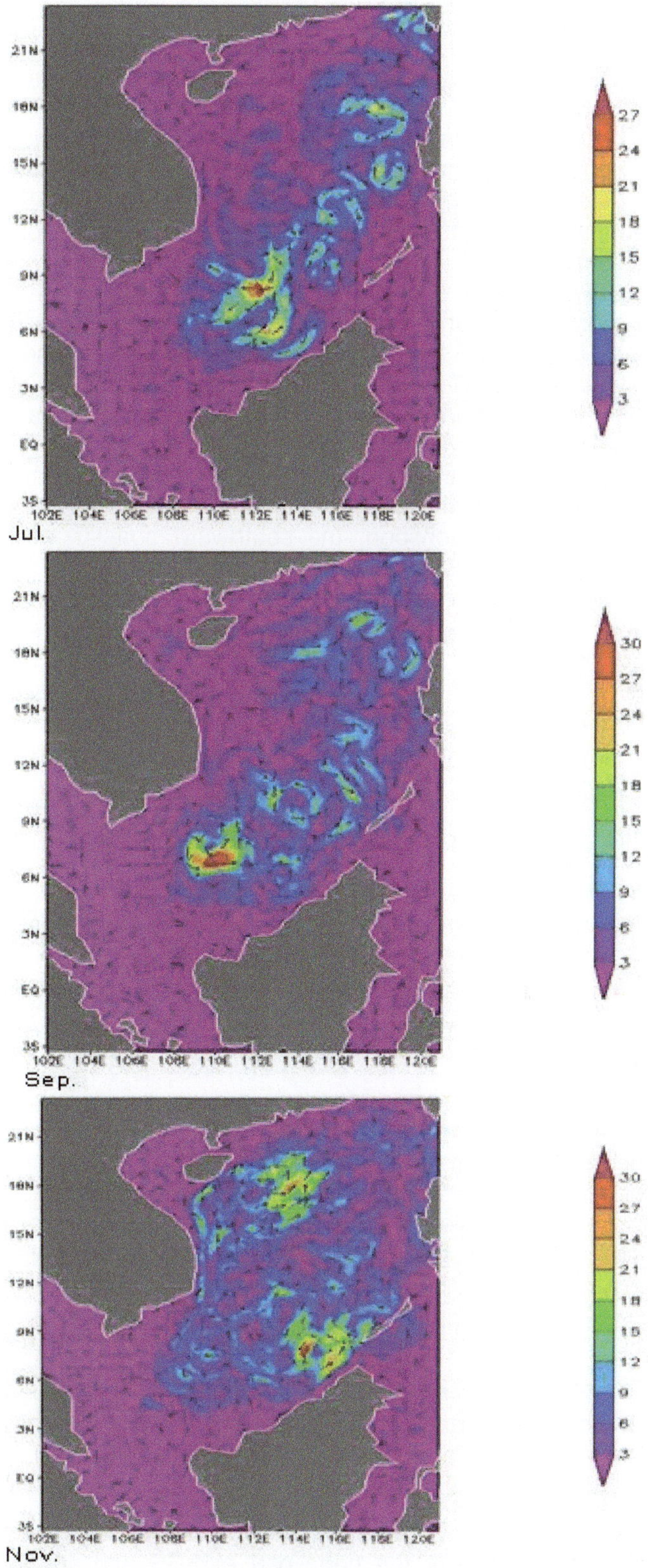

Figure 5. Coriolis-Stokes forcing induced monthly-mean changes of depth-integrated current transport in the mixed layer, $\Delta\vec{T}_c$, over SCS in 2001(m^2s^{-1}) (results of Exp. 3 minus that of Exp. 4) (color indicates the magnitude $|\Delta\vec{T}_c|$ and vector presents the direction).

no CSF, our modeled circulation is a result of the interaction of local wind-driven circulation with topography. The upper layer circulation of SCS is mainly driven by the monsoon circulation. Wyrtki (1961) demonstrated that, there is a distinct seasonal cyclonic gyre in winter and a weak cyclonic gyre in the Northern SCS, but an anti-cyclonic gyre in the southern SCS in summer. In terms of the basin-scale gyre, our modeling results are consistent with that of Wyrtki (1961). Qu (1999) showed similar results by an analysis of the distribution of dynamic height. Liu et al. (2008) found that the barotropic sea surface height has a seasonal distribution similar to that of the sea dynamic height. Another important current in SCS, the Vietnam coastal jet, flows southward during winter monsoon season and reverses in summer, which is also in good agreement with the result of Wyrtki (1961). According to Figure 4, during winter monsoon, the southward coastal jet off Vietnam is displayed as a continuation of the southwestward coastal jet over the continental slope south of China, named the Dongsha Current, since it originates near Dongsha Island. In summer, the northward jet off Vietnam turns eastward into the interior of SCS at about 11°N.

Next, comparing $\Delta\vec{T}_c$ in each odd month with the corresponding current field, Figure 5 provides an insight into the influences of CSF on $\Delta\vec{T}_c$. In January, the magnitudes of $\Delta\vec{T}_c$ are generally between 0 and 5 m^2s^{-1}, and the larger values are distributed over the northern SCS; in the north part of SCS, especially west of the Luzon Strait, and the region to the south of the Vietnam coast, the directions of $\Delta\vec{T}_c$ are generally northward and generally normal to that of the currents there. Knowing that the northeasterly winter monsoon winds are prevailing in this season, this phenomenon can be explained, at least theoretically, by the fact that the flow driven by CSF is normal to the quasi-geostrophic gyre currents. In November, the situation is similar to that of January except that the distribution of large $\Delta\vec{T}_c$ values extends through to south SCS. Also, the maximum value of $\Delta\vec{T}_c$ is larger. From the earlier two winter cases, it can be found that large $\Delta\vec{T}_c$ value centers are always located at the cyclonic/anti-cyclonic gyre region which suggests that CSF plays an important role there. However, some large $\Delta\vec{T}_c$ values also appear in other places. In summer months, both May and July, the circulation is depicted by a weak cyclone in the north part of SCS and an anti-cyclonic in the south due to the actions of southwesterly summer monsoon winds. The large $\Delta\vec{T}_c$ value center is developed in the southern SCS, between 5°N and 9°N in latitude and 110°E and 114°E in longitude, where strong eastward currents exist. No obvious relationship can be found between the current direction and the $\Delta\vec{T}_c$ direction.

WNA Case

For the WNA region, $\Delta\vec{T}_c$, Δthi and Δtem are shown in Figures 3f, g and h, respectively. The majority of $|\Delta\vec{T}_c|$ are less than 6 m^2s^{-1} with a maximum of 27 m^2s^{-1} occurring between Yucatan Channel and Straits of Florida where the Loop Current is located at. It also indicates that, as in the SCS case, the influences of CSF on upper layer current field are stronger in regions with strong quasi-geostrophic currents. The value of annual-mean Δthi generally varies in the range of -20 to 30 m, most of them from -10 to 10 m. Large variability of MLD is found in the Loop Currents region and northern part of Atlantic, where intense currents exist. The annual-mean changes in ML temperature due to CSF are between -1 and 1°C. A great portion of Δtem values fall in the range of -0.4 to 0.4°C. As discussed earlier, mixing intensified (weakened) by CSF decreases (increases) ocean surface temperature by entraining more (less) cold water into the ML leading to the changes in both MLD and temperature.

Monthly-mean ML current fields (including the effects of CSF), for March, June, September and December of 2001, are given in Figure 6, with the western boundary currents (Caribbean Current and Loop Current as well as the Gulf Stream) clearly depicted. In the Gulf of Mexico region, the ring shed off by the Loop Current has a diameter of about 250 km which is consistent with previous model results. We also notice that the core of the ring exists at (89.5°W, 25.5°N) in January and it dissipates at the western boundary of Gulf of Mexico in June. It takes roughly 6 months for the eddy to cross the Gulf of Mexico, corresponding a westward propagation speed of approximately 3 $km \cdot day^{-1}$, which agrees well with the result given by Romanou et al. (2004). Another interesting issue to note is that the Loop Current becomes weaker and weaker while another eddy is developing in the Caribbean Sea south of the Yucatan Channel during the period of June to December.

As mentioned earlier, CSF can influence the upper ocean circulation to a certain degree, but cannot fundamentally change the current pattern. It just serves as a correction to the ocean top boundary condition. The monthly-mean ML $\Delta\vec{T}_c$ in WNA for March, June, September, and December of 2001 are shown in Figure 7. Similar to SCS, the magnitudes of $\Delta\vec{T}_c$ are generally less than 5 m^2s^{-1}. The large ones appear in the area where intense eddy present, such as the Loop Current region and the Gulf Stream region. One possibility is that the dynamic process in the strong eddy/gyre region is

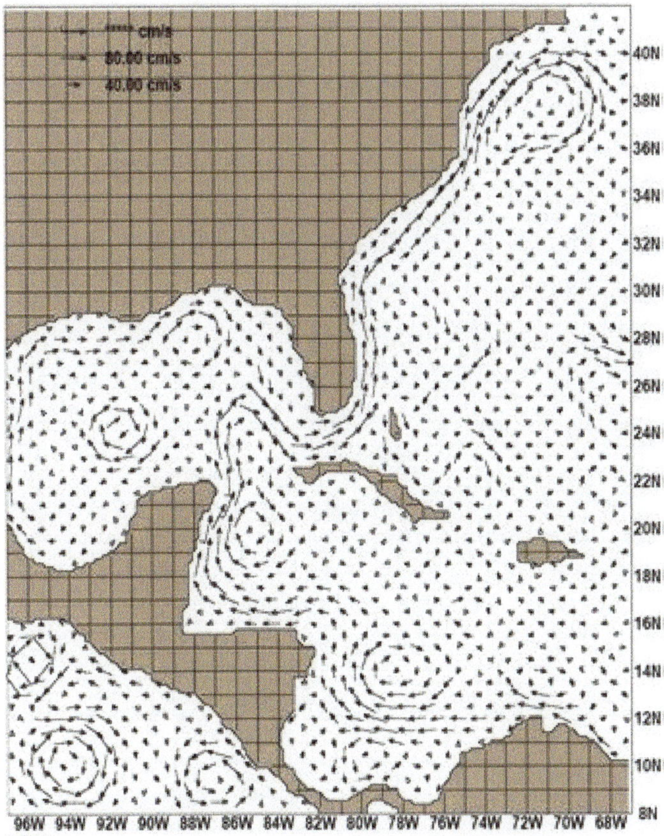

Mar.

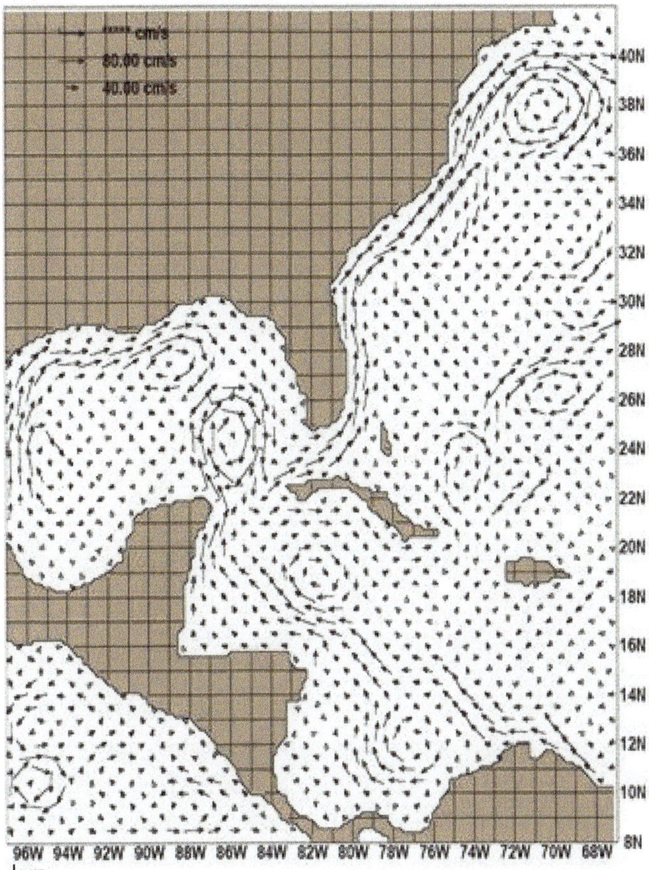

Jun.

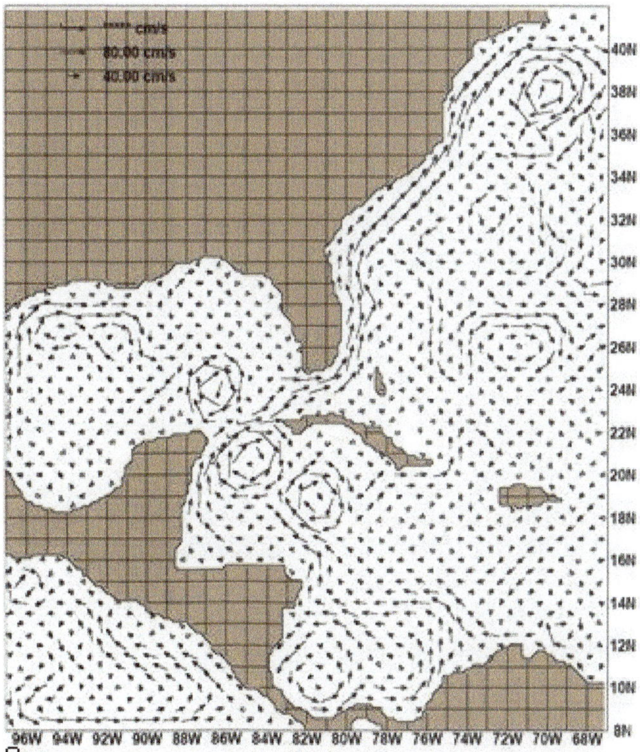

Sep.

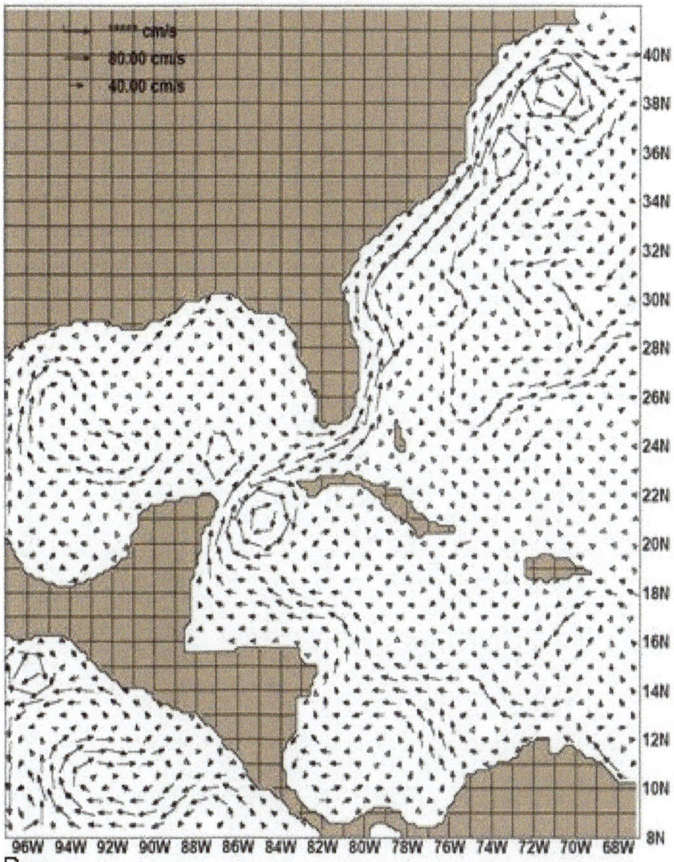

Dec.

Figure 6. Monthly-mean mixed layer current field in 2001over WNA (from Exp. 5, $cm \cdot s^{-1}$).

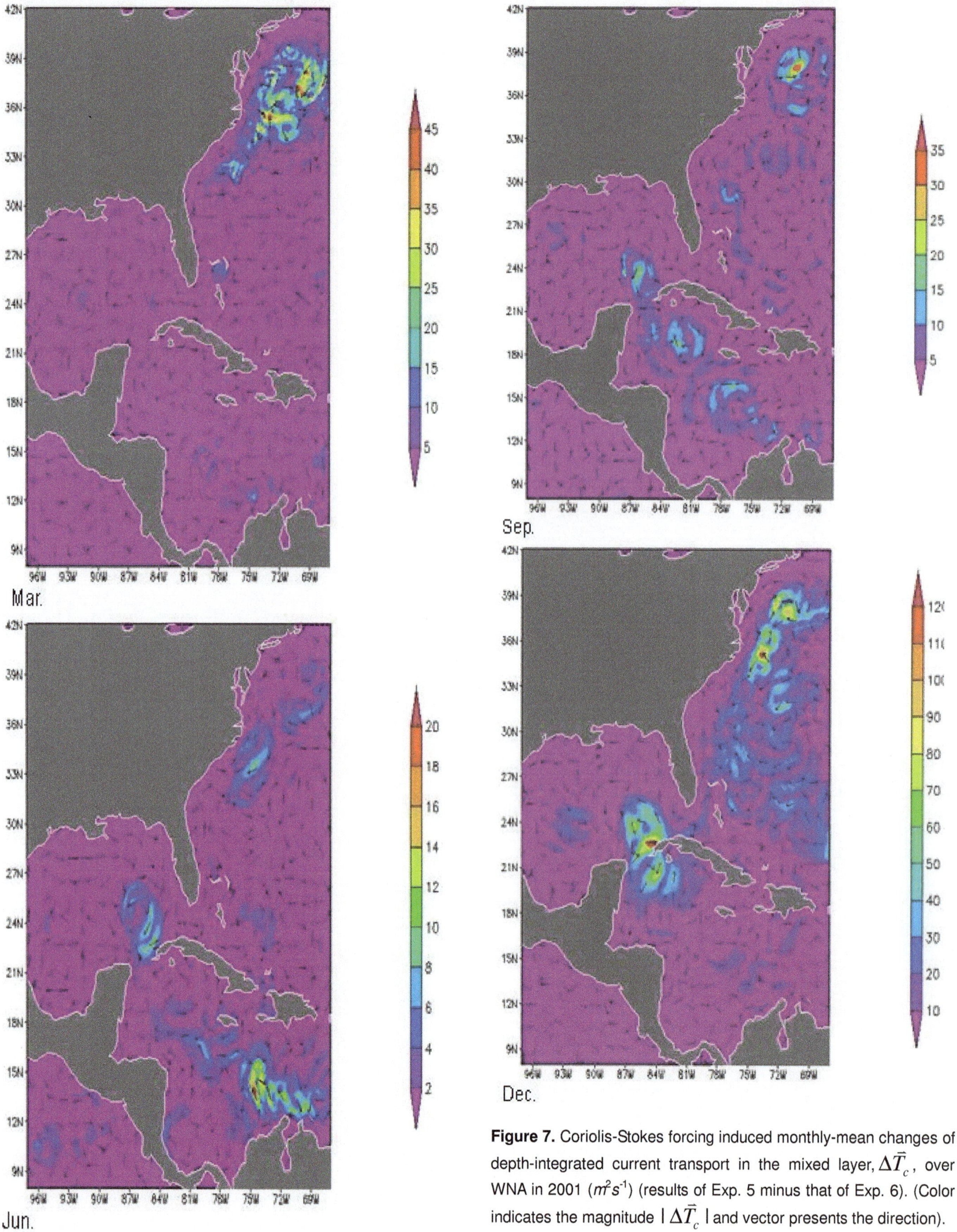

Figure 7. Coriolis-Stokes forcing induced monthly-mean changes of depth-integrated current transport in the mixed layer, $\Delta\vec{T}_c$, over WNA in 2001 (m^2s^{-1}) (results of Exp. 5 minus that of Exp. 6). (Color indicates the magnitude $|\Delta\vec{T}_c|$ and vector presents the direction).

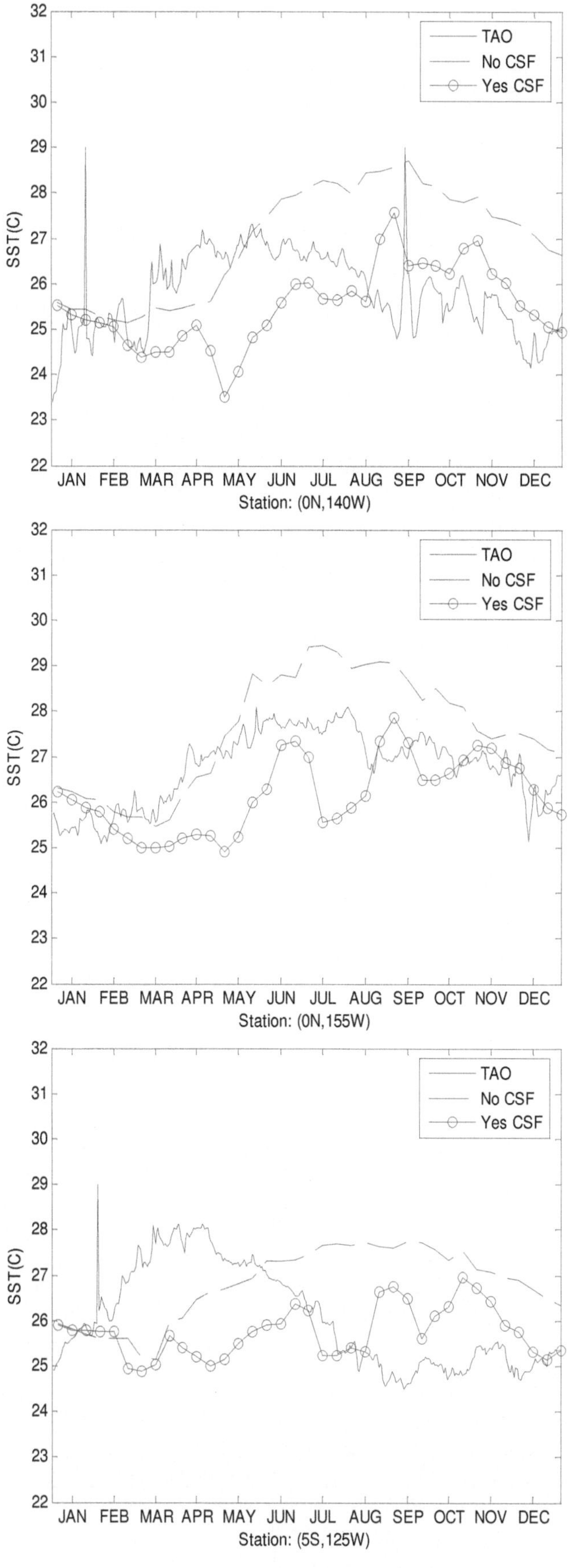

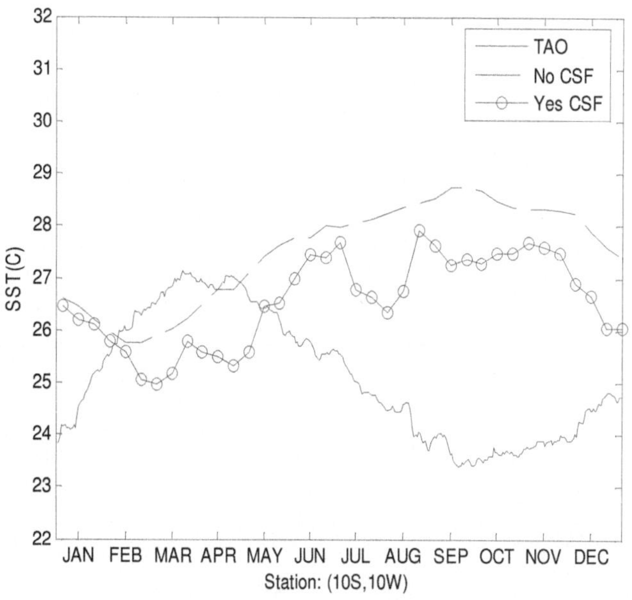

Figure 8. Daily SST time series in 2001from TAO buoys (solid line), and those output from HYCOM simulations with CSF (cycle line) and without CSF (dash line) for 4 stations over Equatorial Pacific.

more sensitive to the boundary forcing. Moreover, the interaction between waves and currents might be more intensive in the strong eddy region. Deng et al. (2009) also demonstrated that the ocean surface currents are more sensitive to wind forcing in the low wind speed region than that in other regions. Since the wind direction is almost westward both in summer and winter over the Caribbean Sea and Gulf of Mexico, there is not much difference in the current structure among seasons in this region. The winds in summer are weaker than that in winter and the values of $\Delta\vec{T}_c$ in summer are also smaller than that in winter (Figure 7), suggesting that the changes in CSF-induced $\Delta\vec{T}_c$ are associated with the changes in wind speed.

VALIDATION AND DISCUSSION

MLD is partly determined by SST. The distribution of annual-mean Δthi and Δtem (Figures 3a, b, d, e, g and h) indicate that increased MLD tends to be accompanied with decreased SST. As an important parameter that can be used to verify the capability of a numerical model in simulating the upper ocean phenomena, SST can be directly measured with high accuracy by buoys. Tropical Atmosphere Ocean (TAO/TRITON) observational array data, which are updated every day, are freely available to the research community. Here, we use SST time series from TAO buoys which are deployed over the equatorial Pacific region to validate our simulation results. Figure 8

Table 2. Monthly-mean mixed layer thickness (for each site, the two lines list the simulation results without and with CSF, respectively), Site 1: (114.5°E, 17.5°N), site 2: (110.5°E, 8.5°N), site 3: (71.5°W, 36.5°N), site 4: (87.5°W, 24.5°N)).

	Unit: *m*	JAN	FEB	MAR	APR	MAY	JUN	JUL	AUG	SEP	OCT	NOV	DEC
Site 1	No csf	93.6	84.4	20.8	19.5	10.9	26.6	19.3	37.2	33.4	38	51.3	106.3
	Yes csf	94.3	86.9	22.8	22.4	12.1	16	18	40.2	46.7	44.8	95.3	119
Site 2	No csf	51.4	49.8	45	36.8	57.4	53.7	78.1	83	52.4	32.4	28.4	36.7
	Yes csf	51	53.2	43.3	37.2	51.3	56.3	76.3	75.1	58.1	38.8	35.2	30
Site 3	No csf	163.5	190.5	159.4	131.8	16.8	12.8	7.8	5.7	13.5	21.2	31.1	65
	Yes csf	163.4	208.6	289.4	125.1	15.5	12.5	7.1	6.5	16.2	23.3	27.8	83.5
Site 4	No csf	59.2	61.7	43.2	28.3	29	29.4	15.8	12.9	9.1	26.7	35.9	35.4
	Yes csf	59.6	61.9	43.4	38.4	19.7	31.9	12.4	13.9	13.2	37.4	32	40.4

displays the comparisons of SST from model simulation and from TAO observation performed on 4 sites. These are located at (0°N, 140°W), (0°N, 155°W), (5°S, 125°W) and (10°S, 10°W), respectively, covering the equatorial Pacific region sparsely. Considering no data assimilation applied in our modeling, the deviations between reproduced SSTs and observations are basically acceptable. All the four cases show that, in general, the results with CSF are decreased and slightly closer to observations comparing to that without CSF. CSF-induced improvements in SST simulation can be appreciated in almost all the seasons except spring.

However, it is currently inappropriate to conclude that the reproduced SST is substantially improved by the inclusion of CSF. Nevertheless, one certainty is that CSF is a necessary physics and is not negligible in upper ocean modeling, which is somewhat indicated by Figure 8. Kantha and Clayson (2004) argued that, the Langmuir cell-induced SST changes are well within the uncertainties in the modeled SST resulting from an imperfect knowledge of the air-sea flues used to drive the mixed layer models. Additionally, in our opinion, the imperfect mixed layer models are also partly responsible. It can also be found from Figure 8 that, the CSF resulted SST variations are quite pronounced during spring/summer periods (more than 1°C), and are relatively small in fall and winter when ML is thicker. Carniel et al. (2005) also found that, the surface wave (though through Langmuir cells and wave breaking) resulted SST differences are around 1°C.

As concluded earlier that CSF exerts more striking impacts on upper layer currents in eddy regions, we step further to discuss the influences of CSF on MLD in eddy regions. Table 2 lists the monthly-mean MLD for four sites which are all located in the eddy regions (site 1 is in north SCS, site 2 south SCS, site 3 north WNA, and site 4 is in Loop Current region). Results show that the ML is shallow in spring/summer and deeper in fall/winter. In other words, it presents a seasonal variability. CSF tends to deepen the ML in every month except July. Previous studies (Kantha and Clayson, 2004; Carniel et al., 2005) found that Langmuir cells deepen the ML because of the increased mixing, and result in the drops of SST. Whereas, as a mediation of top boundary condition, CSF does not always tend to increase the MLD (or decrease the SST) over the whole model domain, but it does in eddy regions. The bigger changes in the eddy regions are probably resulting from the more intensive wave-current interaction reflected by the modulating of top boundary forcing in both momentum equation and ML model.

Conclusions

Ocean wave models and ocean current models are usually run independently, so that the impact of wave-current interactions on the ocean such as CSF is not included. In this study, a one-way coupling system consists of an ocean circulation model (HYCOM) and a wave model (WW3), is used to study the effects of CSF on ocean upper-layer dynamics including currents, temperature and upper-layer mixing. The CSF is calculated by wave variables derived from WW3, and is then imposed to HYCOM as an upper boundary condition. In order to verify the effects of CSF on upper ocean properties are consistent in different regions, experiments configured in three different regions (Global Ocean, SCS and WNA) with different horizontal resolutions (2.5°, 0.083° and 0.17°) are carried out to illustrate the details. The results are summarized as follows.

Only serving as a correction to the upper ocean boundary conditions, CSF can change the structures of current field by interacting with the mean currents as well as releasing TKE. The changes in ML current transport caused by CSF are comparable to wind-driven Ekman transport. The CSF-induced changes in annual-mean ML depth-integrated current transport $\Delta \vec{T}_c$ values are in the range of 0 to 13 m^2s^{-1}, most of which are from 0 to 3 m^2s^{-1} and larger ones appear over the mid-latitude Westerly regions. The annual-mean changes in MLD Δthi are in the range of -15 to 80 m with obvious enhancement in the southern ocean. The annual-mean ML temperature Δtem are from -0.8 to 0.8℃. The magnitude of depth-integrated Stokes transport $\vec{T}_s$ is in the range of 0 to 2.2 m^2s^{-1}, which is in the same order as that of $\Delta\vec{T}_c$. However, for the most part of the global ocean, the directions of $\vec{T}_s$ are totally different from that of $\Delta\vec{T}_c$, except for the southern Westerly region. The CSF intensifies (reduces) the mixing in the ML, thus results in a decrease (an increase) in ocean surface temperature.

Similar to the global case, neither can CSF entirely change the structures of current pattern in the SCS experiments. The upper-layer basin-scale circulation of the SCS, which subjects to the monsoon system, is well simulated. The monthly-mean upper-layer circulation pattern of SCS varies with the monsoon winds. In general, it shows a distinct seasonal behavior which exhibits a strong cyclonic gyre in winter and a weak cyclonic gyre in the Northern SCS but an anti-cyclonic gyre in the southern SCS in summer. The CSF-induced annual-mean $\Delta\vec{T}_c$, Δthi, and Δtem in the ML of SCS vary from 0 to 5 m^2s^{-1}, -10 to 10 m, and -0.2 to 0.25℃, respectively. The monthly-mean $\Delta\vec{T}_c$ in the ML is changing month to month in both direction and magnitude.

The magnitudes of $\Delta\vec{T}_c$ are in the range of 0 to 30 m^2s^{-1}, but mostly within 0 to 3 m^2s^{-1}. The large values appear in the regions of intense gyre, which indicates that the CSF plays a more important role in those regions.

The main features of upper-layer circulation in WNA (Loop current, eddy-shedding and western boundary currents) are successfully simulated in comparison with other important studies. The rings shed episodically by Loop Current have a diameter of about 250 km and a westward propagation speed of $3\,km \cdot day^{-1}$. The CSF-induced annual-mean $\Delta\vec{T}_c$, Δthi, and Δtem in the ML of WNA vary from 0 to 27 m^2s^{-1} (with most less than 6 m^2s^{-1}), -20 to 30 m (with most from -10 to 10 m), and -0.4 to 0.4℃, respectively. The monthly-mean $\Delta\vec{T}_c$ in the ML of WNA also varies monthly in both direction and magnitude. The influences of CSF on $\vec{T}_c$ are most important in winter and least significant in summer. The CSF plays a significant role in regions where strong quasi-geostrophic flow presents.

In conclusion, although wave-induced CSF cannot fundamentally change the pattern of the ocean circulation, it still has a noticeable effect on ocean modeling, while unfortunately most ocean models do not consider such effects. With magnitude less than 5% of that of wind stress, CSF does not always increase the combined top stress; it also decreases the net stress when the wind and CSF are in opposite direction. CSF influences the upper layer properties, such as currents and SST, more pronounced in the eddy regions than in others; and the mixing in the eddy regions tends to be intensified in the presence of CSF, thus results in the increase of MLD. CSF draws important effects on upper layer currents over the Westerlies where strong wind waves present. The comparisons of SST with some available TAO buoy observations show that, the simulated SSTs with CSF are generally improved. However, it is now still hard to say that the model's SST reproduction skill is definitely improved, because the CSF-induced changes in SST are well within the uncertainties of external forcing and because of the imperfect knowledge of the parameterization of turbulent mixing. Nevertheless, the inclusion of CSF in ocean modeling is on the right way to develop a model for better representing the real ocean.

This study uses a one-way coupling which only investigates the effects of waves on currents, not vice versa. In the ocean, swells with wave length of more than 50 *m* commonly exist (indicated by our WW3 simulated results), and their directions are much different from that of the corresponding winds. Typically, Stokes depth is about 5 *m* in the open ocean (when the wave length is about 60 *m*), while under the same condition Ekman depth is much deeper (e.g. in 35°*N*, under a wind speed of 10 ms^{-1}, Ekman depth is ~120 *m*). So, in this context, it is valid to introduce the wave-induced CSF as a top boundary condition of the ocean circulation model. Different from the WNA case where the contributions of CSF to WBC can be found, in global case, the resolution is too coarse to resolve WBC, so we can not properly appreciate the contributions of CSF to WBC, while the effects of CSF on the ocean interior, such as Westerlies and Equatorial Ocean, can be appreciated clearly. Consistent results of experiments that conducted over three different domains (especially the two regional cases) indicate that our simulations are the representative of the real physics. One limitation of this study is that, we adopted 100% of the wind stress to the ocean circulation component, which is the common way used in ocean modeling at present. Actually, there must exist an energy partition depending on sea state between currents and waves; however, this partition still remains unknown. In order to represent the effects of waves here, we introduce the CSF as a correction to the wind stress.

ACKNOWLEDGEMENTS

This research is jointly supported by grants of the National Basic Research Program of China (2007CB816001) and the National Natural Science Foundation of China (40776016, 40906015, and

40906016). The authors also appreciate supports from the National Basic Research Program of China (2005CB422302, 2005CB422307 and 2007CB411806), Great Project of National Natural Science Foundation of China (40490263, 40976005), research foundation form State Oceanic Administration (2011221, 2011224), and research foundation from National Marine Data and Information Service (29110006C).

REFERENCES

Ardhuin F, Jenkins AD (2006). On the interaction of surface waves and upper ocean turbulence. J. Phys. Oceanog., 36: 551-557.
Bleck R (2002). An Ocean General Circulation Model Framed in Hybrid Isopycnic-Cartesian Coordinates. Ocean Modeling, 4: 55-88.
Bye JAT (1967). The wave-drift current. J. Mar. Res., 25: 95-102.
Carder KL, Fanning KA, Betzer PR, Maynard V (1977). Dissolved silica and the circulation in the Yucatan Strait and the deep eastern Gulf of Mexico. Deep-Sea Res., 24: 49-60.
Carniel S, Sclavo M, Kantha LH, Clayson CA (2005). Langmuir cells and mixing in the upper ocean. IL NUOVO CLMENTO, 28: 33-54.
Chu PC, Edmons NL, Fan C (1999). Dynamical Mechanisms for the South China Sea Seasonal Circulation and Thermohaline Variablilities. J. Phys. Oceanogr., 29: 2971-2989.
Cochrane JD (1972). Separation of an anticyclone and subsequent developments in the Loop Current (1969). Texas A & M Univ. Ocean Studies. Eds: L. R. A. Capurro and J. L. Reid, Gulf Publishing Co., Houston, 2: 91-106.
Cooper CG, Forristall Z, Joyce TM (1990). Velocity and hydrographic structure of two Gulf of Mexico warm-core rings. J. Geophys. Res., 95: 1663-1679.
Deng Z, Wu K, Yu T (2007). The wave transport of the eastern area of the Pacific. ACTA OCEANOLOGICA SINICA (Chinese edition), 29(6): 1-9.
Deng Z, Wu K, Zhao D, Yu T (2009). Effects of Wind Waves of the Pacific Westerly on the Eastern Pacific Wave Transport. ACTA OCEANOLOGICA SINICA, 28(1): 83-88.
Fratantoni DM (2001). North Atlantic surface circulation during the 1990's observed with satellite-tracked drifters. J. Geophys. Res., 106: 22,067-22,093.
Hasselmann K (1970). Wave-driven inertial oscillations, Geophys. Fluid Dyn., 1: 463-502.
Huang NE (1979). On Surface Drift Currents in the Ocean. J. Fluid Mech., 91: 191-208.
Ianniello JP, Garvine RW (1975). Stokes transport by gravity waves for application to circulation models. J. Phys. Oceanog., 5: 47-50.
Kantha L, Clayson CA (2004). On the effect of surface gravity waves on mixing in the oceanic mixed layer. Ocean Modeling, 6: 101-124.
Kenyon KE (1970). Stokes transport. J. Geophys. Res., *75*: 1133-1135.
Korvin-Kroukovsky YV (1972). Pure drift current and mass transport in coexistent waves. Deut. Hydro. Z., 25: 1-13.
Large WG, Mcwilliams JC, Doney SC (1994). Oceanic Vertical Mixing: A Review and a Model with a Nonlocal Boundary Layer Parameterization. Rev. Geophysics, 32: 363-403.
Liu Q, Kaneko A, Su J (2008). Recent Progress in Studies of the South China Sea Circulation, J. Oceanography, 64: 753-762.
Mcwilliams JC, Rsetrepo JM (1999). The wave-driven ocean circulation. J. Phys. Oceanogr., 29: 2523-2540.
Mcwilliams JC, Sullivan PP, Moeng CH (1997). Langmuir turbulence in the ocean. J. Fluid Mech., 334: 1-30.
Metzger EJ, Hurlburt H (1996). Coupled Dynamics of the South China Sea, the Sulu Sea, and the Pacific Ocean, J. Geophys. Res., 101: 12331-12352.
Ochoa J, Sheinbaum J, Baden A, Candela J, Wilson D (2001). Geostrophy via Potential vorticity inversion in the Yucatan Channel, J. Mar. Res., 59: 725-747.
Oey LY, Ezer T, Lee HC (2005). Loop Current, Rings and Related Circulation in the Gulf of Mexico: A Review of Numerical Models and Future Challenges. Geophysical Monography Series, 161: 31-56.
Polton JA, Lewis DM, Belcher SE (2005). The role of wave induced Coriolis-Stokes forcing on the wind-driven mixed layer, J. Phys. Oceanogr., 35: 444-457.
Qu T (2000). Upper layer circulation in the South China Sea, J. Phys. Oceanogr., 30: 1450-1460.
Romanou A, Chassignet EP, Sturges W (2004). The Gulf of Mexico circulation within a high resolution numerical simulation of the North Atlantic Ocean. J. Geophys. Res., 109, CO1003, doi: 10.1029/2003CJ001770.
Shaw PT, Chao S (1994). Surface circulation in the South China Sea. Deep-Sea Res., 41: 1663-1683.
Sturges W, Leben R (2000). Frequency of ring separations from the Loop Current in the Gulf of Mexico: A revised estimate, J. Phys. Oceanogr., 30: 1814-1818.
Teixeira MAC, Belcher SE (2002). On the distortion of turbulence by a progressive surface wave, J. Fluid Mech., 458: 229-267.
Vukovich FM (1995). An updated evaluation of the Loop Currents eddy shedding frequency. J. Geophys. Res., 100: 8655-8659.
Wu K, Liu B (2008). Stokes drift-induced and direct wind energy inputs into the Ekman layer within the Antarctic Circumpolar Current. J. Geophys. Res., 113, C10002, doi: 10.1029/2007JC004579.
Wyrtki K (1961). Scientific results of marine investigation of the South China Sea and Gulf of Thailand. NAGA Report, 2: 195.

Removal of Fe(II), Cu(II), Zn(II), and Pb(II) ions from aqueous solutions by duckweed

Jameel M. Dhabab

Department of Chemistry, College Of Science, Al-Mustunsiyria University, Baghdad, Iraq.
E-mail: ihsanalsaimary@yahoo.com.

Duckweed was used to study the adsorption of Fe(II), Cu(II), Zn(II), and Pb(II) ions from aqueous solutions. The experimental conditions which includes; pH, contact time, initial concentration, and loading weight on the removal process were investigated. Bach adsorption was studied at 30°C. The efficiency of adsorption depends on pH value, it increases with the increase of pH value from (2-8) in range time between 60 to 90 min. The most percentage of removal in the main way of loading weighed was 1.5 gm and 50 ml mixed metal ions solution at 50 ppm concentration of each metal ion. The study of isothermal showed that data was confirmed with temkin isotherm model. The removal order was found to be Pb^{2+} < Zn^{2+} < Cu2+< Fe^{2+}. The potential of application for the treatment of solutions containing these heavy metals in multimetal solutions is indicated. The characterization (the infrared) for the surface of duckweed showed the presence of many function groups which can be binding with metal cations.

Key words: Duckweed, heavy metals, adsorption, removal of metals, adsorption isotherm, adsorption from aqueous solution.

INTRODUCTION

The awareness of increasing water pollution implying studies concerning water treatment and the use of removal of heavy metals from industrial wastewater is of primary importance. The use of natural materials for heavy metal removal has become a concern in all countries. Natural materials that are available in large quantities or certain waste from agricultural processes may have potential to use low cost adsorbents, as they represent unused resources, widely available and are environmentally friendly (Afkhami et al., 2008).

To see the decrease of polluted water most researches were concentrated with treatment of heavy metals from industrial wastewater. It uses normal material to removal metals from different sides because it is valid largely in agriculture processes in addition to their low price as adsorbent materials (Deans and Dixon, 1992).

The materials which were used naturally and of low cost include carbonaceous materials, agricultural production, activated carbon and waste by-products which are achieved by previous studies (Kamel et al., 2004; Iihan and Fuat, 2000; Yesim and Izzet, 2005; Singh et al., 2005; Amer et al., 2010; Chi et al., 2009; Choi et al., 2009; Noman and Palanisumy, 2005) and were made as substantive material for some expensive methods. This is so because of the technicalities and price involved in the process of precipitation, ion exchange, solvent extraction, liquid membrane and others to remove heavy metals from industrial water (Garadea et al., 2000; Arpa et al., 2000; Mofa, 1995; Gardea et al., 1996; Abia et al., 2002). So materials of biology origin were used to remove these metals (Sag et al., 2001). All these sources were limited to removing one or two materials. Duckweed - an aquatic plant which grows on water surface, lives in shallow water and is valid in south and the central Iraq.

It is one of the most widely distributed plants in most aquatic areas. This kind of plant exists in almost every sides of rivers in South and the Central Iraq.

Accordingly, this study aimed to investigate the adsorption potential of duckweed for the removal of Fe, Cu, Zn^{2+} and Pb^{2+} ions from wastewater within various experimental conditions. At the same time, this can be considered a way of recycling such environmentally trouble maker materials.

MATERIALS AND METHODS

Duckweed

Duckweed was collected from different clean area far from those which may be industrially polluted. Duckweed was washed by tap water then cut by a small clean cutter, it was then dried at 80°C at about 36 h. Finally, the dried plants were grinded by clean electric mixer then stored in clean plastic page.

Characterization of duckweed

Duckweed was taken by FT.IR spectrophotometer type shimadzu 8400S which work in approximately (4000-400) cm^{-1}, then the sample introduced as KBr (pellet).

Preparation of synthetic wastewater

The mixed metal ions solution from cations (Fe^{2+}, Cu^{2+}, Zn^{2+} and Pb^{2+}) was prepared of Merck – analytical grade stock standard of concentration 1000 ppm. The solution of wastewater was introduced for each metal according to the method of continuous dilution from the sample solution above. The pH of the wastewater was adjusted by using 0.1 M HCl and/or NaOH. The final concentration of metal ions in wastewater was analyzed by atomic absorption spectrophotometer type phoenix – 986.

Bach sorption experiment

Capped tubs containing 50.0 ml of solution test of metal ions with known ion concentration, which was mixed with a known amount of dried untreated plant and the capped tubs were thoroughly mixed, allowing sufficient time for adsorption equilibrium. The mixtures were filtered through filter paper Whatman No. 42 and the metal ions were determined in the filtrate solution by atomic absorption spectrophotometer type phoenix - 986. Each experiment was carried out at temperature room and was repeated two times, and the results are given are averages.

Calculate of removal percentage of metal ions by duckweed

Amount of removal material by duckweed through series of batch investigations were appointed by the following equation:

Removal %= {(C_0-C_f)/ Co} × 100

$$\text{Re } moval \ \% = \frac{C_o - C_f}{C_o} x100$$

Where: C_o and C_f are the initial and equilibrium concentration (ppm) of metal ions in solution respectively.

RESULTS AND DISCUSSION

Duckweed surface characterization

The infrared spectrum was taken for duckweed in (Figure 1) in 3412.01,2926.11 and 2856.67 cm^{-1} which refers to group (-OH), identical alkyl group(-CH_2-) and aldehyde group (-CHO), respectively. Also, the spectrum cleared bands at 1660.77 and 1095.60 cm^{-1} were to be the presence of (C=C) and (C-O), respectively (Stuart, 1996).

The surface IR-characterization of duckweed indicated the presence of many functional groups which were able to bind with the metals cations.

The effect of different operating conditions (contact time, pH, adsorbent loading weight and initial ions concentration) on the removal by adsorption onto duckweed was investigated. The metals removal studies were illustrated graphically in Figures 2, 3, 4 and 5, and showed that their removal was strongly affected by these different operating conditions.

Effect of contact time equilibrium

Figure 2 clarifies the effect of touching time on removal of cation ions (Fe^{2+}, Cu^{+2}, Zn^{2+} and Pb^{2+}). This took 30, 60, 90, 120 and 150 min. It was found that the best time to remove these metal ions was 90 min. The results showed that the removal percentage order at equilibrium was Pb^{2+} (94%) $\gg$ Zn^{2+} (72%) $\gg$ Cu^{2+} (65%) $\gg$ Fe^{2+} (50%). These results are important, as equilibrium time is one of the important parameter for selecting a wastewater treatment system, when the time consumed for wastewater disposal should be considered.

Effect of pH

Effect of pH on process of adsorption movement on duckweed was at a range of (2 - 8) as it is in (Figure 3). It was noticed that the ability of removing iron, copper, zinc and lead ions by adsorbent depends on pH solution and this depends on the ion state and nature of material. In low pH value, binding sites are generally protonated or positively charged (by the hydronium ions). Thus, repulsion occurs between the metal cation and the adsorbent at a higher pH value; binding sites start deprotonating, and makes different functional groups available for metal binding. In general, cation binding increases as pH increases (Forsner and Wittman, 1981; Esposito et al., 2002).

Effect of loading weight of duckweed

It studied the effect of loading weight of duckweed in initial concentration 50 ppm taking the range of the weight (0.5 - 2) gm. It was noticed that the weight of the adsorbent material increase the ability of removal as in Figure 4 which represented the amount of the adsorbent at saturation stage, and also depends on the physical properties of the duckweed (Qadeer and Sohail, 2005).

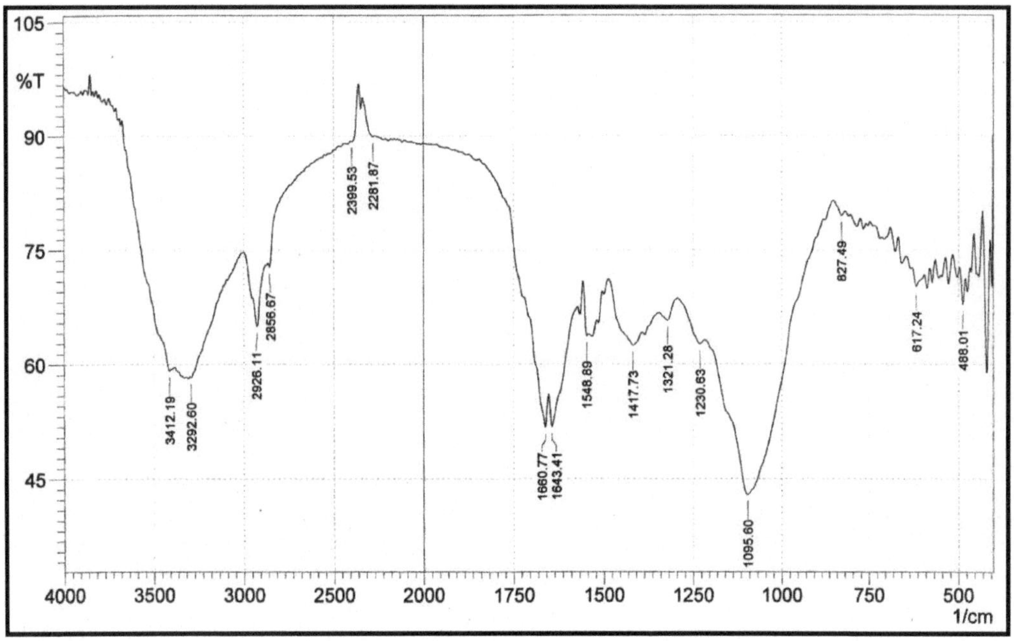

Figure 1. FT.IR spectrum of duckweed.

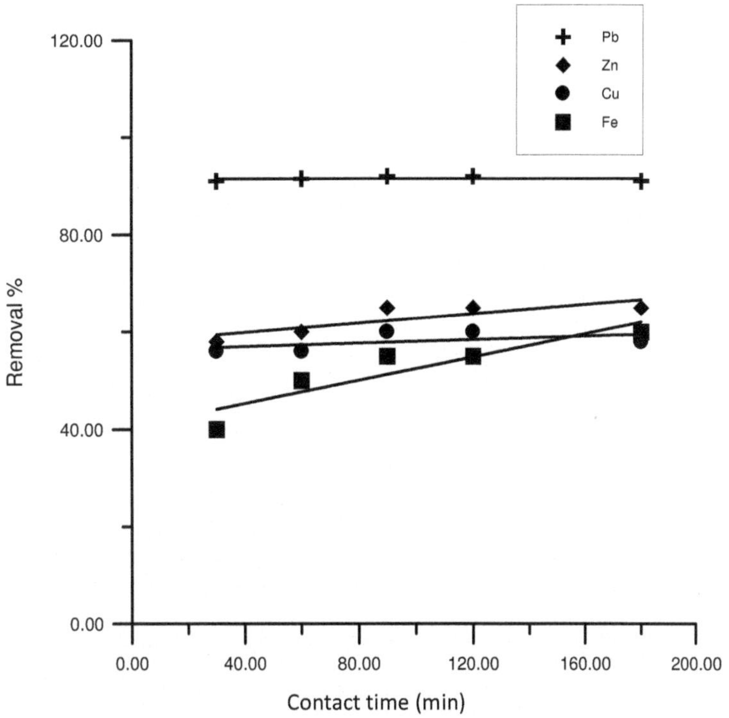

Figure 2. Effect of contact time on the removal metal ions by adsorption on the Duckweed at 50 ppm and 1.5 gm adsorbent weight.

Also, from Figure 4 it can easily be inferred that the percent removal of metal ions increases with the increasing of weight of duckweed; this is due to the greater availability of the exchangeable sites or surface area at higher dose of the adsorbent. These results are in agreement with previous studies on many other adsorbents (Choi et al., 2009; Asubiojo and Ajelabi, 2009).

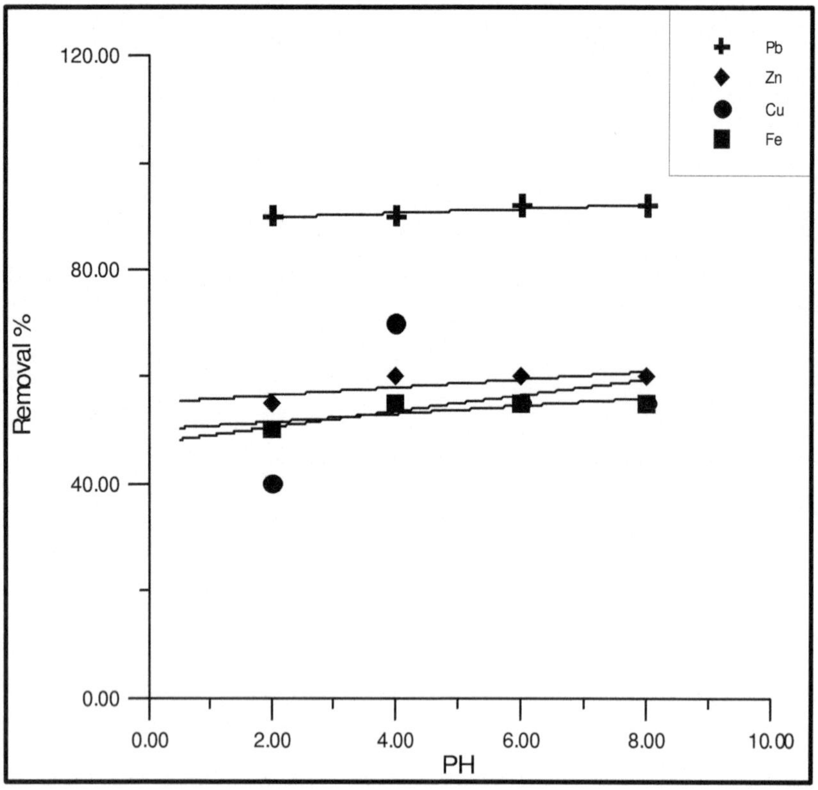

Figure 3. Effect of PH on metal ions adsorption in there mixed solution by duckweed.

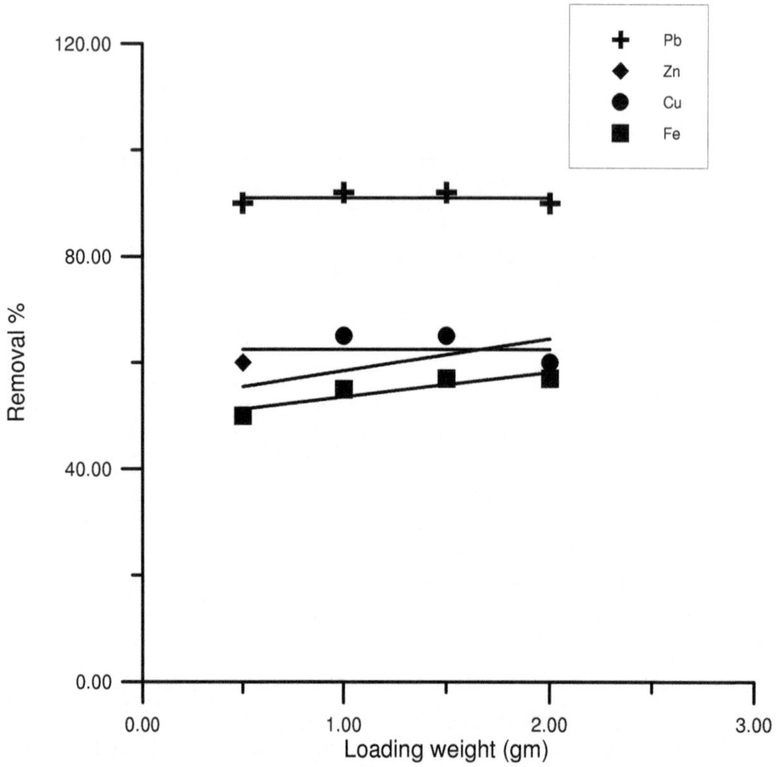

Figure 4. Effect of loading weight of duckweed on metal ions adsorption in mixed solution.

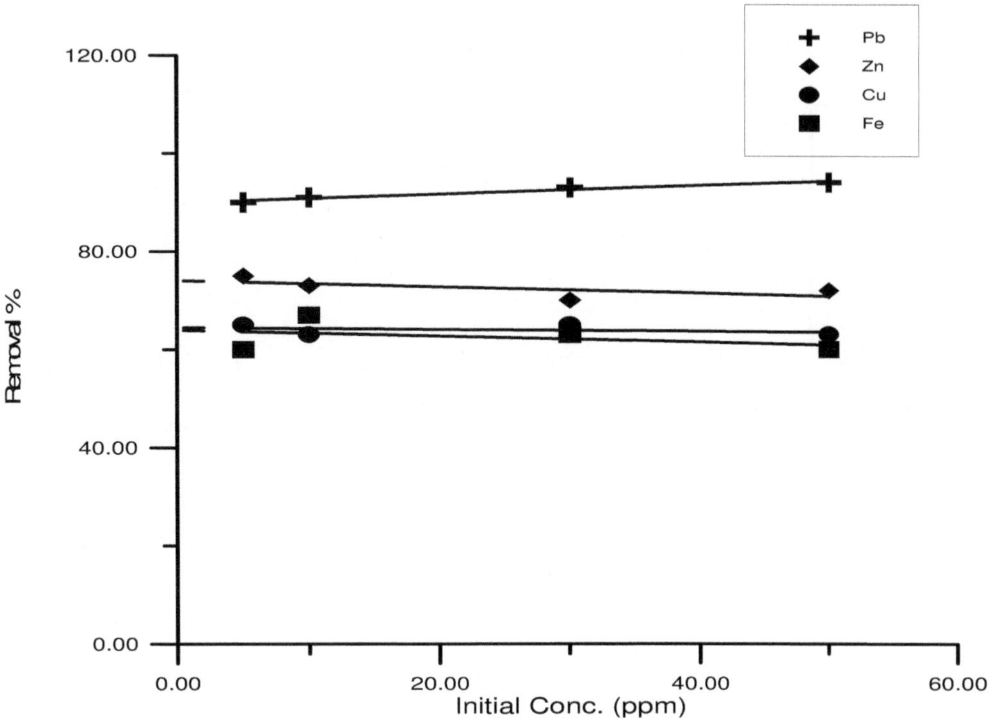

Figure 5. The effect of variation of initial metal ions concentration on to there adsorption using duckweed.

Effect of initial metal ions concentration

Figure 5 clarifies the effect of initial metal concentration on the ability of adsorption by duckweed. It took many different concentrations at a range of (10 - 50) ppm for the mixture of metal ions. The reduction of removal metal ions on duckweed was found with the of increase initial concentration from 10 to 50 ppm. This was due to the increase of metal ions concentration which was completed on the effective site, Also this may be a limited active site for adsorbent material. Therefore, more metal ions were left unabsorbed in solution at higher concentration levels.

Isothermal studies

In the experiment, the information from the removal of metal ions by duckweed was dealt by many adsorption isothermals (langmuir, tekmin and frendlich) which were known. By this information we got, the best isothermal confirmation was tekmin for introducing (chemical adsorption) between the metal ions and duckweed. Tekmin isothermal model was given by the following equation:

$x = a + b \ln c$

where c = concentration of adsorbate in solution at equilibrium(mg/L). x = amount of metal adsorbed per unit weight of adsorbent (mg/g), a and b are constant related to adsorption capacity and intensity of adsorption.

Figure 6 clarifies drawing of x against ln c for Fe (II) on duckweed as example on tekmin model isotherm. This gave a sign of straight line in fixed slope (b). The amount of (b) 11.56 on to duckweed was found.

Conclusions

The result of batch adsorption clarifies the percentage of removal of metal ions from their wastewater solution. Their amounts were between (50 - 94) ppm with experimental conditions as time contact, pH, initial concentration and the weight of loading material. The batch adsorption result showed that removal percentage was Pb^{2+} (94%), Zn^{2+} (72%), Cu^{2+} (65%) and Fe^{2+} (50%). These results are important, as equilibrium time is one of the important parameter for selecting a wastewater treatment system.

It shows that duckweed is active to clean the environment from the wastewater. Also, duckweed is found to be a promising adsorbent for the removal of metal cations from mixed metal ions solution, representing an effective and environmentally clean waste matter.

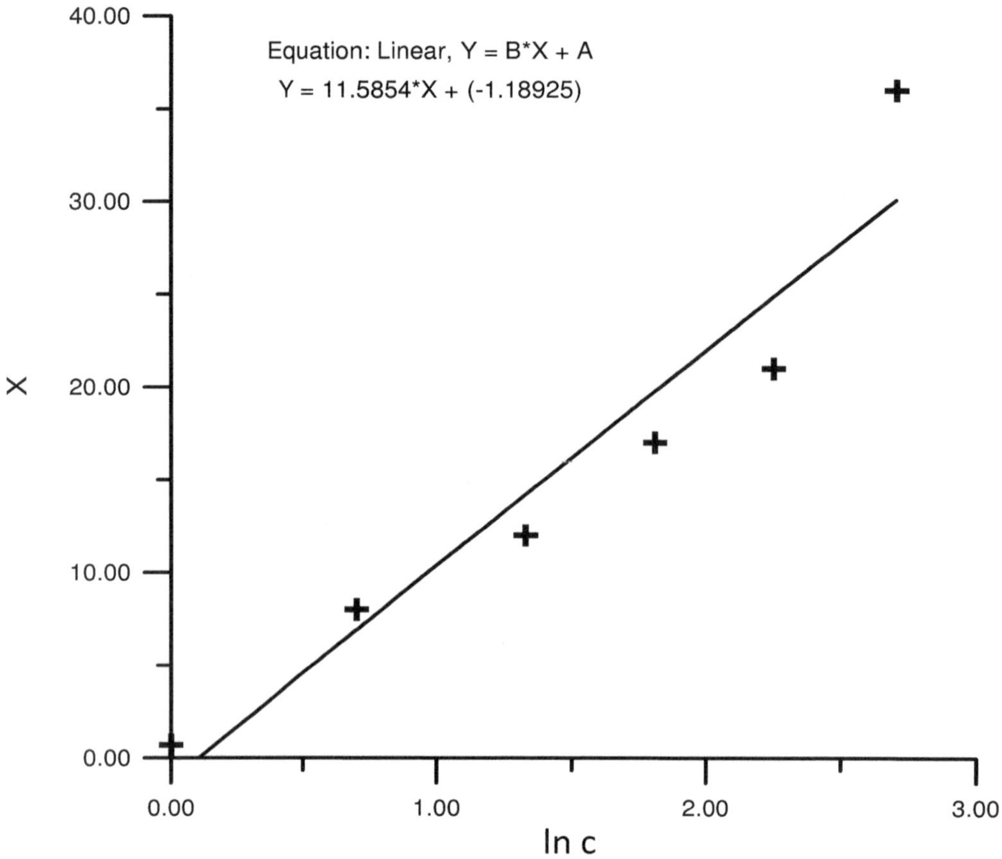

Figure 6. Plot of (x) against Ln (C) for Fe (II) sorption onto duckweed.

REFERENCES

Abia AA, Horsfall MI, Didi O (2002). Studies on the use of agricultural by-products for the removal of trace metals from aqueous solution. J. Appl. Sci. Environ. Manage., 6: 89-95.

Afkhami A., Madrakin T, Karimi Z (2008). Effect of the impregnation of carbon with ethylenediamintetraaceticacid on its adsorption capacity for the adsorption of several metals ions. J. Hazard. Mater., 150: 408-412.

Amer MW, Khalili FI, Awwad AM (2010) Adsorption of lead, zinc and cadmium ions on polyphosphate-modified kaolinite clay. 12(1): 001-008.

Arpa C, Say R, Satiro N, Bekas S, Yurum Y, Genc O (2000) .Heavy metals removal from aquatic systems by Northern Anatoliian smectites; Turk. J. Chem., 24: 209-215.

Asubiojo OI, Ajelabi OB (2009). The removal of heavy metals from aqueous solution by sawdust adsorption. J. Environ. Chem. Toxicol., 91: 883-890.

Chi K, Donghee P, Seung H, Jong M (2009). Removal of cationic heavy metal from aqueous solution by activated carbon impregnated with anionic surfactants, J. Hazard. Mater., 164: 1130-1136.

Choi HD, Cho JM, Xang JS, Lee JX (2009). Influence of cationic surfactant on adsorsoption of Cr(VI) onto activated carbon. J. Hazard. Mater., 161: 1565-1568.

Deans JR, Dixon BG(1992). Uptake of Pb and Cu by novel biopolymers. Water Res., 26(4): 469-472.

Esposito A, Pagnanelli F, Veglio FI (2002). Plant proving there worth in toxic metal. Chem. Eng. Sci., 57: 307-313.

Forsner U, Wittman GT (1981). Metal pollution in the aquatic environment . Springer–verlag, Berlin,Germany, 2nd edition.

Garadea II, Rosa GD, Peralta JR (2000). Use of python filtration technologies in the removal of heavy metals. Pure Appl. Chem., 76 (4): 801-813.

Gardea JI, Tiemanm KI, Gonzaiez IH ,Henning, IA, Townsend MS (1996). Ability of silica–immobilized *Midicago sativa* (alfalfa) to remove copper ion from solution. J. Hazard Mater., 57: 29-39.

Iihan U, Fuat G (2000). Adsorption of some heavy metals ions from aqueous solution by activated carbon and comparison percent adsorption results of activated carbon with those of some other adsorbent. Turk. J. Chem., 24: 291-297.

Kamel MM, Ibrahim MI, Ismail AM, El-Motaleeb MA (2004). Adsorption of some heavy metal ions from aqueous solution by using kaolinite clay. Assoc. Univ. Bull. Environ. Res., 7: 1.

Mofa AS (1995) . Plant proving there worth in toxic metal. Cleanup Sci., 269: 302-305.

Noman SM, Palanisumy K (2005). Removal of heavy metal from industrial wastewater using chitosan coated oil palm shell charcoal. J. Biotechnol., 23: 232-237.

Qadeer R, Sohail A (2005) . kinetic study of lead ion adsorption on active carbon . Turk J. Chem., 20: 51-53.

Sag y, Akcael, Band Kutsal T (2001). Evaluation interpretation and representation of three-metal biosorption equilibria using a fungal biosorbent. Process Biochem., 37: 35.

Singh KK, Rastogi R, Hasan SH (2005). Removal of cadmium from wastewater using agricultural waste rice. Polish J. Hazard Mater., 121(3): 51-58.

Stuart B (1996). Modern infrared spectroscopy. wily.chister.uk.

Yesim K, Izzet K (2005). Removal of cadmium from water using duckweed. Int. Agric. Biol,, 7: 4.

8

Integration of coastal and marine areas into sustainable development strategies: A case study of Africa

Eric O. Odada

Department of Earth Sciences, Marine Geology and Oceanography Unit, University of Nairob, P. O. Box 30197-00100 Nairobi, Kenya. E-mail: eodada@uonbi.ac.ke.

The coastal and marine environments of the African region are uniquely situated to support a variety of activities and to serve diverse human needs for food, transport, and recreation. The pressures from growing populations in coastal zones of Africa, expanding coastal tourism, intensified fisheries, and a large number of other economic activities pose an increasing threat, which jeopardize the quality of these coastal and marine environments. Large-scale destruction of some of Africa's most valuable resources, the coastal forests and mangroves, the lagoons and coral reefs has caused serious degradation of the environment, thus affecting the life of the coastal inhabitants and the economic development of the countries of the African region. The Seychelles Island in the western Indian Ocean for example, was famous for its luxuriant forests and an incredible abundance of wildlife. But many reefs have been mined for coral for the purpose of construction. Mangrove forests on the granitic islands have been raised to the ground or drained and reclaimed. Severe erosion is as a result of this destruction. Many African countries are confronted with serious manpower problems, which are proving to be great impediments in the economic development, especially of their coastal and marine areas. In many cases, the most important cause underlying these problems is the lack of adequate training facilities for the type of manpower required. The teaching of marine science and marine technology in universities of the region is a recent development and in many of the universities, there are as yet no comprehensive study programmes covering the whole spectrum of marine science and technology at the undergraduate or postgraduate level. The situation in regard to research in marine science and technology is equally unsatisfactory. To implement sustainable development of coastal and marine resources of the African region, there is an urgent need to: (1) build human resources by undertaking short-term academic training to strengthen existing capabilities; (2) promote public awareness by producing educational materials on the ecological and socio-conomic contributions of the marine resources and the consequences of unsustainable exploitation (3) organise policy workshops, seminar and/or conferences involving relevant policy and law-makers to increase their understanding of and commitment to the sustainable use of the resources in their coastal and marine areas, and (4) implement integrated coastal zone management programmes by establishing case studies in pilot sites in selected countries of the African region. The object of this paper is to review background information on the present status and trends of coastal and ocean development in Africa, and on the existing capabilities in the region to implement sustainable development strategies. It is intended to lay out the framework and strategy to be used in the design of an integrated management of coastal and marine areas in the region.

Key words: Coastal ecosystems, marine environments, sustainable development, population pressure, environmental degradation, sea level rise, capacity building, management planning.

INTRODUCTION

At the Rio United Nations Conference on the Environment and Development (UNCED), protection of the coastal and marine environments to ensure sustainable use of the natural resources was at the top of the agenda for action (Agenda 21). Under Agenda 21, coastal states should "commit themselves to integrated management and sustainable development of coastal areas and marine environment under their national jurisdiction". The Rio Conference further pointed out the importance of coastal states to develop national policies and management capabilities for integrating the development and management of multi-sectoral activities in coastal and marine areas (UNEP, 1985). Coastal and marine areas are uniquely situated to support a variety of activities and to serve diverse human needs for food, transport, and recreation (UNEP, 1988). Some two-thirds of the world's population lives near the coast and a majority of the world's largest cities are in coastal areas (WWF, 2001a). Ninety per cent of the world's fish catch comes from the continental shelf and upwelling regions, and at present, approximately 10% of the world's total animal protein supply comes from the sea. Again, about 20 per cent of the world's oil production comes from offshore areas and it is estimated that about 70% of the world's ultimate recoverable hydrocarbon resources lie in coastal waters 200 m in depth or less (Mbaki, 1999). Finally, the coastal area is a source of many other raw materials in the form of sand, gravel and a variety of placer minerals such as diamonds. Salt for instance, as an extracted product from the oceans, contributes significantly in food processing and other industrial endeavors. Coastal areas are, therefore, an integral part of the development process in a large number of countries (Ngoile, 1997).

More than half of the world's population lives within 60 km of the shoreline, and in Africa trends in population indicate that this figure could rise to three quarters by the next century (FAO, 1998). The lives and reasonable aspiration for economic advancement of the population depend on these coastal and marine resources (World Bank, 1996b). The shallow tropical coastal areas bordering the African continent have traditionally supported highly productive ecosystems from which fish and other aquatic resources have been harvested. Furthermore, these natural systems provide valuable services in terms of recreation, coastal tourism, marine transportation in addition to their natural protective and defence functions against coastal erosion and flooding (WWF, 2001a). Unfortunately, in many parts of the region, most of the natural resources are over-exploited and the fragile coastal and marine systems severely degraded by unregulated human activities. As a result, the productivity and usefulness of these coastal and marine areas have dramatically diminished and the prospects for sustainable development are greatly jeopardized (Ibe, 1996). There are a number of reasons for this serious situation. In general there is lack of knowledge and understanding of the coastal and marine resources and their interaction with terrestrial and oceanic processes. This lack of information and awareness, together with ineffective coastal and marine planning and management further aggravated the situation (Sain and Knecht, 1998). In this paper, the present status and trends of coastal and ocean development in the African region are examined. The existing capabilities in the region to implement sustainable development strategies are also assessed.

REGIONAL SETTING

The Eastern Africa coastal and marine areas,for example, have some of the world's richest ecosystems containing exclusive coral reefs, lagoons, estuaries and mangrove forests (UNEP, 1998a). The seashores of Somalia, Kenya, Tanzania and Mozambique (Figure 1) are extremely rich in marine habitats and sheer numbers of marine plant and animal species. Within this tropical belt, some of the world's most extensive mangrove forests are found, as well as vast stretches of seagrass beds and several hundreds of kilometers of sand beaches, limestone cliffs, lagoons and subtidally some of the richest coral reefs (Odada, 1993). These ecosystems support a wide diversity of plants and animals and economic benefits derived from the coastal and marine environments are essential for a large part of the population in the countries of the sub-region (Obura et al., 2000).These coastal and marine environments provide coastal population and communities inland with essential requirements in life such as food, building materials, firewood and many other vital resources (UNEP, 1985).

The west and central African coastline for instance, stretches for approximately 8,000 kilometers and is characterized by marked diversity (Allersman and Tilmans, 1993). The climate ranges from desert in the north-the Sahara-through a humid tropical belt, which contains two of Africa's largest rivers, the Niger and Congo, to desert again in the south-the Kalahari. The coastal zone shows equal diversity, ranging from dunes through marshy delta lands and mangroves to Rias with steep cliff's (Ibe, 1996). The continental shelf also varies markedly in width from 70 km or so in the Gulf of Guinea to about 4 km off Angola and Zaire (Figure 2). Even the countries that make up the sub region differ markedly, from small island states like Sao Tome and Principe to large mainland countries like Nigeria and Senegal. Their state of development also differs. Some have economies based essentially on handicraft and subsistence farming, whereas others have quite well developed industries (Awosika et al., 1993)

OBSERVED COASTAL AND MARINE ISSUES AFFECTING THE REGION

Africa is surrounded by four seas-the Atlantic of West and Central Africa, the Mediterranean, the Red Sea and the Indian Ocean/Eastern African maritime zone (WWF, 2001a). Throughout the region, the countries are facing a

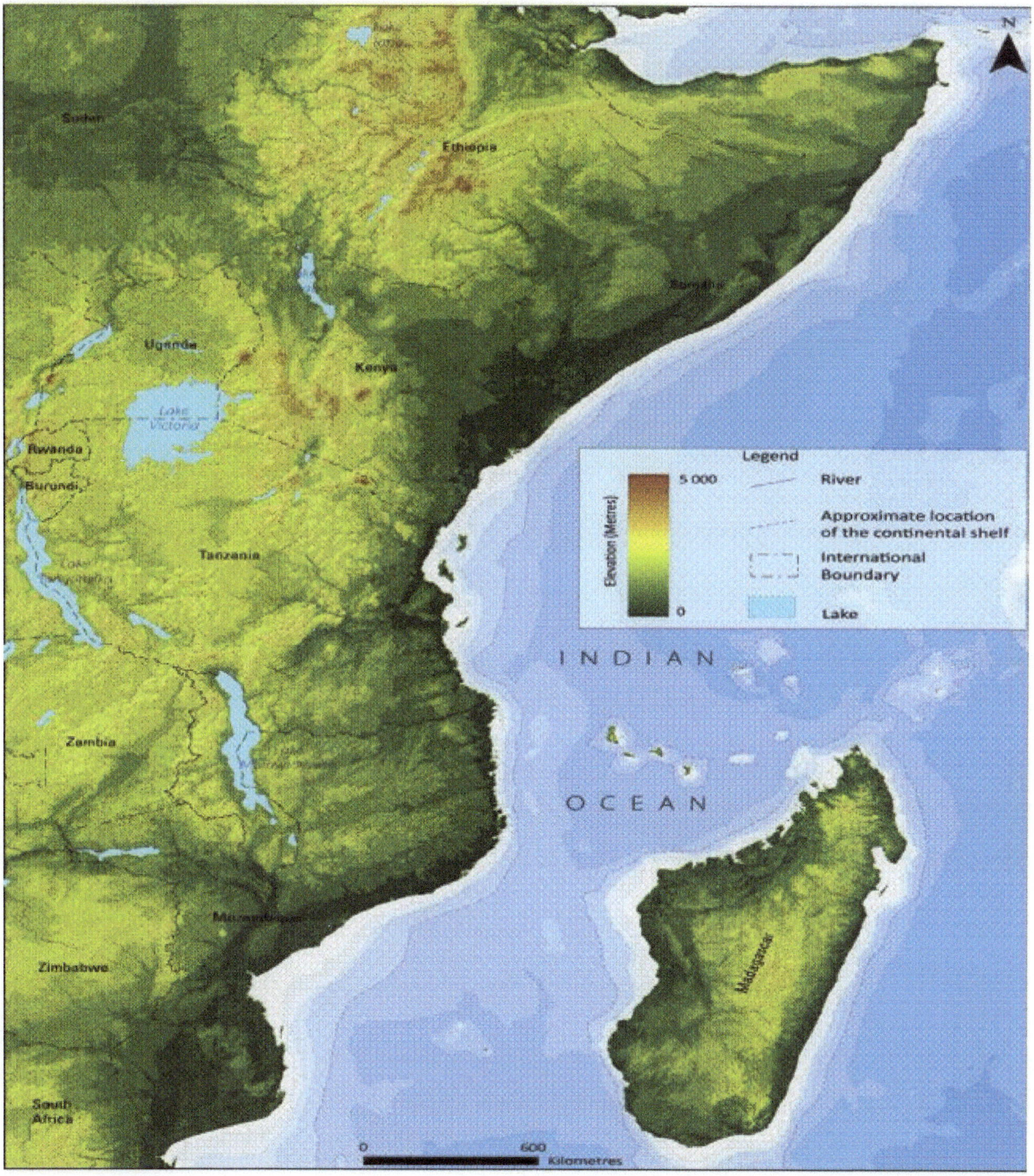

Figure 1. The map of coastal zone of East Africa.

growing number of coastal and marine changes as a result of development and increased population pressure (FAO, 1998). The African marine environment is influenced by unregulated human activities on land and the changes can be viewed as taking place in a series of concentric circles-in the centre, desertification; towards the coasts, deforestation; on the coasts, erosion and pollution of the beaches; on the high seas, over-exploitation of marine resources, dumping of toxic and hazardous wastes and oil spillages (Ibe, 1996). As the population increases, the African coastal and marine environments are assuming greater importance-hence the growing realization that they need to be protected from pollution, coastal erosion, over - exploitation of

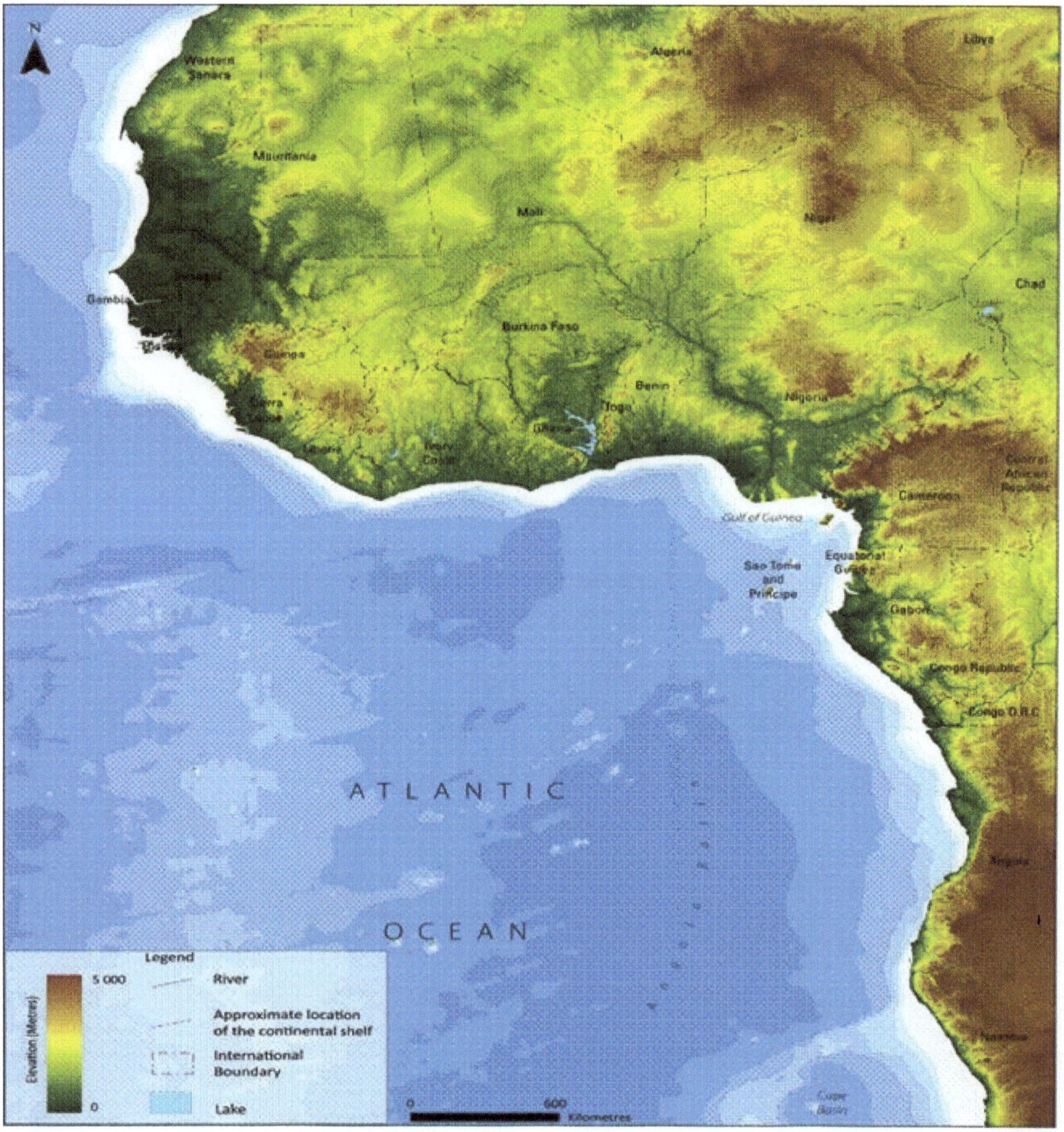

Figure 2. Map of coastal zone along West Africa.

marine resources, etc. through an integrated interdisciplinary and multi-sectoral approach in developing management plans for the coastal and marine areas of the region (Clark, 1996)

Rapid population growth

The coastal areas of a number of countries in the African region are becoming under increasing pressure as a result of population increase and expansion of economic activities such as agriculture, industry, and tourism and infrastructure development (Figure 3). For example, the total population of the countries in the west and central African sub-region is estimated to be around 309 million, of whom about 20% live in the town and villages of the coastal zone (PRB, 2010). Lagos with upwards of 18 million people and 85% of Nigeria's industry, and

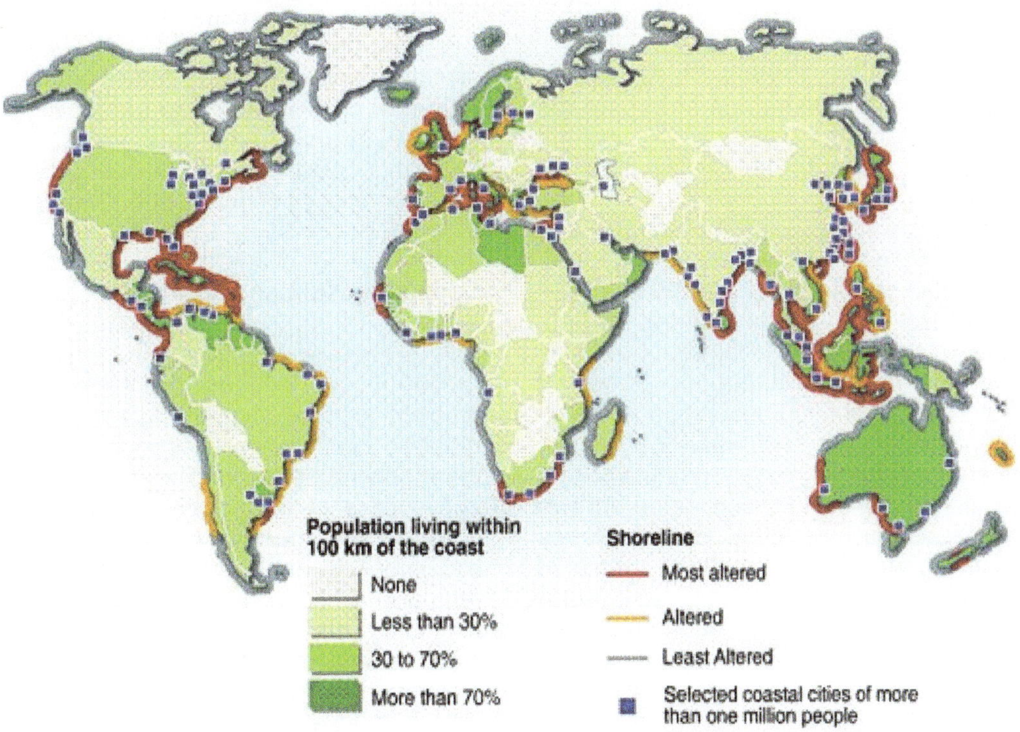

Figure 3. Map showing coastal population and shoreline degradation in Africa and around the world within 100 km of coastline. SOURCE: UNEP/GRID-Arendal (2007).

Accra-Tema with 60% of Ghana's industry are good examples.

In Eastern Africa for example, a total of about 100 million people live in the coastal zone (Figure 3). The country with the most coastal population in the sub-region is clearly Madagascar with about 50.7% of the total population. The mounting concentration of peoples on the coasts of Africa for example has led to substantial increases in the volume of sewage and effluent being discharged into the near shore waters-most of it untreated or only very slightly treated. With sewage discharge taking place in this way there is obvious risk to human health through water contact and through the consumption of sea foods which might be contaminated by sewage organism (FAO, 1996). This potential problem seems to have received little attention in the region, although sporadic outbreaks of human diseases attributed to contact with feacal remains on beaches have been reported (UNEP, 1984). Thus, the rapid growth of population is an urgent problem to be addressed in affecting development in coastal and marine areas of the African region (FAO, 1998).

Environmental degradation

Coastal environmental degradation is a major problem facing many countries in the African region (Figure 3). Sewerage and domestic and commercial discharges in the vicinity of coastal and shallow waters, erosion and siltation, overcutting of forests products such as mangroves on the coastal strips and timber forests inland are mainly the sources of pollution from land (UNEP, 1996b). Poor agricultural practice, lack of agricultural land especially on the islands, over utilization, burning etc and mismanagement and over-exploitation of forest resources have resulted in extensive deforestation and severe soil erosion. This has caused severe siltation with the resultant destruction of coral reefs followed by erosion of the beaches and the destruction of the coastal mangrove and other trees. Examples of this are clearly seen in Madagascar, Comoros and many other parts of the African region (UNEP, 1999a). Development of ports and harbours, coastal construction such as reclamation for airport construction and dredging of the seabed also cause coastal erosion and especially siltation. There has been considerable work of this kind carried out in the African region in the past three decades or so and not enough environmental consideration has been given. For example, extensive dredging has been carried out in the Seychelles within the last decade and siltation of the coral reefs within the marine park and along the east coast of Mahe has been severe (Shah, 1995). In some coastal areas, dynamiting coral reefs as a means of catching fish is widely used. These include Tanzania, Mozambique, and formerly Mauritius, which resulted in serious environmental degradation of their coastal and marine areas (FAO, 1998; UNEP, 1996b).

Sea level rise

The African coastal zone along the Atlantic and Indian Oceans have been more and more vulnerable to sea level rise and other impacts of climate change (Leatherman and Nicholls, 1995). The coastlines of the eastern African sub-region for example, have been retreating inward and seaward as a result of the rise and fall in sea level due to past climate change (Odada, 1991). According to the estimates of the WMO/UNEP Intergovernmental Panel on Climate Change (IPCC, 1998), an average rate of mean global sea level rise could be about 6 cm per decade over the next century. It means that the total sea level rise could attain as much as one metre in one century. In such case, hundred of thousands square kilometres of coastal wetlands and lowlands in the African region could be inundated. African beaches could retreat as much as a few hundred metres and prospective structures may be breached (Smith et al., 1996). Flooding would threaten lives, agriculture, livestock, buildings and infrastructure. Salt water would advance landward into aquifers and up estuaries, threatening water supplies, ecosystems and agriculture in coastal areas of the African region. Given the near certainity of an accelerated rise in sea level, the only hope left for the African government to avoid future chaos is through anticipatory planning and actions (Nicholls and Leinert, 2000).

Development of coastal areas

A number of development activities are leading to major changes in coastal areas of the African region (Ibe, 1996). The most obvious of these are the actual construction of towns with associated industries and the creation or extension of ports and harbour areas. Although, these are confined to a few locations, they are frequently close to areas which could be exploited as tourist centres. For example, at Lagos in Nigeria, Victoria beach has been eroded 2 km inland since the construction of breakwaters (Ibe, 1985) Similar problems were created at the port of Abidjan when the Canal de Vridi was opened in 1950; since then the beach has eroded to the east of the canal and a road has been cut through in the area (Awosika et al., 1993). New or rapidly growing coastal zone activities often lead to the creation or rapid expansion of municipal centres on the coast.

When such rapid expansion occurs, it is extremely difficult for national or local authorities to respond by planning for and providing infrastructure and social services. In Malindi, Kenya for example, the growth of a large tourist-related section has occurred at a traditional small municipal centre with extremely limited services and infrastructure (Odada, 1993). Population growth rates have reached 20% per annum, and authorities are hard pressed to meet basic needs for sanitation, education, and commercial organization. Similar situations do arise wherever new coastal zone activities stimulate extremely rapid growth in new or small existing municipal centres. The social or environmental effects of such development must be considered and dealt with in economic planning for sustainable development of coastal areas (Allersmann and Tilmans, 1993).

Coastal erosion and flooding

This is prevalent problem especially in west and central Africa (Figure 3). The degree of seriousness of the problem and the attempts to mitigate the nuisance and negative economic consequences vary. In some places the problem has reached worrying proportions like in the above case of west and central Africa. Retreat of the coastline with the concomitant flooding cause's hazards by uprooting settlements, destroying agricultural and recreational lands, disrupting harbour and navigational structures and dislodging economic facilities located along coastal towns (Ibe and Quelennec, 1989).

Natural factors for erosion include: storm wave regime with sea level set-up, orientation and nature of the coastline, low relief of the coastal plain, vulnerable sediment budget, narrowness of the continental shelf, presence of off-shore canyons and gullies, global estuatic rise in sea level, etc. In many cases, man's intervention in the natural environment, by the construction of artificial structures on the coastline, mining of beach sand, the location of dams on rivers that normally would supply sediment replenishment to the coastline, the haphazard withdrawal of fluids from coastal aquifers and reservoirs, the destruction of mangroves etc., has served to exacerbate the impact of natural forces (Denis et al., 1995). There is, therefore, an urgent need to react timely and appropriately to coastal erosion and flooding problems in the African coastal areas (Ibe and Quelennac, 1989; Odada, 1993).

Oil pollution

Marine pollution, especially from oil spillage, is a major regional problem and is often aggravated by leakage due to accidents, grounding, harbour operations and discharges from refineries (Portmann, 1998). With the increasing number and size of tankers traveling through the Indian Ocean area for example, fear of oil spills has also increased in the Eastern African subregion (UNEP, 1982). Some paths used by these tankers are shown in Figure 4. In 1981, 355 million tonnes of oil were transported through the main route from Arabian sea to the Far East. Tankers discharge operations in the sub-region are accident prone and many such accidents have been reported in Mombasa, Maputo and Dar es salaam where large areas of mangrove forests were completely

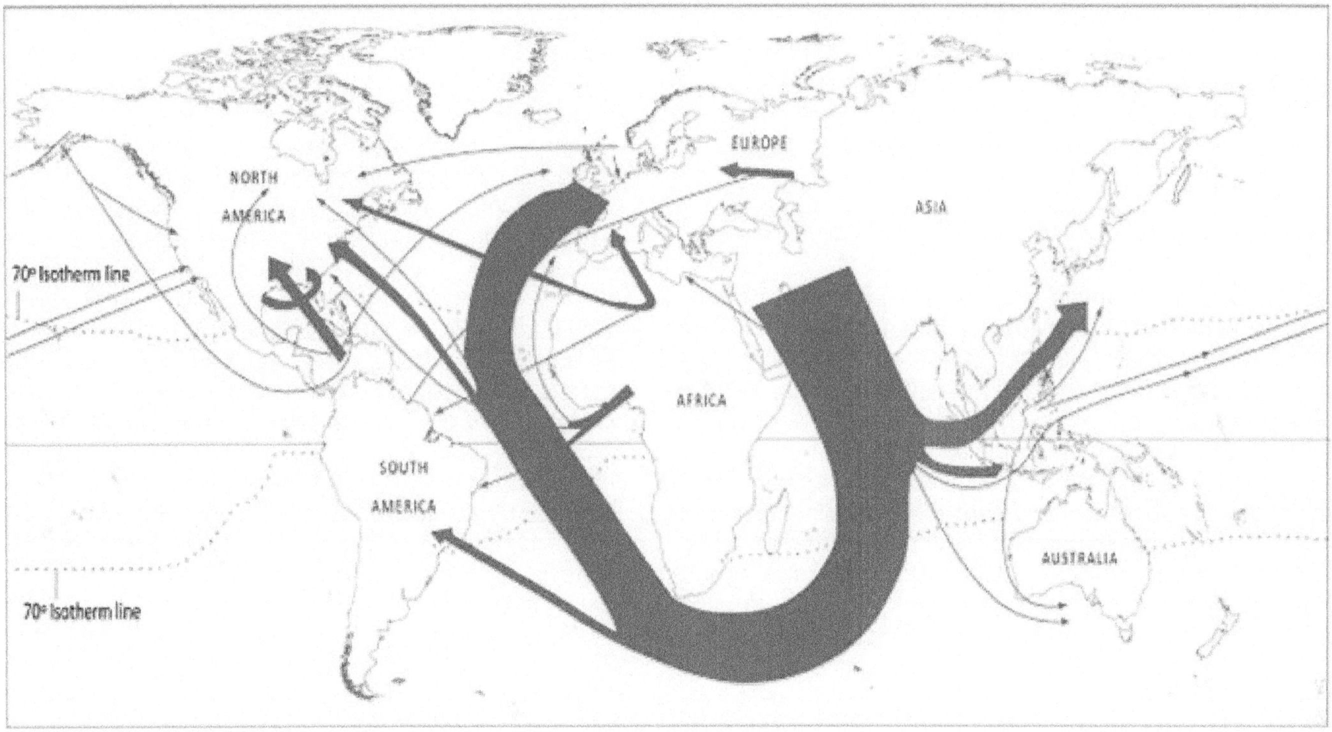

Figure 4. Main routes of oil transportation from the Middle East to the far East and the West around Africa.

destroyed (Munga, 1981). Oil refineries are found in most parts in Eastern Africa, which significantly contributes to oil pollution of the coasts and seas (International Ocean Institute, 2001).

The West and Central African sub-regions exports oil to Europe and America. The coastline lies to the east and is downwind of the main route of oil transport from the Middle East to Europe (Figure 4). The total volume transported annually along the Gulf of Guinea for example has been estimated to be 706 million tonnes (Portmann et al., 1978) and the discharge of tank washings from offshore traffic is a significant source of oil on beaches. However, a more recent investigation has indicated that much of oil found on beaches arises as a result of spills or tank washings discharged from tankers visiting ports in the region although other sources are also important (Portmann et al., 1989). Investigation of pollution in the Ebrie Lagoon (Ivory Coast) by Marchand and Martin (1985) produced a wide range of concentrations of total hydrocarbons in the lagoon sediments (1000 - 24000 mg/kg).

The highest concentrations were associated not with shipping but with industrial and domestic sewage discharges. However, a spill of 400 tonnes of oil at a refinery in 1981 was still clearly detectable at the time of their survey in 1983 (Portmann et al., 1989). There is, therefore, an urgent need not only for the development of national and regional contingency plans to combat oil pollution especially in cases of emergencies, but also for monitoring the levels and effects of pollutants in the Eastern and Western Africa coastal and marine areas (UNIDO, 2000).

Coastal tourism

The people of Africa have for a long time been associated with visitors from the Arabic and Persian nations, Europeans, and other distant lands (Ngoile, 1997). They are friendly and attractive with captivating life styles, customs and traditions, food preparation styles, costumes, artistic expressions. In addition, the coastline of the African region is an area of great physical beauty, rich in living resources. In the Eastern African sub-region for example, Palm fringed beaches of white coral reefs with their wealth of colourful fish, shell and corals. For many countries in the sub-region, coastal tourism is one of the most important sectors of their economies producing foreign exchange (Shah, 1995). Although, available evidence that before the 1980's, the growth of tourism in Africa occurred without significant deterioration in the fragile coastal ecosystem, this is rapidly changing (Odada, 1993). The impacts of tourism on both social and cultural environment and on the natural environment are causing serious concerns in the sub-region. To ensure that the importance of the tourism industry is maintained in the economies of African countries, there is need to develop and maintain environmental policies to govern the industry. Protecting and conservation of tourist attractions must be observed through the formulation of sound laws and regulation governing tourism, especially in the Eastern African sub-region, and elsewhere in the African coastal areas (Okemwa and Wakwabi, 1993).

Coastal agriculture

Agriculture is the main stay of the economy of most African countries. Agriculture contributes between 30 to 60% of their GNP and the majority of the population depends on it for their livelihood (World Bank, 1996b). The population is increasing rapidly with the subsequent increase in food demand. Simultaneously, in most African countries, the land that is available for agriculture is shrinking because of the use of the same land for non-agricultural purposes such as residential accommodation, industry, roads, playfields, hotels, etc., and also due to loss of good agricultural land through soil erosion, salinisation and sodification. As a whole, the net land resources available for agriculture are diminishing while population is rapidly increasing in the African region (Ibe, 1996). The expanding agriculture is having undesirable effects on coastal zone and marine habitats of the African region.

These effects are becoming more and more apparent as pollution increases and ecosystems deteriorate. For example, erosion related to deforestation and unwise agricultural practices is prevalent in the countries of the region (Odada, 1993). In Kenya, silt from rivers is affecting catches of fish, smothering coral reefs and is sulling beaches with serious consequences for fishing and tourism. The effects of pesticide pollution on marine life are now becoming apparent in many countries of Africa and are absorbed into living organisms (Mwaguni and Munga, 1997). The health of humans is threatened by these toxins reaching them through the fish they eat (UNEP, 1989). It is imperative, therefore, that soil conservation measures must be instituted particularly where agriculture is being developed in the coastal areas (UNEP, 1999b).

Over-exploitation of marine resources

In general productivity of African coastal waters is dependent on the extent of the continental shelf, coastal upwelling, mangroves, coral reefs, and run off from rivers (FAO, 1996b). Fisheries in the countries of the region reflect the availability of these physical characteristics. The relatively extensive continental shelves of Madagascar and Mozambique for example, support lucrative shrimp fisheries, while the absences of such areas in island countries make them depend on offshore tuna resources. Shrimp and tuna are the main commodities supporting export ventures in the Eastern African coastal sub-region. For the west and central Africa, the total annual catch of fish in coastal zone is estimated to be about 2.6 million tons per annum (FAO, 1987) about 10% of the coastal population engages in some form of fishing activity. At least 30% is canoe fishing, but larger fishing trawlers account for the bulk of the remaining catch in the sub-region (UNEP, 1989). The most serious problem, however, is over-exploitation of marine resources in most African coastal and marine areas. Fish, shells, beche-demer, dugongs and turtles are all subject to over-exploitation on a massive scale especially in many parts of Eastern and Western Africa, where agricultural land is in short supply and food is scarce. The over-exploitation is due in part to burgeoning human numbers coupled with a shortage of land-based jobs. Kenya's population of over 40 million for example, is increasing at unprecedented rate of 4.3% annually, and is expected to double around the turn of the century, with a consequent increase in demand for land, food, housing, water and social services similar situation prevail in other parts of the African region, straining both terrestrial and marine ecosystems to breaking point (WWF, 2001a). Therefore, solutions need to be found to this resource depletion problem which escalates as populations grow in Africa (FAO, 1997).

Institutional and administrative limitations

The institutional and administrative capacities of the different countries for coastal and marine resources development vary widely in the African region. It ranges from countries with virtually no capabilities to those with a growing capability and considerable resources. There are however, two broad categories of the African coastal states as regards their present level of institutional and administrative capabilities (UNESCO, 1981). The first categories include countries like South Africa, Nigeria and Kenya, in which there are already good infrastructure for the development of coastal and marine areas, in the way of appropriate research and training institutions and other facilities, where the governments are well aware of the importance and crucial role of coastal and marine resources in the development of their economies, and where there are reasonable number of qualified national personnel to undertake an integrated management of coastal and marine areas (Ngoile, 1997; Ibe, 1996). The second category includes countries like Benin, Gambia and Djibouti which for various reasons are at low stage of oceanographic development, where there are not as yet any substantial infrastructure for the development of coastal and marine resources, in the way of research and training institutions and other facilities, and where there are great shortage of trained manpower (UNESO, 1981). In general, the present level of institutional and administrative capacities of practically all African coastal states is low, and far from being adequate in providing a sound and sustainable base for the rational exploitation of coastal and marine resources of these countries. Thus, institutional and administrative capacity building in ocean affairs should be top priority in the African region (Sain and Knecht, 1998).

Awareness and management

There are a number of reasons for this serious situation. In general there is a lack of knowledge and understanding of the coastal and marine resources and their interaction with terrestrial and oceanic processes (UNESCO, 1981). This lack of information and awareness together with ineffective coastal and marine planning and management further aggravate the situation. In the Rio United Nations Conference on the Environment and Development (UNCED) protection of the coastal and marine environments to ensure sustainable use of the natural resources was at the top of the agenda for action (AGENDA 21). Under chapter 17 of Agenda 21, coastal states should "commit themselves to integrated management and sustainable development of coastal areas and the marine environment under their national jurisdiction." UNCED further pointed out the importance of coastal states to develop national policies and management capabilities for integrating the development and management of multisectoral activities in coastal and marine areas (UNEP, 1985; 1988).

DISCUSSION ON INTEGRATION OF COASTAL AND OCEAN AREAS INTO SUSTAINABLE DEVELOPMENT STRATEGIES

In general, in the African region, environmental values and natural resources factors have not always been integrated into national development plans. Development decisions and social trends appear not to have optimized the value of natural resources. Industrial expansion has often been carried out at the expense of the environment. Economic and social development, both in cities and countryside, has tended to deplete natural resources and damage the environment and amenities. Institutions to promote environmental and resource values and to assure that they are taken into account in the governmental decision making are just being created in some countries in the region (World Bank, 1996b; WWF, 2001a; Akpabli, 2000; O'Toole et al., 2001).

Environmental policies

The general situation with respect to the environment and relationships to national development and the expansion of human settlement is exacerbated in the case of the coastal development (UNEP, 1999b). The states of the region have not by large developed explicit policies relating to coastal and marine related development. For example, the West African coastal area has in the recent times acquired a more significant role in the overall economic development of the sub-region (World Bank, 1996b). For the land-locked countries in the sub-region, coastal ports and harbour represent the only natural outlet for their goods and services, both imports and exports. For the coastal states themselves this importance is reflected by the continuing desire to further locate industries within the coastal area to take advantage of the accessibility of the area to the main arteries of transport and communication (UNIDO, 2000).

Environmental concerns have not therefore played a significant role in the development of the coastal and marine areas of the sub-region (WWF, 2001a). Very little efforts has been expended in incorporating socio-environmental concerns in development planning and though widespread problems have generally been discerned in the coastal and marine areas, they have generally been shrugged off as being an inescapable part of the development process. In the African region as elsewhere, the dynamic and interactive nature of coastal and marine resources has meant that absence of adequate planning and management in these areas has been magnified in its effects on marine resources (Wilkinson et al., 1999). Although, serious environmental degradation of the marine environment has not yet occurred in many African coastal and marine areas, disturbing and perhaps irreversible trends are beginning to appear, some in connection with activities practiced outside the coastal zone. The marine and coastal aspects of these problems must therefore, be effectively incorporated into national policy and decision making (UNEP, 1999b).

Legislation and environmental law

Many countries in Africa have formulated regulatory measures for their resources management in coastal and marine areas such as the issuance of permits for fishing, logging and mangrove harvesting. However, most of these measures have proven ineffective for various reasons mentioned earlier (FAO, 1997). Increasingly, the countries of the region are enacting environmental laws that can provide practical frameworks at the national level to implement environmental standards and to regulate activities of enterprises and people in the light of environmental objectives (DEA and T, 1998). At the international level, conventions like the Law of the sea, protocols and agreements such as the UNEP Regional Seas Programme have been providing a basis for co-operation among countries at bilateral, regional and global levels for the management of environmental risks, control of pollution and conservation of natural resources in coastal and marine areas are effectively tackled (UNEP, 1996b). Governments of the African countries should also be encouraged to settle their marine environmental and other related disputes by peaceful means, making use of the existing and emerging agreements and conventions (UNEP, 1988).

Coastal zone management

In the African region a few countries are in the process of developing management plans for their coastal and marine areas (Ngoile, 1997; Ibe, 1996). The Seychelles for examples initiated in 1992 a plan for coastal zone management under the UNEP Eastern African Regional Seas Plans (Shah, 1995). The objectives of this project are to prepare an inventory of the coastal and marine species, the state of coral reefs, mangroves and lagoons, to assess the extent, nature and causes of coastal and marine pollution and also to identify policy and remedial actions. The project components include training, institutional capacity building, workshops, provision of laboratory equipment, etc. It is managed from the Department of Environment that was created in June 1989 under the direct leadership of the President of the Seychelles to solve the environmental problems that stem from a general increase in the population and the rapid development of the island (Shah, 1995).

Tanzania is, as yet, only in the preliminary stages of the development of an integrated coastal zone management programme. In 1991, however, the country began the process of creating a protected area, to be known as the Mafia Island Marine Park (MIMP). This marine park will protect the last pristine coral reef ecosystem found in Tanzania's coastal waters-an area that is important as an economic resource upon which a significant coastal and island population group is quite dependent. The Tanzania government perceives that this project will serve as a preliminary or pilot project providing basic interaction and approaches for the development of Tanzania's integrated coastal zone management (Ngoile, 1997). Coastal and marine resources are valuable assets that can effectively contribute to sustainable development of the African region (Clark, 1996).

Nevertheless, uni-sectoral over-use of some resources has caused grave problems. For example, indiscriminate harvesting of mangroves might have large economic benefits to those countries in the Eastern African sub-region, but have proven detrimental to fisheries, agriculture, and coastal region tourism (Linden, 1993). Similarly, unregulated fishing efforts and the use of destructive fishing methods such as dynamiting have destroyed fish habitats and reduced fish stocks (FAO, 1997). Some African countries have formulated regulatory measures for their resources management such as the issuance of permits for fishing, logging, and mangrove harvesting. However, most of these measures have proven ineffective due to partly enforcement failure, but mainly to lack of support from the communities concerned (FAO, 1998). There is, therefore, an urgent need for an integrated interdisciplinary and multi-sectoral approach in developing management plans for the coastal and marine areas of the African region (Ngoile, 1997).

Institutional and administrative limitations

Practically, all African coastal states are at present making only minimal use of their coastal and marine resources owing to the limitations in the necessary scientific knowledge and technological know-how and to the lack of efficient organizational and administrative machinery (UNESCO, 1981). For example, the shipping and navigation services, which are so vital for the management, control and exploitation of marine resources and for the development of international trade, are still very much in their infancy in most African coastal states (World Bank, 1996a).

Marine technology is also very much underdeveloped in the region. In very few African countries there are marine technology training centres with comprehensive programmes concerning all aspects of marine technology-marine engineering, fishing and fishing gear, boat building and repair, navigation, instrumentation repair and maintenance (including electronic equipment), fish processing and preservation, economic and marketing. Thus, the development of sound training and research programmes and their effective linkages with the production system are basic and most important steps towards enhancing the capability of African states in making full use of their coastal and marine resources (UNIDO, 2000; DEA and T, 2000a).

Basic marine science

Many African countries are confronted with serious manpower problems which are proving to be great impediments especially in the economic development of coastal and marine areas of these countries. In many cases, the most important cause underlying these problems, is the lack of adequate training facilities for the type of manpower required (UNESCO, 1981). Although, the majority of African countries now have national universities and other institutions of higher learning of their own, most of these institutions are young, and many are still facing teething problems concerning adequate staffing, adequate equipment, sound curriculum development, etc. In many of these countries, therefore, the universities are still grappling with the fundamental issues of producing adequate manpower of the vital organs of the civil service requiring high-level personnel, such as public administration, school education, public health and agriculture (Clark, 1996; Ibe, 1996; Sain and Knecht, 1998; World Bank, 1996b). In a considerable number of cases, even these vital sectors of the civil service are still far from being adequately staffed by qualified and experienced nationals. It is understandable, therefore, for the universities in these countries to be pursuing for the present, narrow crash training programmes with the objectives of producing high-level manpower badly needed in key areas of the civil service. In such cases, other needs for manpower development for example training of marine scientists and technologists that would be beyond the capability of most African states working on their own should be met through regional and sub-regional co-operation. Seminars, workshops, conferences, etc. such as those organized by UNESCO-IOC would be needed, useful and appropriate for the development of manpower capable of sound management of coastal and marine areas of the African region (UNESCO, 1981; Mensah, 1994; Linden, 1993).

RECOMMENDED MEASURES FOR INTEGRATING COASTAL AND MARINE AREAS INTO SUSTAINABLE DEVELOPMENT STRATEGIES

The coastal and marine areas of the African region have some of the World's richest ecosystems with extensive coral reefs, lagoons, estuaries, and mangrove forests (FAO, 1996; 1997; WWF, 2001a). These ecosystems support a wide diversity of plants and animals. The economic benefits derived from these areas are essential for the survival of a rapidly growing population (Ibe, 1996). As in many other coastal zones elsewhere, the balance of the ecosystem is threatened by unplanned industrial exploitation of marine resources, shore erosion, expanding coastal tourism, etc (Ngoile, 1997). As stated in Chapter 17 of Agenda 21 of the Earth Summit in Brazil, coastal and marine management must play an important role if resources are to be exploited in a way that ensures sustainable development to a growing population.

To implement sustainable development of coastal and marine areas of the region, therefore, African governments need to (1) develop human resources by undertaking short-term academic training to strengthen existing cap-abilities; (2) promote public awareness by producing educational materials on the ecological and socio-economic contribution of the coastal and marine resources and the consequences of unsustainable exploitation (3) organize policy workshops involving relevant policy and law makers to increase their understanding and commitment to the sustainable use of the resources in their coastal and marine areas, and (4) implement integrated coastal zone management programmes by establishing case studies in pilot sites of coastal and island states in the African region (UNEP, 1999a)

Capacity building

Many countries of the African region are confronted with serious manpower problems, which are proving to be great impediment in the economic development of coastal and marine areas of these countries (UNESCO, 1981). In many cases, the most important causes underlying these problems, is the lack of adequate training facilities for the type of manpower required. Since the requirements for high level manpower in certain areas of marine science e.g., physical, chemical, biological oceanography, aquaculture, etc are not so great in terms of numbers needed by any one country at any one time, existing institutions (e.g. universities) in the region suited for teaching and research in theses areas should specialize as regional or sub-regional training centres by enlarging their facilities to enable the enrolment of students from other member states, wanting to study the subjects of their specialization (UNESCO, 1981).

Oceanographic research should also be carried out on a regional or sub-regional co-operative basis using collectively-operated research vessels which are well equipped and well staffed for all types of oceanographic research and for the on-board training of marine scientific and technical staff (UNESCO, 1981). To start with one or two such vessels should be adequate for each of the sub-regions that is, Eastern and Western Africa. The co-ordination of research on a regional scale, the exchange and dissemination of research information and the storage of research data are important support activities in the protection and development of coastal and marine areas of the African region (International Ocean Institute, 2001).

Public awareness

The African governments need to increase public awareness through education. The public should be informed of long-term ramification of coastal development. At present the general public is generally unaware of problems associated with the development of coastal and marine areas. In Madagascar for example, the wide beaches which previously attracted many tourists to the island have already disappeared (Odada, 1993). In spite of this, many structures are still being constructed on the eroding beaches. Campaigns should be instituted on a national basis to create greater public awareness of national and regional issues in the protection and development of coastal and marine resources of the African region. Education is the principle of protection and development of marine resources should be provided as part of the ordinary educational curriculum at the primary school, secondary school and university levels through training of special instructors or specialized training of general educators, and through seminars and courses offered to general public (UNESCO, 1981).

In the western Indian Ocean for example, marine science association has been formed with the secretariat at the Institute of Marine Science in Zanzibar (Ngoile, 1997). The aims of the Western Indian Ocean Marine Science Association (WIOMSA) are (1) to promote and advance the educational, scientific and technological development of all aspects of marine science in the region, (2) to provide a forum for discussion and dissemination of information and organize meetings, seminars and workshops for the presentation of information, findings and experiences on all subjects related to marine sciences, (3) to encourage the support of marine science research, and the developments and educational activities by government agencies and private sector, and (4) to collect and disseminate scientific, technical and other information on marine sciences (Linden, 1993).

Environmental legislation

Increasingly, environmental legislation has been providing practical frameworks at the national levels to implement environmental standards and to regulate activities in coastal and marine areas of the African region. At the international level conventions, protocols, and agreements have been providing a basis for co-operation among countries at bilateral, regional and global levels for the management of development activities, and control of pollution and conservation of marine resources (UNIDO, 2000).

However, African states urgently need to review and where necessary expand, update or strengthen national legislation and regulation pertaining to the protection and development of the coastal and marine areas (WWF, 2001a). The enforcement of national regulations related to coastal and marine resources protection and development needs to be improved. There is an urgent need to expand the accession to and ratification of international convention, such as the Law of the sea, and institute mechanism at the national level to ensure their application (UNEP, 1988).

Management planning

The value of treating coastal and marine areas as planning entities within the overall framework of national development planning is not always fully recognized in the African region and very often countries lack the administrative and legislative basis for implementing such an approach. Industrial expansion, such as construction of ports and harbours, has often been carried out at the expense of the environment (Clark, 1996). Economic and social development, both in the coastal cities and the coastal zone, has tended to deplete natural resources and damage the environment and amenities. To solve these problems, the countries of the African region need to develop an integrated interdisciplinary and multi-sectoral approach in their management plans for coastal and marine areas. Pilot coastal sites can be selected in one or two countries in each sub-region and intensive interdisciplinary planning programme conducted, involving several resources and scientific personnel from various institutions in those countries, including universities, government agencies and non-governmental organizations (Ibe, 1996). In addition, training courses, workshops and conferences should be organized, and publications and educational materials disseminated as part of the integrated coastal and marine resources management planning (Ngoile, 1997/3).

REFERENCES

Akpabli KR (2000). Integrated Management Of The Gulf Of Guinea Fight Against Coastal Pollution. Paper Presented At The Sustainable Development For Coastal Zones And Instruments For Its Evaluation. October 23–26, 2000. Bremerhaven, Germany.

Allersman E, Tilmans WK (1993). Coastal Conditions In West Africa; Rev. Ocean Coastal Manage., p. 9-240.

Awosika LF, Chidi Ibe A, Schroeder P (1993). Coastlines of West Africa. A Compilation Of Papers Presented At The Coastal Zone '93 Conference, New York. American Association Of Civil Engineers.

Chidi IA (1996). The Coastal Zone And Oceanic Problems Of Sub-Saharan Africa. In; Benneh G,Morgan WB, Uitto JI, (eds.) Sustaining The Future; Economic, Social And Environmental Change In Sub-Saharan Africa. United Nations University Press, Tokyo.

Chidi IA, Awosika LF, Ihenyen AE, Ibe CE, Tiamiyu AI (1985). Coastal Erosion In Awoye And Molume Villages, Ondo State, Nigeria. A Report For Gulf Oil Co. Nigeria Ltd.

Clark JR (1996). Coastal Zone Management Handbook. Boca Raton, FL: Lewis Publishers, p. 94.

DEA and T (1998). Green Paper for a Coastal Policy for South Africa. Department Of Environmental Affairs And Tourism, Pretoria.

DEA and T (1999). State Of The Environment South Africa. Department Of Environmental Affairs and Tourism, Pretoria. http://www.ngo.grida.no/soesa/nsoer.

DEA and T (2000a). The South African Line Fishery: Past, Present, and Future. Department Of Environmental Affairs and Tourism, Pretoria. Wetlands. March–April 1995. Climate Alert, 8: 2.

ESA-ESRIN (1996). Protection of mangrove forests and coastal erosion in Gabon and Nigeria. European Space Agency Publications Division *ERS-SAR* Application, p. 10.

FAO (1996). Fisheries And Aquaculture In Sub-Saharan Africa: Situation And Outlook In 1996. Fisheries Circular No. 922 FIPP/C922. FAO, Rome.

FAO (1997). Review Of The State Of World Fishery Resources: Marine Fisheries. FAO, Rome.

FAO (1998). Coastal Environments Under Threat. FAO Factfile, Available On:Http://Www.Fao.Org/News/Factfile/Ff9804.Htm.

FAO (1999). The Sustainable Fisheries Livelihoods Programme In West Africa. FAO Fisheries Department, Rome.

Ibe AC, Quelennec RE (1989). Methodology for assessment and control of coastal erosion in West and Central Africa, UNEP Regional Seas Reports and Studies. UNEP, p. 107.

ICST (Ivorian Country Study Team) (1996). Vulnerability of Coastal Zone of Côte d'Ivoire to Sea Level Rise and Adaptation Options. Report on the Côte d'Ivoire/USA Collaborative Study on Climate Change in Côte d'Ivoire.

International Ocean Institute (2001). The Coastal And Inshore Marine Environment Of The Western Indian Ocean Region At The Dawn Of The 21st Century.Report Of The International Ocean Institute, Available On: http://www.ioinst.org/reports/report2.htm.

International Rivers Network (1995–2000). The Environmental Impacts Of Large Dams. http://hww.irn.org/basics/impacts.shtml.

IPCC (1995). Climate Change 1995.Impacts,Adaptations And Mitigation Of Climate Change: Scientific And Technical Analysis. Contribution Of Working Group II To The Second Assessment Report Of The Intergovernmental Panel On Climate Change. Edited By Robert T.Watson, Marufu C. Zinyowera,and Richard H.Moss. WMO and UNEP. Cambridge University Press, Cambridge UK.

IPCC (1998). The Regional Impacts Of Climate Change.

Intergovernmental Panel On Climate Change, Geneva.
Leatherman SP, Nicholls RJ (1995).Rising Seas Threaten Cities, Erode Beaches And Drown Wetlands In Key Developing Countries. March–April Climate Alert, 8: 2.
Linden O (1993). Workshop And Policy Conference On Integrated Coastal Zone Management In Eastern Africa Including The Island States. Coastal Management Centre (CMC) Conf. Proc. 1, 371p. Metro Manila, Philippines.
Munga D (1981). Some Observations On Petroleum Pollution Along The Kenya Coast. Pp 290–297 In Proceedings Of The Workshop Of KMFRI On Aquatic Resources Of Kenya, July 13-19.
Mwaguni S, Munga D (1997). Land Based Sources And Activities Affecting The Quality And Uses Of The Marine Coastal And Associated Freshwater Environments Along The Kenyan Coast. (Unpublished Report).
Ngoile MAK (1997). Coastal Zone Issues And ICM Initiatives In Sub-Saharan Africa. Marine and Coastal Management, 37(3): 269-279.
Nicholls RJ, De La Vega-Leinert AC (2000). Report on the SURVAS Expert Workshop on "African Vulnerability And Adaptation To Impacts Of Accelerated Sea-Level Rise" Cairo, 5th–8th November 2000 http://www.Survas.Mdx.Ac.Uk/Publica2.htm#Cairo.
O'Toole MJ, Shannon LV, de Barros Neto V, Malan DE (2001). Integrated Management of the Benguela Current Region:A Framework for Future Development. In B. von Bodungen and RK Turner (ed.) Science and Integrated Coastal Management. Dahlem University Press.
Obura D, Suleiman M, Motta H, Schleyer M (2000). Status Of Coral Reefs In East Africa: Kenya, Mozambique, South Africa And Tanzania pIn:Wilkinson, C. Status Of Coral Reefs Of The World: 2000. Australian Institute Of Marine Science And Global Coral Reef Monitoring Network, Townsville, p. 65-76.
Odada EO (1991). Eroding Kenyan shorelines: The need for scientific appraisal and management planning IOC-SAREC-KEMFRI. Regional Workshop on Causes and Consequences of Sea Level Rise in the Western Indian Ocean and Islands, IOC Workshop Report, p. 7.
Odada EO (1992). Beach erosion:Case studies of the Kenyan coast, Proceed, International Workshop on Rational Use of Coastal Zone UNESCO, Nairobi, Kenya.
Odada EO (1993). The problem of coastal erosion and flooding in Eastern Africa. Background paper presented at the workshop and Policy conference on integrated Coastal zone Management in East Africa, Arusha, Tanzania, April.
Okemwa E, Ntiba MJ, Sherman K (1995). Status and future of large marine ecosystems of the Indian Ocean: A report of the International Symposium and Workshop. Marine Conservation and Development Reports. IUCN, Switzerland, pp. x+201.
Okemwa E, Wakwabi EO (1993). Integrated Coastal Zone Management Of Kenya. In Linden, O. (ed) Workshop And Policy Conference On Integrated Coastal Zone Management In Eastern Africa Including The Island States. Coastal Management Centre (CMC) Conf. Proc. 1, 371p. Metro Manila, Philippines, pp. 193-203.
Okemwa E (1998). Application Of The Large Marine Ecosystem Concept To The Somali Current. In Kenneth Sherman, et al., (eds.), Large Marine Ecosystems Of The Indian Ocean: Assessment, Sustainability, And Management (Oxford: Blackwell Science), pp. 73-99.
Portmann JE (1978). The Gulf of Guinea: Pollution, the need for control and possible mechanisms thereof. Rome, FAO, FAO/UNEP Joint Project No. FP/0503-77-02, p. 55.
Portmann JE, Biney C, Ibe AC, Zabi S (1989). State of the marine environment in the West and Central African region.UNEP Regional Seas Reports and Studies, No. 108. UNEP 1979.
PRB (2010). World Population Data Sheet. Population Reference Bureau. http://www.prb.org/publications/Datasheets/2010/2010wpds.aspx.
Salm RV (1996). The Status Of Coral Reefs In The Western Indian Ocean With Notes On The Related Ecosystems. In: The International Coral Reef Initiative (ICRI) Western Indian Ocean An Eastern African Regional Workshop Report, 29 March– 2 April 1996, Mahé, Seychelles.UNEP Biodiversity/Water) ICRI WIO/EAF WG.1/REP.UNEP 1997.
Shah NJ (1995). Managing Coastal Areas In The Seychelles. Nature And Resources, 31(4): 16-33. UNESCO, Paris.
Shah NJ (1996). Climate Change Poses Grave Danger To African Island States. Splash, 12(1): 14-15, SADC.
Smith JB, Huq S, Lenhart S, Mata LJ, Nemesova I, Toure S (1996). Vulnerability And Adaptation To Climate Change. Environmental Science And Technology Library, Kluwer Academic Publishers.
UNEP (2001). UNEP Mediterranean Action Plan. June 2001 http://www.unepmap.org/.
UNEP (1982). Pollution and the marine environment in the Indian Ocean. UNEP Regional Seas Report and Studies, p. 13.
UNEP (1984). Environmental management problems in resource utilization and survey of resources in the West and Central African region.UNEP Regional Seas Reports and Studies, UNEP, p. 37.
UNEP (1984). The Marine And Coastal Environment Of The West And Central Africa Region And Its State Of Pollution. Regional Seas Programme Reports And Studies No RSRS 46. United Nations Environment Programme, Nairobi.
UNEP (1985). Action plan for the protection, management and development of the marine and coastal environment of the East African region. UNEP Regional Seas Reports and Studies, p. 61.
UNEP (1988). Regional Seas Programme: Legislative Authority. UNEP Regional Seas Reports and Studies, p. 17.
UNEP (1998a). Eastern African Atlas Of Coastal Resources: Kenya. United Nations Environment Programme, Nairobi.
UNEP (1999a). Western Indian Ocean Environment Outlook. United Nations Environment Programme, Nairobi.
UNEP (1999b). Overview Of Land-Base Sources And Activities Affecting The Marine, Coastal,And Associated Freshwater Environments In The West And Central Africa Region. UNEP Regional Seas Reports And Studies, p. 171.
UNEP/GRID-Arendal (2007). Coastal population and altered land cover in coastal zones (100 km of coastline), UNEP/GRID-Arendal Maps and Graphics Library, http://maps.grida.no/go/graphic/coastal-population-and-altered-land-cover-in-coastal-zones-100-km-of-coastline (Accessed 31 August 2010).
UNESCO (1981). Marine Science, and Technology in Africa: Present State and Future Prospects Report No. 14. UNESCO Paris (1981).
UNIDO (2000). Gulf Of Guinea: Water Pollution Control And Biodiversity Conservation (Success Story). UNIDO Media Corner, http://www.unido.org/doc/100452.htmls.
Wilkinson C, Linden O, Cesar H, Hodgson G, Rubens J, Strong AE (1999). Ecological and Socio-Economic Impacts of 1998 Coral Bleaching in the Indian Ocean: An ENSO Impact and Warning of Future Change? Ambio, 28: 188-196.
World Bank (1996b). Towards Environmentally Sustainable Development In West Central Africa. Agriculture And Environment Division, West Africa Department, Africa Region. World Bank, Washington D.C.
WWF (2001a). Africa Programme: Oceans and Coasts. Available on http://www.panda.org/africa/oceans.htm.

9

Spatial pattern of vertical zonation of rocky shore organisms and the influence of ocean exposure at the Islas Cíes (NW Spain)

Jesús S. Troncoso[1] and Jeffrey A. Sibaja-Cordero[1,2*]

[1]Departamento de Ecología y Biología Animal, Facultad de Ciencias del Mar, Universidad de Vigo, Campus Lagoas Marcosende, 36310 Vigo, Spain.
[2]Centro de Investigación en Ciencias del Mar y Limnología (CIMAR), Universidad de Costa Rica, San Pedro, 11501-2060 San José, Costa Rica.

A sampling was conducted around the intertidal coast of the Islas Cíes, (NW Spain) with the goal of determining the variation of the shore landspace, based on vertical zonation profiles of sessile organisms in relationship with the Atlantic Ocean influence. The Western Coast is heavy wave beaten and the Eastern Coast is sheltered. Many species coexist in a limited intertidal space of the sheltered than in the exposed shores. The sheltered coast presents in this way, a mosaic structure of their organisms. On the other hand few species dominated the cliffs of the west coast. In this section, the upper intertidal limit of distribution was raised by the increased of humidity caused by the splash and spray. The winds also favoured the presence at the high littoral level of organisms such as brown algae, lichens and vegetation. Multivariate techniques detected differences in the communities between exposed and sheltered shores as well as between the two main islands. Distribution of species in the intertidal zone was related to the particular geomorphological feature of the coast, such as coves, points or exposed cliffs. Four different landscape intertidal patterns of vertical zonation were represented by each island face. The algae, *Asparagopsis armata* and *Bifurcaria bifurcata* characterized the sheltered North Island. *Pelvetica canaliculata* and *Fucus spiralis* had a greater presence in the sheltered side of the South Island. The fucoids were absent at the wave beaten shore. The lichens, *Verrucaria maura* and *Caloplaca marina*, and the Rhodophyta belt had a higher upper limit of distribution at the exposed sides of the North Island. Bare rock and the crust phase of *Mastocarpus stellatus* had a greater presence at the exposed side of the South Island.

Key words: Intertidal ecology, Galician Coast, ocean exposure, sessile organisms, digital image analysis, multivariate analyses.

INTRODUCTION

Patterns of vertical zonation of intertidal organisms are caused, in general, by the effects of physical stress and biotic interactions (Connell, 1972; Menge et al., 2010). Several belts or assemblages of organisms have been described and delimited in terms of the dominant species (Chapman, 1974). Thus, the high littoral is characterized by snails (Littorinidae), the middle shore is defined by sessile organisms such as barnacles and mussels, and the low littoral is dominated by algae (Ellis, 2003). The principal causes of differences in assemblages were the wave exposure, larval transport, food supply and spatial heterogeneity (Menge, 1976; Menge et al., 1997; Schiel, 2004). In addition, settlement of larval stages depends on the rock type (Fischer, 1981). These factors can change

*Corresponding author. E-mail: jeffro@costarricense.cr.

in their intensity by the coastal geomorphology of rock platforms and cliffs (Bird, 2008). The fauna in temperate rocky shore present truly marine species in the low intertidal zone and semi-terrestrial species in the supra-tidal (Ellis, 2003). De la Huz and Lastra (2008) point out that these two types of organisms respond in a different manner to physical factors in sandy beaches in the Galician shores.

Three-quarters of the world's coastline is composed of cliffs and rocky substrata in general; the majority of inter-tidal ecology studies were, however, developed in shore platforms characterized by sampling accessibility and safe conditions for the researchers (Little and Kitching, 1996; Bird, 2008). The Galician coast is mainly rocky and the cliffs dominate the shore (Pérez-Cirera, 1978; Vilas-Paz et al., 2005). By this reason, few studies were done in wave-exposed rock walls of Galicia (Niell, 1977; 1978; Olabarria et al., 2009). A basic description of the vertical distribution of organisms at the north of Galicia was provided by Pérez-Cirera (1975). Additionally, the rocky shores of the north coast of Portugal present similar assemblages of organisms (Araújo et al., 2005; Pereira et al., 2006).

The southern Galician coast is characterized by the Rías Baixas that are flooded tectonic valleys (deCastro et al., 2000). These mouth of these estuarine systems present islands with intertidal cliffs and rocky shores (Vilas-Paz et al., 2005). The benthic assemblages of these islands have scarcely been studied (Acuña-Castroviejo, 1984; Niell, 1977; Besteiro-Rodriguez et al., 2004) and the vertical zonation of intertidal organisms and their geographically variation are therefore unknown. In this way, this paper describes the vertical zonation of sessile organisms (algae, lichens, mussels and barnacles) along the rocky coast of the two main islands of the Islas Cíes (NW Spain). The aim of this work is to test if the vertical zonation profiles differ in presence, position on the rock and dominance of species according to the oceanic exposure and geographic position within the National Park of the Islas Cíes.

METHODS

Study site

The Ría de Vigo is located at the Atlantic coast of Spain, which is the boundary between the subpolar and temperate regimes of the coastal upwelling system of the Eastern North Atlantic. The seasonal quasi-permanent upwelling occurs from April to October and is driven by the northerly trade winds. The Ría de Vigo is partially enclosed at the mouth by three natural islands, That is, Islas Cíes, which provide some shelter for the inner of the Ría de Vigo (Figure 1). These islands have the status of Galician Natural Parks since 1980, and that of National Park since 2002 (de Castro et al., 2000; Costas et al., 2002; Bernárdez et al., 2006; Diz et al., 2006).

The maximum deep at the south of the Islas Cíes is 53 m and at the north entrance is 30 m. In front of the West Coast of the Islas Cíes the deep is near the 100 m (García-García et al., 2005). The tidal amplitude is of 2 to 3 m during the neap and spring tides as in most of the Galician Rías (de Castro et al., 2000). The two northern islands (Figure 1) are connected at the eastern margin by a 1 km long sand barrier system (Rodas Beach), which confines a shallow lagoon (Costas et al., 2002). The walls of the Islas Cies are made up of granitic rock (García-García et al., 2005; Vilas-Paz et al., 2005). About 20% of the emerged surface corresponds to pronounced cliffs (Acuña-Castroviejo et al., 1984). The western side of the Islas Cíes disrupts the propagation of waves produced during storms. These waves have 7 m height and 16 s of period (Queralt et al., 2002). In Figure 1, the sites 14 to 19 and 29 to 40 have low wave energy and the sites 1 to 13 and 20 to 28 represent a high wave-beaten coast (exposed directly to oceanic influence); this classification follows that by Acuña-Castroviejo et al. (1982) in their study of the sublittoral ecotones and the wave pattern by Queralt et al. (2002).

Methodology

The rocky shore of the south and north islands of the Islas Cíes National Park (NW Spain) were sampled at low spring tide on 28th February and 14th March 2009, respectively. The coast around the islands was divided into 40 sections (Figure 1); those sections were photographed from a boat. In each section, several pictures were taken with a digital camera (7 megapixels). A total of 482 images were taken, each one from an independent point of the rocky shore wall. The pictures encompassed the rock wall from the low sea level until the line of vascular vegetation. The height (~35 cm) of the gull, *Larus cachinans*, captured on first plane in each picture was used for calibrating the spatial scale of images and converting the x-y pixels coordinates to a metric scale. In this way, for each picture in each section the height (y coordinate of the image) above the low tide sea level of each organism was estimated in centimeters by digital image analysis with the "point" command of the UTHSCSA Image Tool, available in the web site of the University of Texas Health Science Center. The height represents the upper and/or the lower limits for marine organisms: algae, mussels, barnacles, and lichens, and the lower limit of terrestrial lichens and vascular vegetation (Sibaja-Cordero, 2008).

Data analysis

The statistical analyses were carried out with the free software PAST (Hammer et al., 2001). For each of the 40 sections the mean height was calculated for each organism and a vertical profile with all organisms was generated. The profiles were categorized whether they are located in south or north islands, western or eastern face and exposed and sheltered to Atlantic Ocean, respectively (Figure 1).

The data matrix contained numerical information of the height above sea level, but in several sections one or more organisms were absent. The vertical profile in these cases has missing data (absence of the organism). In PAST software, the multivariate techniques can handle missing data. The absences were marked with question mark in the data frame (Hammer et al., 2001). The software calculated the Euclidean dissimilarity matrix between sections (vertical profiles) of the shore using a pairwise deletion. This means that when the distance is calculated between two sections (vertical profiles), any variable missing is ignored in the calculation (Hammer et al., 2001). However, the ordination methods as Principal Coordinate Analysis (PCoA) can detect the variable with missing value as a difference, and the profile is plotted apart in the graphic representation. To avoid strong horseshoe effects in the

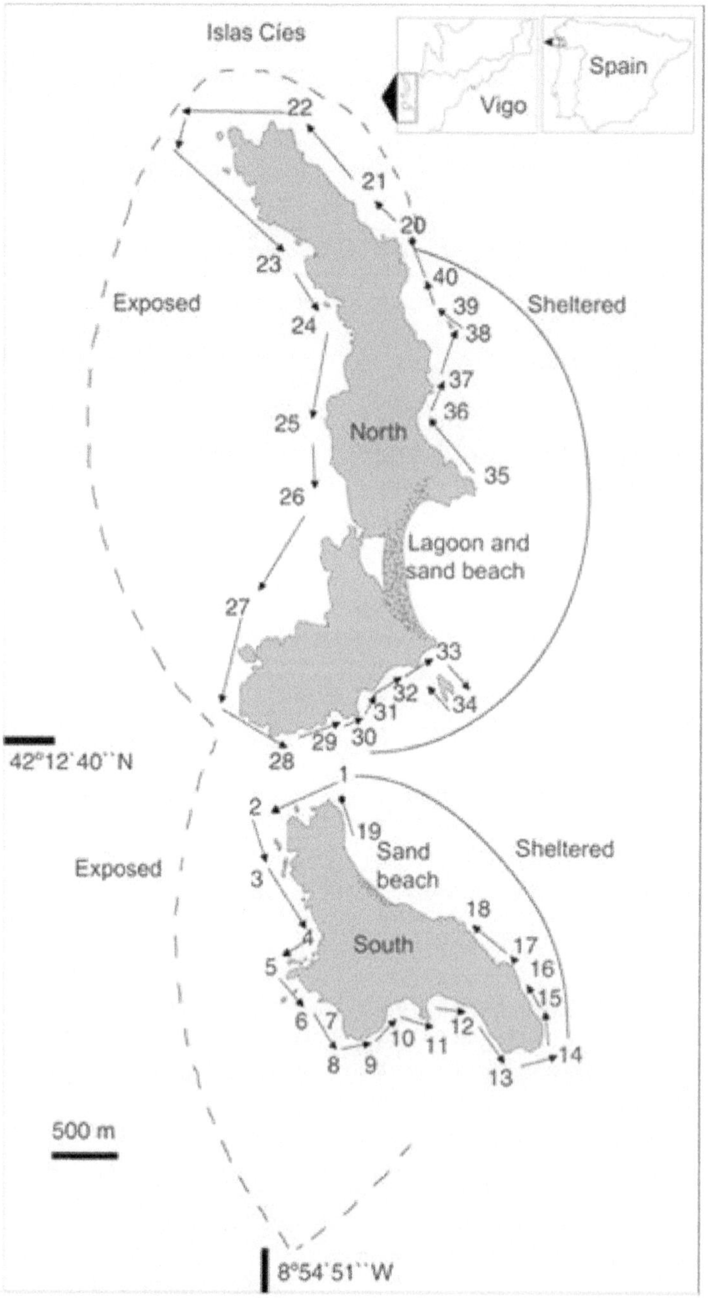

Figure 1. Location of vertical sections of the rocky shore in the Islas Cíes National Park, NW Spain. Classification of sections according to ocean exposure (exposed *vs.* sheltered) is based on Acuña-Castroviejo et al. (1982). Boat transects (GPS) are represented with arrows.

ordination, the mean height of the intertidal limits of each organism by section was transformed with log10, before calculating the Euclidean dissimilarity matrix (Legendre and Gallagher, 2001).

In this way, a Cluster analysis was generated by the paired group clustering method based in the Euclidean distance of the section (vertical profile), to explore the similarity of sections of the shore by their patterns of vertical zonation. Additionally, a PCoA of the same data based in the Euclidean distance was carried out to show the similarity of sections and the organism limit score which indicates where the sessile organisms occur or increase their height on the intertidal (Hammer et al., 2001; Quinn and Keough, 2003). The organism limit score of the PCoA biplot was calculated with the formula in Legendre and Gallagher (2001):

$$\text{Organism limit score}_{jk} = r_{jk}\, s_j / s_k$$

where r_{jk} are the correlation between the data of the organism j and the sections score vector k. Also s_j is the standard deviation for the data of the organism j, and s_k is the standard deviation of sections score vector k (Legendre and Gallagher, 2001). In organisms that occurred in all sections, r was calculated with the Pearson coefficient and when the organisms were absent in one or more sections, the Spearman rank correlation was used with a value of 0 for the absence data.

Finally, the same Euclidean dissimilarity matrix between sections was analyzed by two-way crossed ANOSIM (R) (Clarke and Warwick, 1994). The value of R determines the level of difference; when R = the groups are completely different from each other. The factors tested were the exposure degree of the sections (exposed or sheltered) and the island identity (north or south).

RESULTS

Figure 2 shows examples of the vertical profile found at the exposed and sheltered coast of the Islas Cíes National Park. In general, the first centimeters of the infralittoral presented several species of red algae (Rhodophyta). The dominance of species within this band varies between exposed or sheltered coast. For instance, *Asparagopsis armata* formed dense patches in the sheltered face of both islands and Chlorophyta were mostly found in patches in the sheltered infralittoral. Mussels and barnacles are present above the algae band. Above these bands and in exposed shores, there is an intermittent fringe of bare rock; in the sheltered faces, a fucoid belt occurs below the bare rock fringe. *A. armata* and *Bifurcaria bifurcata* were mainly in the sheltered shore of both islands. These species occurs higher in the intertidal of the sheltered shore of the north island. The high littoral of either exposed or sheltered shores presented the black lichen, *Verrucaria maura* and the orange lichen, *Caloplaca marina.*

The sheltered shore presented more complexity in the vertical profile, that is, there were more species and a more diversified mosaic of bands and patches of organisms. The limits of vertical distribution in the majority of organisms tended to appear higher on the shore at exposed faces for example, that of Rhodophyta, mussels, barnacles, lichens and vascular vegetation (Figure 3). Moreover, the width of the bands for species in the high littoral was compressed in the sheltered coast of both islands (Figure 3).

Cluster analysis showed two major groups of vertical profiles, mostly according to differences in ocean exposure (Figure 4a). The ordination of the PCoA showed a similar pattern with the sections of sheltered coast on one side of the graphic representation and sections of the exposed shore on the opposite side (Figure 4b). The central cluster (Figure 4a) presented profiles of both conditions of ocean exposure. The sections 14, 21, 28, 29 and 40 (Figure 1) are transitional areas between the fully exposed or sheltered condition. The exposed sections 9, 10, and 11, were clustered with the sheltered sections; PCoA ordination showed the same pattern (Figure 4b). These sections are in a small coastal cove (Enseada de Concela) of the south island (Figure 1). The upper intertidal limits of the organisms was lower in these sites (Figure 3) indicating less degree of exposure to the Atlantic Ocean. Sections 33 and 36 were best ordinate in the PCoA by their sheltered condition than in the cluster analysis (Figure 4).

The ANOSIM test showed that the ocean exposure was a significant factor in determining different vertical zonation patterns of sessile organisms ($R = 0.55$, $p< 0.0001$). The organism limit scores in the PCoA (Figure 4B) showed that the invertebrates, red algae and lichens had higher limits of vertical distribution at the exposed shore. On the other hand, *A. armata, B. bifurcata, Fucus spiralis* and *Pelvetia canaliculata* were mainly present in the sheltered coast (Figures 3 and 4). The second coordinate in the PCoA (Figure 4) showed a slightly separation of sections (vertical profiles) according to the island in which they are located. This spatial trend was confirmed by the ANOSIM test ($p = 0.037$) but the degree of difference is low ($R = 0.13$).

The profile in Figure 3 showed than *B. bifurcata* occurs in more sections in the north island. On the other hand, *P. canaliculata* and *F. spiralis* showed positive scores in the second coordinate which suggest these algae are mostly associated (higher height in the intertidal) with the sheltered face of the south island; in fact, both species were present in more sections at this island (Figure 3). The main shared feature among exposed faces was the higher upper limit of vertical distribution of the Rhodophyta band and the black lichen, *V. maura,* but higher in the north island as well as those of the lichen *C. marina* and the vascular vegetation (Figure 3) with positive scores in the second coordinate of the PCoA, near to the most exposed north sections (Figure 4). The crust phase of *Mastocarpus stellatus* and the bare rock band had a higher upper limit of vertical distribution in Sections 2, 3, 5, 8 and 12 of the exposed face of the south island (Figures 3 and 4).

DISCUSSION

The community and their pattern of the vertical zonation in Islas Cíes at the moment of the sampling were similar in the relative position and dominance of the species in the intertidal to that of the temperate rocky shores described for Great Britain (Little and Smith, 1980; Little and Mettam, 1994; Mettam, 1994).

Moreover, the trend of increased vertical distribution of intertidal species at areas with high wave exposure was found in the rocky shore of Islas Cíes; similar patterns have been reported from other intertidal areas (Underwood, 1981). At the exposed coast of both islands, species in high littoral were distributed along the intertidal constituting wide fringes. As expected, the multivariate

Figure 2. Vertical zonation of sessile organisms at the exposed (A) and sheltered (B) rocky coast of the Islas Cíes (NW Spain).

multivariate techniques indicated clear separations among the exposed and sheltered intertidal shores. The vertical profiles of exposed and sheltered shores presented differences in species assemblages, that is, the erect and foliose algae were present mainly in the sheltered shore. Other difference between exposed and sheltered shore was the higher height in the species occurs in the exposed rock wall.

The vertical distribution of littoral species on the sheltered shore of Islas Cíes was similar to those in similar areas in north Portugal (Araújo et al., 2005). The waves were unable to reach high littoral levels and the rock area for settling of marine species was reduced in north Portugal (Araújo et al., 2005). However, more marine organisms occurred in the intertidal of the sheltered coast of the Islas Cíes. The community was more complex in the sheltered face of the south island with more sessile species coexisting in the rock wall. The mid and high littoral of the sheltered face of the south island contained several patches of algae, lichens and bacterial bio-film than in the other island faces. In this way, Wallenstein et al. (2008) found in the Azores islands that the association between organisms and height at which they occur differs geographically (between islands). The changes in other intertidal species may relate to disturbances (herbivores or heavy wave), that reduced the productivity of the most dominant species (Menge and Farrel, 1989).

In the Islas Cíes, the wave patterns influence the intertidal community but in the Azores Islands, the wave exposure is a second factor in regulating the vertical zonation below the substrate ecotones (boulders, cliffs, platforms) (Wallenstein and Neto, 2006).

The two islands in the Azores Archipelago, studied by Wallenstein et al. (2008) did not present sheltered habitats. Moreover, there was some degree of variation between the islands of Santa Maria and São Miguel of the Azores Archipelago (Wallenstein et al., 2008); this pattern is similar to that showed in our study. In the Islas Cíes, the face of the islands with the same degree of exposure showed slightly structural differences in the vertical zonation profiles. The geographic accidents (bays, coves and points) were abundant and the influence of wind and sunlight varies among sites (Vilas-Paz et al., 2005). In this way, diverse microhabitats are present in the Islas Cíes that influence the composition of species and the height of this species in the intertidal. Spatial heterogeneity in Islas Cíes may be a crucial factor in determining the rocky shore communities similar with Wilding et al. (2010) in Scotland, UK and Kostylev et al. (2005) in Hong Kong.

The dominant wind in the Islas Cíes comes from north in summer and from southwest in winter (Vilas-Paz et al., 2005). The winds in other rocky shores also create large

A)

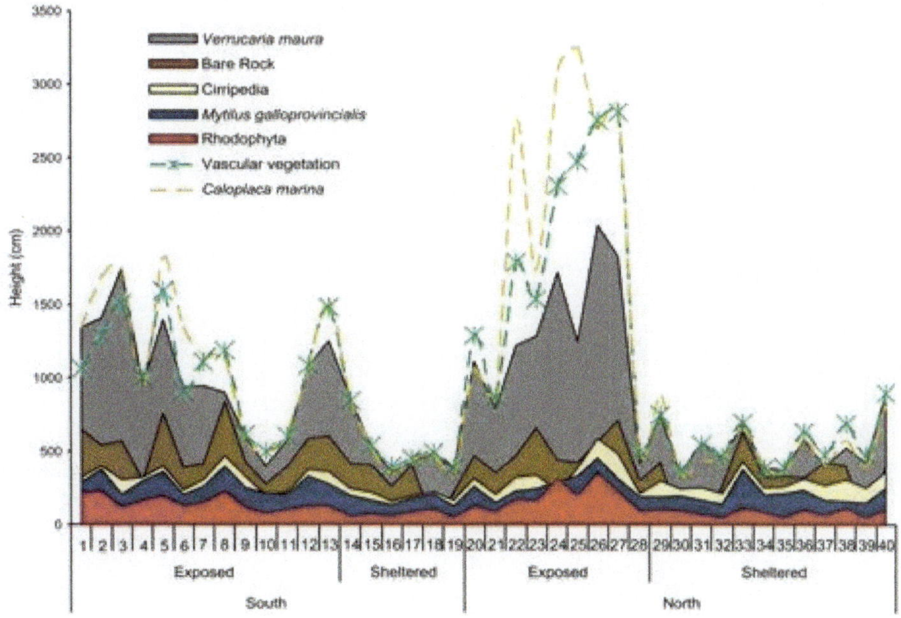

B)

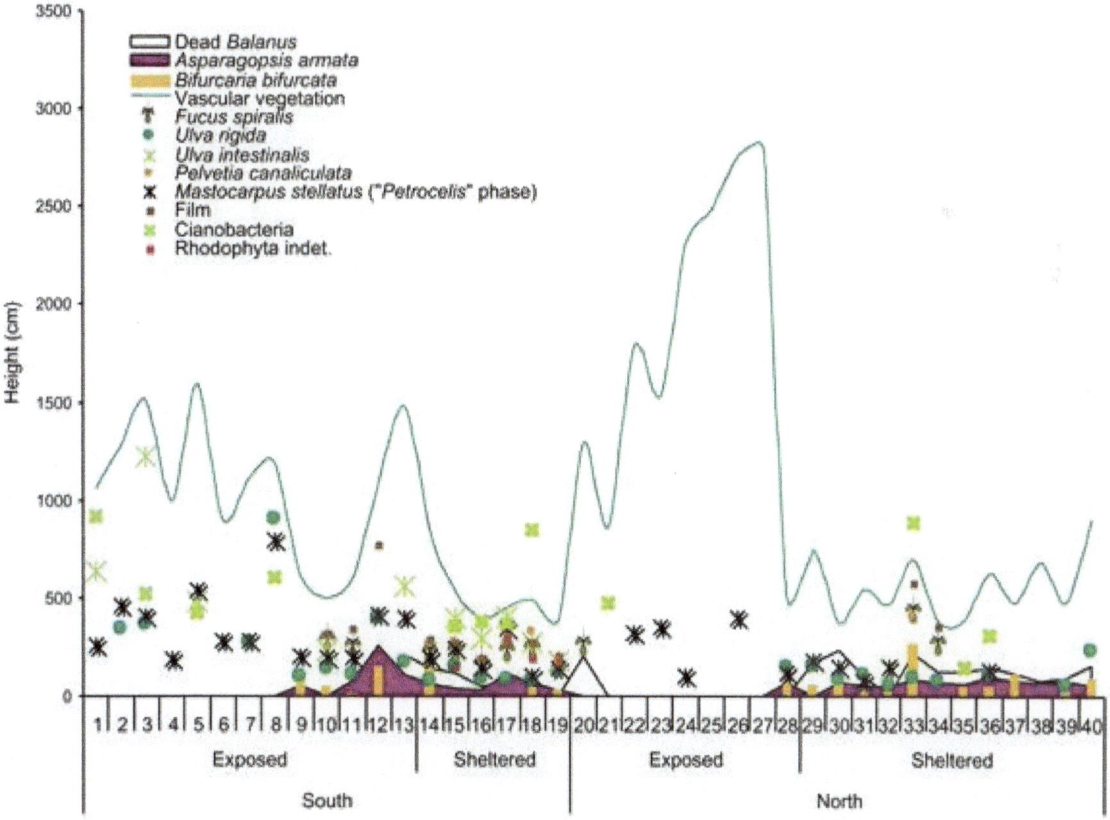

Figure 3. Vertical zonation of sessile organisms (mean height above the low tide of the upper and/or lower limits of distribution) by each section (1 to 40) of the rocky shores at the Islas Cíes (NW Spain, 2009). (A): organisms which form definite bands. (B): organism mainly in interrupted fringes or patches.

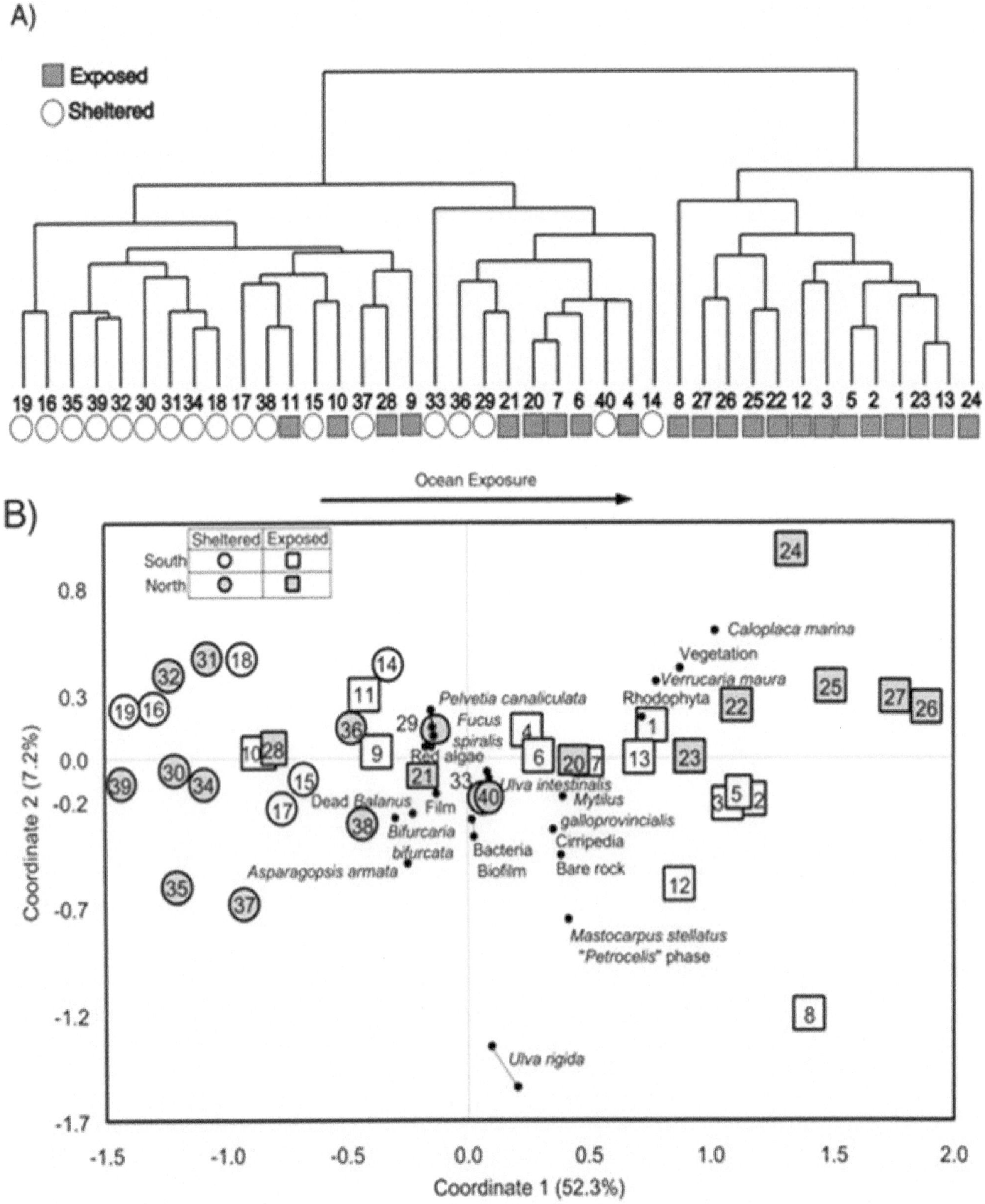

Figure 4. A: Cluster analysis by paired group clustering, and B: Principal Coordinate Analysis, both based in the Euclidean distance matrix of mean height above low tidal level of the rocky shore organisms by section. These analyses showed the similarity of the sections profiles of vertical zonation at the Islas Cíes (NW Spain). The black points in PCoA are the limit score for each organism. A line linked the upper and lower limit score in *P. canaliculata* and *U. rigida*.

large fetch that produce frequent and large waves (Little and Kitching, 1996). The eastern face of the Islas Cíes was sheltered from waves and strong wind astride the year. The sites 20 (Faro do Peito) to 22 (Punta do Cabalo) in north island were directly exposed to winds in summer and this might explain why fucoids were practically absent. Other factor that can explain the reduced population of fucoids in sites 21 to 22 are that in October an outflow current from the Ría of Vigo with high water turbulence reached the north mouth, near the shore of the north island in Cíes Archipelago (Souto et al., 2001).

The sheltered shore profiles were evident from site 14 (Faro do Bicos) to site 19 (Punta Pau de Bandeira) with high presence of fucoids. Niell (1977) found that *P. canaliculata* in Galicia was excluded from wind-beaten coasts. In fact, Jones and Demetropoulos (1968) indicated that *F. spiralis* does not tolerate heavy hydrodynamic conditions. Fucoids occur in the exposed face of the south island but in the Sites 10 and 11 at the Enseada de Concela, which is a cove sheltered from waves and southwest winds in winter.

The spray produced by the wave splash in the north face may help to promote the establishment of some intertidal organisms at higher heights in the rock cliffs (Little and Kitching, 1996). The lichen, *V. maura* reaches higher levels from the exposed site 24 (Punta do Chancelos) to 27 (Punta do Principe) in the north island. Moreover, the mostly terrestrial orange lichen, *C. marina* and the vascular vegetation occurred 24 m above from the low tide limit in this section of the shore. At this height, vascular plants such as *Armenia* and *Crithmum* grow as scattered pillow and can tolerate the high salinities that are present there (Vilas-Paz et al., 2005). The high humidity by the spray in the exposed faces increased the height in the shore of several opportunist species in the Islas Cíes. Patches of *Ulva intestinalis*, cianobacterial and *M. stellatus* occurs higher in this faces of the Archipelago.

At the very exposed intertidal shore in the Ría de Corme e Laxe (A Coruña, Spain; Niell, 1978), the red algae fringe and the mussel band is present from the low tide level to 3 m, which is in agreement with our results in the Islas Cíes. Above the 3 to 4.5 m mark, Niell (1979) also found barnacles and the black lichen, *V. maura*. In the exposed face of the Islas Cíes, the black lichen can, however, reach the 15 m mark above the low tidal level in most of the sites. The orange lichen, *C. marina* in the Ría de Corme e Laxe, occurs between 4.5 to 6.5 m and in the exposed face of the Islas Cíes occurs always above 10 m. The vertical profile of distribution of intertidal organisms at the Ría de Camariñas (A Coruña, Spain; Niell, 1978) is also similar to that of the exposed face of the Islas Cíes, being both characterized because *V. maura* and mussels reach higher intertidal levels than in other areas.

Conclusion

Four different landscape patterns of vertical zonation were found at the moment of the sampling in each of the island faces. The algae, *A. armata* and *B. bifurcata* characterized the sheltered north island. On the other hand, *P. canaliculata* and *F. spiralis* had a greater presence in the sheltered side of the south island. The lichens, *V. maura* and *C. marina*, and the Rhodophyta belt had a higher upper limit of distribution at the exposed sides of the north island. Bare rock and the crust phase of *M. stellatus* had a greater presence at the exposed side of the south island. The wave force and dominant winds were possibly factors in modeling the community structure. Moreover, transition zones, geographical accidents and shore slope, also can contribute to explain the differences in assemblages of species and the height in the rocky intertidal for each species. For similarities with previous studies some degree of extrapolation of general aspects of the community (position on the rock and dominance of the common species) can be made for zones that have not been sampled previously in North Atlantic Shores.

ACKNOWLEDGMENTS

We are grateful to our colleagues Marcelo Rodrigues Gómez, Cristian Aldea and Patricia Esquete of the Adaptations of Marine Animals Laboratory and ECIMAT, Universidad de Vigo for help with the field work. The second author received a partial financial support from the Ministerio de Ciencia y Tecnología (MICIT) and Consejo Nacional para Investigaciones Científicas y Tecnológicas (CONICIT) of the Government of Costa Rica.

REFERENCES

Acuña-Castroviejo R, Durán-Neira C, Rodríguez-Solórzano M, Sanjuán-López A (1984). Campañas de estudio del macrobentos infralitoral rocoso en el parque natural de las islas Cíes (NW de España). II. Actas IV Simp. Ibér. Estud. Benthos Mar., 1: 287-305.

Araújo R, Bárbara I, Sousa-Pinto I, Quintino V (2005). Spatial variability of intertidal rocky shore assemblages in the northwest coast of Portugal. Estuar. Coast. Shelf Sci., 64: 658-670.

Bernárdez P, Francés G, Prego R (2006). Benthic-pelagic coupling and postdepositional processes as revealed by the distribution of opal in sediments: The case of the Ría de Vigo (NW Iberian Peninsula). Estuar. Coast. Shelf Sci., 68: 271-281.

Besteiro-Rodriguez MC (2004). Evaluación del Impacto en las comunidades y especies de interés comercial y ecológico en las zonas costeras. Periodo Primaveral. 2003. Informe Final. Programa Nacional: Acción Urgente Prestige-Recursos Naturales. Ministerio de Ciencia y Tecnología. Galicia.

Bird E (2008). Coastal geomorphology: an introduction. 2ª ed. John Wiley and Sons. West Sussex, England. p. 411.

Chapman ARO (1974). The Ecology of Macroscopic Marine Algae. Ann. Rev. Ecol. Evol. Syst., 5: 65-80.

Clarke KR, Warwick RM (1994). Change in Marine Communities: An Approach to Statistical Analysis and Interpretation. Bourne, England, p. 144.
Connell JH (1972). Community interactions on marine rocky intertidal shores. Ann. Rev. Ecol. Syst., 3: 169-192.
Costas S, Alejo I, Alcántara-Carrió J (2000). Human Influence on the Evolution of Islas Cíes Sand Barrier (NW Spain). In: EUROCOAST (ed). Littoral 2002, The Changing Coast. EUROCOAST/EUCC. Porto, Portugal. pp. 283-287.
De La Huz R, Lastra M (2008). Effects of morphodynamic state on macrofauna community of exposed sandy beaches on Galician coast (NW Spain). Mar. Ecol., 29:150-159.
deCastro M, Gómez-Gesteira M, Prego R, Taboada JJ, Montero P, Herbello P, Pérez-Villar V (2000). Wind and Tidal Influence on Water Circulation in a Galician Ria (NW Spain). Estuar. Coast. Shelf Sci., 51: 161–176.
Diz P, Francés G, Rosón G (2006). Effects of contrasting upwelling–downwelling on benthic foraminiferal distribution in the Ría de Vigo (NW Spain). J. Mar. Syst., 60: 1–18.
Ellis DV (2003). Rocky shore intertidal zonation as a means of monitoring and assessing shoreline diversity recovery. Mar. Poll. Bull., 46: 305-307.
Fischer R (1981). Bioerosion of basalt of the Pacific coast of Costa Rica. Senckenb. Mar., 13:1-41.
García-García A, García-Gil S, Vilas F (2005) Quaternary evolution of the Ría de Vigo, Spain. Mar. Geol., 220: 153–179.
Hammer Ø, Harper DAT, Ryan PD (2001). Past: Paleontological Statistics Software Package for Education and Data Analysis. Palaeontol. Electron., 4: 1-9.
Jones WE, Demetropoulos A (1968). Exposure to wave action: measurements of an important ecological parameter on rocky shores on Anglesey. J. Exp. Mar. Biol. Ecol., 2: 46-63.
Kostylev VE, Erlandsson J, Ming MY, Williams GA (2005). The relative importance of habitat complexity and surface area in assessing biodiversity: Fractal application on rocky shores. Ecol. Complexity, 2: 272–286.
Legendre P, Gallagher ED (2001). Ecologically meaningful transformations for ordinations of species data. Oecologia, 129: 271-280.
Little C, Kitching JA (1996). The Biology of Rocky Shores. Oxford, England. p. 240.
Little C, Mettam C (1994). Rocky shore zonation in the Rance tidal power basin. Biol. J. Linnean Soc., 51: 169-182.
Little C, Smith LP (1980). Vertical Zonation on Rocky Shores in the Severn Estuary. Estuar. Coast. Shelf Sci., 11: 651-669.
Menge BA (1976). Organization of the New England Rocky Intertidal Community: Role of Predation, Competition, and Environmental Heterogeneity. Ecol. Monogr., 46: 355-393.
Menge BA, Daley BA, Wheeler PA, Dahlhoff E, Sanford E, Strub PT (1997). Benthic-pelagic links and rocky intertidal communities: Bottom-up effects on top-down control? Proc. Natl. Acad. Sci. USA, 94: 14530-14535.
Menge BA, Farrell TM (1989). Community structure and interaction webs in shallow marine hard-bottom communities: test of an environmental stress model. Adv. Ecol. Res., 19:189-262.
Menge BA, Foley MM, Pamplin J, Murphy G, Pennington C (2010). Supply-side ecology, barnacle recruitment, and rocky intertidal community dynamics: Do settlement surface and limpet disturbance matter? J. Exp. Mar. Biol. Ecol., 392: 160–175.
Mettam C (1994). Intertidal zonation of animals and plants on rocky shores in the Bristol Channel and Severn Estuary-the northern shores. Biol. J. Linnean Soc., 51: 123-147.
Niell FX (1977). Distribución y zonación de las algas bentónicas en la facies rocosa del sistema intermareal de las Rías Bajas Gallegas. Invest. Pesq., 41: 219-237.
Niell FX (1978). Catálogo florístico y fenológico de las algas superiores y cianofíceas bentónicas de las Rías Bajas Gallegas. Invest. Pesq., 42: 365-400.
Olabarria C, Rodil IF, Incera M, Troncoso JS (2009). Limited impact of *Sargassum muticum* on native algal assemblages from rocky intertidal shores. Mar. Environ. Res., 67: 153–158.
Pereira SG, Lima FP, Queiroz NC, Ribeiro PA, Santos AM (2006). Biogeographic patterns of intertidal macroinvertebrates and their association with macroalgae distribution along the Portuguese coast. Hydrobiologia, 555:185–192.
Pérez-Cirera JL (1975). Notas sobre la vegetación ficológica bentónica de la Ría de Cedeira (NO. de España). Anales Inst. Bot. Cavanilles, 32: 161-171.
Pérez-Cirera JL (1978). Notas sobre la vegetación bentónica del litoral de la Península Iberica. I. Las comunidades de *Ceramium shuttleworthianum* (Kutz.) Rabenh. y *Callithamnion granulatum* (Ducluz.) C. Ag. de zonas batidas del Atlántico. Anales Inst. Bot. Cavanilles, 35: 5-28.
Queralt S, Rodríguez-Fernández N, Bernabeu AM, Rey D, Vilas F (2002). Características morfosedimentarias de las playas de la Ría de Vigo: influencia de las condiciones hidrodinámicas. Geogaceta, 32: 211-214.
Quinn GP, Keough MJ (2003). Experimental Design and Data Analysis for Biologists. Cambridge Univ., Cambridge. p. 537.
Schiel DR (2004). The structure and replenishment of rocky shore intertidal communities and biogeographic comparisons. J. Exp. Mar. Biol. Ecol., 300: 309–342.
Sibaja-Cordero JA (2008). Vertical zonation in the rocky intertidal at Cocos Island (Isla del Coco), Costa Rica: A comparison with other tropical locations. Rev. Biol. Trop., 56 (2): 171-187.
Souto C, Fariña-Busto L, Alvarez E, Rodríguez I (2001). Wind and tide current prediction using a 3D finite difference model in the Ría de Vigo (NW Spain). Sci. Mar., 65 (Suppl 1): 269-276.
Underwood AJ (1981). Structure of a rocky intertidal community in New South Wales: patterns of vertical distribution and seasonal changes. J. Exp. Mar. Biol. Ecol., 51: 57-85.
Vilas-Paz A, Gamallo-Liste B, Framil-Barreiro J, Bonache-López J, Sanz-Ochoa K, Lois-Silva M, Toubes-Porto M (2005). Atlantic Islands of Galicia National Terrestrial-Maritime Park. Galicia. Visitor`s Guide. Autonomous Body for National Parks. Galicia. p. 303.
Wallenstein FFMM, Neto AI (2006). Intertidal rocky shore biotopes of the Azores: a quantitative approach. Helgol. Mar. Res., 60: 196-206.
Wallenstein FM, Neto AI, Álvaro NV, Santos CI (2008). Algae-based biotopes of the Azores (Portugal): spatial and seasonal variation. Aquat. Ecol., 42:547–559.
Wilding TA, Palmer EJL, Polunin NVC (2010). Comparison of three methods for quantifying topographic complexity on rocky shores. Mar. Environ. Res., 69: 143–151.

10

Insights on the environmental risk of metallic accumulation in sediments of Santos Harbor, Santos Estuarine System, Southeastern Brazil

Siqueira G. W.[1] and Aprile F. M.[2]*

[1]Laboratório de Planejamento e Desenvolvimento de Fármacos – LPDF, Campus Universitário do Guamá. Av. Augusto Corrêa n. 1, Universidade Federal do Pará, Belém, PA, # 66075-100, Brasil.
[2]Laboratório de Estudos de Ecossistemas Amazônicos - LEEA, Universidade Federal do Oeste do Pará, Campus Rondon. Av. Marechal Rondon s/n Caranazal, Santarém, PA, # 68040-070, Brasil.

The present study aims to evaluate the Santos harbor environmental health, the most important port of the South American, where anthropic activities are associated to metallic contamination (Al, Cr, Fe, Co, Ni, Cu, Zn, Cd and Pb). Enrichment factor, geo-accumulation index and geochemistry mobility were applied. The results showed increases in the levels of the metals with indicative of moderately to extremely pollution between the estuary and industrial park. The environment risk can become potentially high due to the transfer of metals to species of the trophic net.

Keywords: metal contamination, bioaccumulation, organic source, mangrove, estuary.

INTRODUCTION

The revitalization program of the Brazilian ports includes the expansion of the Port of Santos, the main hub for moving agricultural and industrial products in Brazil, located in the most industrialized and populous South American Port of Santos is inserted into a of the last areas of the country's coastal rainforest, surrounded by mangrove vegetation, which suffers strong pressure for the expansion of cities and industries around. The result is a continuous and intense forest clearing, with irregular occupation of areas of permanent protection, and production of waste and waste from urban and industrial, which contaminate soils, sediments and waters of the estuarine system.

Several domestic and industrial activities in the Santos Estuarine System (SES), such as shipping activities, refinery and fossil-fuel combustion effluents, discharge of untreated sewage by non-point sources, the input of urban drainage waters, irregular domestic and industrial landfills, among others (Medeiros and Bícego, 2004; Abessa et al., 2005; Luiz-Silva et al., 2006, 2008; Umbuzeiro et al., 2006; Bícego et al., 2006; Hortellani et al., 2008), have raised the risk of environmental degradation, especially due to the accumulation of hydrocarbons, organic matter and metals in the sediments. The present study aims to evaluate the Santos harbor environmental health due to diverse anthropic activities associated to discharge of metals (Al, Cr, Fe, Co, Ni, Cu, Zn, Cd and Pb) in the local.

*Corresponding author. E-mail: aprilefm@hotmail.com.

Description of the study area

Santos harbor (Figure 1), located in the SES in southeastern São Paulo State - Brazil, is economically the most important harbor of South America with one vital Cubatão Industrial Park – CIP located in the Cubatão River, which consists of approximately 1100 industries (Martins et al., 2007), including steel mill, fertilizer, chemical, petrochemical and metallurgical industrial areas and also cereals storage. The region still includes the cities of Santos, São Vicente and Cubatão, with approximately 874,000 inhabitants (CETESB, 2008), where the main economic activities are tourism, navigation and fishing. In the last forty years the port of Santos was one of the major examples of pollution and

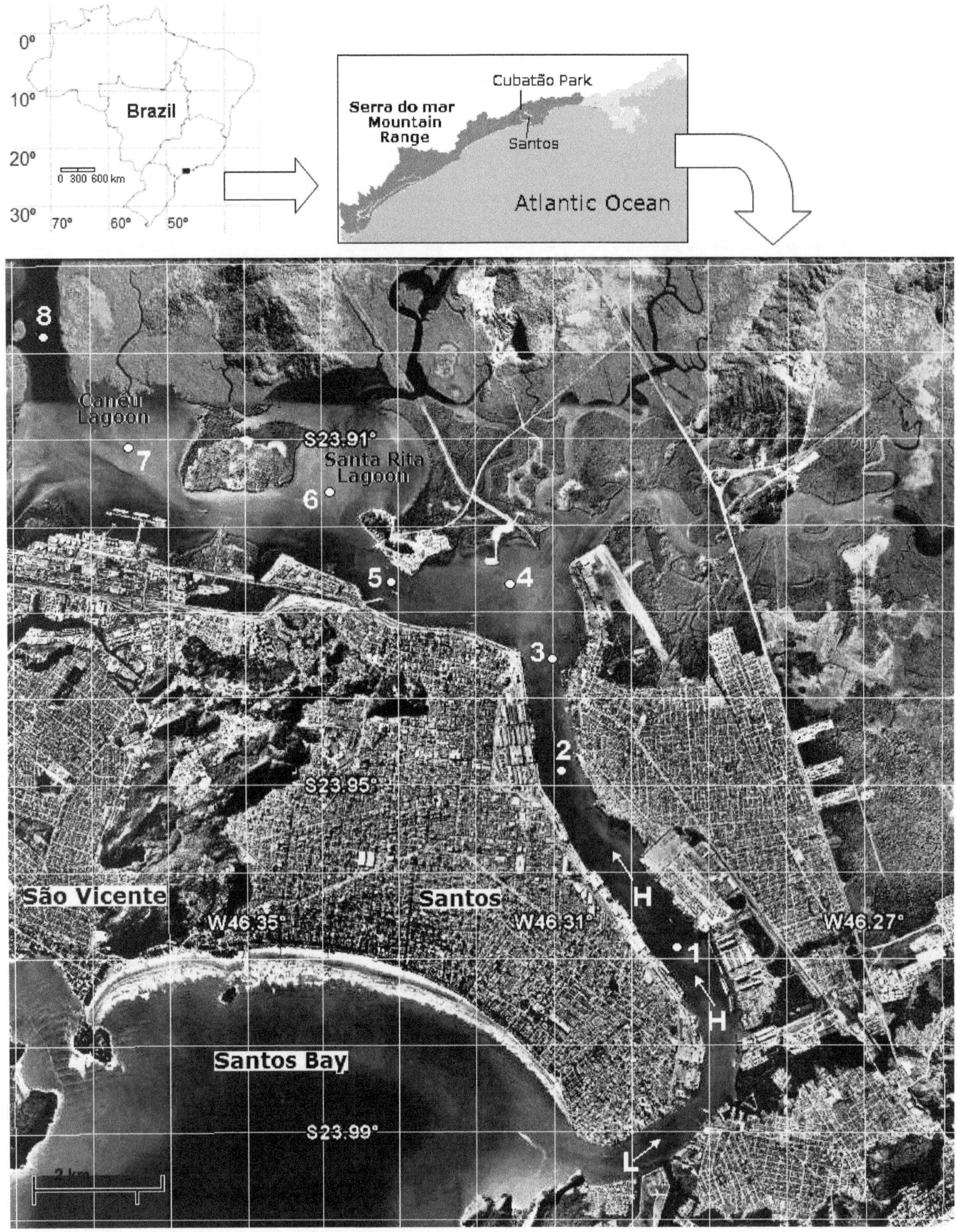

Figure 1. Santos Harbor in Santos Estuarine System with the sampling sites (Source: Google Earth 2010). Arrows: L=low and H=high tidal.

degradation of the environment, in special for the mangrove and estuary. Studies developed by Abessa et al. (2005) and Luiz-Silva et al. (2008) showed that the extensive and uncontrolled urbanization around the harbor area affects the environment quality, because several pollutants are emptied daily in the water system.

Table 1. Environmental risk factors to the heavy metals in sediments of Santos harbor.

	EF	I_{geo}	%BA	GM*		EF	I_{geo}	%BA	GM*
Sc	<1	0	<6	L	Ni	<1-5	3-6	30-90	H
Al	<1–2	0–1	15-20	L	Cu	<1–2	1	30–60	M
Cr	1–10	1–4	10–50	L-M	Zn	3–10	2-3	35–55	M–H
Fe	1–9	0–5	25-50	L-M	Cd	<1-3	6-7	40-70	M-H
Co	1-5	1-3	20-40	L	Pb	1-20	2–3	10–70	H

EF= Enrich factor; I_{geo}= geo-accumulation index; BA= bioavailable fraction of the metals; GM= geochemistry mobility. *L=low; M=medium; H=high.

The domestic sewage that outfalls from the cities, around Santos Harbor, represent a substantial contribution of organic matter to the water column and sediments, while the industrial effluents and emissions come mainly from the CIP and represent primary source of metals.

MATERIALS AND METHODS

Eight sampling sites of sediment were selected along the port system (Figure 1) seasonality in the last ten years in order to determine metallic concentrations (Al^{13}, Cr^{24}, Fe^{26}, Co^{27}, Ni^{28}, Cu^{29}, Zn^{30}, Cd^{48} and Pb^{82}). The sediment fractions were homogenized and passed through 0.063 mm sieve, and then the samples were digested with (HNO_3 + HCl)-$HClO_4$-HF (Merck 2.5:1:10) in an airtight Teflon system, heated up by six hours for 100°C, and diluted to 100 ml (APHA, 2005). A weak acid digestion was also accomplished being used of tubes of polyethylene of 50 ml with HCl 0.1 mol.L^{-1} for 24 h with continuous agitation. The product of that weathering was left in rest by 24 hours for sedimentation, later centrifuged and filtrate in Whatman 44 filter paper. The extracts obtained were transferred for 25 ml volumetric balloons with HCl (1%) + Milli-Q water solution. The metallic concentrations were determined by flame atomic absorption spectrometry using Perkin-Elmer AAS 3300 with air-acetylene flame. Quantification of heavy metal concentrations was based upon calibration curves involving standard solutions prepared from a commercial stock solution (trace metals 1 ICM-411H in 5% HNO_3, Radian International LLC) and use of standard seawater reference materials (CASS 1-4) from the National Research Council of Canada, and Estuarine Sediment (nº1646a) from NIST Standard Reference Materials. The detection limits were 1.00 Al; 0.01 Cr, Fe, Ni and Pb; 0.002 Co; 0.001 Cu; 0.03 Zn and 0.005 Cd mg mL^{-1}.

Normalization is a procedure very common to detect and to quantify irregular concentrations of metals in sediment samples. The extent of sediment contamination was assessed using the EF and geo-accumulation index (I_{geo}). According to EF technique metal concentrations were normalized to the textural characteristic of sediments. The widely used element to normalize the metals in sediments is the Al (Woitke et al., 2003; Selvaraj et al., 2004). Other option usually adopted is the Fe as the element to normalize (Aprile and Bouvy, 2008). Unfortunately, both elements have problem of variation in its levels in Santos Estuary to be used as element to normalize the metals concentrations. Abessa et al. (2005) showed there to be variability in the Al concentrations to the Santos coast, while Martins et al. (2007) and Luiz-Silva et al. (2008) showed that Fe concentrations in the Santos Estuary are historically influenced by anthropogenic inputs from metallurgical industries and fossil fuel combustion. Therefore, in this work we chose the aluminasilicates-silicate (Sc) because we believe this is more suitable to normalize the metals concentrations in the SES, besides the silicates are the predominant contents of costal-marine sediments, and the main constituent of the thin mud where sediments are associated to trace metals. The EF expression is defined according to Woitke et al. (2003) as:

$$EF = \frac{(x/Sc)_{sediment}}{(x/Sc)_{crust}} \tag{1}$$

where: X/Sc is the ratio of the metals (X) to the aluminasilicates-silicate (Sc). The I_{geo} values for the metals studied were calculated using the Muller's (1979) expression:

$$I_{geo} = \log_2\left(\frac{C_n}{1.5B_n}\right) \tag{2}$$

where: C_n is the measured content of element 'n', and B_n is the geochemistry background.

The bio-available fraction of the metals is the main environmental risk for aquatic ecosystems. Metals determined are considered potentiality bio-available, and an increase of the oxygen levels in the water column allows a remobilization of the metallic elements to the trophic net (Förstner and Wittmann, 1983). The percentage of bio-available metals in the sediments describe by Siqueira (2003) was determined by the expression:

$$\%BA = \frac{MC}{TC}100 \tag{3}$$

here: MC is the concentration of the mobile phase and TC the total concentration of the metal in the samples.

RESULTS AND DISCUSSION

The EF, I_{geo}, percentage bio-available (%BA) and geochemistry mobility (GM) of the metals are summarized in Table 1. The aluminasilicates (Sc) showed low mobility with BA below 6% indicating a conservative behavior for the system, unlike the elements Al and Fe that were not conservatives. Considering the EF, all the analyzed metallic elements presented additional source of metallic contribution for the compartment sediment, besides the natural source originating from of the rocks. The elements Al, Cu and Cd had modest contribution local. However, the elements Co and Ni, and

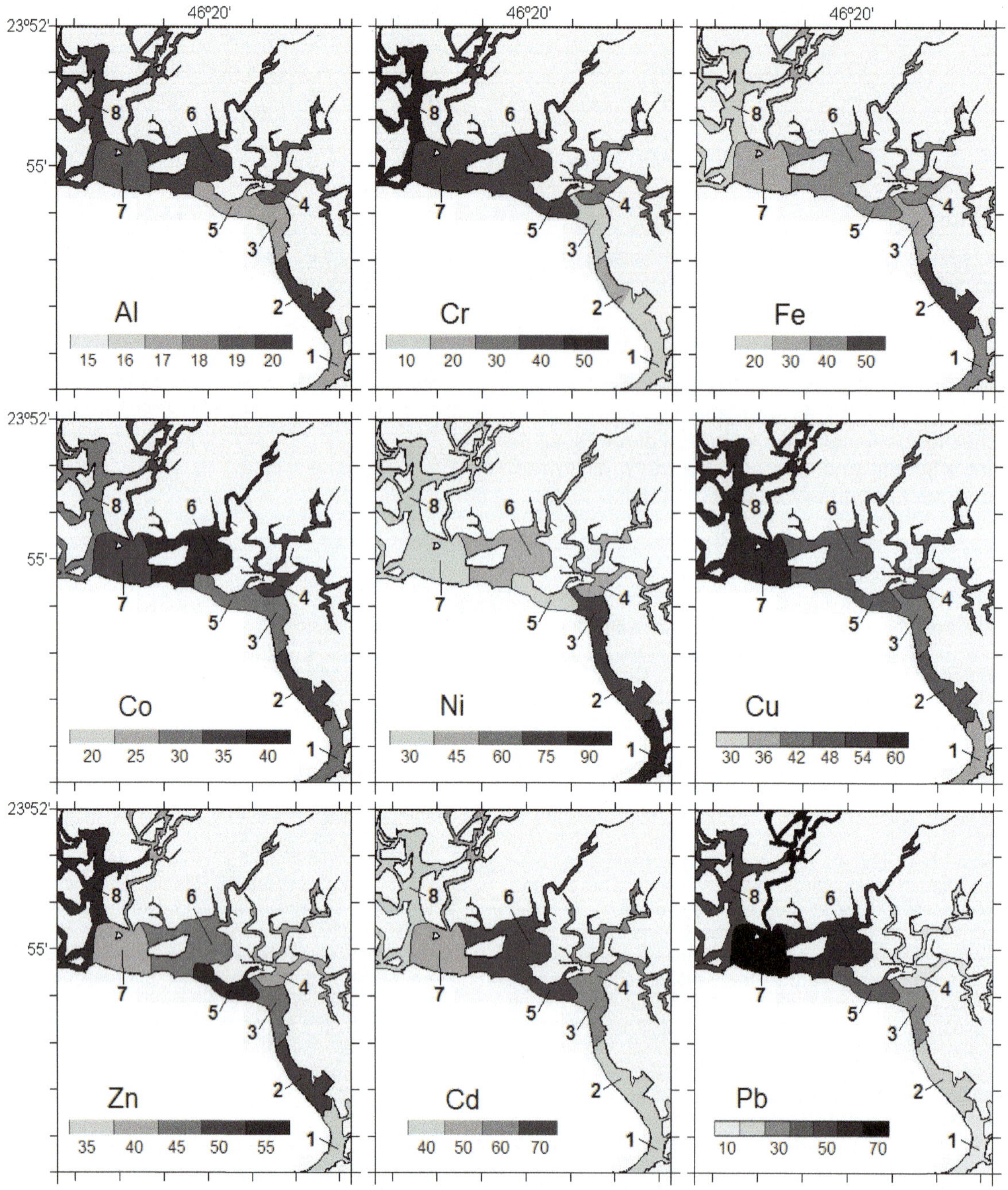

Figure 2. Environmental risk analysis in the Santos harbor based in the percentage of bio-available of heavy metals in sediments.

more intensely Cr, Fe, Zn and Pb presented strong evidences of metallic contamination of the sediments for external discharges, typically originating of anthropogenic inputs from local industries. The BA fraction of heavy metals in sediments is showed in Figure 2. Cr and Fe presented from low to medium geochemistry mobility with medium BA fraction of until 50%. On the other hand, Pb, Cd and Ni presented elevate mobility with medium to high BA fraction of until 70% for Pb and Cd and 90% for Ni, mainly close of the container's area of the port zone, in Santos channel. That means to say that the three metallic elements are very mobile in the aquatic environment, and they are, generally, enriched in sediments leached of the subtropical areas. The estuarine system also has some contribution in the mobility and transport of those metallic elements, especially due to marine dynamic and the action of the high tides in the Santos bay (Figure 1).

The I_{geo} index demonstrated that the quality of the sediment is variable depending on the type of metallic contribution and of the site of discharges. Al and Cu had low I_{geo} indicating environment not polluted, while the Zn and Pb had indicative of moderately polluted; Cr strongly polluted, and Fe, Ni and Cd extremely polluted in some sites between the estuary and the industrial park. That is an indicative of metallic contamination by anthropic activities in the region, and the spatial variability indicates that the contamination is punctual in special in the whole port zone in association with the industrial waste from Cubatão Industrial Park (located the West of the site 8, Figure 1). Metals with magnetic properties, such as Pb and Zn, occur as contaminants in the effluents of petrochemical plants, as a waste product of oil refining into the Santos Harbor. Most part of the chemical wastes from anthropogenic activities, which including toxic organic and inorganic elements, contribute to the degradation of aquatic environments. This is particularly important for the estuarine systems that are constantly affected by multiple contamination sources.

Particularly interesting it is the geochemistry behavior of the Pb, with high mobility and EF, Pb is connected to the structures of the silicates in the collected samples, what represents a serious problem in periods of drainage of the main channel of the harbor, when the (re)suspension of the sediments elevate the Pb levels potentially in the water column. High sediment deposition rates content Pb-Fe could reflect anthropogenic activities vicinity to the port zone. Pb is toxic for the whole aquatic biota and although that element is not considered one of the metals of larger geochemistry mobility in the aquatic environment, there are still considerable evidences showing the bio-available of the Pb associated to sediments for species provide by the sediment compartment. The metals can be associated to the organic matter (complex or colloidal forms) or cations, precipitated and/or co-precipitated with oxi-hydrate and hydroxides of Fe and Mn, as well as absorbed in the fine sediments as silt and clay particles (Förstner and Wittmann, 1983). The bioaccumulation of the Zn in aquatic organisms depends on the trophic levels, and it is largely distributed in the benthonic and pelagic organisms, especially in the carnivorous species.

A decreasing gradient was observed in the Cu levels from the channel to the estuary (Figure 2), indicating absence of stocks of Cu in the sampling sediments. Cu is present in industrial waste effluents, sanitary embankments and sewage discharged into the SES. The main hazard of the Cu for the aquatic organisms is in the fact that the insoluble form can be easily transformed in soluble for the simple alteration of the pH in waters with low salinity (Förstner and Witmann, 1983), allowing that the metallic element can be incorporated in the tissue of the aquatic organisms very fast and in high levels.

The increase of the salinity in the channel due to entrance of water from the sea, in high tides periods, it blocked the solubility of most part of the metals studied. However, in the low tides as well as in the periods of strong precipitation, when waters of low salinity prevail in the channel, it is possible to determine the presence of metallic elements in its soluble forms, with largest mobility and bioavailable in the aquatic organisms.

The results showed high environmental risk for the elements Zn and Pb, and a situation of alert for the Cu. The intense leaching process that occurs in the Santos Harbor is due to the transport of sediments from the highest areas (Serra do Mar Mountain Range), as well as deforesting activities, and erosion of soils from the mangroves. Both ocean and inland-related exploitations have produced environmental impacts, such as pollution and coastal erosion (Harari and Camargo, 2003), and the suspended sediments are transported by the channel precipitating close of the seaport area. As a consequence, the channel is dredged periodically, revolving the sediments of the bottom and liberating metallic elements to the water column. Thus, this process is highly harmful for the environment because it is possible metals reacting with the organisms of the trophic net. Sediments may accumulate contaminants in concentrations higher than those observed in the water column, producing negative effects to the benthic biota and to the organisms that feed on the benthos or on the sediment (Cesar et al., 2007).

Although the local environment quality has improved in the last decade in relation to the chemical parameters and toxicity due to low enforcement actions, the sediment still is contaminated by metals. Non-stop activities of sediment dredging from the central channel of the port allow the (re)suspension of sediments with metallic elements and it is useful to be absorbed by the biota (bioaccumulation process). The accumulation of metals in benthonic organisms, phytoplankton, zooplankton, and fish is dangerous to the traditional communities of fishermen that bring fish to consumption into the city of Santos, São Vicente and Guarujá. Thus, the environmental risk can become potentially high due to the transfer of metals to humans as well as to top species on the highest levels of the trophic net.

REFERENCES

Abessa DMS, Carr RS, Rachid BRF, Sousa ECPM, Hortelani MA, Sarkis JE (2005). Influence of a Brazilian sewage outfall on the toxicity and contamination of adjacent sediments. Mar. Pollut. Bull., 50: 875-885.

APHA - American Public Health Association, AWWA - American Water Works Association, WPCF - Water Pollution Control Federation(2005). Standard Methods for the Examination of Water and Wastewater. American Public Health Association, Maryland.

Aprile FM, Bouvy M (2008). Distribution and enrichment of heavy metals in sediments at the Tapacurá river basin, Northeastern Brazil. Braz. J. Aquat. Sci. Technol., 12(1): 1-8.

Bícego MC, Taniguchi S, Yogui GT, Montone RC, Silva DAM, Lourenço RA, Martins CC, Sasaki ST, Pellizari VH, Weber RR (2006). Assessment of contamination by polychlorinated biphenyls and aliphatic and aromatic hydrocarbons in sediments of the Santos and São Vicente Estuary System, São Paulo, Brazil. Mar. Pollut. Bull., 52: 1784-1832.

Cesar A, Choueri RB, Riba I, Morales-Caselles M, Pereira CDS, Santos AR, Abessa DMS, DelValls TA (2007). Comparative sediment quality assessment in different littoral ecosystems from Spain (Gulf of Cadiz) and Brazil (Santos and São Vicente estuarine system). Environ. Int., 33: 429-435.

CETESB - Companhia de Tecnologia de Saneamento Ambiental (2008). Report of quality of the littoral waters of the São Paulo State 2007. São Paulo, CETESB. pp.537 (in Portuguese)

Förstner U, Wittmann M (1983). Metal Pollution in the Aquatic Environment. Berlim, Springer-Verlag, p. 486.

Harari J, Camargo R de (2003). Numerical simulation of the tidal propagation in the coastal region of Santos (Brazil, 24ºS 46ºW). Cont. Shelf. Res., 23: 1597-1613.

Hortellani MA, Sarkis JES, Abessa DMS, Sousa ECPM (2008). Assessment of metallic element contamination in sediments from the Santos - São Vicente Estuarine System. Quim. Nova, 31(1): 10-19 (in Portuguese).

Luiz-Silva W, Machado W, Matos RHR (2008). Multi-Elemental Contamination and Historic Record in Sediments from the Santos-Cubatão Estuarine System, Brazil. J. Braz. Chem. Soc., 19(8): 1490-1500.

Luiz-Silva W, Matos RHR, Kristosch GC, Machado W (2006). Spatial and seasonal variability of the concentration of trace elements in sediments of the Santos-Cubatão Estuarine System (SP). Quim. Nova, 29: 256-263 (in Portuguese).

Martins CC, Mahiques MM, Bícego MC, Fukumoto MM, Montone RC (2007). Comparison between anthropogenic hydrocarbons and magnetic susceptibility in sediment cores from the Santos Estuary, Brazil. Mar. Pollut. Bull., 54: 240-246.

Medeiros PM, Bícego MC (2004). Investigation of natural and anthropogenic hydrocarbon inputs in sediments using geochemical markers. I. Santos, SP—Brazil. Mar. Pollut. Bull., 49: 761-769.

Muller G (1979). Schwermetalle in den sediments des Rheins-Veranderngen seitt 1971. Umschan, 79: 778-783.

Selvaraj K, Ram Mohan V, Szefer P (2004). Evaluation of metal contamination in coastal sediments of the Bay of Bengal, India: geochemical and statistical approaches. Mar. Pollut. Bull., 49: 174-185.

Siqueira GW (2003). Study of the levels of heavy metals and other elements in surface sediments of Santos Estuarine System (Baixada Santista/São Paulo) and Amazon Continental Shelf (North continental edge of Brazil). Thesis, Oceanography Institute, São Paulo., p.386 (in Portuguese).

Umbuzeiro GA, Kummrow F, Roubicek DA, Tominaga MY (2006). Evaluation of the water genotoxicity from Santos Estuary (Brazil) in relation to the sediment contamination and effluent discharges. Environ. Int., 32: 359-364.

Woitke P, Wellmitz J, Helm D, Kube P, Lepom P, Litheraty P (2003). Analysis and assessment of heavy metal pollution in suspended solids and sediments of the river Danube. Chemosphere, 51: 633-642.

Study on Mg^{2+} removal from ammonium dihydrogen phosphate solution by a novel emulsification extraction process

JianHong Luo, Jun Li*, Yang Jin and DongSheng Zheng

Department of Chemical Engineering, Sichuan University, Chengdu, Sichuan, 610065, People's Republic of China.

The emulsification extraction process (EEP) is a novel method for separating solutes from aqueous solution, and for its high extraction ratio, the EEP presents huge potential for extraction from dilute solution. In this paper, the EEP is successfully generated with mixing solutions containing aqueous and organic phases at high speed. The extraction of Mg^{2+} from ammonium dihydrogen phosphate ($NH_4H_2PO_4$) solution is studied by using kerosene as a solvent, di(2-ethylhexyl)phosphoric acid (D2EHPA) as an extractant. To study the extraction ratio and advantages of the EEP process in the removal of Mg^{2+}, various parameters - D2EHPA volume fraction, reaction temperature, phase volume ratio, initial pH of $NH_4H_2PO_4$ solution, stirring speed, stirring time were studied and optimized. The results show that Mg^{2+} in $NH_4H_2PO_4$ solution can be effectively removed by EEP process. An extraction ratio of more than 90.1% is attained at the optimized parameters and superior-grade $NH_4H_2PO_4$ (MAP) can be obtained by two levels of extraction.

Key words: Emulsification extraction process, di(2-ethylhexyl)phosphoric acid, $NH_4H_2PO_4$, Mg^{2+}.

INTRODUCTION

MAP, which is the common name of $NH_4H_2PO_4$ is used as flame retardant and drip-irrigation fertilization, which needs highly pure MAP mainly manufactured with thermal-process phosphoric acid in the past. The cost of thermal-process phosphoric acid is very high. The yellow-phosphorus manufacturers are closed because of the pressure from the energy consumption and environment protection. The cost of yellow phosphorus, as a basic raw material of thermal-process phosphoric acid is becoming higher and higher. So the low cost of wet-process phosphoric acid (WPA) is gradually being paid attention in recent years. However, there are some undesirable impurities (Fe^{3+}, Al^{3+}, Mg^{2+}) in WPA. They will lower the quality of MAP products.

To get the superior grade MAP, WPA should be purified. Improving the pH of the solution, usually between 4 to 4.5, can remove most of the metal ions, but there still are some Mg^{2+} which can cause formation of troublesome water-insoluble substance in the following concentration and crystallization process. The main ingredients of water-insoluble substance are magnesium-containing phosphate. Therefore, the Mg^{2+} must be removed, before concentrating the neutralized MAP solution. Several methods based on solvent extraction (McCullough, 1976; Lo et al., 1983; Bradford, 1977; Kenneth et al., 1972; Miki et al., 1997; Mohammad et al., 1997; Hannachi et al., 2007; Wang et al., 2000) are used to remove Mg^{2+} in phosphorus chemical industry; however, conventional solvent extraction suffers from two disadvantages. It needs a mixing-setting stage, and it requires a high phase volume ratio of solvent/water below which extraction is poor.

*Corresponding author. E-mail: ljh00308033@163.com.

Furthermore, because of the nature of the mixing, sometimes there is the undesirable possibility of the formation of a third colloidal phase which is difficult to eliminate. Faced with these problems, a novel technique of EEP can solve them efficiently. The idea of EEP (the emulsification extraction pocess) originated from the emulsification phenomena which is often produced in conventional solvent extraction process. However, it is well known that the emulsification can form an enormous interfacial area as in Figure 1, which can lower the phase volume ratio of solvent/water and where a transfer of solute from one phase to another can occur very rapidly with a minimum energy requirement and shorten the equilibrium time of extraction process. So, how to make full use of the advantages of emulsification phenomena is considered in this work. In this study, the removal of Mg^{2+} from ammonium dihydrogen phosphate with EEP is investigated. The aim of this work is to experimentally study the effects of various factors on the extraction ratio using EEP.

MATERIALS AND METHODS

Materials and equipment

The solvent used in this work is kerosene. D2EHPA is employed as an extractant produced by Luo yang Zhongda Chemical Company (China) (AR grade). Pure water is produced by Aquapro making-water machine (ABZ1-1001-P) in our laboratory. Emulsification generator using a high-speed stirrer (JRJ-300-I) produced by Shanghai Specimen and Model Factory (China).

Parameters that could affect the EEP process

To study Mg^{2+} removal efficiency and advantages of the EEP (the emulsification extraction pocess) process, it is necessary to optimize various parameters that could affect the process. The parameters to be optimized are the D2EHPA volume content, the reaction temperature, the phase volume ratio, the stirring speed, the stirring time, the phase volume ratio and the initial pH of $NH_4H_2PO_4$ solution.

Extraction experiment conditions

Solvent: kerosene, D2EHPA volume content: 50%, phase ratio (aqueous/organic) = 5:1, stirring speed: 4500 r/min, initial Mg^{2+} concentration: 600 mg/L, initial pH of ammonium dihydrogen phosphate: 4.5, stirring time: 60 s, reaction temperature: 55℃.

Analysis

The concentration of Mg^{2+} is determined by atomic absorption spectrophotometry (GF3000).

RESULTS AND DISCUSSION

Extraction ratio (E) is defined as follows:

$$E = \frac{M_{(i)}^{Mg^{2+}} - M_{(r)}^{Mg^{2+}}}{M_{(i)}^{Mg^{2+}}} \times 100\% \quad (1)$$

E represents the efficiency of EEP; $M_{(i)}^{Mg^{2+}}$: Mole of Mg^{2+} in initial solution, mol;

$M_{(r)}^{Mg^{2+}}$: Mole of Mg^{2+} in the raffinate, mol.

Effect of D2EHPA volume fraction (%)

Increasing D2EHPA concentration in solvent phase will increase the amount of extractant, so the numbers of free extractant taking participate in the extraction reaction will also increase. However, when the D2EHPA concentration increases to a certain value, the extraction ratio (E) remains almost unchanged, because the extraction reaction reaches equilibrium. Therefore, the extraction ratio (E) increases as shown in Figure 2.

Effect of phase ratio

The phase ratio (aqueous/organic) has a significant effect on extraction ratio (E) and entrainment. This effect is studied by changing the phase ratio (aqueous/organic) from 5:1 to 1:1. The results presented in Figure 3 clearly show that, a phase ratio of 1:1 gives the best extraction of Mg^{2+}. The reason (Wang et al., 2000) for this is that, for the fixed Mg^{2+} concentration in solution, increasing phase ratio in extraction process will enhance the amount of solvent and extractant.

Effect of stirring time

The extraction of Mg^{2+} by emulsification extraction and solvent extraction are presented in Figure 4. Figure 4 indicates that the extraction ratio (E) of Mg^{2+} is improved with the increase of stirring time. The equilibrium time for the emulsification extraction of Mg^{2+} is observed to be 100 to 120 s, which is far less than the solvent extraction equilibrium time of about 30 min. What is more surprising is that, the extraction ratio (E) of Mg^{2+} by emulsification extraction is higher than solvent extraction. This is because EEP (the emulsification extraction process) make the reaction of Mg^{2+} with D2EHPA more complete than solvent extraction. Therefore, this means that the equilibrium time for emulsification extraction needs 120 s to ensure completion.

Effect of initial pH of ammonium dihydrogen phosphate solution

The results of the experiments are shown in Figure 5.

Figure 1. Samples of emulsion liquid are examined by a microscopic camera.

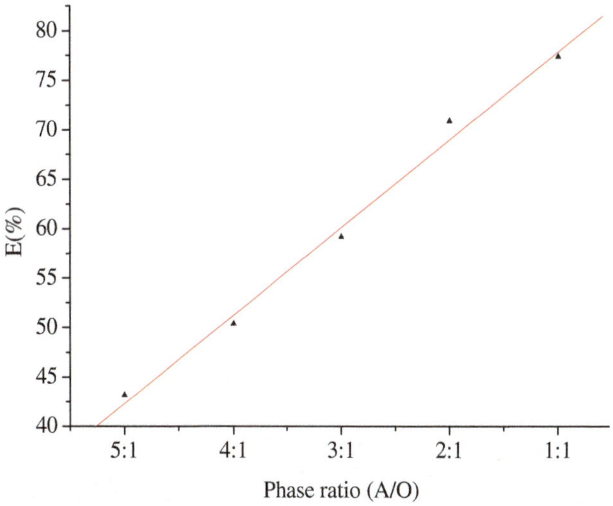

Figure 3. The extraction ratio (E) versus the phase ratio (A/O).

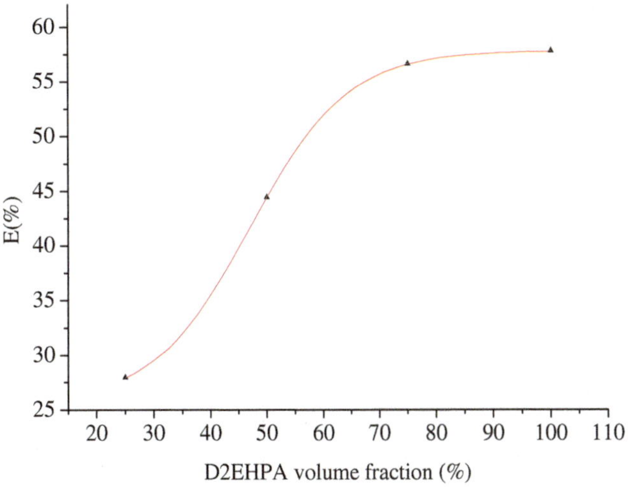

Figure 2. The extraction ratio (E) versus the D2EHPA volume fraction (%).

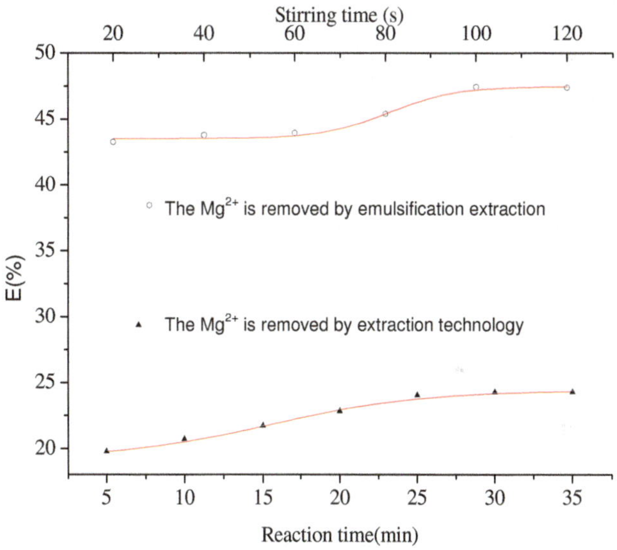

Figure 4. The extraction ratio (E) versus the stirring time.

The extraction of Mg^{2+} reach a maximum at the pH 4.5 where the formation of an ion-pair in ammonium dihydrogen phosphate media is favorable. It was observed that, with a further increase in pH to 5.0, Mg^{2+} extraction decreases gradually. At higher pH, there is a possibility of hydrolysis of the Mg^{2+} ion-pair complex. The lower values obtained at higher acid concentrations (pH<3.5) are due to competition between the extractable Mg^{2+} and $H_2PO_4^-$ which predominates in ammonium dihydrogen phosphate media at low pH.

Effect of reaction temperature

From Figure 6, the extraction ratio (E) increases as the temperature rises up, which shows that the extraction of Mg^{2+} with D2EHPA is endothermic.

Effect of stirring speed

As the stiring speed increases, a smaller globule size will lead to a larger transfer interfacial area between the feed and the solvent phase. This increased transfer area allows the extraction to occur at a higher rate. However, Figure 7 shows that the emulsification speed does not affect the extraction ratio (E) too much when the stiring speed reaches 4500 r/min. Therefore, the optimum value for stiring speed is found to be 4500 r/min.

Examination

A kind of practical wet-process phosphoric acid is

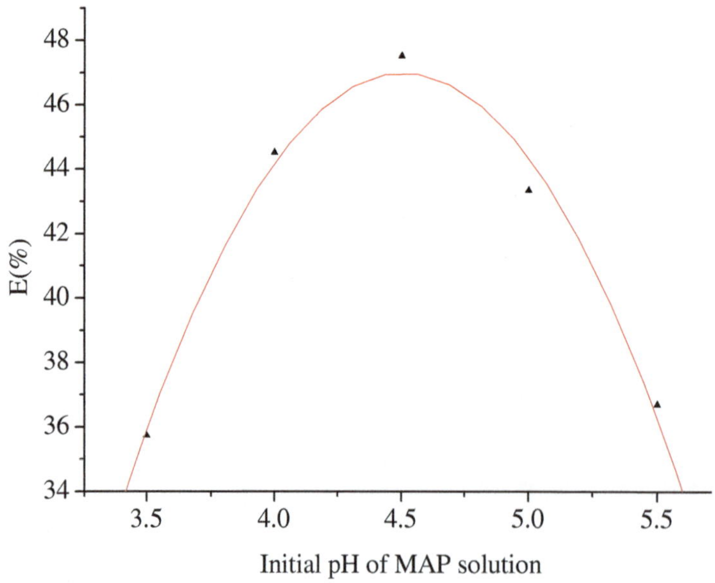

Figure 5. The extraction ratio (E) versus the initial pH of ammonium dihydrogen phosphate solution.

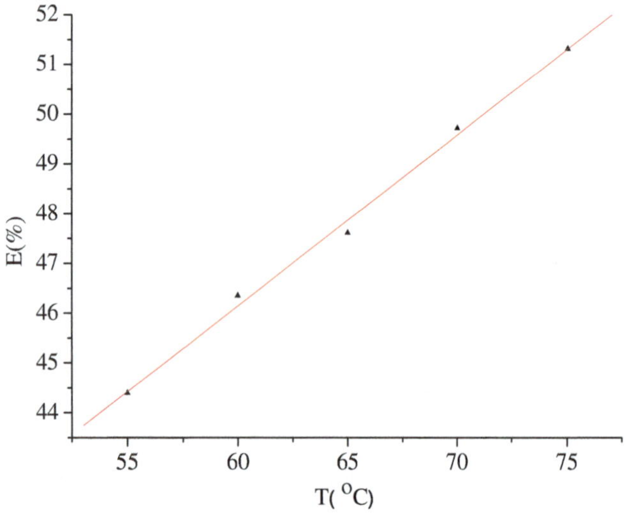

Figure 6. The extraction ratio (E) versus the reaction temperature.

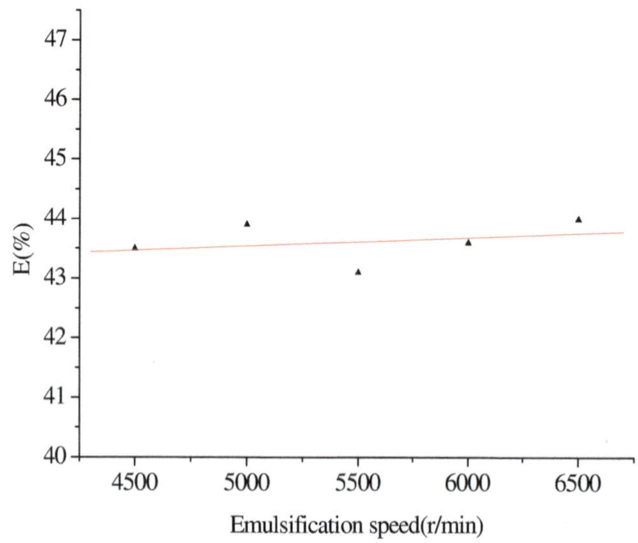

Figure 7. The extraction ratio (E) versus the emulsification speed.

neutralized to pH=4.5 with ammonia and filtered. The neutralized solution is then extracted under the aforementioned optimal technology conditions. The superior grade MAP is produced by the following concentration and crystallization process using the extracted solution. As Table 1 shows, the food grade MAP can be obtained.

Conclusion

Based on the results of this research on the removal of Mg^{2+} from ammonium dihydrogen phosphate solution with EEP (the emulsification extraction process), the following specific conclusions can be drawn:

1. EEP treatment can be a more effective method for the removal of Mg^{2+} from ammonium dihydrogen phosphate solution than solvent extraction.
2. The optimized parameters affecting the process are as follows: The D2EHPA volume fraction: 75%, the stirring time: 100s, the reaction temperature: 65℃; the phase volume ratio: 1:1; the stirring speed: 4500 r/min, and the initial pH of $NH_4H_2PO_4$ solution: 4.5.

Table 1. Composition of $NH_4H_2PO_4$ product.

Item	N (%)	P_2O_5 (%)	Fe^{3+} (%)	Mg^{2+} (%)	Al^{3+} (%)	Heavy metal (Pb) %
Mass fraction	≥12	≥60.5	≤0.0003	≤0.0010	≤0.0002	≤0.0005
Item	**As (%)**	**F^- (%)**	**SO_4^{2-} (%)**	**pH**	**H_2O (%)**	**Water-insoluble substance (%)**
Mass fraction	≤0.0090	≤0.0085	≤0.0020	4.5-4.8	≤0.2	≤0.05

ACKNOWLEDGEMENTS

The authors gratefully acknowledge the financial support of this work by eleventh five year national key technology R&D program (No.2008BAE58B01) and the New Century Excellent Talents of Ministry of Education (No. NCET-07-0577), the People's Republic of China.

REFERENCES

McCullough JF (1976). Phosphoric acid purification: comparing the process choices, Chem. Eng., 83(26): 101-103.

Lo TC, Baird MH, Hanson C (1983). Handbook of Solvent Extraction. Wiley. New York, 3: 12.

Bradford (1977). Method for purification of phosphoric acid high purity polyphosphoric acid. JP, p. 4,053,564.

Kenneth A, Williams La Mirada, Calif, David R (1972). Extraction process for purification of phosphoric acid. JP:, pp. 3, 694, 153.

Miki M, Hiroshi S, Isao K (1997). Formation of Reverse Micelles and Extraction of Aluminum and Beryllium by Mixture of Bis (2-Ethylhexyl) Phosphoric Acid and Dinonylnaphthalene Sulfonic Acid. J. Chem. Eng. Japan. 30(4): 706-711, 712-717.

Mohammad I, Ei-Khaiary (1997). Separation and Purification of Technology, 12: 13-16.

Hannachi A, Diamonta H, Chtaraa C, Ratela A (2007). Purification of wet process phosphoric acid by solvent extraction with TBP and MIBK mixtures. Separation Purif. Technol., 55: 212-216.

Wang YD, Cheng M, Xu LL, Dai YY (2000). Removal of Phenol from Dilute Solutions by Predispersed Solvent Extraction. Chin. J. Chem. Eng., 8(2): 103-107.

12

Macrobenthos of shelf zone off Dhamara estuary, Bay of Bengal

Debasish Mahapatro[1]*, R. C. Panigrahy[2], S. Naik[2], S. K. Pati[3] and R. N. Samal[1]

[1]Chilika Development Authority, Bhubaneswar, Odisha, India.
[2]P.G. Department of Marine Sciences, Berhampur University, Berhampur -760007, Odisha, India.
[3]Zoological Survey of India, WRC, Pune, India,

The present study is an attempt to know the status of the macrobenthos at the shelf region off Dhamara Estuary. A total of 1870 individulas per square meter of macrobenthic organisms were encountered during the study period. Population density and species diversity was found to be higher as depth increased. Polychaeta emerged as dominate macrobenthic group. Among dominant benthic macroinvertebrates *Nereis*, *Nepthys*, *Capitella*, *Owennia*, *Prionospio*, *Gammarus*, *Ampelisca*, *Tellina*, *Donax*, *Dentallium*, *Echinus* etc were predominant in the study region. Environmental parameters like salinity, pH and dissolved oxygen exhibited strong correlation with population density of macrobenthic organisms. Factors 1 and 2 pooled from principal component analysis represented 64.09% of the total variability. The noteworthy aspect of the present study is higher value of population density marked at deeper regions with heterogeneous sediment than shallow depth.

Key words: Shelf region macrobenthos, correlation coefficient, PCA.

INTRODUCTION

Ocean covers about 71% of the earth's surface and is covered by continental land mass. A continental shelf is the moderately shallow region of seabed surrounding a continent which could be extended away from the exclusive economic zone (EEZ) measuring about 200 nautical miles (360 km) from the shoreline. In tropical regions, the continental shelf zone covers about 30% of the total geographical area and 40% of open sea (Alongi, 1990). These regions are the most productive areas in the world for vast exploitable fishery and petroleum hydrocarbon resource hence supports the socio-economic condition of the country.

In recent days, the stress over the continental shelf region is found to be higher due to over-fishing, excessive bottom trawling for demersal fishery, dredging activities, drilling operations for exploration of petroleum hydrocarbon and natural gas resources including accidental and occasional seepage of oil and natural gas from ships and cargo vessels etc. Other anthropogenic influence from radioactive pollutants, heavy metals, pesticides, insecticides, hydrocarbons, PPEs, toxic chemicals are also adding additional stress to this highly productive ecosystem.

Therefore, it is the need of the hour to monitor the status or health of such sensitive ecosystem using suitable indicator. Macrobenthic organisms are well known indicator of environmental stress due to their sedentary habitat (Dauer and Corner, 1980). They posses active participation in benthic-pelagic trophic relation (Boesch et al., 1977; Simboura et al., 1995) and shows direct link with other organisms of higher trophic groups in marine food web (Snelgrove, 1998). Snelgrove (1998) highlighted regarding the noteworthy features of benthic macroinvertebrates like significant contribution towards metabolism of pollutants, nutrient cycling,

*Corresponding author. E-mail: dsmahapatro@gmail.com.

Abbreviation: PD, Population density (Number per square meter) salinity (ppt); **DO,** dissolved oxygen (mg/L); **WT,** water temperature (℃); **D,** dominance; **H',** shannon's diversity index; **J,** evenness; **d,** Margalef's species richness.

dispersion metabolism etc while Gerlach (1978) and McIntyre (1977) stated that benthic organisms provide food for larger epibenthic fauna. The feeding habits of demersal fishes is largely depended on benthic organisms as the chief source of nourishment (Ansari et al., 1977), while Parulekar et al. (1982) strongly argued to use them as a basic tool to access demersal fishery resource.

The shelf region of Bay of Bengal provides exploitable pelagic and demersal fishery resource. Thus, it needs to be monitored as priority basis. Benthic research in the shelf region of Bay of Bengal is few (Harkantra et al., 1982; Mahapatro, 2006) as compared with the shelf region of Arabian Sea of west coast of india (Parulekar and Wagh, 1975; Parulekar et al., 1976; Harkantra et al., 1980; Joydas and Damodaran, 2001; Joydas, 2002; Joydas and Damodaran 2009; Jayaraj et al., 2007, 2008).

The main objective of the present study was to provide comprehensive information regarding the status of benthic macroinvertebrates along with sediment texture analysis and physico-chemical parameters affecting significantly different macrobenthic groups of the continental shelf zone of north East Bay of Bengal, off Dhamara Estuary.

MATERIALS AND METHODS

Study area

The Bay of Bengal is as highly influenced by tropical climate and regarded as the 64th large marine ecosystems (LME) in the world. It measured as moderately productive (Class II) type of LME, that is, 150 to 300 g of carbon is produced per square meter per year located at 6°N and 80° E to 22°N and 80° 94°E (Mark, 2010). The Bay is situated at the monsoon belt and thus receives high fresh water from rainfall and discharges from major river systems (Dwivedi, 1993; Aziz et al., 1998). Recent seismic surveys indicated the presence of hydrocarbon deposits in the sea bed of Bay of Bengal over 300,000 km^2 area. The continental shelf region off Dhamara Estuary of Bay of Bengal falls in that region where the petroleum hydrocarbon deposit was noticed. Therefore the present study area confides to the shelf region off Dhamara Estuary.

Dhamara Estuary is located at 20° 80' North and 86° 90' East. Four stations were named as D1, D2, D3, and D4 straight to the offshore region from the Dhamara River Estuary (Figure 1) and positioned by GPS. The sampling stations were located at a depth range from 20 to 60 m. Dhamara Estuary is the integral part of the Bhitarkanika mangrove forest and Gahirmatha sea turtle rookery endowed with fragile sandy beaches. These beaches are well known for mass nesting activity of endangered olive ridley sea turtles (*Lepidochyles olivacea*). The shelf region off Dhamara Estuary is a potential fishing ground for both pelagic and demersal fishery.

Sampling periodicity

Three offshore cruises were carried out for collection of hydrological, biological and sediment parameters. The first cruise program was held in February 2005, the second and the third cruise was in March 2005 and in April 2005, respectively. The entire cruise program was carried out by the help of research vessel "Sagar Paschimi" provided by NIOT, Chennai, India.

Sampling methodology

Environmental parameters such as salinity (ppt), dissolved oxygen (mg/L), pH and water temperature (°C) were measured in duplicate from four stations during three cruise programs. All the hydrographical parameters were analyzed according to standard procedures given by Strickland and Parson (1972).

Sediment samples were collected in triplicate by the help of a grab sampler having surface area of 0.25 m^2. The first sample was kept for sediment texture analysis. The sediment texture analysis was made according to Krumbein and John (1938). Other two sediment samples were wet-sieved on-board through a sieve having mesh size of 0.5 mm (Birkett and McIntyre, 1971). The remains over the sieve were collected carefully and preserved in with 5 to 7% neutralized formalin-Rose Bengal-sea water mixture. Rose Bengal is a dye which makes the macrobenthic organisms red or pink in color, so that they can be easily collected from the sediment. The collected macrobenthic samples were identified to their possible lower taxonomic level and grouped into five major groups such as polychaetes, bivalves, gastropods, crustacea and others group. The others group contains the organisms belonging to nematodes, tardigrada, gastrotricha, echinoderms and the representatives of some minor phyla. The population density of macrobenthos was calculated by the counting of individual organisms collected per grab and then converted in to number per square meter area. Station wise relative abundance was carried by taking the mean values of all three cruises.

Statistical analysis

In order to find out significant relation between environmental parameters like salinity, pH, dissolved oxygen and water temperature with population density and some macrobenthic groups like polychaets, crustaceans, bivalves, gastropods with others group with sediment composition like, sand , silt and clay, the Pearson's correlation coefficient was calculated. Correlation coefficient measures the strength of a relationship between variables by putting emphasis on the degree to which mean value of the variables differ from each other. To further validate the findings of correlation analysis, principal component analysis (PCA) was also carried out. PCA is an important statistical tool which helps to get the relation between non-correlated factors which are in linear combinations of the original variables. This statistical tool helps to quickly visualize and analyze the possible correlations between N variables in a 2 dimensional space where N refers to number of variables. The statistical analysis was carried out by XLSTAT statistical software. To find out the community structure, univariate analyses were carried out to the biological data. The diversity indices studied were Margalef's species richness index 'd' (Margalef, 1968), Pielou's eveness index 'J" (Pielou, 1966) , and the Shannon species diversity index 'H"(Shannon and Weaver, 1963) and species dominance 'D'. All these analysis were carried out by PRIMER (Plymouth Routines in Multivariate Ecological Research, Version 5) as proposed by Clarke and Warwick (1994).

RESULTS

Hydrographical analysis

In the present study, the higher value of salinity was

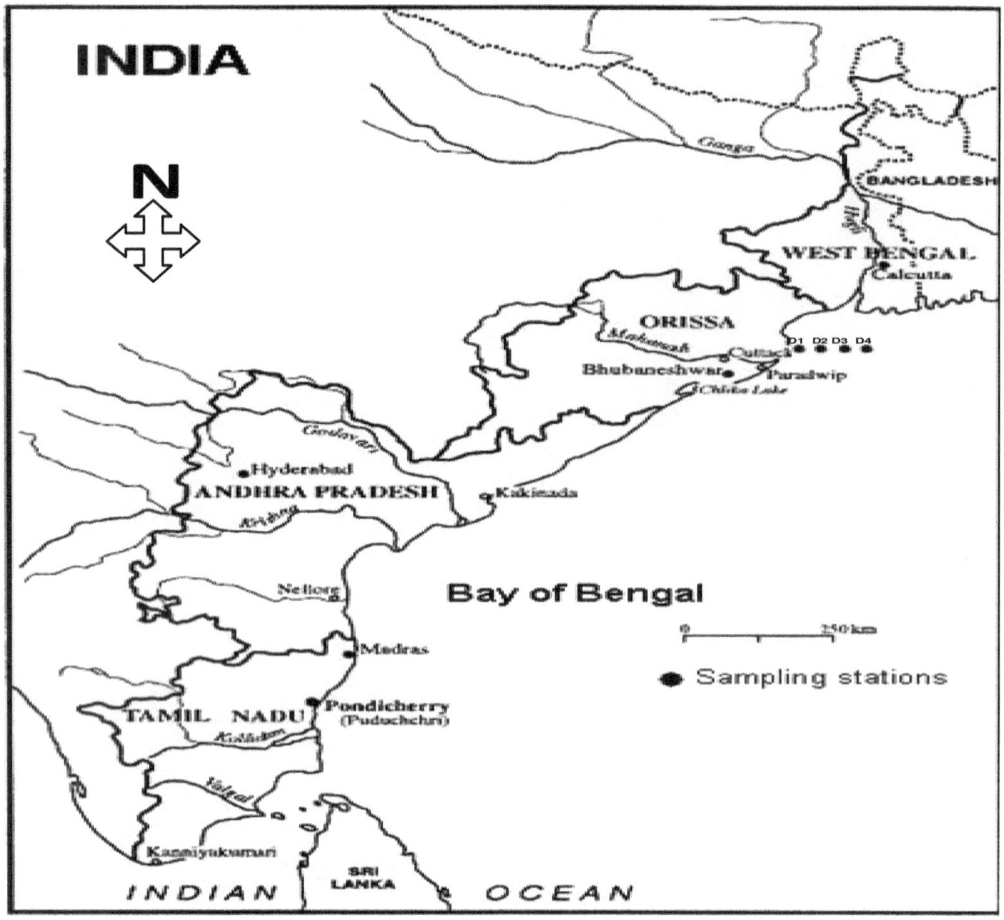

Figure 1. Explains India, Orissa state, Dhamara estuary and the sampling stations located at the northeastern region of Bay of Bengal. This region is becoming crucial for potential petroleum hydrocarbon resource.

observed in April 2005, that is, 35.09 ppt followed by March and lower at February 2005 having 31.4 ppt. Dissolved oxygen was found higher in April 2005 possessing 5.7 mg/L followed by February, 2005 (4.8 mg/L). Lower value was observed at March 2005 as 4.41 mg/L. pH value was higher at April 2005 (8.52) followed by March (8.28) and February 2005 (8.18). Water temperature was higher in April 2005 containing 29.8°C followed by March 27.8°C and February 2005 having 25.9°C. The mean and standard deviation of all environmental parameters are given in Table 1.

Diversity indices

The Shannon's diversity index H' was showed increasing trend as depth increases. The diversity value found higher during April 2005 possessed 2.75 followed by March (2.15) and lowest in February 2005 as 1.86. The dominance 'D' was found higher at February 2005 and lower at April 2005. Evenness J was higher in March 2005 followed by February 2005 and April 2005, while species richness was higher from February 2005 to April, 2005 (Table 1).

Sediment texture analysis

The results of sediment texture analysis are given in the Figure 2. During the study period, there was no great variation in the sediment. Broadly, the sediment can be categorized as mixed or heterogeneous type, that is, the combination of sand, silt and clay. During February 2005, lower content of clay fraction in the shelf sediment was marked and increased negligible towards April, 2005.

Population density

The population density of macrobenthos per station during three months is given in Figure 3 while the Figure 4 represented the mean and standard deviation at each station. In the first cruise program held on February, 2005, D2 showed higher values of macro benthic density

Table 1. Physico-chemical and diversity indices during February to April, 2005.

	February 2005		March 2005		April 2005	
Variables	**Mean**	**Std. dev**	**Mean**	**St. dev**	**Mean**	**St. dev**
PD	75	41.43	61.25	46.26	312.50	233.11
Salinity	31.94	0.65	33.55	0.00	35.09	0.61
pH	8.16	0.07	8.28	0.00	8.52	0.00
DO	4.8775	0.13	4.41	0.52	5.07	1.16
W.T.	25.9	1.4	27.8	0.5	29.5	1.2
D	0.2229	0.12	0.13	0.06	0.10	0.04
H'	1.863	0.51	2.15	0.52	2.75	0.36
J'	0.888	0.09	0.98	0.02	0.87	0.08
d	3.0088	1.13	3.68	1.38	5.95	1.63

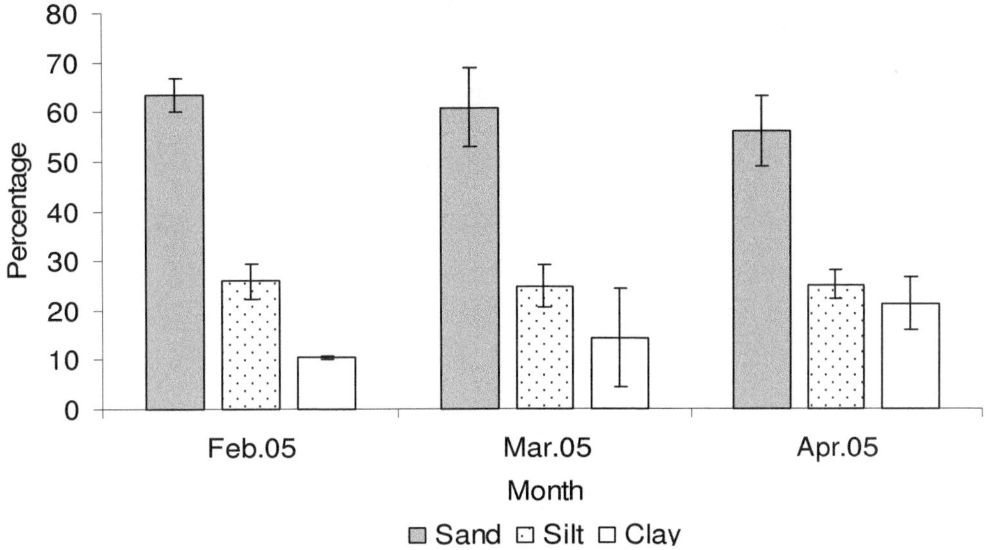

Figure 2. Mean values and standard deviation (error bars) of sand, silt and clay fraction within February to April, 2005.

while in the second cruise, it was D1. In third cruise program station, D4 showed highest values of macrobenthic density (Figure 4). The mean of population density observed during entire period was given in the Figure 5 which shows the mean value of macrobenthic density at first station D1 was 140±70.5) m^2, in D2 it was 108±58.3 m^2, in third station D3 it was 91±107.2 m^2 while in the last station D4, it was 258±349 m^2. Values in parenthesis showed standard deviation. The average population density per station was estimated as 149 No/m^2.

Relative abundance of macrobenthos

The relative abundance (mean) of macrobenthos at four stations is given in Figure 5. Among four stations, D1 showed polychaeta as the dominant group contributed 49% followed by D4 sharing 34%. In D2 and D3, the contribution of polychaeta was 20 and 22%, respectively. Crustacean become dominant in the First station D1 having 23% followed by D4 contributing 20% while low contribution was found at D2 and D3 having 3 and 6% respectively. Bivalves which showed highest contribution in D2 contributed 32% followed by D1 having 21%. D3 and D4 Possessed 16 and 8% only. Major contribution of gastropods was observed in D3 contributing 34% and lowest value at D1 sharing 1% only. The second station D2 and fourth station D4 showed equal contribution 28% to the total macrobenthic bulk. Other group exhibited major sharing at D3 having 22% followed by D2 17% followed by D4 10% and least values at D1 6% only.

Dominant macrobenthic organisms

Dominant macrobenthic organisms encountered during the study period were *Nereis*, *Nepthys*, *Goniada*, *Sabella*,

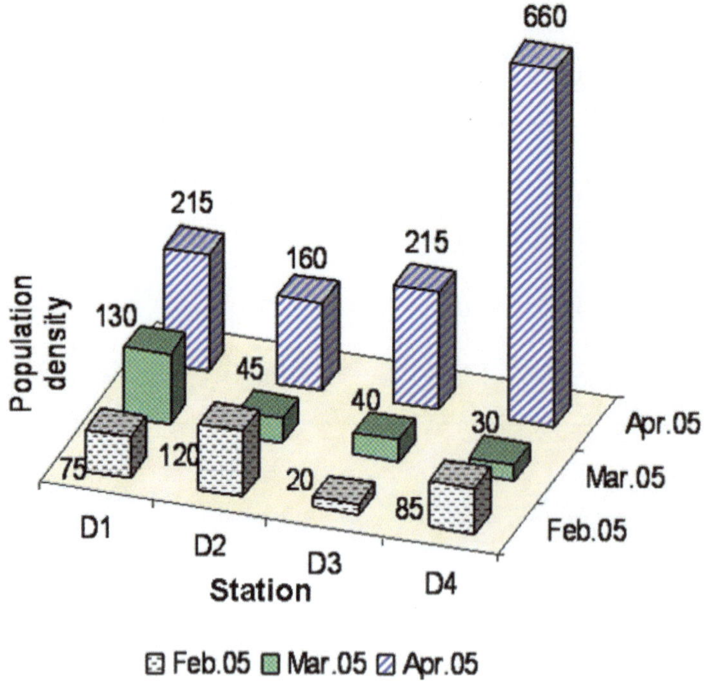

Figure 3. Population density of three months at different stations.

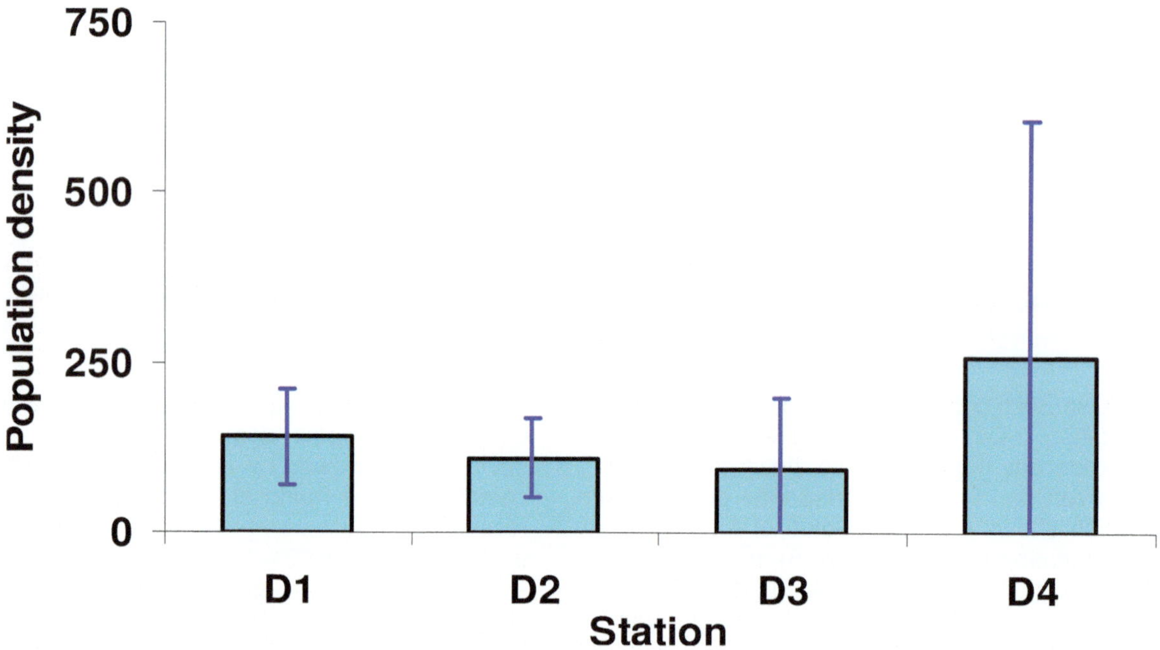

Figure 4. Mean and standard deviation of macrobenthic density at four satiations.

Clymenella, *Glycera*, *Myriochele*, *Phyllodoce*, *Cossura*, *Capitella*, *Axiothelaa*, *Owennia* from polychaeta group. Among bivalves, *Tellina*, *Donax*, *Thyasira*, *Tagelus*, *Anadora*, *Bathyarea*, etc. While among gastropods, the common species were *Apporahis*, *Buccinum*, *Hyanssa*. *Retusa*, *Dentallium*, *Bursa* etc. Crustacean representatives were dominated by gammarid amphipods, *Gammarus*, *Ampelisca* and *Ampithoe*, Harpecticoid copepod, stomatopods, *Echinus* (Sand dollar), Sipunculids, nematodes, Cumacea, Gastrotrichs, Tardigrada, etc. The images of macrobenthic organisms are given in Figure 8.

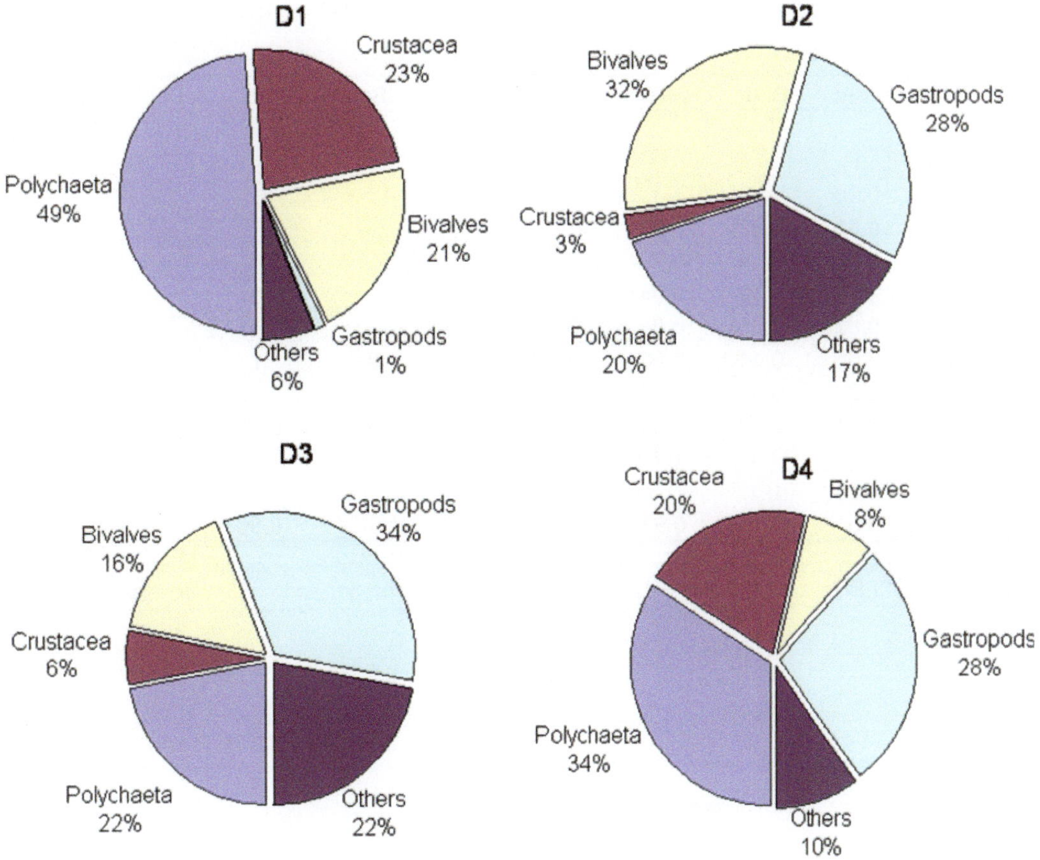

Figure 5. Relative abundance of macrobenthic groups of four stations.

Statistical analysis

Pearson's correlation coefficient

The Pearson's correlation coefficient (Table 2) explains about the possible relationship between the population density, macrobenthic groups and environmental parameters. During the present study, 13 parameters were taken such as population density, polychaeta, bivalves, gastropods, crustaceans, others, salinity, pH, dissolved oxygen, water temperature, sand, silt and clay.

Population density was significantly correlated with salinity, dissolved oxygen and pH ($p<0.05$). Polychaetes and gastropod are significantly correlated with salinity ($p<0.05$). Bivalve was significantly interrelated with water temperature, while dissolved oxygen becomes more essential for gastropod density ($p<0.05$). Crustacean did not show any statistically significant relation with any of the parameter.

Principal component analysis

Figure 6 represented the eigenvalues of different factors. Eigenvalues provides the information related to the quality of the projection of the variables from higher to lower number. Each eigenvalue represents a factor which is a linear combination of the initial un-correlated variables. In the present study (where N=13), the first eigenvalue was calculated as 6.219 which represented 47.84% of variability. Similarly, the second eigenvalue was equal to 2.11 represented 16.24% of the variability. Thus the eigenvalue of first two factors represented 64.09% of the total variability. This represented that the first two factors have the highest variability as compared to others. Figure 6 showed the scree plot where X-axis represents the projection of the factors and Y-axis represents the corresponding eigenvalue.

Figure 7 represents the correlation circle where first factor (F1) represented 47.84% of the variability at X-axis; second factor (F2) represented 16.25% of variability at Y. Variables nearer and closer to each other represented positive correlation, while located on opposite side of the centre represented negative correlation while those variables found closer to the center have no relationship. Silt and others posses no strong correlation in the correlation circle. Table 3 represents squared cosines values of variables in relation to the respective factor axes. Values closer to 1 were significantly related with the variables. In the present

Table 2. Pearson's correlation coefficient showing the strength of relationship between variables. Values in bold are significant at 95 % confidence limit.

Variable	PD	Poly.	Biva.	Gast.	Crust.	Oth.	Salin.	pH	DO	W.T.	Sand	Silt	Clay
PD	**1**												
Poly.	**0.90**	**1**											
Biva.	0.33	0.20	**1**										
Gast.	**0.89**	**0.64**	0.37	**1**									
Crust.	**0.92**	**0.85**	0.18	**0.75**	**1**								
Oth.	0.20	0.18	-0.28	0.04	0.21	**1**							
Salin.	**0.69**	**0.62**	0.56	**0.68**	0.51	0.01	**1**						
pH	**0.63**	0.55	0.44	**0.68**	0.47	-0.14	**0.92**	**1**					
DO	**0.60**	0.37	0.15	**0.74**	0.45	-0.09	0.12	0.18	**1**				
W.T.	0.32	0.21	**0.68**	0.39	0.14	-0.02	**0.86**	**0.75**	-0.10	**1**			
Sand	-0.45	-0.22	0.07	**-0.65**	-0.30	-0.39	-0.49	-0.47	-0.38	-0.40	**1**		
Silt	-0.06	0.20	-0.17	-0.30	-0.22	0.17	-0.04	-0.06	-0.13	-0.11	0.29	**1**	
Clay	0.57	0.51	0.17	**0.61**	0.47	-0.02	**0.81**	**0.90**	0.04	0.56	-0.48	-0.08	**1**

Values in bold are different from 0 with a significance level alpha=0.05. Abbreviation: PD =population density, Poly.=Polychaeta, Biva.= Bivalves, Gast..= Gastropod, Crust..=Crustacea, Oth.=Other, Salin.=Salinity, DO=Dissolved oxygen and W.T.=Water temperature.

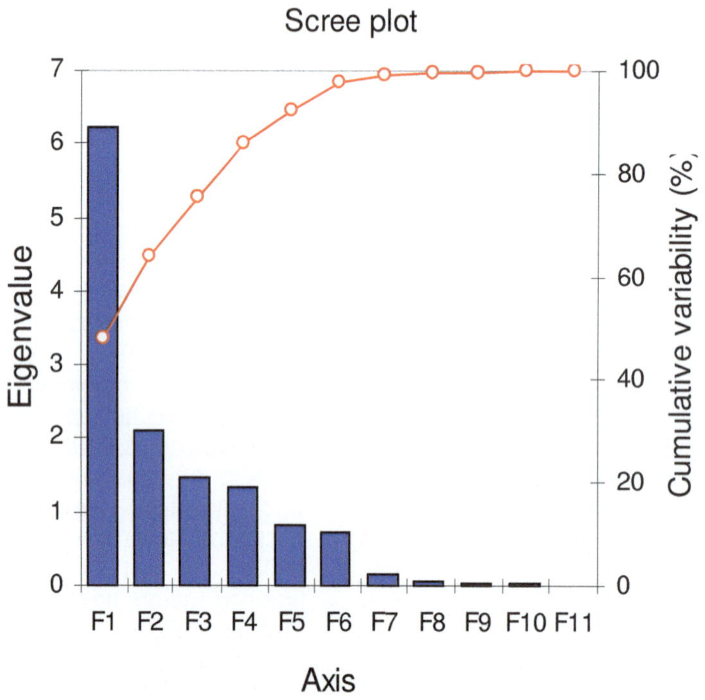

Figure 6. Scree plot showing factor with Eigenvalue.

study, salinity and pH (F1 axis), water temperature and dissolved oxygen (F2 axis) emerged as principal components that were remarkably putting impact over macrobenthos.

DISSCUSSION

The population density of macrobenthos of shelf region of north-eastern part of Bay of Bengal is comparable with Harkantra et al. (1982), Vijaya Kumar et al. (1991) and Mahapatro (2006) who studied the offshore region off Dhamara Estuary. The population density of macrobenthos was lower as compared to the Arabian sea of west coast of India as reported elsewhere (Joydas, 2002; Joydas and Damodaran, 2001). This is in concurrence with the work of Parulekar et al. (1982) who studied the benthos of the Indian seas and opined that

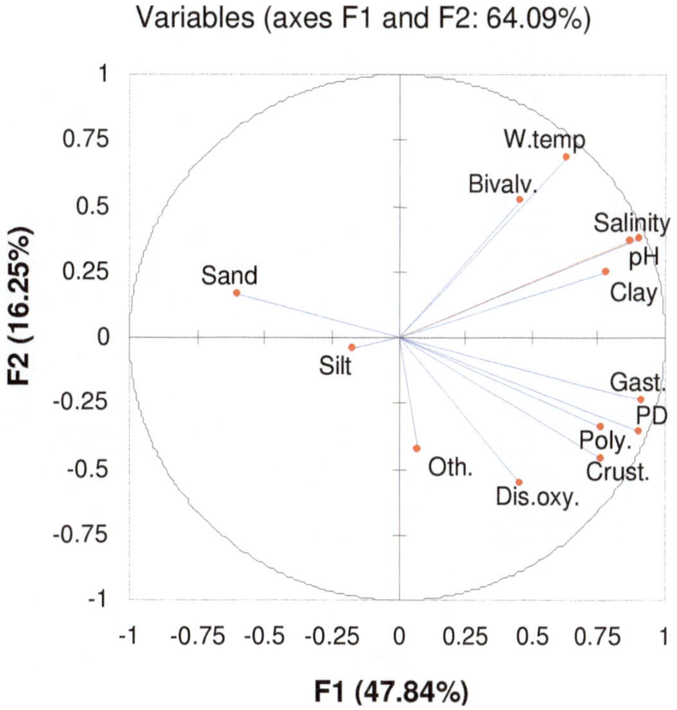

Figure 7. Correlation circles of principal components in two axis.

Table 3. Shows squared Cosine values of different parameters pooled out against Different factors up on X and Y axis.

	Squared cosines of the variables				
	F1	**F2**	**F3**	**F4**	**F5**
Population density	**0.82**	0.13	0.00	0.04	0.01
Polychaeta	**0.58**	0.11	0.05	0.19	0.00
Bivalves	0.21	**0.27**	0.15	0.09	0.25
Gastropods	**0.83**	0.06	0.07	0.01	0.00
Crustaceans	**0.59**	0.21	0.00	0.03	0.00
Other	0.01	0.18	**0.43**	0.12	0.23
Salinity	**0.81**	0.14	0.03	0.00	0.00
pH	**0.76**	0.14	0.01	0.00	0.06
Dissolved oxygen	0.21	**0.31**	0.27	0.00	0.00
W. Temperature	0.40	**0.47**	0.01	0.02	0.07
Sand	0.36	0.03	0.01	**0.51**	0.00
Silt	0.03	0.00	**0.39**	0.32	0.00
Clay	**0.61**	0.06	0.06	0.01	0.19

Values in bold correspond for each variable to the factor for which the squared cosine is the largest.

the Bay of Bengal posses lower values of macrobenthic density as compared with Arabian Sea and also he pointed out the reasons of lower values are due to fresh water flux during monsoon, unprecedented storm surges, and tropical cyclones. Again, out flux from large riverine systems such as the Ganges, the Bramhputra, the Mahanadi, the Godavari, the Krishna and the Cauvery carries huge quantity of silt borne freshwater (Dwivedi, 1993; Aziz et al., 1998). The riverine systems and the environmental parameter together might have influenced the phytoplankton productivity. It is proved that phytoplankton productivity has significant influence on benthic productivity (Ingole et al., 2010) which further strengthen the fact that surface water chlorophyll have

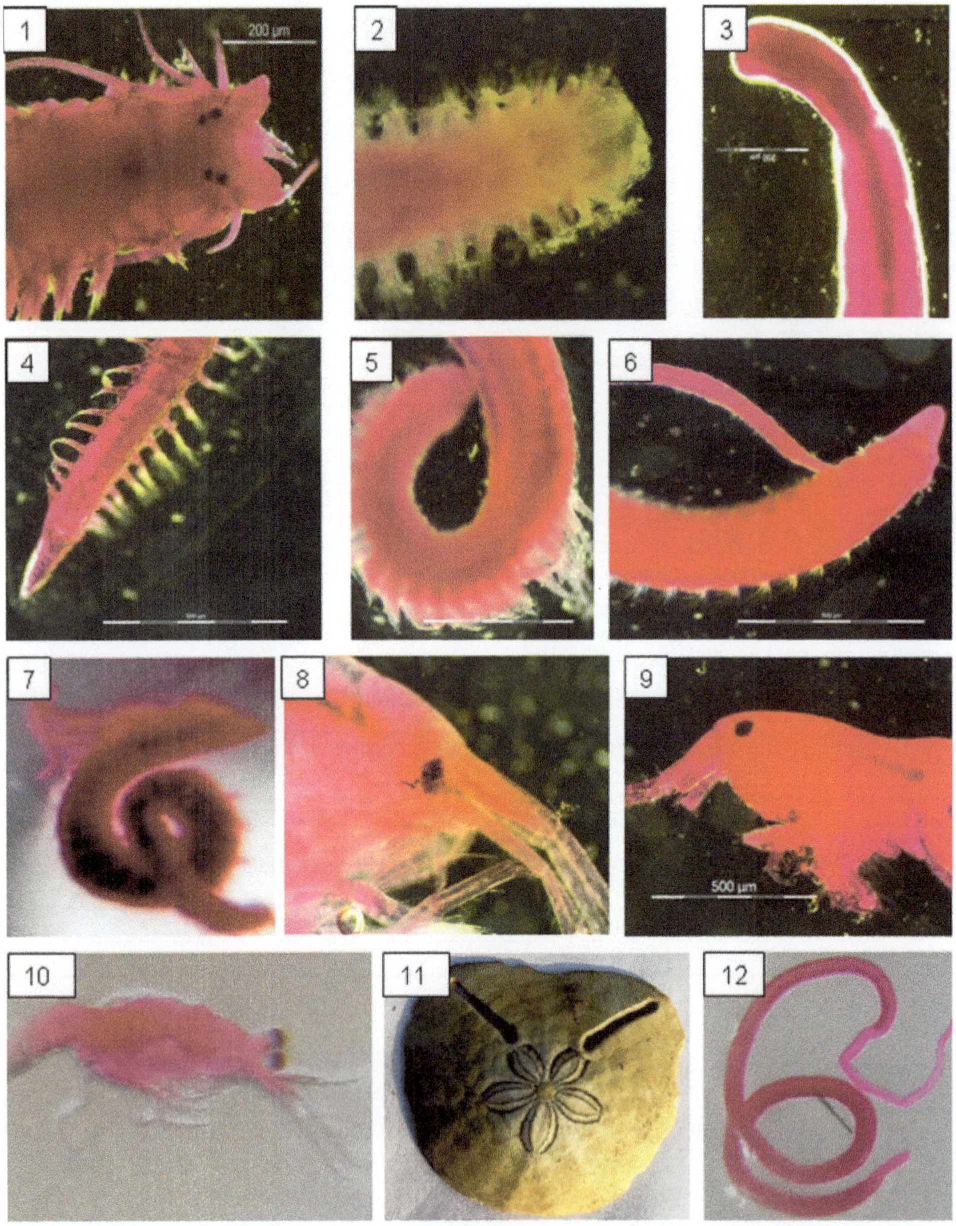

Figure 8. Images of macrobenthos of shelf region off Dhamra estuary. 1. *Nereis sp.*; 2. *Nepthys sp.*; 3. *Capitella sp.*: 4. *Goniada sp.*; 5. *Prionospio sp.*; 6. *Cossura sp.*; 7.*Cirratulus sp.*; 8. *Gammarus sp.*; 9. *Ampelisca sp.*; 10. *Mysis sp.*; 11. Sand dolar, 12. Nematoda.

major authority over macrobenthic productivity and helped for benthic pelagic coupling (Snelgrove, 1998) while Pavithran et al. (2009) found significant correlation between surface chlorophyll-a with macrobenthic density and sediment organic matter in Central Indian Ocean Basin. In Arabian Sea, the surface water chlorophyll was found higher than Bay of Bengal as reported by Prasanna et al. (2002) and Madhuparatap et al. (2003). This was due to upwelling, less disturbance from cyclone and storm surges and a consistent environmental condition unlike Bay of Bengal.

So far, the present study was concerned with the salinity value along with dissolved oxygen, water temperature, macrobenthic density and species richness, was increased from February to April 2005 which showed the strong relationship between macrobenthic organisms with environmental (Table 2). There was a clear evidence of increasing rate of population density and deeper regions than the shallow depth. Because in shallow depth regions, high turbidity, often fluctuation of environmental parameters and more fishing pressure disturbs habitat condition, while in deeper region favoring condition is existed due to stable environmental parameters. Similar results were obtained by Ingole et al. (2010) in the shelf region of Arabian Sea.

Salinity is a major influencing parameter for the abundance and distribution of marine organisms. A constant range of salinity favors the proliferation of macrobenthic organisms. Vizakat (1991) reported that recolonisation of benthos with increased salinity value at Konkan West coast of India. He further opined that in tropical waters, salinity is the most influencing parameter over benthic productivity. Same observation was made in the present study from the squared cosine table (Table 3) which is pooled from the principal component analysis that showed salinity and pH as major guiding factors. Harkantra and Parulekar, (1991) suggested that no single factor could be considered as an ecological master factor that totally responsible for community structure of macrobenthos. This is rightly evidenced from the present study that four environmental parameters such as salinity, pH, dissolved oxygen and water temperature were significantly affecting the macrobenthic density off Dhamara Estuary (Correlation Table 2). In contrast to this, northwest shelf of Arabian Sea exhibited none of the physico-chemical parameter that significantly correlated with the macrobenthos (Jayaraj et al., 2007, 2008). This corroborate that the macrobenthos off Dhamara Estuary of Bay of Bengal was significantly influenced by physico-chemical parameters as compared to Arabian Sea. Harkantra et al. (1982) studied the shelf region of Bay of Bengal stated that other than physico-chemical parameters, sediment texture and community interaction are the possible cause of low values of macrobenthic density. The results of granulometric analysis revealed that there was no great variation in the texture composition of the study region. Whereas the community interaction such as predation, competition, recruitment and mortality are such factors which can significantly alter the numerical density of macrobenthos (Mahapatro, 2006; Harkantra and Parulekar, 1991).

Perusal of literature at Indian context revealed that polychaeta was a major component of macrobenthos at the shelf region of east coast and west coast of India (Harkantra et al., 1980; Harkantra et al., 1982, Ingole et al., 2002, Mahapatro, 2006; Joydas and Damodaran 2009; Jayaraj et al., 2007; 2008; Ingole et al., 2010). In the present study, polychaeta group emerged as a major stuff among all other macrobenthic groups. The station wise distribution of relative abundance of different macrobenthic groups during the study period exposed that the first station D1 and the last station D4 possessed higher contribution of polychaets in the shelf waters of Bay of Bengal. This supports the fact that deeper areas of the Bay of Bengal posses higher density of macrobenthos is in agreement with Harkantra et al. (1982) (Joydas and Damodaran, 2009; Pavithran et al., 200; Ingole et al., 2010).

From the study of PCA, three principal components were pooled out, such as salinity (F1), Water temperature (F2) and silt (F3). The variability of first two factor was 64.09% of the total variance. Salinity along with pH (F1) was found most significant among other variables having good relation with population density, specifically with polychaeta, gastropods, and crustaceans, while F2 scores correspond to higher values of water temperature, dissolved oxygen and bivalves. These parameters commonly favor high colonization of macrobenthic organisms (Jayaraj et al., 2007, 2008; Ingole et al., 2009, 2010). Under low turbidity, fair transparency and optimal temperature favors high photosynthetic rate. This causes high dissolved oxygen content favoring good colonization of benthic organisms.

Conclusion

It is concluded from the above study that multiple environmental parameter exhibited significant impact over the macrobenthic organisms in the shelf region of Bay of Bengal. Certain parameters such as salinity, water temperature, dissolved oxygen and pH were considerably influencing over macrobenthic density. No great variation was exhibited in the sediment composition. The role of community interaction could not be ruled out completely regarding the lower values of macrobenthic density. Polychaetes was emerged as major macrobenthic group during the present study, which can act as an indicator of environmental stress. In the future, more study should be carried out regarding the physico-chemical status along with macrobenthos in the shelf region of Dhamara Estuary in general and macrobenthic polychaets in particular to assess demersal fishery and to evaluate natural and anthropogenic stress.

ACKNOWLEDGEMENTS

Authors are grateful to Head of the Department, P.G. Department of Marine Sciences, Berhampur University for financial assistance and laboratory facilities. Authors are also thankful to the scientist of Estuarine Biological Centre, Gopalpur-on-sea of Zoological Survey of India for identification of samples.

REFERENCES

Alongi DM (1990). The ecology of tropical soft-bottom benthic ecosystems. Oceanogr. Mar. Biol. Annu. Rev., 28: 381-496.

Ansari ZA, Parulekar AH, Harkantra SN, Ayyaapan Nair (1977). Shallow water macrobenthos along central west coast of India, Mahasagar, 10: 123-127.

Aziz Ahmad HB, Luqueman A, Atapattu A, Chullasorn S (1998). Regional stewardship for sustainable marine resources management in the Bay of Bengal. In: K. Sherman, E. Okemwa and M. Ntiba (eds), "Large Marine Ecosystems of the Indian Ocean: Assessment, Sustainability, and Management." Blackwell Science. ISBN: 0632043180.

Ingole BS, Sautya S, Sanitha S, Singh R, Nanajkar M (2010). Macrofaunal community structure in the western Indian continental margin including the oxygen minimum zone Marine Ecology 31, doi:10.1111/j.1439-0485.2009.00356.x.

Birkett L, McIntyre AD (1971). Treatment and sorting of samples. In: Methods for the study of marine benthos, IBP Handbook No. 16: Blackwell Sci. Publ. Oxford and Edinburgh, p. 157.

Boesch DF, Kraeuter JN, Serafy DK (1977). Distribution and Structure of Communities of Macrobenthos on the Outer Continental Shelf of the Middle Atlantic Bight;1975-1976 investigations. Virginia Institute of Marine Science Special Report in Applied Marine Sciences and Ocean Engineering No.175.

Clarke KR, Warwick RM (1994). Change in marine communities: An approach to statistical analysis and interpretation. Plymouth Mar. Lab., p. 144.

Dauer DM, Corner WG (1980). Effect of moderate sewage input on benthic polychaetes population; Estu. Coa and Shelf Sc., 10: 335-362.

Dwivedi SN (1993). "Long-Term Variability in the Food Chains, Biomass Yield, and Ocenaography of the Bay of Bengal Ecosystem," in Kenneth Sherman, et al. (eds.), Large Marine Ecosystems: Stress, Mitigation, and Sustainability (Washington, D.C.: American Association for the Advancement of Science, 1993), pp. 43-52. ISBN: 087168506X.

Gerlach SA (1978). Food chain relationship in subtidal silty sand marine sediment and the role of meiofauna in stimulating bacterial productivity; Oecol., 33: 55-69.

Harkantra SN, Rodreigues, CL, Parulaekar AH, (1982). Macrobenthos of the shelf off northeastern Bay of Bengal. Indian J. Mar. Sci., 11: 115-121.

Harkantra SN, Parulekar AH (1991). Interdependence of environmental parameters and sand dwelling benthic species abundance: a multivariate approach. Indian J. Mar. Sci., 20: 232- 234.

Harkantra SN, Ayyappan Nairs, Ansari ZA, Parulekar AH (1980). Benthos of shelf region along the west coast of India. Indian J. Mar. Sci., 9: 106-110.

Ingole BS, Sivadas S, Nanajkar M, Sautya S, Nag A (2009). A comparative study of macrobenthic community from harbours along the central west coast of India. Environ. Moni. Asses., 154, 135–146.

Ingole BS, Rodrigues N, Ansari ZA, (2002). Macrobenthic communities of the coastal waters of Dabhol, west coast of India. Indian J. Mar. Sci., 31(2): 93-99.

Jayaraj KA, Jayalakshmi KV, Saraladevi K (2007). Influence of environmental properties on macrobenthos in the northwest Indian shelf. Environ. Moni. Asses. , 127: 459-475, DOI: 10.1007/s10661-006-9295-5.

Jayaraj KA, Josia J, Dinesh Kumar PK (2008). Infaunal macrobenthic community of soft bottom sediment in a tropical shelf. J. Coast. Res., 24: 708-718.

Joydas TV, Damodaran R (2009). Infaunal macrobenthos along the shelf waters of west coast of India, Arabian Sea. Indian J. Mar. Sci., 38 (2): 2009, 191-204.

Joydas TV, (2002). Macrobenthos of the shelf waters of the west Coast of India. Cochin, India: Cochin University of science and Technology, Doctoral thesis.

Joydas TV, Damodaran R (2001). Macrobenthic polychaetes along the shelf waters of the west coast of India. (Paper presented at IAPSO/IABO Ocean Odyssey Conference held at Mar Del Plata).

Krumbein P, John FJ (1938). Manual of Sedimentary Petrography Appleton Century Crafts. New York, p. 549.

Madhupratap M, Gauns M, Ramaiah N, Prasanna Kumar S, Muraleedharan PM, DeSousa SN, Sardessai S, Muraleedharan S (2003). Biogeochemistry of the Bay of Bengal: Physical, chemical and primary productivity characteristics of the central and western Bay of Bengal during summer monsoon 2001. Deep_sea Res.(II), 50: 881-896.

Mahapatro D (2006). Studies on the macrobenthos off the Dhamara River mouth, Orissa Coast. M.Phil Dissertation, Berhampur University, Berhampur, Orissa, India

Margalef R (1968) Perspectives in Ecological Theory. University of Chicago Press, Chicago, p. 111.

McGinley M, (2010). "Bay of Bengal large marine ecosystem". In: Encyclopedia of Earth. Eds. Cutler J. Cleveland (Washington, D.C.: Environmental Information Coalition, National Council for Science and the Environment).<http://www.eoearth.org/article/Bay_of_Bengal_large_ marine_ ecosystem> .

McIntyre AD (1977). Effect of pollution on inshore benthos: *In*: Ecology of marine benthos (sd.) Coull B.C. (University of South Carolina press), pp. 301-312.

Parulekar AH, NAIR SA, Harkantra SN, Ansari ZA (1976). Some quantitative studies on the benthos off Bombay. Mahasagar, 9(1/2): 51-56.

Parulekar AH, Wagh AB (1975). Quantitative studies on benthic macrofauna of northeastern Arabian Sea shelf. Indian J. Mar. Sci., 4: 174-176.

Parulekar AH, Harkantra SN, Ansari ZA (1982). Benthic production and assessment of demersal fishery resource of Indian seas. Indian J. Mar. Sci., 11: 107-114.

Pielou EC (1966). Species diversity and pattern diversity in the study of ecological succession. J. Theor. Biol., 10: 372-383.

Prassana kumar S, Muraleedharan PM, Prasad, TG, Gauns M, Ramaiah N, De Souza SN, Sardesai S, Madhupratap M (2002). Why is the Bay of Bengal less productive during summer monsoon compared to the Arabian Sea? Geophys. Res. Lett., 29: 881 http: //drs. nio.org. /drs /bitstream / 2264 /111/1/Geophys_res_Lett_29_88-1.pdf.

Pavithran S, Ingole BS, Nanajkar M, Raghukumar C, Nath BN, Valsangkar AB (2009). Composition of macrobenthos from the Central Indian Ocean Basin. J. Earth Syst. Sci. 118, No. 6, December, pp. 689–700.

Shannon CE, Weaver W (1963). The Mathematical Theory of Communication. University of Illinois Press, Urbana, Illinois,144 pp.

Snelgrove PVR, (1998). The biodiversity of macro-faunal organisms *In*: marine sediments. Biodiv. Conser., 7: 1123– 1132.

Strickland JDH, Parsons TR (1972). A practical handbook of seawater analysis, vol 167. Bulletin of Fisheries Research Board of Canada, Ottawa. 1–310. pp

Vijaya kumar R, Ansari ZA, Parulekar AH (1991). Benthic fauna of Kakinada Bay and Back water, east coast of India. Indian J Mar Sci. 20: 195-199.

Vizakat L, Harkantra SN, Parulekar AH (1991). Population ecology and community structure of subtidal soft sediment dwelling macroinvertebrates of Kankan, west coast of India. India J. Mar. Sci. 20:40-42.

Simboura N, Zenetus A, Panayotides PP, Makra A, (1995) Changes in benthic community structure along an environmental pollution gradient. Mar. Poll. Bull.30:470-474.

Seasonal and spatial variations of dissolved carbon and nitrogen in the Santos estuarine system, Southeastern Brazil

Fabio Aprile[1]* and Gilmar W. Siqueira[2]

[1]Laboratório de Estudos de Ecossistemas Amazônicos - LEEA, Universidade Federal do Oeste do Pará, Campus Rondon. Av. Marechal Rondon s/n Caranazal, Santarém, PA, # 68040-070, Brasil.
[2]Laboratório de Planejamento e Desenvolvimento de Fármacos – LPDF, Universidade Federal do Pará, Campus Universitário do Guamá. Av. Augusto Corrêa n. 1, Belém, PA, # 66075-100, Brasil.

We investigated the concentrations of nutrients carbon and nitrogen in Santos Estuarine System (SES). The SES is located in the region of Baixada Santista (Santos, Brazil) and is ecologically and economically important due to biological diversity, high levels of metal pollution and sewage, and the presence of the port considered the largest and most productive in Latin America. For the preparation of the nutrient flow model, limnological and oceanographic data were obtained in water and sediment of 27 stations distributed in SES between 2000 and 2008. Significant contributions especially of organic carbon and nitrogen were observed from the communities living around the estuary and the Santos bay. Equations were developed to explain the flow of nutrients based on the type estuary, tidal flows and channel. The SES was classified as a partially mixed system. The results showed that the system has a conservative behavior to salinity and non-conservative to carbon and nitrogen.

Key words: Carbon, nitrogen, modeling, salinity, nutrients flow.

INTRODUCTION

The carbon cycle is the most complex of all biogeochemical cycles, and more than 97% of carbon present in ocean water is present as dissolved inorganic, especially as CO_2, HCO_3^- and CO_3^{2-}, called total dissolved inorganic carbon (DIC) and total CO_2. Levels of dissolved organic carbon (DOC) in marine and continental waters are associated with the decomposition or oxidation process, which contains organic compounds like amino acids, monosaccharides, polysaccharides, humic substances and proteins. The nitrogen exists in many forms due to the large number of oxidation states can take. Nitrogen can be found in the aquatic environment as nitrate (NO^{3-}), nitrite (NO^{2-}), ammonium compounds (NH_3 and NH_4^+), nitrous oxide (N_2O) and organic forms (urea, amino acids and peptides).

Biogeochemical cycles of carbon and nitrogen in water are mainly associated with the tides, the abiotic and biotic processes, climatic conditions and associated ecosystem function (Cox et al., 2002). The knowledge about the nature and cycling of nutrients in estuarine systems allows for better estimates of carbon transport to the coastal zone and ocean (Pettine et al., 1998; Trefry et al., 1994). Estuaries are important routes for transfer of dissolved and particulate materials from the mainland to the coastal zone and ocean. The flow of nutrients into the estuaries is so intense and dynamic, the result is a strong interaction between the biotic and abiotic system, resulting in a high local biodiversity. The rivers carry and distribute nitrogen in estuaries, whose source can be natural or anthropogenic depending on the degree of

*Corresponding author. E-mail: aprilefm@hotmail.com.

Abbreviations: SES, Santos estuarine system; **DIC,** dissolved inorganic carbon; **DOC,** dissolved organic carbon; **DIN,** dissolved inorganic nitrogen; **DON,** dissolved organic nitrogen; **POC,** particulate organic carbon; **POM,** particulate organic matter; **PON,** particulate organic nitrogen.

preservation of the surrounding forest, of the intensity of erosion and leaching processes, as well as of the impact resulting from anthropogenic activities. In local and regional scales there are indications of an increase in levels of C, N and P from rivers to coastal and ocean (Howarth et al., 1996; Meybeck, 1993), and this increase in concentration of nutrients has contributed to an increase in the eutrophication process and degradation of coastal marine ecosystems, especially in estuarine areas.

It becomes necessary to measure the mass balance and the transformation of materials through groundwater, rivers, estuaries and coastal waters, in order to estimate the reservoirs of nutrients and understand how to process the flow paths as well as the advancement of degradation of coastal waters. Thus, we sought to apply all the information gained from the experiment in a flow model, whose objective was to answer the behavioral tendency of the ecosystem studied, as well as serve as a representative for other similar systems, especially in relation to geomorphology and aspects biogeochemical. Among the various models used for nutrient flow, the mass balance model is one of the most highly regarded when it comes to quantify the concentrations of the elements involved. In this case, some protocols for the construction of the database are suggested, in which case the methodological protocols used LOICZ - Land-Ocean Interactions in the Coastal Zone, which propose a typology of consistent database, and allows for comparison with databases various types of coastal and marine environments (Gordon Jr. et al., 1996; Smith and Crossland, 1999; Smith, 2002; Jorgenson and Brown, 2005; Le Tissier et al., 2006; Kannen and Burkhard, 2009; Lantuit et al., 2009; Streletskaya et al., 2009), was designed to estimate the flows of C, N and P in coastal areas around the world. According to Smith (2002), the role of the coastal oceans in the global C-N-P cycle is one of the major and most challenging questions that LOICZ is evaluating.

Through this research, we sought to observe and better understand the transport and mass balance of carbon and nitrogen by water and sediment compartments. For that it was researched the capture of those elements by the rivers that make up the mesh of the Santos estuarine system to the edge of the continental shelf ocean.

MATERIALS AND METHODS

Study area

Santos Estuarine System - SES (23°50'S and 24° S; 46°15'W and 46°30'W) is located in the Baixada Santista, coast of São Paulo State, Southeastern Brazil. This region is ecologically and economically important because of the great biological diversity originating from the Atlantic Forest biome, the presence of mangrove strands, whose role both in species diversity as in nutrient cycling is already well known, and high levels of pollution and contamination, coming mainly from the port and industrial complex, which together produce solid waste and effluents, including sewage, which are deposited in the channels of SES. The main beaches of southeastern Brazil, in terms of population density, are located exactly in the vicinity of the SES, in the municipalities of São Vicente, Santos and Guaruja. There is also the Port of Santos, the largest and most active port in Latin America, with intensive loading and unloading of goods and products, being the great center of runoff from Latin America to the world and vice versa. It should also highlight the industrial city of Cubatão, where are placed heavy industries, formed mainly by steel and metallurgy (Figure 1).

The Cubatão city has an industrial area remarkable, with some types of industries related to production of oil, production of manure and steel. As a result of intense industrial activity, much of the waste material are released in the channels that make up the mesh of the SES, found that mesh heavy metals, organic compounds, petroleum hydrocarbons and other (Braga et al., 2000).

The location of the port of Santos, as well as the industrial complex of Cubatão is strategic because it is close to the most developed region, industrialized and populous of the South America, the city of São Paulo. However, because the port was built in a mangrove area, where the sediment load is very high and therefore the rate of sedimentation, it is necessary that the main channel of the harbor is dredged annually. As a result, the material retained in the surface sediments returns to the water column, increasing the possibility of assimilation and consequent contamination of aquatic organisms, especially filter feeders such as shellfish.

Analytical procedures and model construction flow

Limnological and oceanographic data used to develop the flow model were obtained between the years 2000 and 2008 in 27 sampling stations distributed within the SES as follows: four stations near of industrial complex of Cubatão, six stations in Santos channel, eight stations in São Vicente channel, and nine stations in the Santos bay (Figure 1). Sampling followed the local seasonal pattern, and the sampling was conducted twice a year with total n = 416 samples to gauge the influence of ocean currents and the hydrological cycle in the estuarine system. All analytical procedures follow international protocols for the collection, preservation and analysis described in Strickland and Parsons (1968), Grasshoff et al. (1983) and Wetzel and Likens (2000). The methodological procedures for the determination of the parameters under investigation in the water column (surface and bottom) and surface sediment (up to 0.15 m) are summarized below. Water temperature (°C), pH and salinity (%) were measured with probes for multiparameter water quality WTW and YSI; density (g/cm^3) was calculated from the data of temperature and salinity of the water; free CO_2 and total CO_2 were obtained by titration with control of pH, using NaOH to raise the sample pH to 8.3 and HCl to reduce the pH to 4.3; total alkalinity (TA meq/l) and alkalinity carbonate (CA meq/l) were obtained by calculation from the potentiometric titration method in the analysis of total CO_2 and free CO_2, and calculated by Equations 1 and 2 (Parsons et al., 1984; Grasshoff et al., 1999). The dissolved inorganic carbon (DIC mg/l), also defined as dissolved CO_2 content, was calculated from the total amount of CO_2 per mole transformation according to Equation 3. The dissolved organic carbon (DOC mg/l) was determined by oxidation with potassium dichromate 1N and acid solution of [3 g Ag_2SO_4 + 1000 ml H_2SO_4 + 20 ml H_3PO_4 85% + 0.4 g NaF] and 1 ml of diphenylamine indicator solution, and titration was done according to determinations described by Aprile (2001), with a solution of iron sulfate and 0.5 N ammonia and detection limit of 0.001 mg/l. Dissolved inorganic nitrogen (DIN mg/l) represents the sum of the inorganic fractions in filtered samples (Equation 4), and ammonium (NH_3 mg/l) was determined by reaction with phenol and hypochlorite in alkaline solution catalyzed by sodium nitroprusside, and detection limit of 0.05 mg/l; nitrite (N-NO_2^- mg/l) was determined

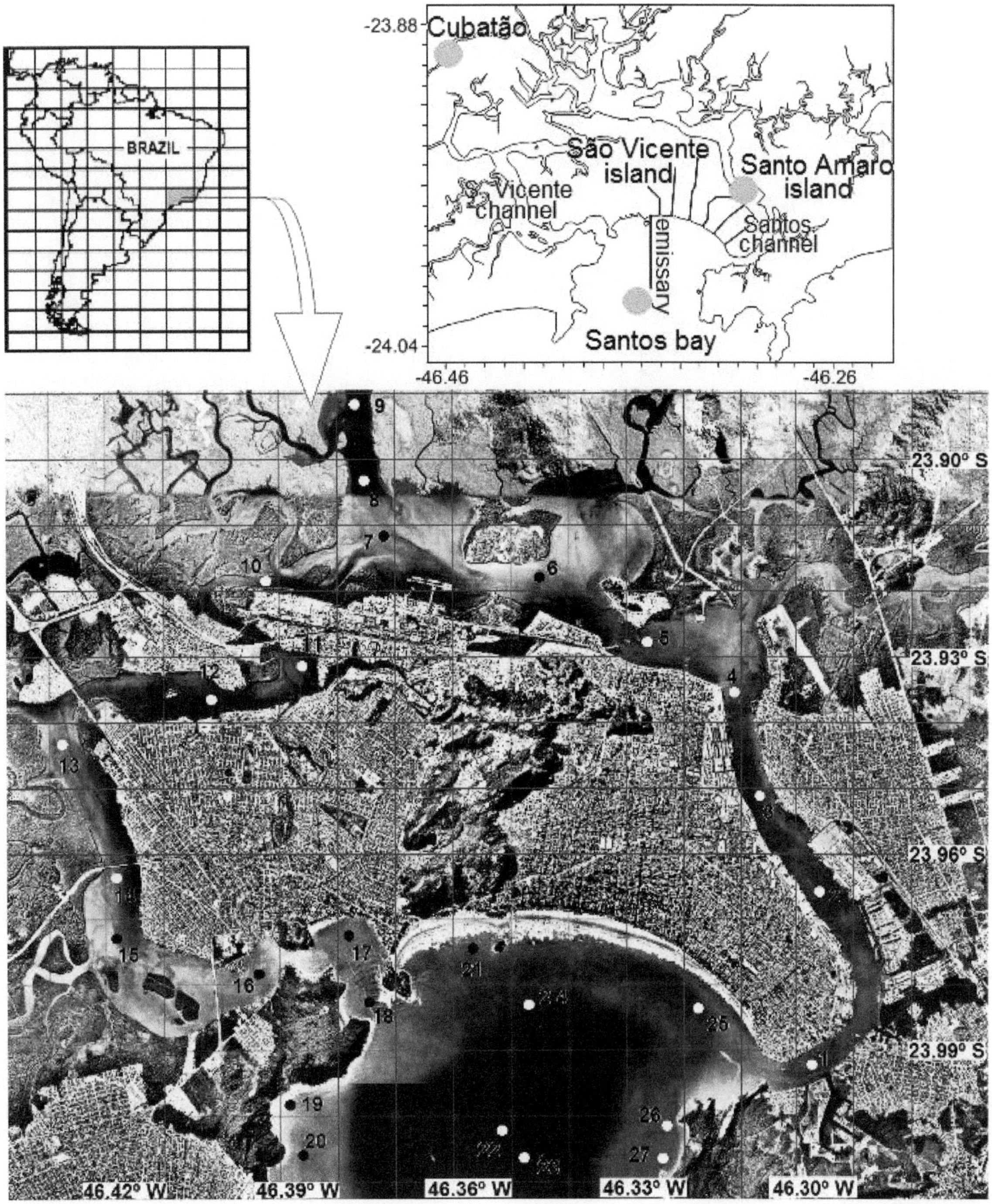

Figure 1. Santos estuarine system (Southeastern Brazil) with the location of sampling stations, and indication of the areas potentially polluting in the smaller map (gray circle) for the industrial Park Cubatão, Port of Santos and outfall.

in the presence of sulfanilamide prepared in acid solution and N-naphthyl ethylenediamine dichloro in alcohol solution with a detection limit 0.005 mg/l, and nitrate ($N\text{-}NO_3^-$ mg/l) was determined in cadmium column with reduction to nitrite in filtered and buffered samples with ammonium chloride, and detection limit 0.009 mg/l. Dissolved organic nitrogen (DON mg/l) was determined in samples filtered by the classical method of Kjeldahl distilling in the presence of 46% NaOH and H_3BO_3, and detection limit of 0.005 mg/l. The particulate organic matter (POM ug/g) in sediments was obtained by the method of oxidation with hydrogen peroxide (H_2O_2 30%) in hot plate at 80°C. The particulate organic carbon (POC ug/g) was calculated by equation 5, which shows the relationship between POC and POM, based on experiments to decomposition in tropical marine environment. The particulate organic nitrogen (PON ug/g) in sediments was determined by the Kjeldahl method of extraction, with a pre-digestion acid (H_2SO_4) of the samples with gradually increasing the temperature to 300°C.

Equations applied in analytical procedures:

$$TA = \left[\left(\frac{V_{HCl}.N_{HCl}}{V_{sample} + V_{HCl}} \right) - \frac{1}{10^{pH}.fH^+} \right] \times K \times 10^3 \Rightarrow TA = \left(2.5 - 1100 \times \frac{1}{10^{pH}.fH^+} \right) \quad (1)$$

Where TA is total alkalinity ($mEq.L^{-1}$), V is volume of HCl and of sample, N is the normality of HCl, and K is the correction factor of the sample.

$$CA = TA - A \quad (2)$$

Where A is the dissociation constant of boric acid in mEq/l obtained in calculation table.

$$\sum CO_2 = CA \, x \, F_t \quad (3)$$

Where F_t is the correction factor for temperature and salinity obtained in calculation table.

$$DIN = [NH_3] + [NO_3^-] + [NO_2^-] \quad (4)$$

Where [X] is the concentration of element X in the sample.

$$POC = \frac{POM}{K_x} \quad (5)$$

Where K_x = 1.9 is a factor of proportionality in the tropical marine environment, and represented the average value of 2.1 suggested by Aprile (2001) and 1.7 of Siqueira et al. (2003) for the region.

The results were statistically analyzed using multivariate analysis, Pearson correlation analysis, considering $p < 0.05$ level of significance, and linear regression: a) to examine the degree of affinity between environmental variables; b) identify the correlation between the variables and the existing local human activities, and c) establish a pattern of carbon and nitrogen flow between compartments within the Santos Estuarine System. Isolines maps were constructed using Surfer 9.0 (Golden Software Inc.) and interpolation of data on maps was made using the kriging method.

The use of historical data was crucial to improve the mathematical model, which is much more accurate as they studied the event in a longer period of time. In this research, we chose to apply the model of mass balance. The mass balance models are considered functional tools, specially developed for decision-making, aiming at solving complex problems associated with the dynamics of an ecosystem and the various interactions between the biotic and abiotic systems. The simulation was established from historical data and based on protocols to flow of nutrients established by LOICZ - Land-Ocean Interactions in the Coastal Zone (Gordon et al., 1996). Among the main characteristics of this method are: the ability to work with secondary data, the minimum requirement of data, the broad applicability and uniformity of methodology and information on the process flow of nutrients (C, N and P).

RESULTS AND DISCUSSION

The isolines of organic and inorganic carbon and nitrogen to the water and sediment are presented in Figure 2. The concentrations of DIC (Figure 2A) in the water ranged between 2.67 and 32.16 mg/l, with average 24.39 ± 3.62 mg/l. The higher isolines were found in Santos channel between stations 5 and 8, and in the eastern part of the Santos bay between stations 25 and 27. There was a wide variation in the concentration of DIC in vertical profile (surface and bottom), this phenomenon is explained by the fact that surface waters contain between 10 and 15% less DIC than deep waters, a difference in the procedure of photo-assimilation carbon by living organisms, and by the streams of falling particles of dead organic matter and carbonate shells into the sediment (Grasshoff et al., 1999). DOC levels (Figure 2B) in the water ranged from 0.10 to 21.53 mg/l with an average 9.61±6.29 mg/l. The high variation of DOC, identified by the high standard deviation reveals a spatial variability of the types of solid waste and liquid effluents released in the SES. The organic shapes are mainly originated from domestic sewage spills in the region, as shown by the high values found at station 22, located at the outlet of the outfall, which ruled in the Santos bay untreated domestic sewage associated with urban stormwater in large part the cities of Santos and São Vicente. With respect to nitrogen forms, levels of DIN (Figure 2C) ranged between 0.04 and 4.71 mg/l, and average 0.61±0.59 mg/l. The higher concentrations were found in the area under the influence of the industrial complex of Cubatão, between stations 8 and 9, where predominated levels of ammonium and nitrite, the most common forms of nitrogen in non-natural water or municipal wastewater. The concentrations of DON (Figure 2D), the same way as with organic carbon, showed a pattern of a tendency to concentrate the highest levels of organic nitrogen in the vicinity of the outfall exit, between the stations 19, 20 and 22 (output emissary), 23, 26 and 27, all stations located in the Santos bay. DON concentrations in water ranged between 0.01 and 2.29 mg/l, with average 0.31±0.33 mg/l. Comparatively, the levels of carbon and nitrogen organic and inorganic were always significantly higher in the bottom waters than surface waters, due to the higher density in the deeper layers. In general, the Santos bay concentrated the highest levels of organic compounds, especially NH_3 forms, which represent a strong pollution by domestic sewage. In general, the nitrogen in the organic matter varies with the type of organism. According to Braga et al. (2000), autotrophic organisms require inorganic nitrogen; ammonia ions on the one hand, nitrite and nitrate or molecular nitrogen on the other hand, and in some cases organic nitrogen. In coastal and estuarine systems, variations in concentrations or overload of nitrogen are an important factor in the evolution of ecosystems.

Concentrations of particulate organic carbon (Figure 2E) in surface sediments ranged between 900 and 38,638 µg/g, with average 19,658±10,548 µg/g, whereas

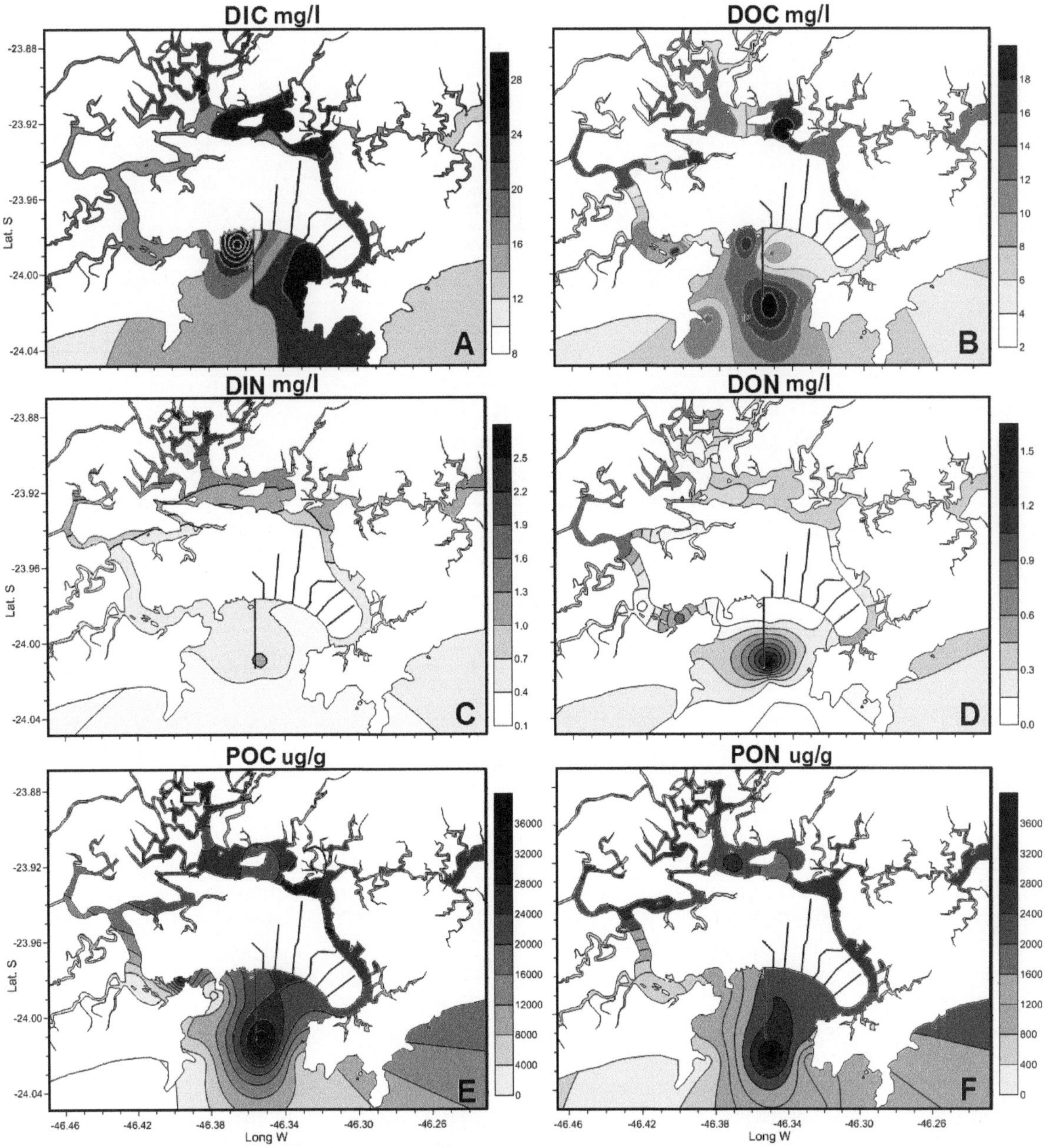

Figure 2. Isolines of carbon and nitrogen organic and inorganic to the water (DIC, DOC, DON and DIN) and sediment (POC and PON) compartments in Santos estuarine system.

the concentrations of particulate organic nitrogen (Figure 2F) ranged between 200 and 3550 µg/g, with average 1875±912 µg/g. The high standard deviations identified are explained by high variation in sediment composition and texture. In principle, higher levels of POC and PON were found on stations with predominantly muddy sediments rich in organic matter associated with clay, while the lowest levels were highly correlated with the stations with a predominance of sandy sediments and sandy clay. In general, in the sediment compartment the isolines concentrations of POC and PON tended to have higher levels in the Santos bay, near the outfall (station 22), and in the Santos channel where it is concentrated all the port activity loading and unloading of ships. The area under the influence of the industrial complex of Cubatão also showed higher values of POC and PON as a result of industrial and urban activities, such as the Village of Cubatão, an area of concern with various issues of public sanitation, including the treatment of disabled sewers.

Considering that salinity has a conservative behavior in coastal systems, and based on linear regression analysis shown in Figure 3, it was concluded that both total carbon (TC), result of adding the inorganic (DIC) and organic

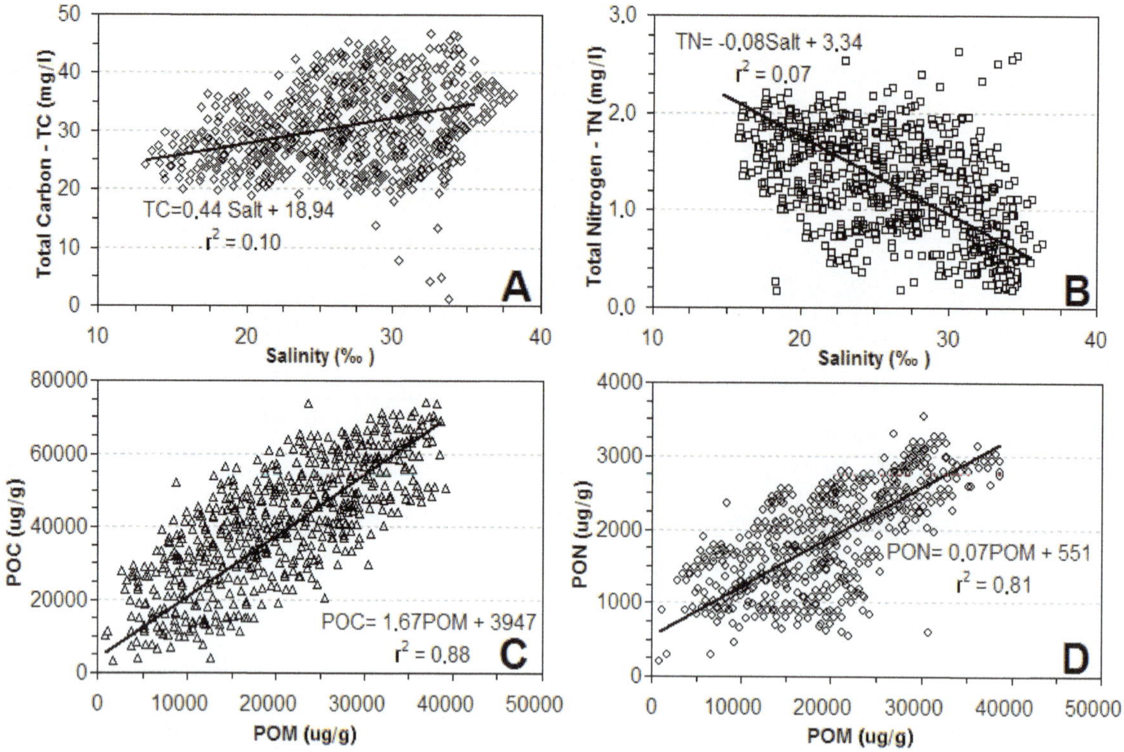

Figure 3. Linear regression analysis for correlation coefficient with $p < 0.05$ between salinity and total carbon (A) and salinity and total nitrogen (B) both the aquatic compartment, and correlation between organic matter and organic carbon (C) and organic matter and organic nitrogen (D) both in the sediment compartment.

(DOC) fractions, and total nitrogen (TN), the sum of DIN and DON fractions, did not have a conservative pattern in the aquatic system. The investigation of TC and TN revealed that its concentrations entering in the SES annually through of the Santos and São Vicente channels were of 1400 and 40 mg/l, respectively, while the levels of these elements that leave the system, by the connection between the Santos bay and the Ocean, were lower that 300 and 10 mg/l, respectively. This means that the SES accumulates carbon and nitrogen, which once not absorbed by the biota is stored in the sediment. According to the mass balance, the station 22, located at the outlet of the outfall, is permanent source of carbon and nitrogen, and depending on the tidal flow, this station can be considered as input (V_x) during the high tidal mark or as output (V_y) during the low tidal mark of TC and TN to the SES.

The correlation coefficients r^2 between TC and salinity and TN and salinity were low 0.10 and 0.07 respectively (Table 1, Figure 3A and B), indicating there is a variation in concentrations among these elements depending on the seasonality and of the flux of tides. Possibly, one part of the nutrients is assimilated by the intense local biota, and other part is added to the sediment, especially the muddy sediments present in the Santos and São Vicente channels. In general, the correlation among nutrients was weak in water, demonstrating independence between variables. The dissolved inorganic nitrogen and total nitrogen were highlights also with a strong positive correlation. The correlation coefficients determined for the regression analysis between POC and POM, and between PON and POM were significant with r^2 values 0.88 and 0.81 respectively (Table 1, Figure 3C and D), indicating a strong influence of particulate organic matter (POM) on the levels of carbon and nitrogen organics. In fact, in the channels where the concentration of organic matter was higher the levels of POC and PON were always high. An exchange may be occurring between the sources of marine and terrestrial organic matter prior to its disposal, caused by hydrodynamic parameters.

It is known the ability of organic matter to precipitate ionic elements through the mechanism of adsorption or complex formation. Because of this characteristic, and as a result of higher concentrations of organic matter are associated with muddy sediments, the port area (Stations 2 to 4) becomes the main area for adsorption of pollutants and contaminants in the sediments of the SES, particularly by metallic elements. The organic matter content in the Santos and São Vicente channels had a very similar behavior, while in the Santos bay this presented a heterogeneous behavior, in part due to the flow of tides that in the stations more distant from the outfall runs washing sediments seasonality, in each cycle of high tide. One factor that may have influenced the

Table 1. Pearson's correlation analysis* for the variables determined in the water column (surface and bottom) and sediment from the SES.

Variable	r (Pearson)	r^2	*p*	Variable	r (Pearson)	r^2	*p*
DICxDOC	0.2000	0.0400	0.0110	TCxDON	0.1712	0.0293	0.0297
DICxTC	0.8270	**0.6840**	0.0000	TCxTN	0.2315	0.0536	0.0032
DICxDIN	0.1804	0.0326	0.0219	TCxSalt	0.3163	0.1001	0.0000
DICxDON	0.1003	0.0101	0.2045	DINxDON	0.0277	0.0008	0.7263
DICxTN	0.2006	0.0402	0.0107	DINxTN	0.8240	**0.6790**	0.0000
DICxSalt	0.1166	0.0136	0.1401	DINxSalt	-0.1965	0.0386	0.0124
DOCxTC	0.8904	**0.7928**	0.0000	DONxTN	0.3067	0.0940	0.0001
DOCxDIN	0.2109	0.0445	0.0073	DONxSalt	0.1242	0.0154	0.1160
DOCxDON	0.1527	0.0233	0.0528	TNxSalt	-0.2646	0.0700	0.0060
DOCxTN	0.1654	0.0274	0.0358	POCxPOM*	0.9384	**0.8805**	0.0000
DOCxSalt	0.3279	0.1075	0.0000	POCxPON*	0.7794	0.6074	0.0000
TCxDIN	0.2565	0.0658	0.0010	POMxPON*	0.9010	**0.8118**	0.0000

*Variables for sediment; significant coefficients of correlations are marked in **bold** ($p < 0.05$), N= 416 and GL= 414.

results is the geomorphology of the region, with dozens of tiny channels and connections meander areas, always surrounded by mangrove vegetation. Sampling stations located in areas more sheltered from the action of tides showed low flow in the bottom, facilitating the accumulation of dissolved organic matter in sediments.

The organic pollutant loads from industrial sources exceeded the organic home-grown in the SES. According to Braga et al. (2000), the waters of Santos bay are contaminated by various industrial effluents, port waste dumps and domestic, those are released in the estuarine region and to come to bay through the Santos and São Vicente channels.

Carbon and nitrogen are two of the principal components of the organic matter. C/N ratios in sediments have aided to determine spatial and temporal variations of organic matter and lignin phenols in sources of organic matter to lakes (Mahapatra et al., 2011), as well as it is an alternative to identify fresh deposition of C and N from water column to sediment compartment. Plankton is rich in nitrogen forms and thus the C/N ratio is low, changing approximately between 5 and 9, including the deposits of organic matter in those organisms. Terrestrial vascular plants and their derivates in sediments have C/N ratios above 15, and high C/N ratios, above 24, can be considered as the mixing of terrestrial and autochthonous organic matter in sediments (Mahiques et al., 1999; Burone et al., 2003; Mahapatra et al., 2011), sometimes with contribution of macrophyte materials, implying, in this case, an increase of the remains partially destroyed of mangrove forests, with high content of plant fibers of cellulose and lignin. It is assumed that the mangrove vegetation is a strong supplier of nutrients, especially carbon to the water column and, consequently, to the sediment. Table 2 shows the range of variation of the rates C/N in the subareas that comprise the SES. The molar rate C/N had high variation in the aquatic compartment, between 7 and 106 with an average 46±26, while the rates for the sediment compartment varied between 1 and 22 with an average 10±4. The high amplitude of variation, especially in the aquatic compartment, indicated a general instability in the concentrations of carbon and nitrogen. The stations located in São Vicente channel and in part of the Santos bay were the main responsible for these high rates. This pattern of concentration of nutrients appears to be a consequence of the following factors: 1) marked reduction of the nitrogen fixed by excessive consumption in the sediment by benthic organisms, 2) growth of carbon compared to nitrogen by deposition of waste material from terrestrial organic matter of the basin drainage of Baixada Santista. Especially in the Santos bay because of the link with the sea, the biological material of marine origin is enriched in nitrogen when compared with terrestrial material. This probably caused a C/N ratio lower. The diagenetic processes of organic carbon change the rates C/N, which became higher.

Hydrodynamic model and nutrient budgets

To understand the processes governing the biogeochemical cycles in estuarine and coastal areas is necessary to know the local processes that control the flow of C, N and P. In areas partially or totally rich of mangrove and of the estuary, the dynamics of nutritional flow is intense and very fast. The processes of decomposition and releasing of organic matter are continuous and the ionic charge of elements released is very large, making these ecosystems among the most dynamic and intense in nutrient cycling. The marine biological productivity in surface waters is mainly controlled by the availability of macroelements, particularly nitrogen and phosphorus, and probably is not influenced

Table 2. C/N rates in water and sediment compartments of the SES.

Area	Rate C/N water	Rate C/N sediment
Santos Channel	16-90	9-14
Cubatão	7-33	10-12
São Vicente Channel	20-90	1-22
Santos Bay	18-106	4-14

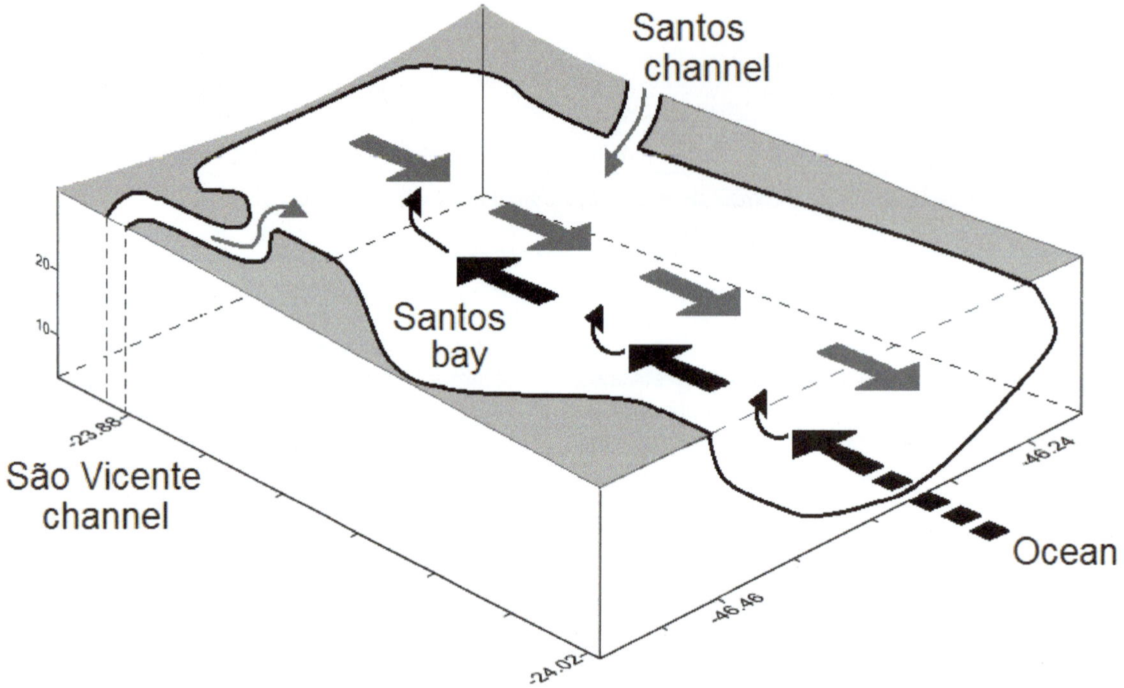

Figure 4. Simplified model of freshwaters and marine waters circulation in the Santos estuarine system.

by changes in inorganic carbon. For this reason, the flow of organic material is often not included in the models used to simulate the response of the oceans to increase anthropogenic discharges.

Equations were applied to the results to understand the fluxes based on the type of estuary, tides and channels, taking into account the density gradients that drive the patterns of circulation and mixing in the estuarine system. The SES was classified as a partial mixture system of water, which receives a supply of dense waters by the tide and layers of freshwater less dense and rich in organic material coming from the channels that border the swamp vegetation (Figure 4). In Santos bay, the tidal influence is greater, besides the flow from the channels can not dominate the circulation pattern, which in this case, occurs most often in a counterclockwise direction, partially removing the water bodies of the SES and discharging them into the ocean. Thus, the renewal of the waters of the SES is partial as these waters are carried by currents to the sea each tidal cycle. There is also a smaller circulation that occurs in the phases of low tide, also in counterclockwise, just inside the Santos bay.

The salinity gradient showed that the Santos channel at the port of Santos, has a greater influence of marine waters, regardless of what stage of high or low tide, shows stratification from station 1 to station 7. The water circulation occurs with marine waters flowing into the bay by the deeper layers, and freshwaters flowing in the bay by the more superficial layers. The São Vicente channel, for their localization, and morphological and morphometric characteristics, presents most part of the hydrological cycle a predominance of freshwaters coming from the rivers that supply the estuarine system. In fact, analyzing the degree of stratification in this channel, it is noted that from station 17 toward the interior of the estuary to station 9, the salinity drops to levels of brackish waters. Thus, the physical stratification imposed by the density is reduced gradually over the water column, contributing to the water mixture. The horizontal gradient of salinity in the bay is maintained by diffusive

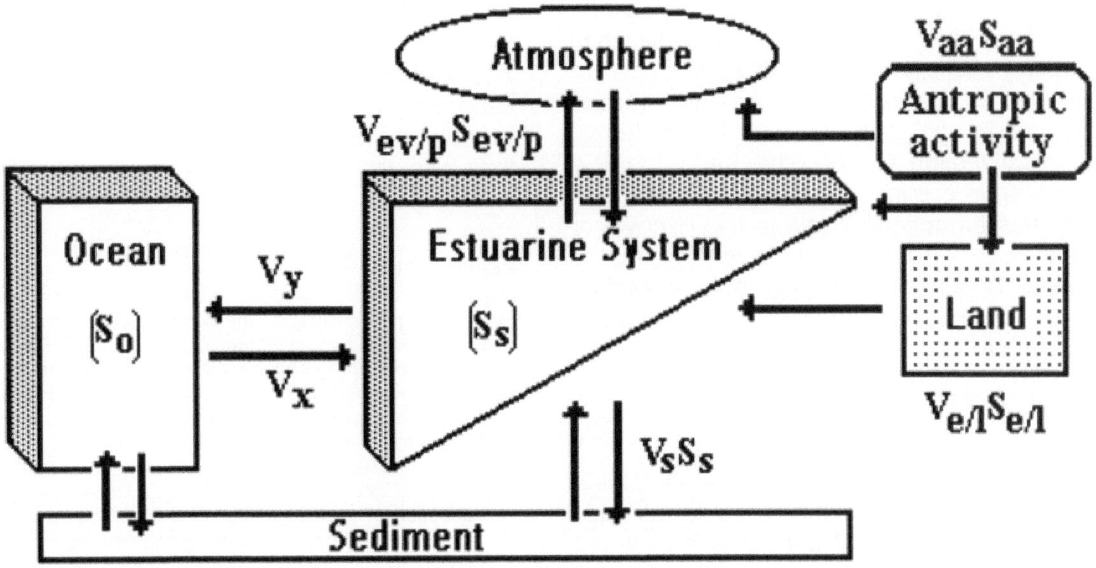

Figure 5. Flow model of carbon and nitrogen for the Santos estuarine system - SES, which is based on the analysis of mass balance of nutrients and conservative pattern of salinity in the system.

flux of salts to the sea due to the mix by the action of tides and winds.

Figure 4 shows schematically the flow of marine waters and freshwaters circulating inside the estuarine system. From this model, the circulation of water masses, and based on nutrient data, we developed a flux model of C and N based on the suggestions of Gordon et al. (1996) for an area previously defined. Furthermore, we took into consideration, for developing the model, the residence time of water in the system and routes of entry and exit of the estuary, including the atmosphere, land adjacent eroded or leached, and coastal and ocean systems. The sediment compartment was considered simultaneously as a reservoir and source of carbon and other elements, including metallic elements to the water. Thus, we have the following Equation 6 of conservation of mass:

$$\frac{dM}{dt} = \left(\Sigma In_i - \Sigma Ou_i\right) + \left(\Sigma P - \Sigma C\right) \Leftrightarrow \frac{dM}{dt} = 0 \qquad (6)$$

Where In_i is the inflow to the system; Ou_i is the outflows of the system, P is the production inside the system, and C is considered consumption (decomposition/oxidation and respiration).

A solution derived from Equation 6 and using the system flow shown in Figure 5 for the inputs of nutrients in the system (In_i) can be explained by the equation:

$$\sum In_i = V_{e/l}S_{e/l} + V_{ev/p}S_{ev/p} + V_sS_s + V_x + V_{aa}S_{aa} \qquad (7)$$

Where V is the nutrient load; and S is the charge of dissolved ions; *e/l* represents the nutrients from adjacent soil transported by the erosion and leaching processes; *ev/p* represents the evaporation and precipitation processes; S is the sedimentary flux; x is the nutritional flux from ocean to the estuary, and y is the nutritional flux from estuary to the ocean.

Equation 7 can be rearranged in order to estimate the anthropogenic activity in the region (V_{aa}), which can be used for any material in the water. We replace the flow of the ocean to the SES (V_x) in Equation 7, simplifying the expression to determine the total load on the system input (Equation 8):

$$V_x = \frac{V_rS_r}{S_0 - S_s}, \text{ and } S_r = \frac{S_0 + S_s}{2} \text{ we have} \Rightarrow$$

$$\sum In_i = V_{e/l}S_{e/l} + V_{ev/p}S_{ev/p} + V_sS_s + V_{aa}S_{aa} + \frac{V_r(S_0 + S_s)}{2(S_0 - S_s)} \Leftrightarrow$$

$$\sum In_i = V_iS_i + \frac{V_r}{2}\frac{(S_0^2 - S_s^2)}{(S_0 - S_s)^2} \qquad (8)$$

Where V_r and S_r are part of the residual volume of nutrients and ions in the system.

A solution based on the flow system shown in Figure 5 for the outputs of nutrients in the system (Ou_i) can be explained by Equation 9:

$$\sum Ou_i = V_sS_s + V_y \Leftrightarrow \sum Ou_i = V_sS_s + V_rS_r \qquad (9)$$

Of all the sources of input to the SES via anthropogenic, $V_{aa}S_{aa}$ is definitely the most important by volume, which is almost constant of discharges of effluents rich in inorganic (industrial sources) and organic (urban sources) compounds, as well as its high potential pollution and contamination. On the other hand, the flow or pulse of tide that is coming in and circulates through the Santos bay has an important role in the removal of excess waste transported by the connection of Santos and São Vicente channels, and also by the outfall, responsible for important contribution of organic compounds to the bay.

Conclusion

Within the study period, the volume of water circulated inside the SES was governed by the hydrological cycle (rainfall) with a strong influence of the pulse of tidal, not just in the bay, but also in all the large and small channels that make up the mesh of the estuarine system. The salinity gradient was conservative for all SES, while the variation in concentration of nutrients depended on much of the area of direct influence, considered in this research as the source of supply of nutrients, and the area of indirect influence, defined as the area that removes or modifies the concentration of the elements C and N in the water column and sediment: Could not find a conservative pattern of nutrients to the estuarine system as a whole, but the inorganic forms remained constant in certain parts of SES, especially in the region under direct influence of the industrial complex of Cubatão and port of Santos (Stations 3 to 9). In this case, the DIC has always remained at high concentrations in the water column in the space-time (Figure 2A), which indicates that there is in these stations one or more continuous sources of dump material rich in inorganic carbon. On the other hand, the Santos bay presented the highest variations in nutrient concentration along the current flow, a fact explained by local ocean dynamics. The data for the Santos bay has shown that it differs from Cubatão region and port area due to the tendency to conserve dissolved nutrients in the suspension. The effluent from the Cubatão city associated with solid waste in the port area are transported to the sea through the Santos and São Vicente channels, in particular the first, while in the bay and areas dominated by mangrove vegetation, the load of pollutants and contaminants is kidnapped, degraded and diverted to the food chain.

This study allowed understanding better the dynamics of nutrient flow due to the local hydrological cycle, especially under the influence of the pulse of tides. Thus, the results added to other work already developed in the region, may contribute to the management of industrial areas (industrial complex of Cubatão and port of Santos) and urban areas (Santos and São Vicente cities), to reduce pollutant loads and contaminants in the estuarine system.

ACKNOWLEDGEMENTS

The authors extend their thanks to colleagues at the Laboratory of Nutrients (LABNUT) of the Oceanographic Institute of USP, especially Professor Elisabete Braga for their help and support with some chemical analysis.

REFERENCES

Aprile FM (2001). Study of the dynamic and mass balance modelo of the carbon in Santos Estuarine System, São Paulo – Brasil. Pos-doctoral report, Oceanograph Institute, São Paulo University, p.129.

Braga ES, Bonetti CH, Burone L, Bonetti Filho J (2000). Eutrophication and bacterial pollution caused by industrial and domestic wastes at the Baixada Santista Estuarine System – Brazil. Mar. Pollut. Bull., 40(2): 165-173.

Burone L, Muniz P, Pires-Vanin MAS, Rodrigues M (2003). Spatial distribution of organic matter in the surface sediments of Ubatuba Bay (Southeastern – Brazil). Anais da Academia Brasileira de Ciências, 75(1): 77-90.

Cox P, Friedlingstein P, Rayner P (2002). Modelling climate – Carbon cycle feedbacks: A cross disciplinary collaboration priority. Glob. Change News Lett., 49: 12-14.

Gordon Jr DC, Boudreau PR, Mann KH, Ong JE, Silvert WL, Smith SV, Wattayakorn G, Wulff F, Yanagi T (1996). LOICZ Biogeochemical Modelling Guidelines. LOICZ Rep. Stud., 5: 1-96.

Grasshoff K, Kremling K, Ehrhardt M (1999). Methods of Seawater Analysis. Weinheim: Wiley VHC Verlag, 3rd. Edition, p. 600.

Howarth RW, Billen G, Swaney D, Townsend A, Jaworski N, Lajtha K, Dowling JA, Elmgren R, Caraco N, Jordan T, Berendse F, Freney J, Kudeyrof V, Murdoch P, Zhao-Liang Z (1996). Regional nitrogen budgets and riverine N and P fluxes for the drainages to the North Atlantic Ocean: natural and human influences. Biogeochemistry, 35(1): 75-139.

Jorgenson MT, Brown J (2005). Classification of the Alaskan Beaufort Sea coast and estimation of carbon and sediment inputs from coastal erosion. Geo-Mar. Lett., 25: 69-80.

Kannen A, Burkhard B (2009) Integrated Assessment of Coastal and Marine Changes Using the Example of Offshore Wind Farms: the Coastal Futures Approach. GAIA, 18(3): 229-238.

Lantuit H, Rachold V, Pollard WH, Steenhuisen F, Ødegård R, Hubberten HW (2009). Towards a calculation of organic carbon release from erosion of Arctic coasts using non-fractal coastline datasets. Mar. Geol., 257: 1-10.

Le Tissier MDA, Buddemeier R, Parslow J, Swaney DP, Crossland CJ, Smith SV, Whyte HAY, Dennison W C, Hills JM, Kremer HH (Eds.) (2006). The role of the coastal ocean in the disturbed and undisturbed nutrient and carbon cycles - A management perspective. LOICZ, Geesthacht, Germany, p. 44.

Mahapatra DM, Chanakya HN, Ramachandra TV (2011). C:N ratio of Sediments in a sewage fed Urban Lake. Int. J. Geol., 3(5): 86-92.

Mahiques MM, Mishima Y, Rodrigues M (1999). Characteristics of the sedimentary organic matter on the Inner and Middle Continental Shelf between Guanabara bay and São Francisco Sul, Southeastern Brazilian. Margin. Cont. Shelf. Res., 19: 775-798.

Meybeck M (1993) Carbon, nitrogen and phosphorus transport by world rivers. Am. J. Sci., 282: 401-450.

Parsons TR, Maita Y, Lalli CM (1984). A manual of chemical and biological methods for seawater analysis. Pergamon Press, Michigan, 173 pp.

Pettine M, Patrolecco L, Camusso M, Crescenzio S (1998) Transport of carbon and nitrogen to the northern Adriatic Sea by the Po River. Estuar. Coast. Shelf Sci., 46: 127-142.

Siqueira GW, Aprile FM, Mahiques MM, Braga ES (2003). Determinação da matéria orgânica em sedimentos de fundo dos estuários de Santos/São Vicente e Baía de Santos - SP/Brasil. In: III Congresso Brasileiro de Pesquisas Ambientais e Saúde, 2003, COPEC, São Paulo, pp. 131-136.

Smith SV (2002). Carbon-nitrogen-phosphorus fluxes in the coastal zone: the global approach. Glob. Change News Lett., 49: 7-11.
Smith SV, Crossland CJ (1999). Australasian Estuarine Systems: carbon, nitrogen, and phosphorus fluxes. LOICZ Rep. Stud., 12: 182.
Streletskaya ID, Vasiliev AA, Vanstein BG (2009). Erosion of sediment and organic carbon from the Kara Sea coast. Arct. Antarct. Alp. Res., 41: 79-87.
Trefry JH, Metz D, Nelson TA, Trocine RP, Eadie BJ (1994). Transport of particulate organic carbon by the Mississippi River and its fate in the Gulf of Mexico. Estuaries, 17: 839-849.
Wetzel RG, Likens GE (2000). Limnolgical analysis. W.B. Saunders Co., Philadelphia, p. 357.

14

Notes on three new records of scleractinian corals from Andaman Islands

Tamal Mondal[1]*, C. Raghunathan[1] and Ramakrishna[2]

[1]Zoological Survey of India, Andaman and Nicobar Regional Centre, National Coral Reef Research Institute, Haddo, Port Blair-744 102, Andaman and Nicobar Islands, India.
[2]Zoological Survey of India, Prani Vigyan Bhawan, M- Block, New Alipore, Kolkata-700 053, India.

Andaman group of islands being a part of India (megadiverse country) shows a great deal of ecological diversity both in terrestrial and marine ecosystems. Coral Reefs are the building blocks of marine ecological niche for the sustainable existence of marine biodiversity and advocates the other associated faunal communities for their betterment, as well as their way of life towards the future generation. Three scleractinian corals *Montipora gaimardi* Bernard, 1897, *Podabcia lankaensis* Veron, 2000, *Podabacia sinai* Veron, 2000, are identified as the new record to Indian water from this group of islands of Andaman Archipelago. These three corals belong to Acroporidae and Fungiidae families. The present study describes the taxonomical features of the identified scleractinians based on their morphology and corallites structures.

Key words: Scleractinia, Andaman Islands, Montipora, Podabacia.

INTRODUCTION

Andaman Sea is a semi-enclosed basin connected by the three main channels to the Bay of Bengal on one side and to the Pacific on the other sides. The Preparis channel, on the north, which is 285 km wide, the ten degree Channel, which is about 150 km wide, separating the Andaman groups from Nicobar groups, whiles the Great Channel, south of Great Nicobar Island is about 189 km wide. Large part of the bottom contained hard rock, a series of ridges and depressions, which include Alcock Sea Mount and Narcodum- Barren Basis (Swell, 1925). Andaman Sea is gifted with rich legacy of corals and their associated fauna (Rao, 2010). Coral reefs of the world cover an estimated area of 600000 km^2 (Smith, 1978; Kleypas, 1977). Over half of these (54%) coral reefs lie in the Mediterranean Indian Ocean (Scheer, 1985). The corals of the Andaman Islands belong to the Indo-West Pacific faunal province (Hoeksema and Dai, 1992). The coral reefs of Andaman and Nicobar Islands are the biodiversity hot spot of India (Jeyabaskaran, 1999). The importance of coral reef ecosystems may be seen in their numerous ecological, aesthetic, economic and cultural functions. Coral reefs, in particular, are critical habitat that supports diversity of both residential and migratory fauna species especially for those considered to be endangered and vulnerable. The structure of a reef provides homes and food for many types of plants, fish and invertebrates (Nelson, 1999). Coral reefs are the largest structures made by living things and exist as extremely productive ecosystems in tropical and sub-tropical areas of the world. These ecosystems are packed with the highest densities of animals to be found anywhere on the planet. Thronging with life, they rival even the tropical rainforests in terms of diversity (Mark, 2000). Coral reefs, in particular, are critical habitat that supports diversity of both residential and migratory faunal species especially for those are considered to be endangered and vulnerable. Coral and rocky reefs constitute are considered as important ecosystems in our planet, being their astonishing diversity, productivity, abundance and beauty are some of the main characteristics (Goldman and Talbot, 1976) According to recent studies, coral reefs constitute one of the country's main marine assets (Schleyer et al., 1999;

*Corresponding author. E-mail: t_genetics@yahoo.com

Rodrigues et al., 1999). The present paper deals on the taxonomical features of the three identified scleractinian corals and their key characters as these are noted as record to Indian Water.

MATERIALS AND METHODS

The survey was made at several sites (Figure 1) of Andaman group of islands during the month of January to May, 2010 by using self-contained underwater breathing apparatus (SCUBA) diving and snorkeling. Line intercept transect and quadrate methods (English et al., 1996) were applied to investigate the diversity and distribution of the corals. During SCUBA diving, species recording was made by underwater digital photography (Sony - Cyber shot, Model-T900, marine pack, 12.1 megapixels) for detailed identification. Identification was done by *in situ* observation and photographs in conjunction with Veron (2000), Wallace (1999) and Sheppard (1987). Specimens were also collected to study their morphological and taxonomic characters. After completion of the examination of taxonomic features, all the collected specimens were registered and deposited in the National Zoological Collection of Zoological Survey of India, Andaman and Nicobar Regional Centre, Port Blair.

RESULTS

During the present study, three species of scleractinian corals were identified as new record to Indian Waters. One species belonged to the genus *Montipora* (Family-Acroporidae) and two species belonged to genus *Podabacia* (Family- Fungiidae). On the basis of taxonomical features of their corallites structure, the following description was made.

Montipora gaimardi (Bernard, 1897)

Place of record in Andaman and Nicobar Islands: Surumai Dikri (Latitude 11°25.504'N and Longitude 92°40.301'E), Rutland Island, Andaman. The sample was collected from a depth of 4.5 m of the surveyed area (Figures 2 and 3).

Material examined: One sample was examined. The length was 102 mm, width was 86 mm and height was 69 mm, Reg. No. ZSI/ANRC- 4737.
Description: Colonies have thick encrusting bases and branches which may be open or form a compact clump. The coenosteum forms a complex of fine ridges. The coarallites are in and between these ridges.
Key character: No axial corallite, corallites < 2 mm diameter, columella absent, branches with basal structures...Genus *Montipora,* Colony dominated by branching growth form, branches thick, branches not submassive, colonies without explanate bases, *Montipora gaimardi.*
Colour: Brown or green, some times with pale branch tips.
Similar species: *Montipora cactus* and *Montipora australiensis,* which have corallites in and between coensteum ridges.
Occurrence: Common.
Distribution: Australia, Indonesia, Irian Jaya, Japan, Philippines, Solomon Islands and Tonga.
Cites appendix: II
EC regulation: B
IUCN red list category and criteria: Vulnerable, 2010

Podabacia lankaensis (Veron, 2000)

Place of record in Andaman and Nicobar Islands: Rail Island (Latitude 12°59.033'N and Longitude 92° 54.137'E), Andaman. The sample was collected from a depth of 7 m (Figures 4 and 5):

Material examined: One material was examined. It is slightly oval in structure. Radius of the polyp is 7.9 cm in longest side and 7.7 cm in shortest side, Reg. No. ZSI/ANRC- 4738.
Description: Colonies are loosely attached or free-living, encrusting or laminar, and unifacial, with irregularly lobed margins and an irregularly contorted surface. There is no central corallite. Peripheral corallites are inclined towards the plate margins. Septo-costae are similar to those of *Halomitra.*
Key character: Colony are attached to substrate,colony mostly explanate...Genus *Podabacia:* Colony irregularly contorted *Podabacia lakaensis.*
Colour: Grey-brown white margins.
Similar species: This species combines various characters of *Lithophyllon, Podabcia* and *Halonitra.* Corallites are smaller, less regular and more inclined on the colony surface than are those of *Podabcia motuporensis.*
Occurrence: Rare.
Distribution: Sri Lanka.
Cites appendix: II
EC regulation: B.
IUCN red list category and criteria: No entries found.

Podabacia sinai (Veron, 2000)

Place of record in Andaman and Nicobar Islands: Surumai Dikri (Latitude 11°25.504'N and Longitude 92°40.301'E), Rutland Island, Andaman. The sample was collected from a depth of 6.5 m (Figures 6 and 7).

Material examined: One material was examined. It is slightly oval in one side and other side is irregular in shape. Radius of the longest side is 8.5 cm. and shortest side is 7.3 cm, Reg. No. ZSI/ANRC- 4739.
Description: Colonies are attached, laminar, unifacial and may form tiers. Plates have usually lobed margins and commonly have radiating costal ridges. Corallites are small and not inclined towards the plate margins. Septo-costae are similar to those of *Halomitra.*
Key character: Colony attached to substrate, colony

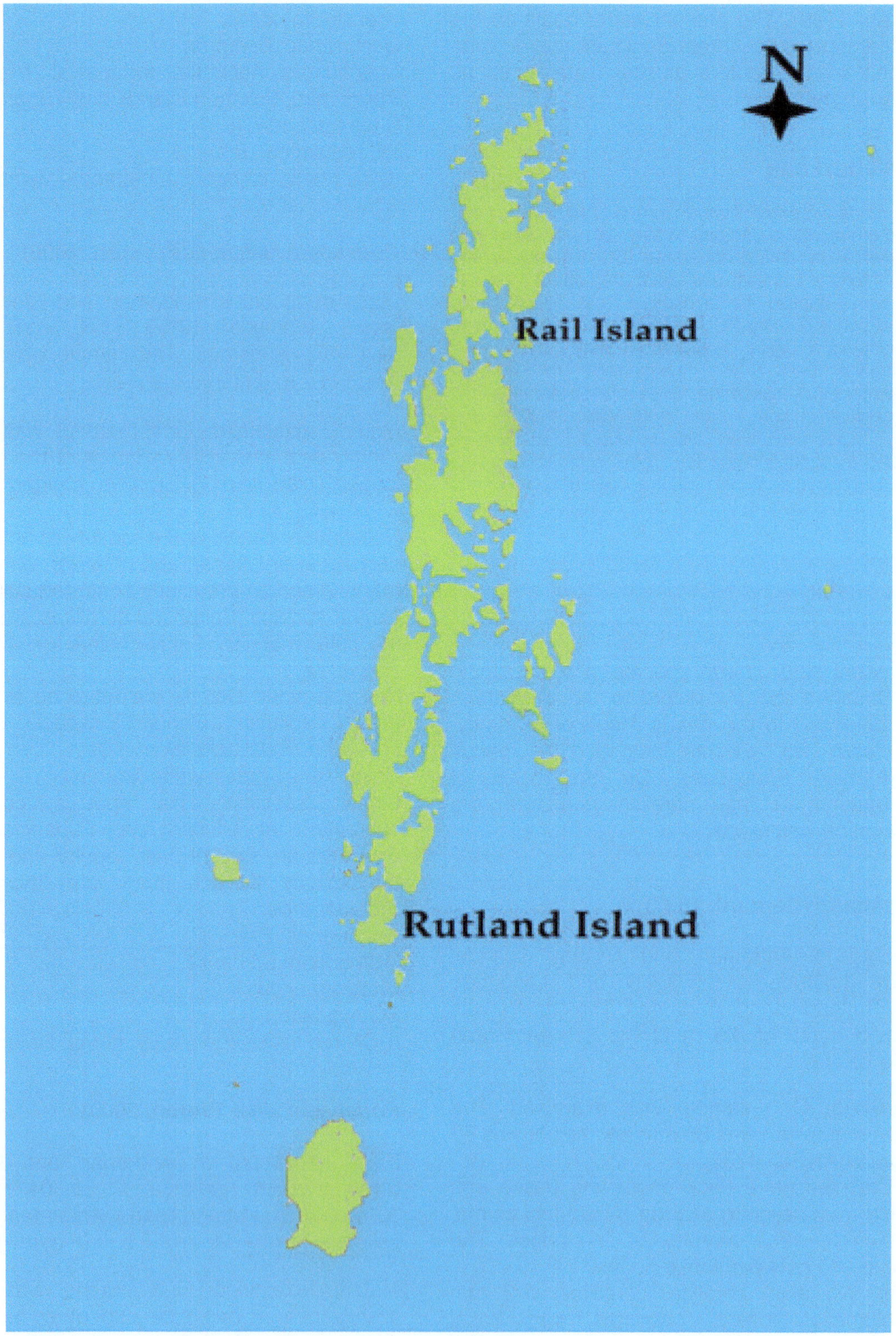

Figure 1. Study areas of Andaman Island".

mostly explanate...Genus *Podabacia:* Colony not contorted, colony with lobed margins *Podabcia sinai.* Colour: Pale brown with white margins.

Similar species: *P. motuporensis:* Is very similar, but have smaller corallites and finer costae. Podabacia crustacean: Forms larger colonies and has larger

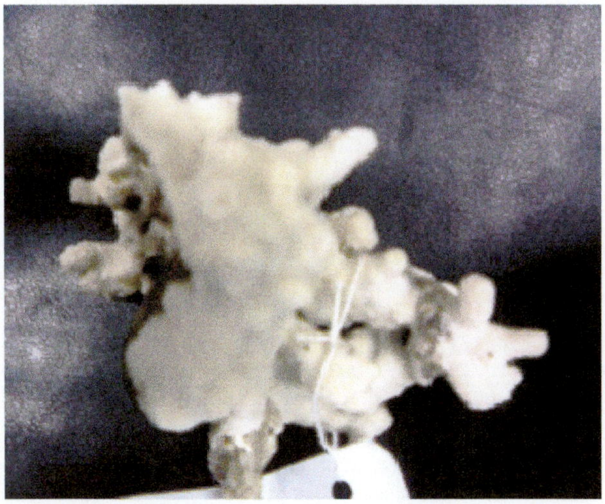

Figure 2. *Montipora gaimardi* (Bernard, 1897).

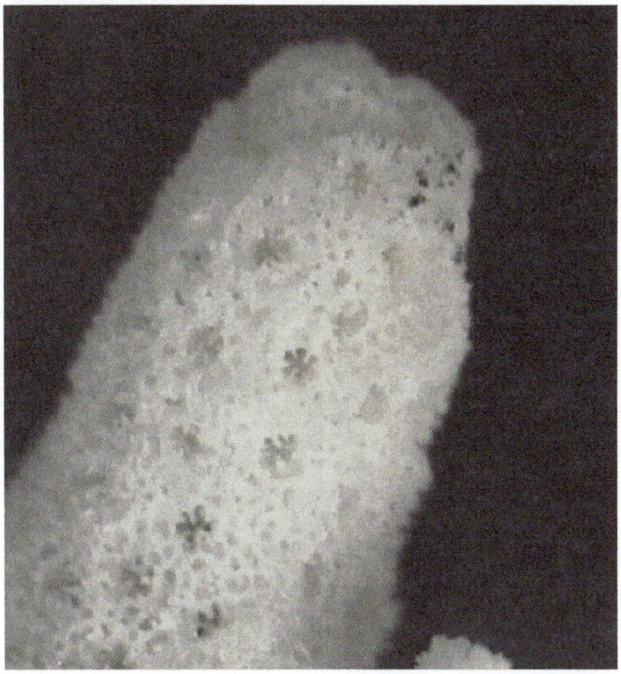

Figure 3. Corallites of *Montipora gaimardi.*

Figure 4. *Podabcia lankaensis* (Veron, 2000).

Figure 5. Septal arrangement of *Podabcia lankaensis.*

corallites.
Occurrence: Uncommon.
Distribution: Egypt.
Cites appendix: II
EC regulationand criteria: Data deficient.

DISCUSSION

Coral reefs are some of the most productive ecosystems, providing habitat for numerous species and serving important ecological functions (Precht, 2006). Coral reefs support extraordinary biodiversity although they are located in nutrient-poor waters. On the basis of a taxonomical character, three scleractinian corals are identified from this group of islands as the new record to Indian water. Pillai (1983) listed 135 coral species from Andaman, and found that Andaman Islands were less diverse (31 genera with 82 species) than the Nicobar Islands (43 genera with 103 species). Presence of these corals in the mentioned region is the indicator of

Figure 6. *Podabacia sinai* (Veron, 2000).

Figure 7. Septal arrangement of *Podabacia sinai* (Veron, 2000).

progressiveness of marine biodiversity. The Government of India and UNDP GEF field mission (2001 diving studies) reported a total 198 species of scleractinian coral from Andaman groups of island of which 111 are supposed to be new records to India (on verification with other studies only 94 species are found to be new records and this also include some non-scleractinian corals). Venkataraman et al. (2003) described 228 species of corals belonging to 58 genera and 15 families. The presence of the corals will increase the associated faunal communities on the selected places (Licuanan, 2004). 223 species of scleractinian have been reported from the Andaman Islands (Turner et al., 2009). Later Tamal et al. (2010, 2011) recorded seven and five scleractinian corals from Andaman and Nicobar Islands as an indicator of increasing Scleractinian database. This identification of scleractinian corals will be helpful for the preparation of conservation and management action plan for the scleractinian corals of Andaman Islands. Further surveys are required in near future to get more apprehensive data on different types of species.

REFERENCES

Bernard HM (1897). Catalogue of the Madreporarian corals of the British useum (Natural History). III. The genus *Montipora*. The genus Anacropora-London.

English S, Wilkinson C, Baker V (1996). Survey Manual for Tropical Marine Resources. Australian Institute of Marine Science, Townsville.

Goldman B, Talbot FH (1976). Aspects of the Ecology of Coral Reef Fishes. In: Jones, O. A. & R. Endean (editores). Biology and Geology of Coral Reefs., New York, Academic Press, 3(2): 125-154.

Hoeksema B, Dai CF (1992). Scleractinian of Taiwan. II. Family Fungiidae (including a new species). Bull. Zool. Acad. Sin., 30: 201-226.

Jeyabaskaran R (1999). Report on Rapid Assessment of coral reefs of Andaman and Nicobar Islands. GOI/UNDP/GEF Project on Management of Coral Reef Ecosystem of Andaman and Nicobar Islands. Published by Zoological Survey of India, Port Blair, p. 110.

Kleypas J(1977). Modeled estimates of global reef habitat and carbonate production since last glacial maximum. Palaeoecanography, 12: 535-545.

Licuanan WY (2004). New records of stony corals from the Philippines previously known from peripheral areas of the Indo-Pacific. Raffles Bull. Zool., 52(2): 285-288.

Mark D, Spalding, Corinna R, Edmund PG (2000). World Atlas of Coral Reefs University of California Press, Berkeley Los Angeles, London.

Nelson V (1999). State of Coral Reefs Cambodia. Environment Coastal Zone Project, MoE/Danida, Phnom Penh, Cambodia.

Pillai CGS (1983). Structure and genetic diversity of recent scleractinian of India. J. Mar. Bio. Assoc. India, 25: 78-90.

Precht WF (2006). Coral Reef Restoration handbook, Laylor and Francis Group, 6pp.

Rao DV (2010). Field Guide to Coral and Coral Associates of Andaman & Nicobar Islands, pp. 1-283.

Rodrigues MJ, Motta H, Pereira MAM, Gonçalves M, Carvalho M, Schleyer M (1999). Reef Monitoring in Mozambique. I: The Monitoring Programme and 1999 Report, Maputo, MICOA/IIP, p. 57.

Scheer G (1985). The distribution of coral reefs in Indian Ocean with a historical review of its investigation. Deep Sea Research, Part A, 31: 885-900.

Schleyer MH, Obura D, Motta H, Rodrigues MJ (1999). A Preliminary Assessment of Coral Bleaching in Mozambique. South African Association for Marine Biological Research, Unpublished Report, 168: 12

Sewell RBS (1925). Temperture and salinity of the costal waters of the Andaman Sea. Mem. Asiat. Soc. Bengal, 9(1): 133-198.

Sheppard CRC (1987). Coral species of the Indian Ocean and adjacent seas: a synonymised compilation and some regional distribution patterns. Atoll Res. Bull., p. 307.

Smith SV (1978). Coral reef area and the contribution of reefs to processes and resources of the world oceans. Nature, 273: 225-226.

Tamal M, Ragunathan C, Sivaperuman C, Ramakrishna (2010). Identification of Seven Scleractinian Corals from Andaman and Nicobar Islands as New Record to IndianWater. Proc. Zool. Soc., 63 (1): 61-66.

Tamal Mondal, Raghunathan, C, Ramakrishna. (2011) New Record of Five Scleractinian Corals from Rutland Island, South Andaman Archipelago. Asian J. Exp. Biol Sci., 2(1): 114-118.

Turner JR, Vousden D, Klaus R, Satyanarayana CH, Fenner D, Venkataraman K, Rajan PT, Subba Rao NV, Alfred JRB, Ramakrishna, Raghunathan C (2009). Coral Reef ecosystem of Andaman Islands - Remote Sensing and Rapid Site Assessment Survey. Rec. zool. Surv. India, Occ. Paper No., 301: 1-132.

Venkataraman K, Satyanarayan CH, Alfred JRB, Wolstenholme J (2003). Handbook on Hard Corals of India, pp. 1-266.

Veron JEN (2000). Corals of the World. Aust. Inst. Mar. Sci., pp. 1-3.

Wallace CC (1999). Staghorn Corals of the world. CSIRO Publications, Melbourne, p. 421.

Litter decomposition dynamics associated with cashew nut plantation in coastal habitat of Orissa, India

T. Panda[1]*, P. K. Pani[2] and R. B. Mohanty[3]

[1]Department of Botany, S.N.College, Rajkanika, Kendrapara- 754220, Orissa, India.
[2]Department of Botany, Agarpara College, Agarpada, Bhadrak, Orissa, India
[3]Department of Botany, N.C.Autonomous College, Jajpur, Orissa, India.

Being a key process in nutrient and carbon cycling in coastal ecosystems, decomposition of litter in coastal sand dunes of Orissa was studied in relation to abiotic and biotic factors. Microbial isolation was performed using standard procedures and decomposition of leaf litter was studied using litter bag technique. Higher microbial populations were encountered in soil with cashew plantations compared to barren sand dunes, corresponding with the fluctuation of prevailing temperature and moisture. Micro-fungi of both soils showed a positive correlation with soil moisture but were negatively correlated with soil temperature. A total of 129 species of fungi belonging to 68 genera were enumerated of which sand dune with cashew plantation had a share of 45 genera, 114 spp. and barren sand dune contributed 51 genera and 112 spp. Sand dune without vegetation revealed 32 restricted spp. while *Anacardium* plantation soil revealed 34 spp. *Trichoderma, Aspergillus* and *Penicillium* are predominant occurrence and have been found to be effective decomposers in the present study. Rates of litter loss and carbon dioxide output followed the same trends as the fungal numbers. Shannon- Wiener index and richness were higher in soil with *Anacardium* plantation than the barren sand dune. Maximum fungal population density was observed in the rainy season followed by winter and summer months. The rate of decomposition was controlled by rainfall, soil temperature, soil moisture, relative humidity and fungal count of soil. This research will help in monitoring the isolation, successful introduction and management of decomposing fungi and their importance in ecosystem service could be a valuable tool for understanding the long term effects of changes in soil condition of the forest floors in coastal sand dunes of Orissa, India.

Key words: Decomposition rate, diversity indices, fungi, abiotic factors, soil respiration.

INTRODUCTION

Orissa, one of the eastern coastal states of India has 480 km long barren coast line of sand dunes. Presently, monoculture plantations of *Casuarina equisetifolia* L. in the outer belt and *Anacardium occidentale* L. in the inner belt have been created along coastline to prevent wind blast and erosion of sand dunes. Although, it has solved the problem to some extent, but the litter decomposition dynamics of *Anacardium* species is yet to be studied. Plant litter decomposition is considered as complex and important factor in controlling both vegetation structure (Liu et al., 2004) and ecosystem function (Xiong and Nilsson, 1999). In particular, litter decomposition plays an

*Corresponding author. E-mail: taranisenpanda@yahoo.co.in.

important role in nutrient cycling and organic matter turnover within ecosystems (Smith and Bradford, 2003) and is important determinant for maintaining the biosphere; it also performs unique and indispensable activities on which larger organisms including humans depend.

However, the rate of decomposition is influenced by environmental factors and also by physico-chemical properties of leaves of the species studied and decomposing organisms present in the soil (Vesterdal, 1999; Wedderburn and Carter, 1999; Smith and Bradford, 2003). Detritus decomposition also encompasses interactions among many types of organisms and environmental variables (Reshi and Tyub, 2006), each effecting quality of the litter as a substrate for the subsequent decomposers (Berg et al., 2003). Such an understanding has been largely obtained from litter decomposition studies of natural forests and grasslands (Malaisse et al., 1975; Swift et al., 1981; Bargali et al., 1993; Joshi et al., 1993; Singh et al., 1993; Gartner and Cardon, 2004; Singh et al., 2004; Tyub and Reshi, 2008). Not withstanding this importance, very few attempts have been made to study the decomposition of manmade ecosystems in coastal sand dunes (Panda, 2010).

Hence, the present study examines the decomposition rate using litter bag technique in coastal sand dunes of Orissa, India. Besides, relationship between litter mass loss and biotic and abiotic factors such as microbial community structure and different groups of microbes, the effects of plantation on occurrence, distribution, diversity and succession of micro flora, soil respiration rate, soil temperature and soil moisture, were worked out for better understanding and management of the important process of litter decomposition in coastal sand dunes.

MATERIALS AND METHODS

The study site covered 65km of coastline along the Bay of Bengal at 6 to 8 m elevation above mean sea level in Ganjam district of Orissa with geographical position of 19°15'N and 84°50'E. The climate of the region is monsoonic and maritime. Air temperature ranges from 37°C in summer to 13°C in winter with an annual average rainfall of approximately 130 cm. The soil structure is coarse and varies from very fine sand (0.05 cm) to small pebbles (1.5 cm). The soil profile is aeolin and geogentic in character (Panda, 1998). The height of the planted *A. occidentale* trees ranged from 2.57 m to 3.28 m with diameter from 16.8 to 21.2 cm at breast height. The planted stand age was 6 to 8 years old with an average density of 700 trees/ha.

Some of the unproductive uplands and coastal sand dunes in this region were extensively covered by *C. equisetifolia* and Cashew (*A. occidentale*) plants. Cashew plantation at the inner belt of the site covered an area about 1500 h extending 4 to 5 km with a width of 250 to 450 m including *Casuarina* plantation of 30 to 40 rows covering 15 to 20 m along the coast of the sea (in the outer belt of the site) that form a shelter belt cum wind break vegetation. *A. occidentale* is one of the most popular farm forestry trees in the coastal lands of Orissa.

Two sites each of about one hectare were selected for the present investigation for a period of two years. First site was with 6 to 8 years old plantation of *A. occidentale* without any undergrowth and the second site was a big patch of sand dune situated adjacent to *Anacardium* plantation comprising few grasses only. Soil samples in triplicate were collected randomly from two sites in sterilized test tubes at monthly intervals by inserting sterile glass tubes at selected point at a depth of 0.3 cm for microbial analysis. Each tube was allowed to contain 10 to 15 g soil and stoppered tightly.

Samples were brought to the laboratory in sterilized polythene packets along with identification tags in sealed condition. The samples were temporarily stored in an ice chest prior to isolation of microbes. The fungi were isolated by dilution plate (Waksman, 1927) and soil plate (Warcup, 1950) techniques using potato dextrose agar medium. Fungi were identified according to Sarabhoy (1983). Soil moisture was measured by oven dry method and soil temperature by using a soil thermometer. Soil metabolism was studied by the alkali absorption method (Witcamp, 1966). Litter bag technique was used to measure decomposition of leaf litter. Freshly collected litter (only leaves) weighing 40 g were taken in nylon cloth bags (25 × 25 cm, 0.25 mm mesh size) and were placed in the field at both sites. Total of 80 bags were placed in the study plots at each site of which three bags were removed at monthly intervals. The bags were carefully tapered to remove adhering soil particles. Its contents were oven dried at 80ºC and weighed. Rates of litter loss were determined by the percent remaining of the original. The dry weight of litter on successive sampling dates was used to study the kinetics of litter decay using single negative exponential model (Olson, 1963):

$$(M_0/M_t) = e^{-kt}$$

Where, M_0 is the initial mass of the litter and M_t is the litter mass remaining at time (t days), k represents the rate constant at which litter is disappearing.

Data analysis

The following indices of diversity were calculated based on species level identification (Krebs, 1989; Ludwig and Reynolds, 1988).

Shannon –Wiener index $H = -\sum_{i=0}^{s} Pi \ln Pi$

Where Pi is the proportion of the individual found in the ith species, ln denotes natural logarithm and H is the Shannon –Wiener index.

Simpson's index $D = \sum_{i=0}^{s} (Pi)^2$

Where Pi is the proportion of the individual found in the ith species and D Simpson's index.

Evenness index (E) = $H/\ln S$

Where H is the Shannon –Wiener index of diversity, S total number of species and ln is the natural logarithm.

Jacquard's index $S_{ab} = S_{AB} / (S_A + S_B - S_{AB})$

Where S_{AB} is the number of species shared by two locations (A and B), S_A the total number of species in location A and S_B the total number of species in location B. S_{ab} is the extent of similarity between the species in location A and B.

Table 1. Decomposition of litter with biotic and abiotic variables (Mean± standard error) in coastal habitat of Orissa.

Variable	Soil type	Rainy 1st Yr	Winter 1st Yr	Summer 1st Yr	Rainy 2nd Yr	Winter 2nd Yr	Summer 2nd Yr
Litter loss (g)	Soil A	1.96 ± 0.67	1.26 ± 0.4	0.59 ± 0.45	1.78 ± 0.41	0.81 ± 0.21	0.47 ± 0.05
	Soil B	0.92 ± 0.25	0.68 ± 0.07	0.37 ± 0.03	0.78 ± 0.15	0.69 ± 0.11	0.34 ± 0.045
% weight remaining	Soil A	90.9 ± 4.07	70.9 ± 1.4	63.9 ± 0.9	52.6 ± 3.4	38.6 ± 1.2	32.95 ± 0.74
	Soil B	95.5 ± 1.8	86.2 ± 1.0	81.6 ± 0.58	76.2 ± 1.45	67.8 ± 0.95	63.3 ± 0.49
K value $x10^{-5}$	Soil A	178 ± 0.001	139 ± 0.0008	76 ± 0.001	279 ± 0.001	165 ± 0.0003	116 ± 0.0001
	Soil B	79 ± 0.0005	63 ± 0.0001	38 ± 0.002	83 ± 0.0003	81 ± 0.0002	40 ± 0.0001
Fungal numbers (cfu/g dry soil)	Soil A	74 ± 9.7	73 ± 4.0	53 ± 5.1	72 ± 8.5	69 ± 1.5	50 ± 4.0
	Soil B	61.5 ± 7.5	57.5 ± 5.7	43.5 ± 5.0	61 ± 5.8	57.5 ± 6.3	40.3 ± 5.1
Soil temperature(°C)	Soil A	30.4 ± 0.65	26.5 ± 0.62	32.5 ± 1.64	30.5 ± 0.8	27.1 ± 0.72	33.5 ± 1.41
	Soil B	34 ± 0.5	32.2 ± 0.19	36.4 ± 1.5	33.3 ± 0.52	32 ± 0.15	36.9 ± 1.8
Soil moisture (%)	Soil A	2.1 ± 0.56	1.14 ± 0.54	0.72 ± 0.18	2.05 ± 0.42	0.89 ± 0.28	0.68 ± 0.16
	Soil B	1.17 ± 0.21	0.62 ± 0.17	0.2 ± 0.035	1.3 ± 0.27	0.61 ± 0.14	0.29 ± 0.1
Relative humidity (%)	Soil A	82 ± 1.48	77.3 ± 2.06	72.8 ± 3.8	86.8 ± 1.1	77.8 ± 1.5	62.5 ± 3.06
	Soil B	79.3 ± 1.5	67.3 ± 3.5	71.3 ± 2.8	82.5 ± 2.9	70 ± 1.8	55.3 ± 1.7
Rainfall(mm)	Soil A	190.3 ± 67	111.1 ± 151	36.4 ± 29.7	195 ± 96.3	87.93 ± 124	20.6 ± 12.9
	Soil B	190.3 ± 67	111.1 ± 151	36.4 ± 29.7	195 ± 96.3	87.93 ± 124	20.6 ± 12.9
Respiration (mg of $CO_2 m^{-2} hr^{-1}$)	Soil A	218 ± 13	163 ± 14	155 ± 14	230 ± 7.5	206 ± 3.8	139 ± 16.5
	Soil B	135 ± 12.5	106 ± 18.5	85 ± 2.5	123 ± 10.0	127 ± 5.5	90 ± 7.4

Richness index (Margalef, 1963) $R = S\text{-}1/\ln N$

Where S is the total number of species and N is the sampling number.

RESULTS AND DISCUSSION

Because of the similar process of succession, coastal sandy belts throughout the globe are structurally similar.But they differ in several functional properties like biomass production, cycling of matter, productivity, litter fall, and litter decomposition. Coastal sandy soils have poor structural stability, low microbial community and poor nutrient holding capacity (Parrotta, 1999; Sall et al., 2003). Organic matter decomposition is the main source of ecosystem energy in these soils and plays a major role on soil plant relationship (Lavelle and Spain, 2001; Diallo et al., 2005). Soil moisture and total fungal numbers were significantly higher in site A than in site B (Table 1). Micro fungi of both soils show positive correlation with soil moisture but are negatively correlated with soil temperature (Soil A moisture $r=+0.767$ $p < 0.001$, temperature $r=-0.643$ $p < 0.001$; Soil B moisture $r=+0.822$ $p < 0.001$; temperature $r=-0.737$ $p < 0.001$). Plantations are often linked to a strong increase of soil microorganisms of the sand dunes (Manlay et al., 2000; Panda, 2009). The qualitative and quantitative differences of microbial populations at two sites indicate that surface vegetation as well as abiotic factors influences micro fungal inhabitants of the soil.

The result of the present study revealed that the mass loss of litter occurred exponentially with time. During present investigation striking differences in decomposition rate was recorded between the two soil types. Maximum decomposition was at Cashew plantation and minimum at sand dunes (Table 1). Maximum decomposition has been recorded during rainy followed by winter and summer months both in plantations and sand dunes (Pant and Tiwari, 1992; Manlay et al., 2004). The high rate of decomposition in rainy season attributed to the suitable moisture, rainfall and micro-fungal population. Similar observations have also been made earlier

Table 2. Pearson's correlation matrix of litter loss (g) with different biotic and abiotic variable in coastal habitat of Orissa.

Variable	Soil type	Soil temperature	Soil moisture	Fungal Number	Relative Humidity	Rainfall	Respiration	K value
Litter loss	Soil A	-0.223	+0.913***	+ 0.837***	+0.573**	+0.518**	+ 0.544**	+0.865***
	Soil B	-0.555**	+858**	+0.882***	+0.412*	+0.605**	+0.837***	+0.952***

Significant level at * P<0.05, **P<0.01, *** P<0.001

(Singh and Joshi, 1982; Osborne and Maucauley, 1988; Okeke and Omaliko, 1992; Wedderburn and Carter, 1999; Sarjubala and Yadav, 2007). Much lower rate of decomposition during summer may be due to paucity of soil water and low microbial load resulting from low rainfall (Austin and Vitosek, 2000). The mass remaining in litter bags decreased with time in both the sites. In case of soil with *Anacardium* plantation, percent mass remaining was 32.95% (Table 1) and in case of sand dune it was 63.3% (Table 1). Percent weight remaining of two sites differed significantly (t test 2.59< p 0.05). The correlation matrix between loss of leaf litter in gram and measured biotic and abiotic variables during different months was significantly correlated with fungal numbers, rainfall, soil moisture, soil respiration rate, k value and relative humidity but negatively correlated with temperature (Table 2).

A distinct pattern of fungal community structure was observed in both the samples during the study period. A total of 129 species of fungi belonging to 68 genera were enumerated of which sand dune with Cashew plantation had a share of 45 genera, 114 spp. and barren sand dune contributed 51 genera and 112 spp. Sand dune without vegetation revealed 32 restricted spp. while *Anacardium* plantation soil revealed 34 spp. Eighty fungal species were detected common to site without vegetation or with *Anacardium* plantation. The percentage composition and rank abundances of different fungal species fluctuated in both the soil types (Table 3). *Trichoderma, Aspergillus* and *Penicillium* are predominant occurrence and have been found to beeffective decomposers in the present study. Such an observation draws support from the studies of Chapin et al. (2002), Osono and Takeda (2002), and Panda (2010). They stated that fungi are the main ecosystem engineers of terrestrial dead plant material and the same holds true for the present observation as well because the rate of decomposition correlates well with total number of colony forming units. Other studies also revealed predominance of these fungi in various Indian soils (Rai and Kumar, 1988; Panda et al., 2007, 2008).

The extent of decomposition (% weight remaining) in the present study (Figure 1) is significantly lower than that reported previously (Pandey, 1986; Mamtaz et al., 1991; Singh et al., 1993; Panda 2010). It is because of low microbial load of the soil (Table 1). The value of k (Table 1) determined in the present study is in well agreement with Das and Ramakrishna (1985). Shannon's diversity index is reasonably high correspondingly the dominance value is low (Table 4). The D value and H value in soil A indicates many species with maximum diversity. The evenness index varies from 0.869 to 0.881 an indication that species were fairly evenly distributed. The soil A, have the highest species richness where as soil B shows lowest richness (Table 4). Mycoflora of plantation soil was found to be richer than the soil without plantation (Similarity index value 0.55). Many studies have reported an increase in rate of weight loss in relation to temperature soil/air, soil moisture, microbial load and rainfall (Gupta and Lekha, 1989; Pant and Tiwari, 1992; Cortez, 1998; Neher, 1999; present study) but none had paid heed the importance of relative humidity during decomposition process. However, in the present study significant correlation between litter loss and relative humidity indicates its role in decomposition process. Carbon dioxide output (Table 1) showed similar trends of fluctuation that closely followed those of the fungal numbers (Soil A r=+ 0.599 p<0.01; Soil B r= + 0.861 p<0.001) and rate of decomposition (Soil A r=+ 0.544 p<0.01; Soil B r= + 0.837 p<0.001). It is similar to the findings of Behera and Pati (1986), Panda et al. (1996) and Panda (2010). The slow rate of carbon dioxide evolution and decomposition at site A might be due to low moisture, high temperature and with low microbial population in comparison to site B.

The results of our study in coastal sand dunes of Orissa demonstrate the importance of plant litter decomposition and soil microorganisms in controlling the biological soil properties. The present investigation clearly brings out that the plantation significantly influences the rate of decomposition in coastal sand dunes and soil fungi are the dominant decomposers. Besides, soil moisture and relative humidity are the important predictors of the decomposition of *A. occidentale* litter. Thus, vegetation cover of the sand dunes should be protected for proper management of the biodiversity in Orissan coastal belts. This research will help in monitoring and management of decomposing fungi and their importance in ecosystem service could be a valuable tool for understanding the long-term effects of changes in soil condition of the forest floors in coastal sand dunes of Orissa, India.

Table 3. Percentage contribution and ranks of some dominant fungi isolated from coastal habitat of Orissa.

Fungi	Soil with *Anacardium* plantation(Soil A)			Soil without vegetation(Soil B)		
	No. of colony	%	Rank	No. of colony	%	Rank
Absidia butleri	45	5.32	4	14	1.99	21
A. glauca	23	2.72	10	-	-	-
A. spinosa	17	2.01	20	-	-	-
Alternaria alternata	-	-	-	13	1.85	22
Aspergillus awamori	57	6.74	1	56	7.98	1
A. flavus	24	2.84	9	24	3.2	8
A. fonsecaceus	21	2.48	11	-	-	-
A. fumigatus	25	2.96	8	28	3.99	6
A. luchuensis	18	2.13	16	18	2.56	14
A. niger	49	5.79	3	43	6.12	2
A. terreus	16	1.89	21	19	2.71	13
Chaetomium homopilatum	14	1.66	25	22	3.13	10
C.murorum	13	1.54	27	-	-	-
Cladosporium cladosporoides	20	2.36	12	18	2.56	15
C. oxysporum	16	1.89	22	15	2.14	19
Curvularia eragrostidis	-	-	-	27	3.85	7
C. lunata	15	1.77	23	17	2.42	16
C.pallescens	-	-	-	12	1.71	23
Drechslera australiensis	-	-	-	16	2.28	17
Fusarium species	20	2.36	13	16	2.28	18
Mucor species	13	1.66	26	-	-	-
Penicillium citrinum	44	5.2	5	30	4.27	4
P.javanicum	32	3.78	7	39	5.56	3
P.minio-leuteum	19	2.25	15	22	3.13	11
P. nigricans	18	2.13	17	11	1.57	24
P. oxalicum	15	1.77	24	20	2.85	12
P. rubrum	18	2.13	18	15	2.14	20
P.rugulosum	19	2.25	14	-	-	-
P. verruculosum	52	6.15	2	30	4.27	5
Rhizopus nigricans	18	2.13	19	-	-	-
Trichoderma viride	44	5.2	6	23	3.28	9

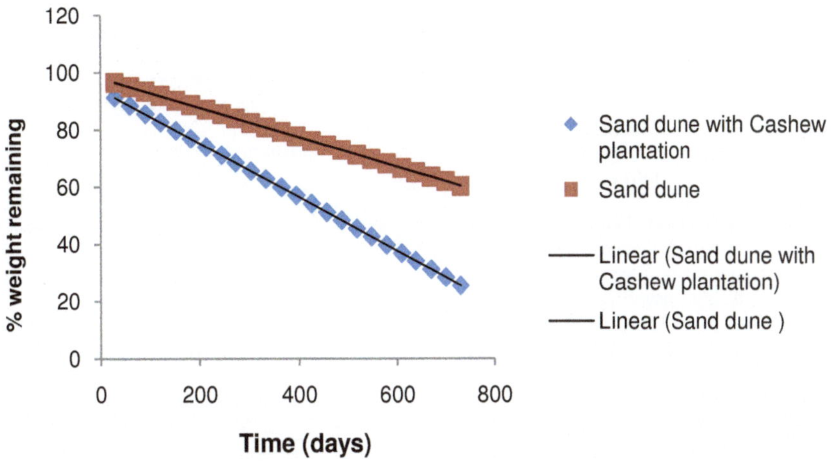

Figure 1. Percentage weight remaining with time (days) in coastal habitat of Orissa.

Table 4. Dominance, diversity, evenness and richness indices of fungi in coastal habitat of Orissa.

Sites	D	1-D	H	E	R
Site without vegetation	0.0323	0.968	3.718	0.881	21.08
Site with *Anacardium* plantation	0.032	0.968	3.744	0.869	23.28

D= Simpson dominance index, H= Shannon diversity index, E= Evenness, R= Richness.

REFERENCES

Austin AT, Vitosek PM (2000). Precipitation, decomposition and decomposability of *Metrosideros polymorpha* in active forest of Hawaii. J. Ecol., 88: 129-138.

Bargali SS, Singh SP, Singh RP (1993). Pattern of weight loss and nutrient release from decomposing leaf litter in an age series of eucalyptus plantations. Soil Biol. Biochem., 25: 1731-1738.

Berg B, De Santo AV, Rutigliano FA, Fierro A, Ekbohm G (2003).Limit values for plant litter decomposing in two contrasting soils.:Influence of litter elemental composition.Acta Oecologia, 24: 295-302.

Behera N, Pati DP (1986).Carbon budget of protected tropical grassland with reference to primary production and total soil respiration. Revue D'Ecologie Et De Biologie Du Sol., 23(2): 167-181.

Chapin FS, Matson PA, Mooney HA (2002).Principles of terrestrial ecosystem ecology. Springer Verlag, NewYork.

Cortez J (1998). Field decomposition of leaf litters: Relationship between decomposition rates and soil moisture, soil temperature and earthworm activity. Soil Biol. Biochem., 30: 783-793.

Das AK, Ramkrishanann PS (1985). Litter dynamics in Khasi Pine (*Pinus* kesiya Royle ex Gordon) of North East India. For. Ecol. Manage., 10: 135-153.

Diallo MD, Guisse A, Niane AB, Sall S, Chotte JL (2005). In situ effect of some tropical litters on N mineralization. Arid Land Res. Manage., 19: 1-9.

Gartner TB, Cardon ZG (2004).Decomposition dynamics in a mixed species leaf litter. Oikos, 104: 230-246.

Gupta SR, Lekha A (1989).Decomposition and carbon turn over in grassland, forest and agricultural systems in a dry subhumid regions.p 131-168. In: J. S. Singh and Gopal B (Eds) Perspectives in ecology.Jagmander Book Agency, New Delhi.

Joshi SR, Sharma GD, Mishra RR (1993). Microbial enzyme activities related to litter decomposition near a highway in a subtropical forest of northeast India. Soil Biol. Biochem., 25: 304-306.

Krebs CJ (1989).Ecological methodology. New York: Harper and Row Publisher.

Lavelle P, Spain AV. (2001). Soil Ecology. Dordrecht, The Netherlands: Kluwier Academic Publishers, p. 654.

Liu C, Westman CJ, Berg B, Kutsch W (2004).Variation in litterfall climate relationships between coniferous and broadleaf forests in Eurasia. Global Ecol. Biogeogr., 13:105-114.

Ludwig JA, Reynolds JF (1988).Statistical Ecology. New York: John Willey. pp. 85-143.

Malaisse F, Goffinet R, Mousset MM (1975).Litter fall and litter breakdown in Maimbo. Tropical Ecological Systems. In: Golly FB and Medina E, eds. Trends in terrestrial and aquatic research. Springer New York, pp. 137-152.

Mamtaz S, Har DK, Mishra RR (1991).Decomposition of maize (*Zea mays* l) crop residues. J. Indian Bototacal Society. 70: 135-138.

Manlay RJ, Cadet P, Thioulouse J, Chotte JL (2000). Relationship between abiotic and biotic soil properties during fallow periods in Sudanian zone of Senegal. Appl. Soil Ecol., 14: 89-101.

Manlay RJ, Masse D, Chevallier T, Smith AR , Friot D, Feller C (2004). Post fallow decomposition of woody roots in the Western African Savanna. Plant Soil, 260: 123-136.

Margalef R (1963).On certain unifying principles in ecology. Am. Natl., 97: 357-374.

Neher DA (1999). Soil community composition and ecosystem processes. Comparing agricultural ecosystems with natural ecosystems. Agroforestry Esytems. 45: 159-185.

Okeke AI, Omaliko CPE (1992). Leaf litter decomposition and carbon dioxide evolution of some agro forestry fallow species in southern Nigeria. For. Ecol. Manage., 50: 103-116.

Olson JS (1963).Energy storage and the balance of producers and decomposers in ecological systems. Ecol. 44:322-331

Osborne JL, Macauley BJ (1988).Decomposition of *Eucalyptus* leaf litter. Influence of seasonal variation in temperature and moisture conditions. Soil Biol. Biochem., 20: 369-375.

Osono T, Takeda H (2002).Comparison of litter decomposing ability among diverse fungi in a cool temperate deciduous forest in Japan. Mycologia, 94: 421-427.

Panda T (1998).Study of microfungi and bacteria of coastal sand dunes of Orissa stabilized with uniculture plantation of *Casuarina.* Ph.D. Thesis, Magadh University, Bodh-Gaya, India. p. 18.

Panda T. (2009).Diversity of sac fungi in coastal sand dunes of Orissa, India. J. Mycol. Plant Pathol., 39(1): 94-98.

Panda T (2010).Role of fungi in relation to litter decomposition associated with *Casuarina equisetifolia* L. in coastal sand dunes, Orissa, India. Int. J. Biodiversity Management Ecosystem Services and Sci. (In Press).

Panda T, Mohanty RB, Prasad BK (1996). Soil fungi and bacteria of coastal sand dunes of Orissa in relation to soil respiration. J. Bihar Bot. Society, 5(1-2): 11-15.

Panda T, Panda B, Mishra N (2007). A comparative study of Penicillia from soil, leaf, litter and air in a coastal sandy belt of Orissa. J. Phyto. Res., 20(2): 335-336.

Panda T, Mishra N, Pradhan FC (2008). Distribution of Aspergilli in early stages of succession in coastal sand dunes of Orissa J. Phytol. Res., 21(2):289-291.

Pant SC, Tiwari SC (1992). Litter fall and litter decomposition in a Montake Oak forest of Garhwal Himalaya. Trop. Ecolol., 33: 103-109.

Pandey PK (1986) Litter production and decomposition, mineral release and biochemical diversity of four forest stands (*Shorea robusta,Tectona grandis*, *Eucalyptus* sp. and *Pinus roxburghii*) at FRI demonstration area. Ph. D Thesis Garhwal University, Srinagar.

Parrotta JA (1999). Productivity, nutrient cycling and succession in single- and mixed-species plantations of *Casuarina equisetifolia, Eucalyptus robusta* and *Leucaena leucocephala* in Puerto Rico. Forest Ecol. Manage., 124(1): 45-77.

Rai B, Kumar A (1988). Microbial decomposition of a forest leaf litter as influence by certain edaphic factors. J. Indian Bot. Society, 67: 18-26.

Reshi Z, Tyub S. (2006).Detritus and decomposition in ecosystems.New India Publishing Agency, New Delhi.

Sall SN, Masse D, Reversat FB, Guisse A, Chotte JL. (2003).Microbial activities during the early stage of laboratory decomposition of tropical leaf litters: the effect of interactions between the litter quality and exogenous inorganic nitrogen. Biol. Fertility Soils, 39(2): 103-111.

Sarabhoy AK (1983). Advanced Mycology. New Delhi: Today and Tomorrow Printers and Publishers.

Sarjubala AD, Yadava PS (2007).Wood and leaf litter decomposition of *Dipterocarpus tuberculatus* Roxb. in a tropical deciduous forest of Manipur, North East India. Curr. Sci., 93(2): 243-246.

Singh O, Sharma DC, Rawat JK (1993). Production and decomposition of leaf litter in Sal, Teak, *Eucalyptus* and Polar forests in Uttar Pradesh. Indian Forester, 119: 112-12.

Singh R, Joshi MC (1982). Studies on decomposition of root and litter materials in sand dune regions of Narhar near Pilani, Rajasthan. Annals Arid Zone, 21(3): 157-161.
Singh RK, Dutta RK, Agarwal M (2004). Litter decomposition and nutrient release in relation to atmospheric deposition of S and N in a dry tropical region. Pedobiologia, 48: 305-311.
Smith VC, Bradford MA (2003). Litter quality impacts on grassland decomposition are differently dependent on soil fauna across time. Applied Soil Ecol., 24: 197-203
Swift S, Russell A, Perfect TJ (1981). Decomposition and mineral nutrient dynamics of plant litter in a regenerating bush-fallow in sub-humid tropical Nigeria. J. Ecol., 69: 981-995.
Tyub S, Reshi Z (2006).Decomposition of mixed litter in a Kashmir Himalayan grassland. Int. J. Ecol. Environ. Sci., 34(1): 19-28.
Vesterdal L (1999). Influence of soil type on mass loss and nutrientrelease from decomposing foliage litter of beech and Norway spruce. Canadian J. For. Res., 29: 95-105.
Waksman SA (1927). Principles of soil microbiology. Baltimore: Williams and Willikins Co. p. 897.
Warcup JH (1950). The soil plate method for isolation of fungi from soil. Nature, 166: 117-118.
Wedderburn ME, Carter J (1999). Litter decomposition of four tree types for use in Silvipastoras systems. Soil Biol. Biochem., 31: 455-461.
Witcamp M (1966). Rate of CO_2 evolution from the forest floor. Ecology. 47: 492-494.
Xiong S, Nilsson C (1999).The effect of plant litter on vegetation: a meta analysis, J. Ecol., 87: 984-994.

16

Water circulation modeling study upon the construction of al-Dannat development at the Halfmoon Bay, Saudi Arabian Gulf

Kaushik Sasmal*, Hari Warrior and Subhendu Maity

Department of Ocean Engineering and Naval Architecture,Indian Institute of Tsechnology Kharagpur, Kharagpur - 721302, India.

The present study aims to understand the flow field near the coast of Halfmoon Bay located in the Arabian Gulf arising due to creation of an artificial island by al-Dannat. The study was performed using the state-of-art hydrodynamic model, Princeton ocean model (POM), applied for a detailed analysis of flow conditions, temperature and salinity distributions after the creation of this artificial island. These hydrodynamic conditions are felt to be critical in assessing stagnation of water, which may eventually lead to eutrophication process. The computed maximum current speeds are in the order of 0.03 ms^{-1}, which suffices to produce reasonable ventilation to the Arabian Gulf even after the construction activity.

Key words: Ocean circulation, modeling, stagnation.

INTRODUCTION

The Arabian Gulf, in the Southwest Asian region, is an extension of the Indian Ocean located between Iran and the Arabian Peninsula. It is historically and commonly known and referred as the *Arabian Gulf* or simply *The Gulf.* This is a shallow basin that separates the Saudi mainland from Bahrain and has many good fishing grounds, extensive coral reefs, and abundant pearl oysters, but its ecology has come under pressure from industrialization, and in particular, petroleum spillages during the recent wars in the region of which the 1991 Gulf war is quite notable. The Gulf is a semi-enclosed marginal sea that is exposed to an arid subtropical climate. It is located between latitudes 24°N and 30°N and is surrounded by deserts.

The study area is located in the Kingdom of Saudi Arabia having an outlet into the Arabian Gulf. The Halfmoon Bay region, located in the Arabian Gulf where the Dannat development is proposed (Figures 1a to b) is a region of relatively quiet and stagnant water. The development of an island in this region can modify the existing flow field and hence the importance of this study.

It is mandatory to have an environmental impact assessment study to understand the impacts arising due to construction activity. It is expected that the effects due to coastal reclamation can change the flow field in the study domain. The dredging activity is limited to a width of 70 to 80 m without altering the banks. The present study involves a detailed analysis of the flow velocity and the resultant temperature/salinity distribution after the construction activity. The aforementioned conditions will determine the residence/flushing time of water surrounding the island having direct implications on the water quality. This deciphers the fact that residence time resulting from release of contaminants into this body of water from the mainland through near-shore waters into the Gulf can be pre-determined. Hence this study is critical to understanding the water quality and eutrophication process.

The computations were performed for a very small domain (having size of 1.45 × 1.45 km) hereafter referred to as the 'simulation domain' (Figure 1b). The focus of this work is to determine the penultimate flow field arising due to the construction activity, and hence details relevant to dredging and dredged sediments are not within the scope of this study. Hence, quantification of physical oceanographic processes including the overall circulation features; temperature and salinity distribution fields were thereby affected due to the fact that this new

*Corresponding author. E-mail: kaushik.sasmal@gmail.com.

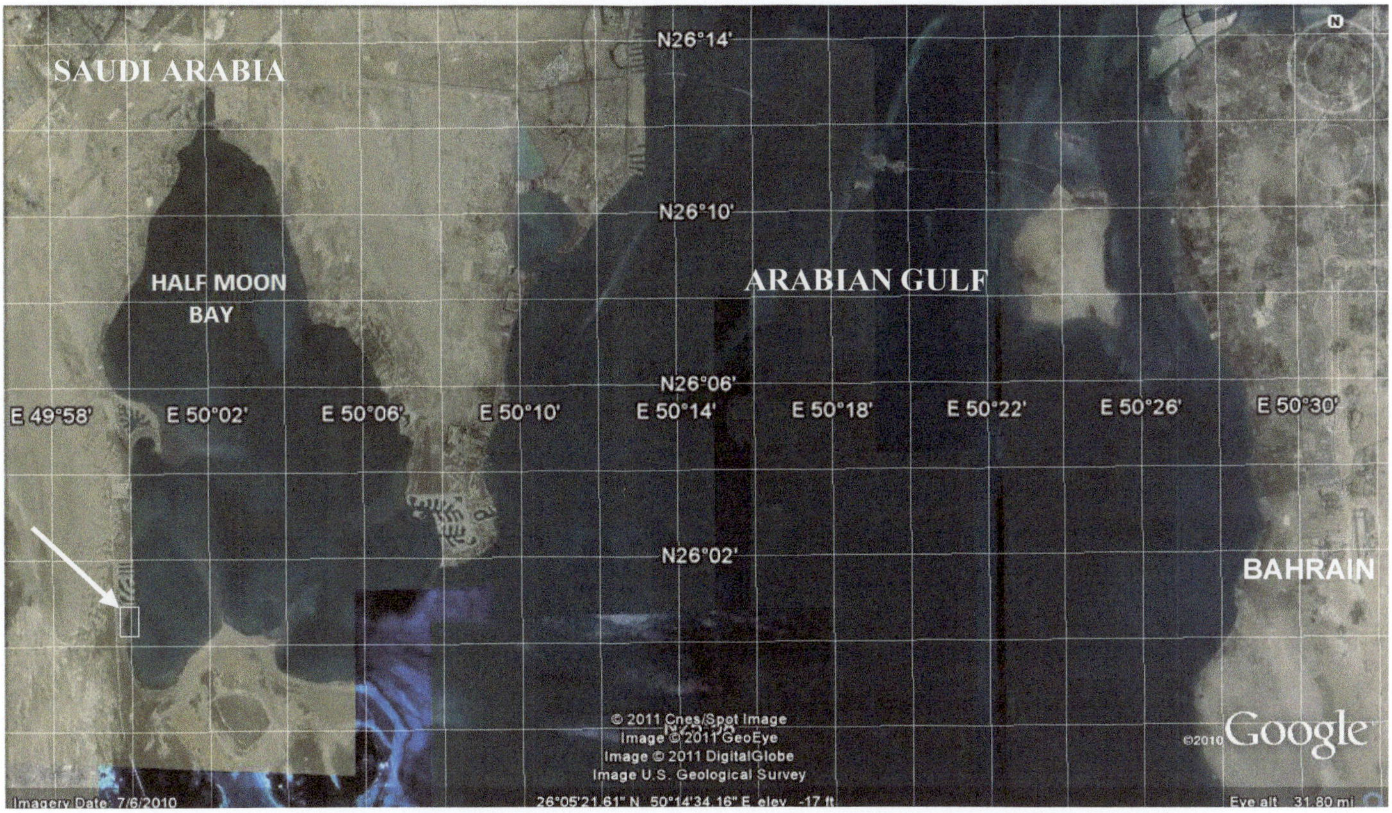

Figure 1a. Location of Halfmoon Bay. The arrow points to the domain of construction and the site for the present study. The coastal water where the island will be constructed is about 4 to 5 m deep. The development of this island is initiated by the Dannat development which involves dredging and refilling the dredged sediments for the island expected off Halfmoon Bay. A comparatively smaller curved island (shown in Figure 1b) is planned for Commercial Township and recreational purposes within close vicinity of Dannat Development Agency.

Figure 1b. Domain for simulation as proposed. 1 cm in x – direction = 242 m; 1 cm in Y- direction = 191 m.

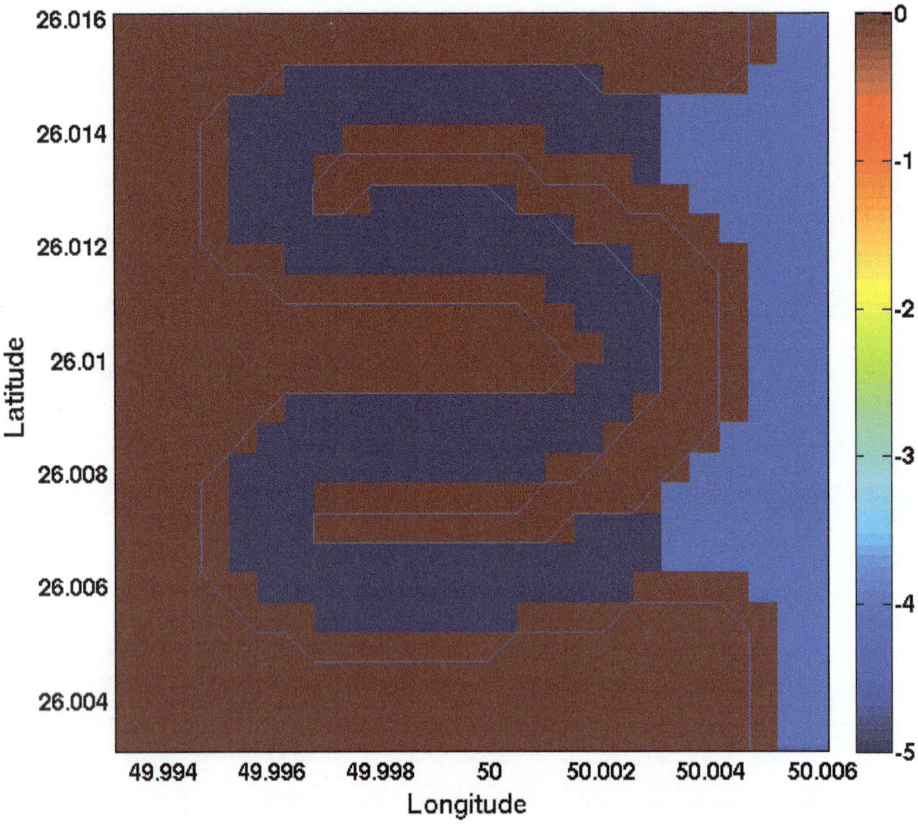

Figure 2a. Bathymetry of the simulation domain for model runs. The deep blue and the sky blue colours indicate depths of 5 and 4 m, respectively and the brown colour indicates zero depth or land acoording to the construction proposal.

construction was given more importance.

METHODOLOGY

Physical background

Bathymetry

The bathymetric details essential for model runs is obtained from the digitized bathymetry map provided by the King Fahd University of Petroleum and Minerals (KFUPM), Kingdom of Saudi Arabia. The simulation domain covers the near shore having water depths less than 6 m. These shallow depths can be modeled using the Princeton Ocean Model (POM, description given under the heading, hydrodynamic model). The bathymetric details are shown in Figure 2a where the water depths are expressed in meters.

Surface winds

The stress resulting from surface wind forcing is an essential pre-requisite for POM runs. In this context, meteorological wind data for the simulation domain was obtained from satellite observations of QuikSCAT (Quick Scatterometer). The final product from QuikSCAT comprises gridded scalar wind speed, zonal and meriodional components of surface winds over the global oceans. In the present study, daily winds for the entire year during 2009 were extracted for the geographic location (50°E, 26°N). As seen, the spatial extent of the simulation domain is very limited in size, hence the extracted wind information from QuikSCAT can be assumed as the representative wind for the simulation domain. The wind vector represented as a polar diagram (wind-rose) for the entire year of 2009 is shown in Figure 2b. The extracted wind field from QuikSCAT for the aforementioned location is then provided as input to force POM model.

The winds in the simulation domain are typically north westerly winds associated with large scale sand-storms. These winds, referred to as 'Shamal', by Perrone (1981), prevail throughout the year and are usually stronger during the day than at night. The intensity of these winds is strong during summer months and less common during the winter periods. The wind data used for the present study also show the dominant direction as north-west (Figure 2b) followed by the south-west direction.

Temperature and salinity

The initial conditions required for POM runs require diagnostic fields such as temperature and salinity for the simulation domain. These fields are obtained from the state-of-art World Ocean Atlas (WOA05), which provides fields of temperature and salinity at standard water depths on monthly, seasonal and annual time scales for the world oceans. The WOA05 is a set of objectively analyzed 1° × 1° spatial grid climatology; in the context to this study, the derived temperature and salinity fields for the geographic

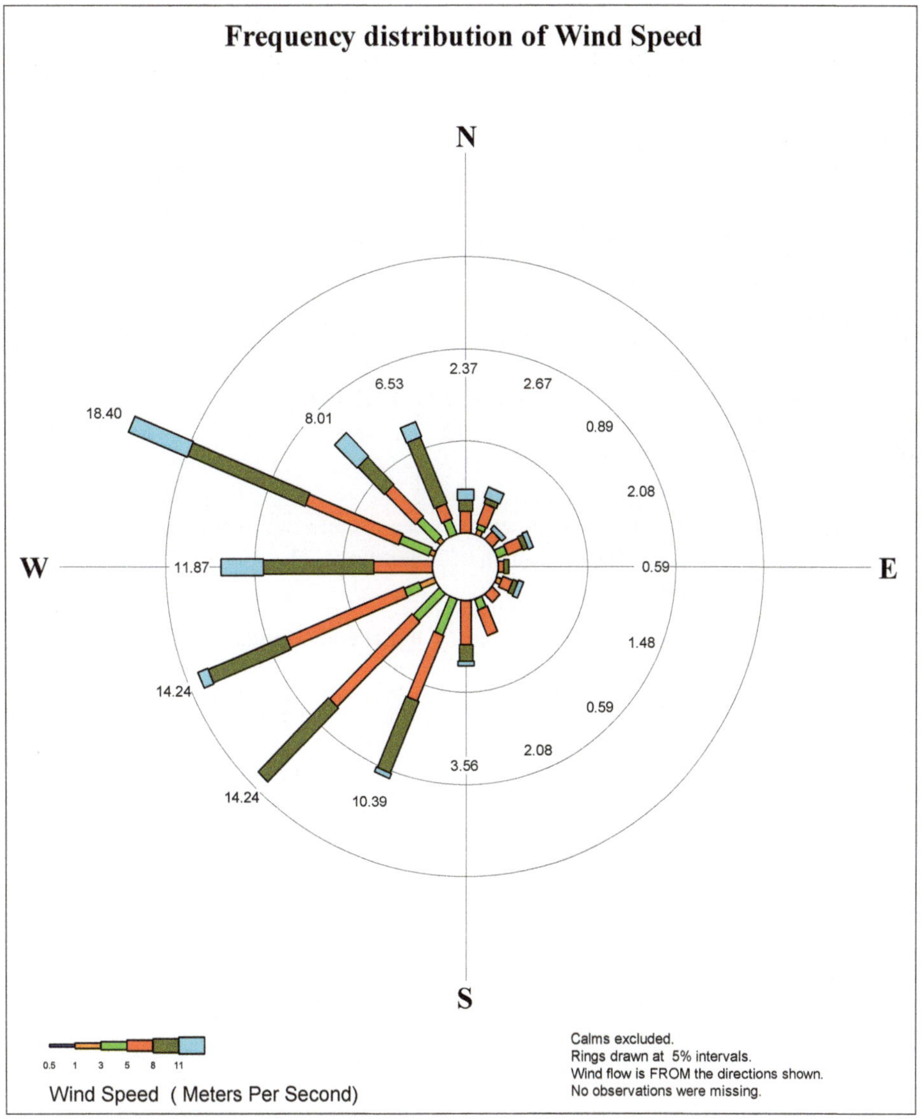

Figure 2b. Annual distribution of wind field in 2009 expressed as Wind Rose over Dannat (simulation domain).

location 50°E, 26°N was used representing the simulation domain. The temperature and salinity data used for the present study is from the monthly climatology of WOA05.

DYNAMICAL ASPECTS

The conservation principles of volume and salinity can be used for the estimation of residence/flushing times in semi-enclosed water bodies. In this context, key parameters such as the flow field, temperature and salinity are essential pre-requisite. An exhaustive study of water circulation at the construction site is also required to understand the water quality characteristics. Since the water depth at simulation domain is limited to less than 6 m, the prevalent currents which control the fluid motion can be attributed due to wind and tides. The depth averaged case for currents can be assumed in the simulation domain due to very shallow depths. In case of wind driven currents, the Ekman depth will be the total water depth (due to depth limitation) and Ekman transport governs the fate of transport mechanism. The intensity and mixing by tidal currents can be pivotal in context to the water quality. Hence, a comprehensive understanding of wind and tidal currents is important to the near-shore dynamics in the simulation domain.

HYDRODYNAMIC MODEL

Model description

The model used for this study is the Princeton ocean model (POM). This model has the capability to simulate three dimensional flow fields relevant for this study. It is a community general circulation numerical model used for nowcasting and forecasting of ocean parameters. The applicability of POM includes wide spectrum of oceanic problems such as circulation and mixing processes in estuaries, semi-enclosed seas, coastal seas and open oceans. The

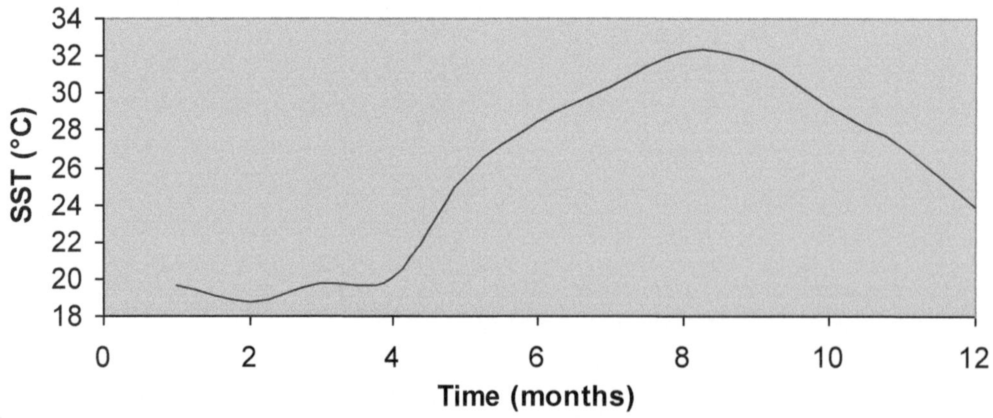

Figure 3a. Time series of SST (°C).

model is based on primitive equations, sigma coordinate system having sophisticated level of turbulence mixing schemes. The recent version also includes an option for wet and dry grids accounted from varying water levels near-shore. The governing equations are discretized in a finite-difference grid. The original version of POM was developed at Princeton University (Blumberg and Mellor, 1987) with collaborative efforts as reported in Herring and Patchen (1994). The version of POM used in the present study is POM2k developed by Mellor (Mellor, 2003). The Blumberg-Mellor model (later termed as POM) includes additional features such as free surface to handle tides, sigma terrain following coordinates to handle complex topographies in shallow regions, curvilinear grids to better represent coastlines, and turbulence scheme called Mellor-Yamada scheme (Mellor and Yamada, 1974) to handle vertical mixing. In the early 1980s, POM model was used for estuarine applications such as the Hudsen-Raritan estuary (Oey et al., 1985a, b) and the Delaware Bay (Galperin, 1990 a, b). The first attempts to use a sigma coordinate model for basin-scale problems started with a coarse resolution model for the Gulf of Mexico (Blumberg and Mellor, 1981).

Governing equations

The model equations are usually expressed in a flux form. This is desirable in terms of computational power to separate the vertically integrated equations (external mode) from the vertical structure equations (internal mode). This method, known as mode-splitting, permits the calculation of the free surface elevation with no priori restriction on computation time. This is achieved by solving the velocity transport separately from three-dimensional estimation of the velocity and thermodynamic variables. The time integration is performed using a leap-frog two level time scheme. This method has been found to be extremely efficient and stable in calculating the surface mixed layer and also rigid boundary layer as in case with land boundaries. The imposed condition of free sea surface in the model permits the dynamic change of water level elevation due to tides.

Model domain and grid resolution

The simulation domain is shown in Figure 2a. The horizontal spatial extent is approximately 1.45 × 1.45 km having depth less than 6 m. The total number of grid points along the horizontal are 26 × 26 with spatial resolution of 56 m. A rectilinear coordinate system is used in the horizontal and 16 sigma levels were specified in the vertical direction.

Initial and boundary conditions

The inputs required to initialize the model are bathymetry, surface winds, temperature and salinity fields with appropriate boundary forcings. In this case study, the forcing at open boundary includes water level variations, temperature and salinity. The free surface forcing is accounted from wind stress, flux of heat and moisture. In the present study, the Blumberg-Kantha radiation type scheme (Blumberg and Kantha, 1985) is used for open boundary condition.

RESULTS AND DISCUSSION

Temperature and salinity

The model computed sea surface temperature (SST) and sea surface salinity (SSS) are shown in Figures 3a and b, respectively. The occurrence of maximum SST is noticed during the month of August (32.2°C) with a net difference of more than 10°C during the winter months (≈ 18.75°C). The sea surface salinity (SSS) is shown in Figure 3b which reveals that maximum surface salinity occurs during the month of January and minimum during June. This temperature and salinity distribution decides the fate of the biological population there. There are about 11 profiling stations (designated from A – K) in the simulation domain (Figure 4a).

Surface circulation

The surface circulation features for different months are shown in Figures 4b to e, respectively for January, April, July and October. As seen from these figures, the surface currents are oriented along eastward and north-eastward directions at various segments of the simulation domain. The color in these Figures indicates the temperature pattern (in °C) and a particular color for each month indicates uniformity of thermal field within the study domain.

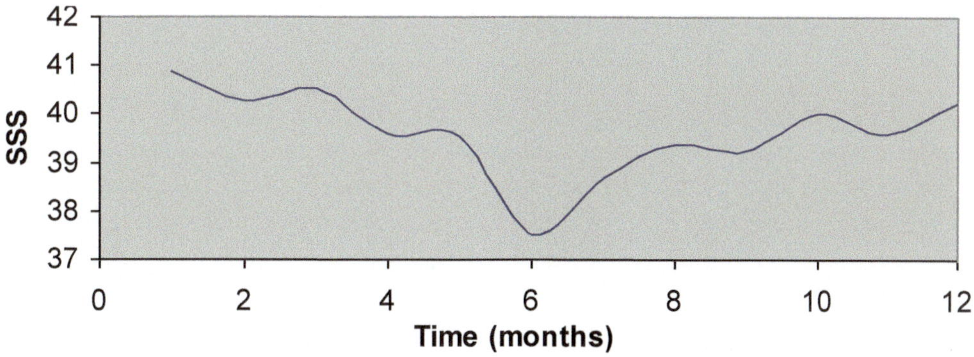

Figure 3b. Time series of SSS (‰).

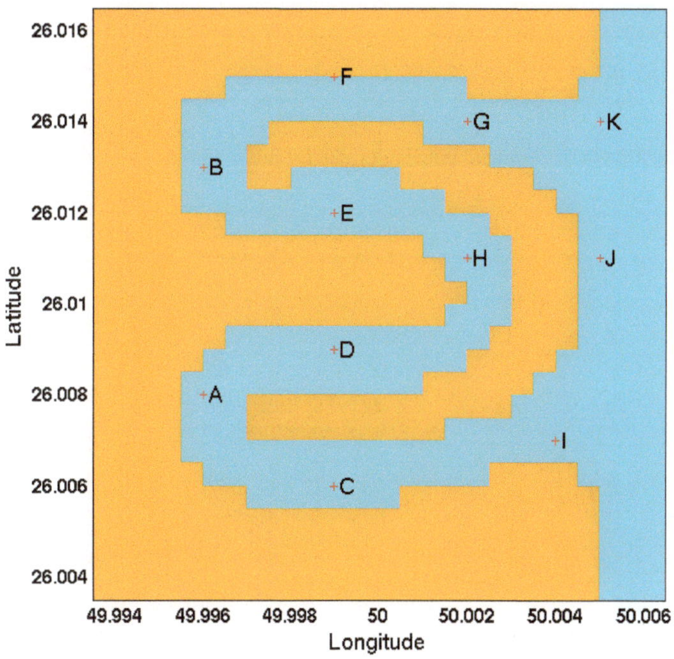

Figure 4a. Locations of the eleven profiling stations.

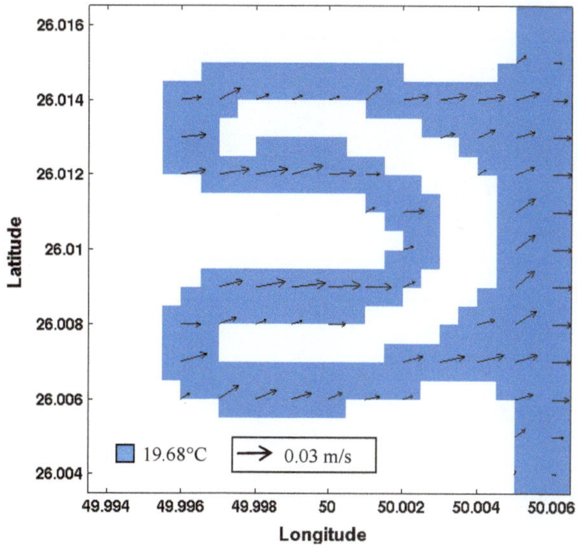

Figure 4b. Surface circulation for the month of January.

Vertical velocity profile of currents

The current velocity profiles (magnitudes as a function of water depth) at eleven profiling stations (Figure 4a) near the site are shown in Figures 5a to d. The advantage of such profiling leads to an understanding of the vertical distribution of velocities. As can be seen from these figures, the vertical profiles show difference among the 11 locations.

A comprehensive analysis of the velocity profiles (Figures 5a to d) shows that the domination current directions are either eastward or northward. This may not lead to stagnation of water mass in the construction site. The magnitudes of surface currents are shown as velocity contours for four months in Figures 6a to d. The

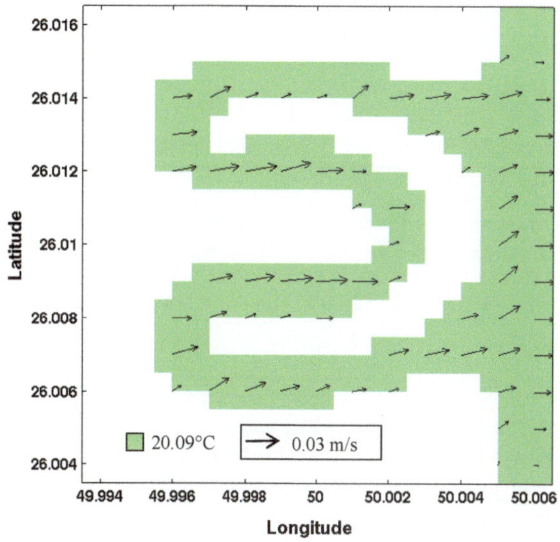

Figure 4c. Surface circulation for the month of April.

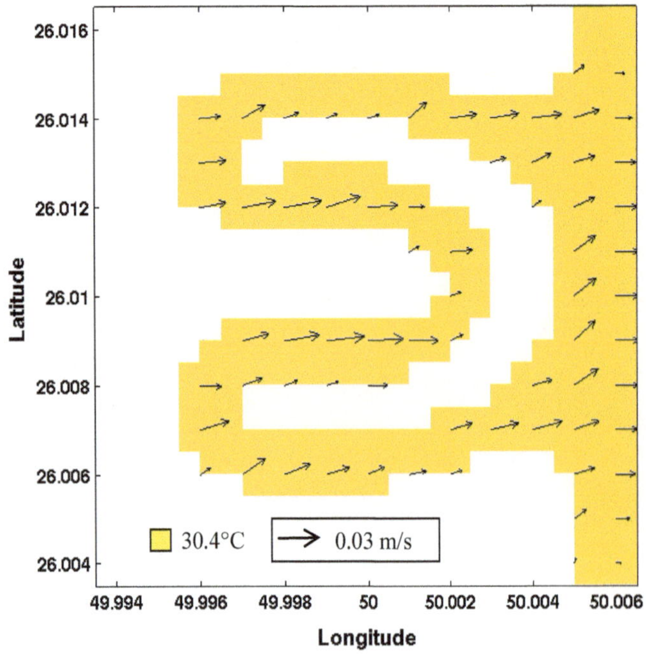

Figure 4d. Surface circulation for the month of July.

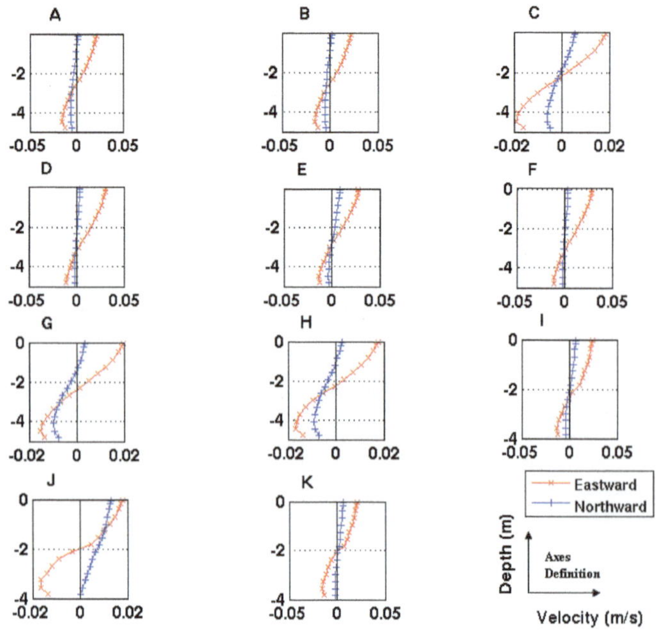

Figure 5a. Vertical velocity profile for the month of January.

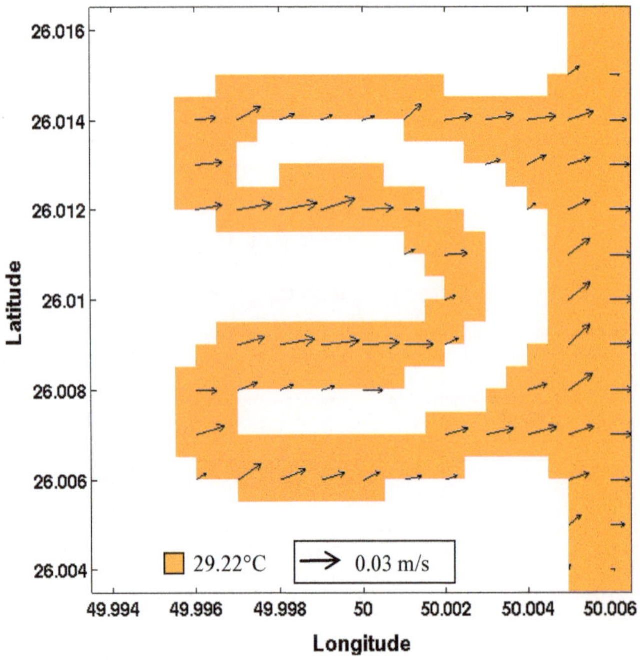

Figure 4e. Surface circulation for the month of October.

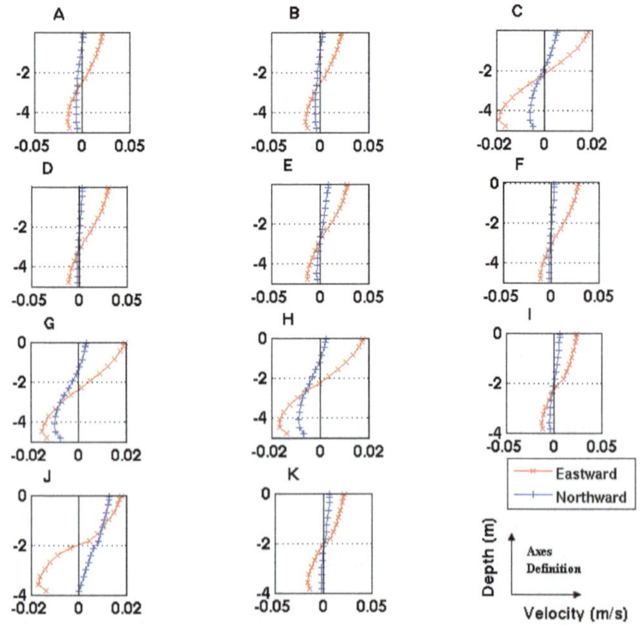

Figure 5b. Vertical profile of velocity for the month of April.

variability of current magnitude are seen along the 11 representative locations, however the seasonal difference is marginal. More details are tabulated and shown in Table 1. The dominance of eastward component is seen compared to the northward component suggesting that the net flow will not result in water stagnation.

Conclusions

The primary goal of the present study was to investigate the net flow field of waters near-shore Dannat area of Halfmoon Bay, Kingdom of Saudi Arabia. After the re-filling process of dredged sediments, an artificial island is expected near the coast for various activities. The circulation aspect relevant to this change of natural

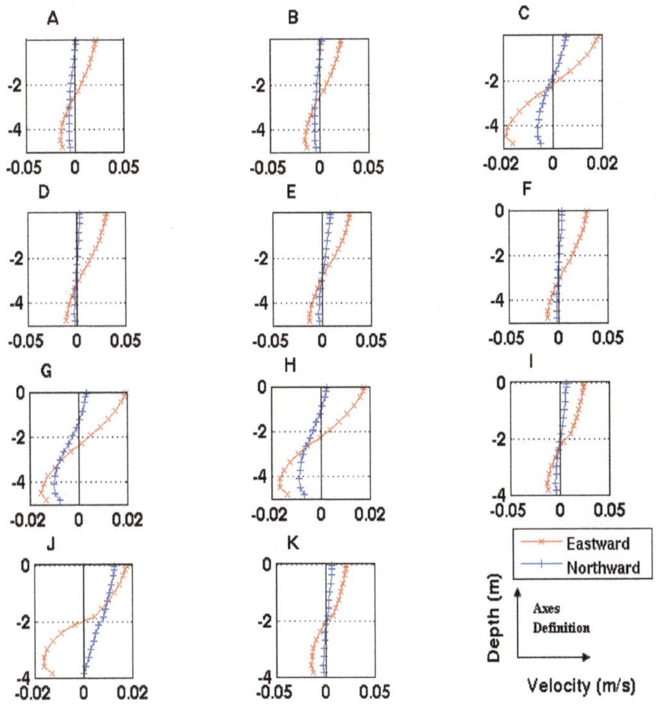

Figure 5c. Vertical profile of velocity for the month of July.

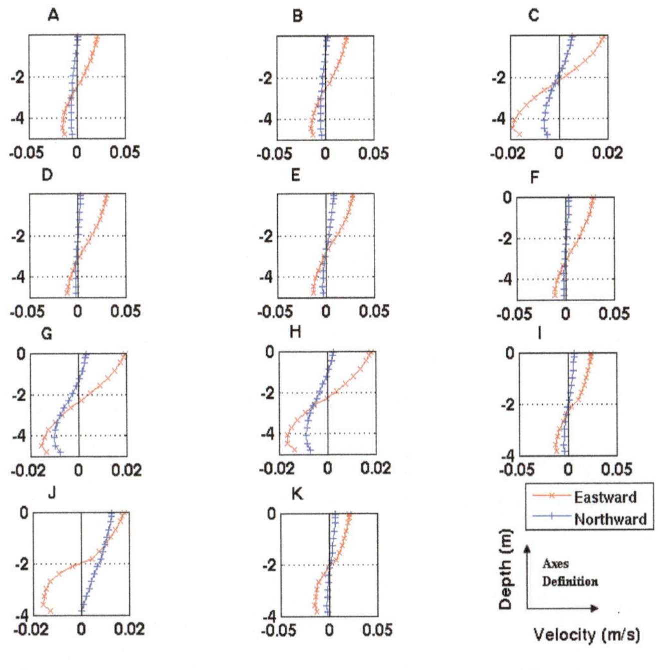

Figure 5d. Vertical profile of velocity for the month of October.

system arising from re-filling and construction activity has been studied. The simulated current velocities are in the order of 0.03 m/s. The persistence of such currents along with net eastward transport direction can produce reasonable circulation to prevent water stagnation in the

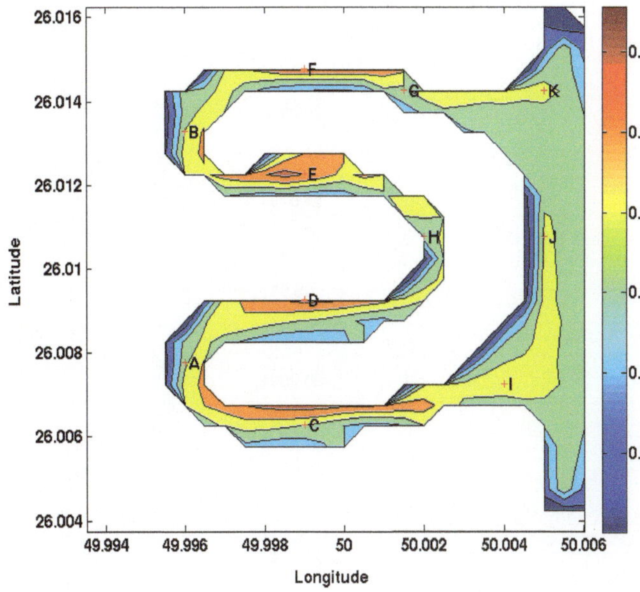

Figure 6a. Magnitude of surface current for the month of January.

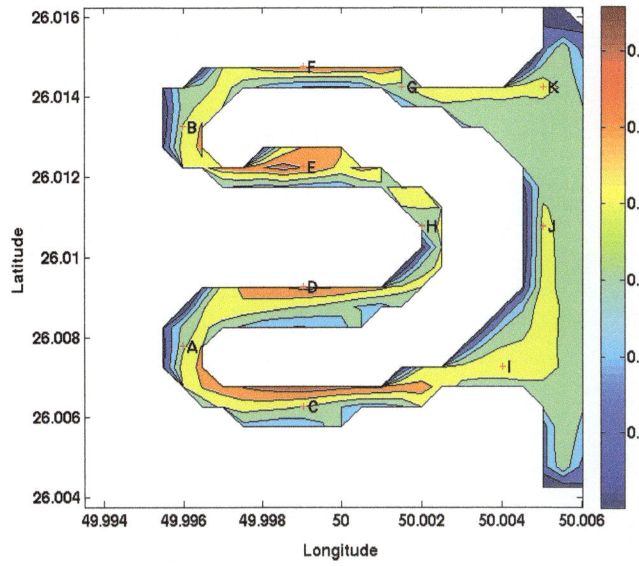

Figure 6b. Magnitude of surface current for the month of April.

eventuality of construction activity at Halfmoon Bay, Kindgom of Saudi Arabia.

ACKNOWLEDGEMENTS

Authors thank Mrs. Sikha Dasgupta of Indian Institute of Technology Kharagpur, for preparing the wind-rose plot. Authors are also grateful to the anonymous reviewers for their useful comments and suggestions in improving the manuscript.

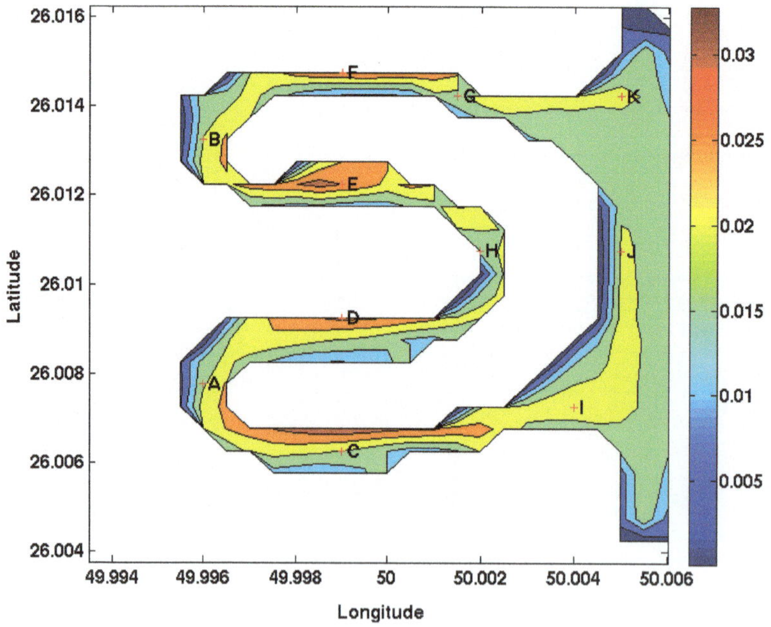

Figure 6c. Magnitude of surface current for the month of July.

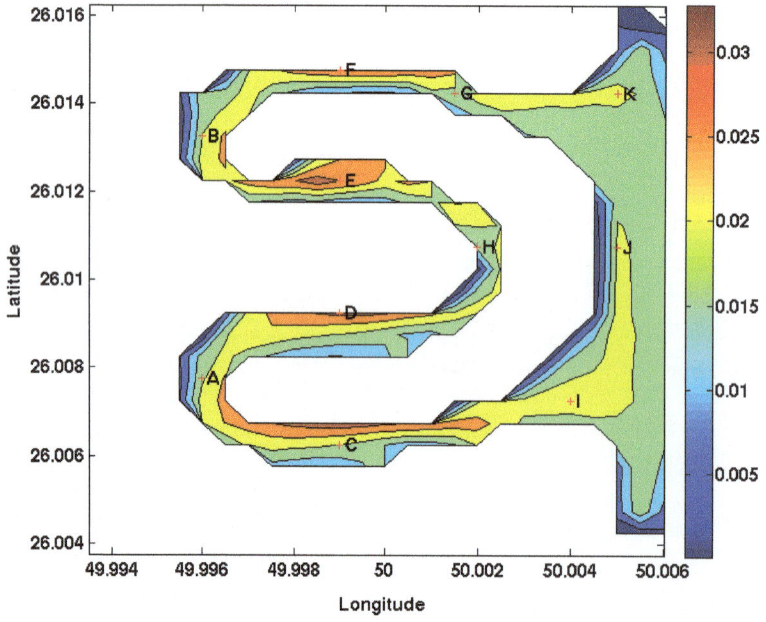

Figure 6d. Magnitude of surface current for the month of October.

Table 1. Computed surface flow field at four different months.

Month	Current-magnitude (m/s)		U-component (m/s)		V-component (m/s)	
	Minimum	Maximum	Minimum	Maximum	Minimum	Maximum
January	0.00	0.0328	-0.0238	0.0322	-0.0126	0.0155
April	0.00	0.0329	-0.0238	0.0323	-0.0127	0.0155
July	0.00	0.0328	-0.0238	0.0322	-0.0127	0.0155
October	0.00	0.0328	-0.0238	0.0323	-0.0126	0.0156

REFERENCES

Blumberg AF, Kantha LH (1985). Open boundary condition for circulation models. J. Hydraulic Eng., 111: 2.
Blumberg AF, Mellor GL (1981). A numerical calculation of the circulation in the Gulf of Mexico, Dynalysis of Princeton.
Blumberg AF, Mellor GL (1987). A description of a three-dimensional coastal ocean circulation model, in Three-Dimensional Coastal Ocean Models, edited by N.Heaps, American Geophysical Union, Washington D.C., 4: 208.
Galperin B, Mellor GL (1990). A time-dependent, three-dimensional model of the Delware Bay and River system. Part 1: Description of the model and tidal analysis. Estuar. Coast. Shelf Sci., 31: 231-253.
Galperin B, Mellor GL (1990). A time-dependent, three-dimensional model of the Delware Bay and River system. Part 2: Three-dimensional flow fields and residual circulation. Estuar. Coast. Shelf Sci., 31: 255-281.
Herring HJ, Patchen RC (1994). Modelling circulation of the Southern California Bight: final report.
Mellor GL (2003). Users guide for a three-dimensional, primitive equation, numerical ocean model (June 2003 version), Prog. in Atmos. and Ocean Sci., Princeton University, p. 53. [Available online at http://www.aos.princeton.edu/WWWPUBLIC/htdocs.pom/.]
Mellor GL, Yamada T (1974). A hierarchy of turbulence closure models for planetary boundary layers. J. Atmos. Sci., 31: 1791-1806.
Oey L-Y, Mellor GL, Hires RI (1985a). A three-dimensional simulation of the Hudson-Raritan estuary. Part I: Description of the model and model simulations. J. Phys. Oceanogr., 15: 1676-1692.
Oey L-Y, Mellor GL, Hires RI (1985b). A three-dimensional simulation of the Hudson-Raritan estuary. Part II: Comparison with observation. J. Phys. Oceanogr., 15: 1693-1709.
Perrone TJ (1981). Winter shamal in the Persian Gulf, Tech. Rep., Naval Environ. Predict. Res. Facil., Monterey, Calif., pp. 79-06.

17

Numerical simulation of seasonal variations in circulations of the Bay of Bengal

Sourav Sil and Arun Chakraborty*

Centre for Oceans, Rivers, Atmosphere and Land Sciences (CORAL), Indian Institute of Technology, Kharagpur, India.

The seasonal variations in circulation at different layers of the Bay of Bengal (BOB) basin have been studied using multilayered regional ocean modeling system (ROMS). The model is simulated with climatological comprehensive ocean-atmospheric data set (COADS) wind and thermal forcing until annual cycle is reached. The northward flowing western boundary current (WBC) during spring and southward flowing east India coastal current (EICC) during autumn are developed due to the formation of anticyclonic and cyclonic Eddies respectively. The influence of Malacca Strait is prominent during winter season. The north equatorial current in the southern basin during winter to spring and southwesterly monsoon current (SMC) during summer are well simulated by the model. At 100 m depth current pattern along the Indian coast is similar to surface pattern in all the seasons. The 500 m depth current depicts the western boundary current in the winter season only. Many Eddies are simulated along the western coast of the basin during summer. The 1000 m current pattern indicates the existence of different Eddies in the basin for longer time. A strong anticyclonic Eddy persists near the head of Sri Lanka for almost every season. Andaman Sea is driven generally by one Eddy in the north and another in the south.

Key words: Comprehensive ocean-atmospheric data set (COADS), western boundary current (WBC), east India coastal current (EICC), Bay of Bengal (BOB), regional ocean modeling system (ROMS), southwesterly monsoon current (SMC).

INTRODUCTION

The Bay of Bengal (BOB) is a three side bounded tropical small ocean basin situated in the northern part of the Indian Ocean and highly influenced by semiannually reversing monsoon winds. A number of studies mentioned that the geographical location and the unique wind forcing make the circulation very complex. Using ship drift data, Cutler and Swallow (1984) showed the reversibility of this surface current, that is there is clockwise circulation from February to July and reverse for the rest of the year. Then, the complex circulation was analyzed by researchers with hydrographic, satellite observation and numerical model as well. Legeckis (1987) first showed the existence of western boundary current (WBC) from satellite infrared image of sea surface temperature in the BOB. He found the similarities of this current with the well-known Gulf Stream as it carries warm water along the east coast of India during February. Sanilkumar et al. (1997) showed existence of WBC using MICOM STC profiles. Shetye et al. (1993) showed the western boundary current of the seasonal subtropical gyre in the BOB. Suryanarayana et al. (1993), Shetye et al. (1996) and Babu et al. (2003) investigated the circulation feature using hydrographic dataset. Somayajulu et al. (2003) and Gopalan et al. (2000) showed the seasonal and interannual variations of surface circulation of the BOB using satellite observations. Potemra et al. (1991) with Isopycnic layered model, Yu et al. (1991) with linear reduced-gravity model, McCreary et al. (1993) with two and half layer thermo-dynamics model and Paul et al. (2009) using multilayered modular ocean model showed the semi-annual variability of the upper layer circulation.

As a result of the typical wind forcing and geographical location they suggested that higher resolution numerical model study is required to understand the mesoscale feature of the BOB. In order to accomplish this, a high resolution-modeling setup has been done for the BOB (4 to 24°N and 76 to 100°E) using regional ocean modeling system (ROMS) to analyze the circulation features for the basin (Figure 1a), which has very shallow coastline and there is sharp slope in bathymetry from south to north. The coastal

*Corresponding author. E-mail: arunc@coral.iitkgp.ernet.in

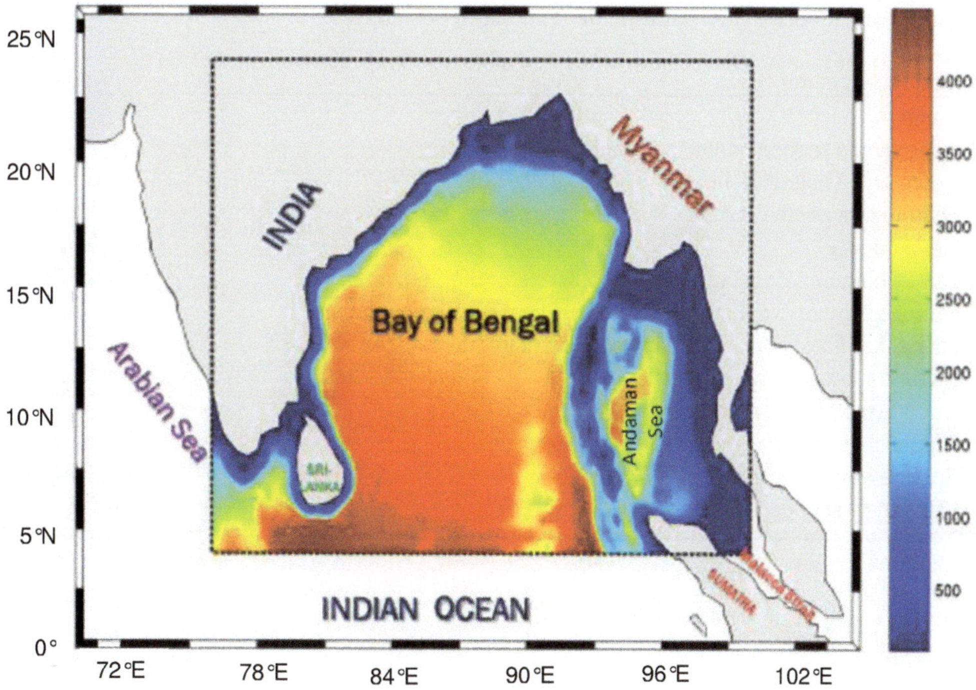

a

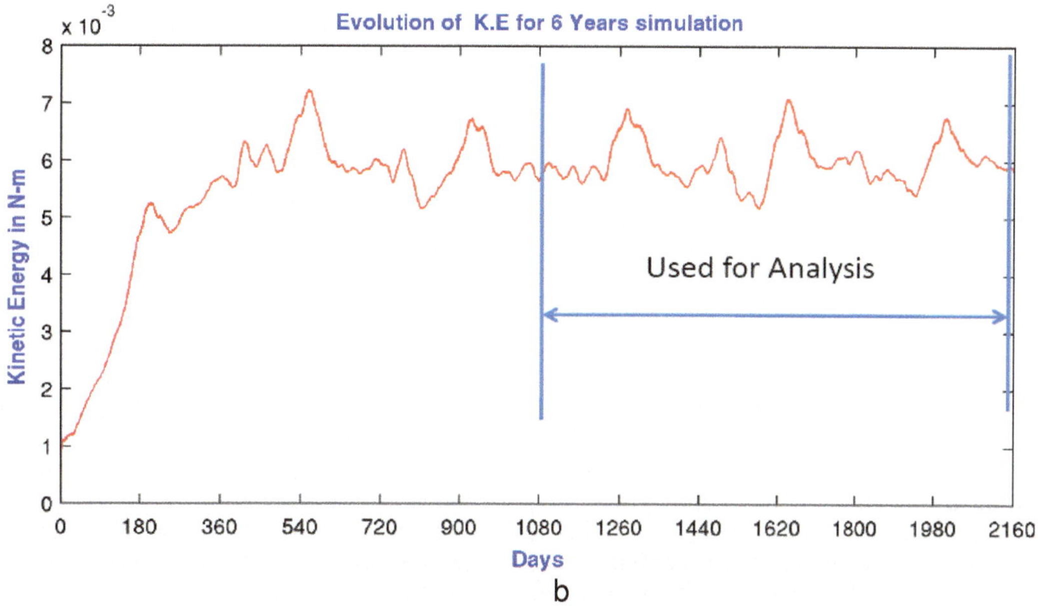

b

Figure 1. (a) Domain of study with its bathymetry (in m). Western side of the basin is deeper compared to AnS. Being shallower around the AnS, the area seems like a trench. (b) Evolution of kinetic energy (N-m) with time (days). Last three years are used for analysis.

region is very shallow (<100 m) while the southern part is very deep (>3500 m). The Palk Strait (PS) between Sri Lanka and India is shallower than 100 m. The Malacca Strait (MS) is also very shallow which connects the BOB and the South China Sea. The eastern part of the BOB basin is known as the Andaman Sea (AnS) which looks like a trench with shallow surroundings and very deep center. Chakraborty and Gangopadhyay (2010) showed that the difference of water mass characteristics of the AnS and the rest of the BOB is due to its bathymetry. The circulation feature of the AnS is not yet studied due to non-availability of datasets, but the model simulation can provide some idea about the circulation feature in this region. In this paper, study on the circulation pattern for different seasons from

Table 1. Parameters with their values used for model configuration.

Model parameter	Value
Depth	0-5500 m
s-coordinate surface control parameter (theta_s)	7.0
s-coordinate bottom control parameter (theta_b)	0.1
Thermocline depth (Tcline)	10 m
Mean density	1025 kgm^{-3}
Lateral harmonic constant mixing coefficient for momentum	4000 m^2s^{-1}
Lateral biharmonic constant mixing coefficient for momentum	100 m^4s^{-1}
Lateral harmonic constant mixing coefficient for tracer	1000 m^2s^{-1}
Linear bottom drag coefficient	10^{-4} ms^{-1}
Quadratic bottom drag coefficient	0 ms^{-1}
Slipperiness variable	1(no slip)
Sponge Layer Thickness	10^5m
Viscosity in sponge layer	800 m^2s^{-1}

ROMS simulation is presented. A materials and methods section is given. The validation of the model with observational datasets was performed by Sil et al. (2010) and Sil and Chakraborty (2011). The models simulating currents for different seasons and for different levels are discussed as results and discussion and finally conclusions.

MATERIALS AND METHODS

Brief description of the model

ROMS is a free-surface; terrain-following, primitive equations ocean model widely used by the scientific community for a diverse range of applications (Haidvogel et al., 2000; Lermusiaux et al., 2001; Wilkin et al., 2005; Chakraborty and Gangopadhyay, 2010; Sil et al., 2010; Sil and Chakraborty, 2011). The model is discretized in the vertical over variable topography using a stretched terrain-following coordinate and can be used for water bodies like coastal basin, estuary, river, sea and open ocean. The primitive equations are evaluated using boundary-fitted orthogonal, curvilinear coordinates on a staggered Arakawa C grid.

There are 287 and 248 grid points in zonal and meridional directions respectively corresponding to 10 km resolution with 32 vertical levels in the model. The northern, eastern and most of the western boundary are closed. The southern boundary and part of the western boundary are open. On the open boundaries, the temperature and salinity values are relaxed to match the Levitus monthly climatology. Vertical turbulent mixing is based on the K-profile parameterization (KPP) vertical mixing scheme. Chakraborty and Gangopadhyay (2010) showed from Brunt-Vaisala frequency distribution that the vertical level should be discretized with more levels in the mixed layer region. Therefore, the model parameters are chosen accordingly (Table 1).

Description of experiment and data used

Bathymetry data was derived from the ETOPO2 (2 min) gridded dataset (National Geophysics Data Center, 1988). Surface forcing functions were drawn from the comprehensive ocean-atmosphere data set (COADS) climatology of Da Silva et al. (1994). These forcing fields included surface wind stresses, heat and freshwater fluxes, and heat flux sensitivity to sea surface temperature. The latter is used to represent the feedback between model SST and the surface heat flux (Da Silva et al., 1994). In the open boundary (east, west and south) temperature and salinity are relaxed to Levitus monthly climatology.

The earlier defined climatological forcing parameters are used for interannual model simulation. The surface seasonal wind forcing transfers the energy to the deeper layers through the vertical advection term in the momentum equations of the model. This vertical advection along with the vertical mixing (KPP) is the main contributor to the deeper circulation in the model. The volume integrated kinetic energy shows that annual cycle is reached after three years of simulation (Sil et al., 2010; Sil and Chakraborty, 2011). The seasonal climatological features are studied for the next three years simulation (Figure 1b).

RESULTS AND DISCUSSION

Seasonal wind pattern over the BOB

Similar to Somayajulu et al. (2003), the seasons are selected on the basis of wind pattern as shown in Figure 2. The northeasterly winds during the winter (Figure 2a) and southwesterly winds during the summer (Figure 2c) flow over the BOB with maximum intensity between 5 to 14°N. Clockwise wind pattern with maximum intensity along the western boundary of the basin during the spring (Figure 2b) and anticlockwise wind circulation in the southern BOB but clockwise wind circulation in the northern BOB (north to 14°N) during the autumn (Figure 2d) prevail over this region. Although the model has 32 terrain following layers but for our analysis we have discussed only three layers; 10 m representing surface, 500 m representing the bottom of the thermocline and deep zone as 1000 m.

Circulation at 10 m

In the winter, the model simulating 10 m current shows that strong north equatorial current flows westward (Figure 3a). The current which enter to BOB basin from MS joins with the above current system and form a jet. This strong north equatorial jet at around 6°N reflects at the Sri Lankan coast

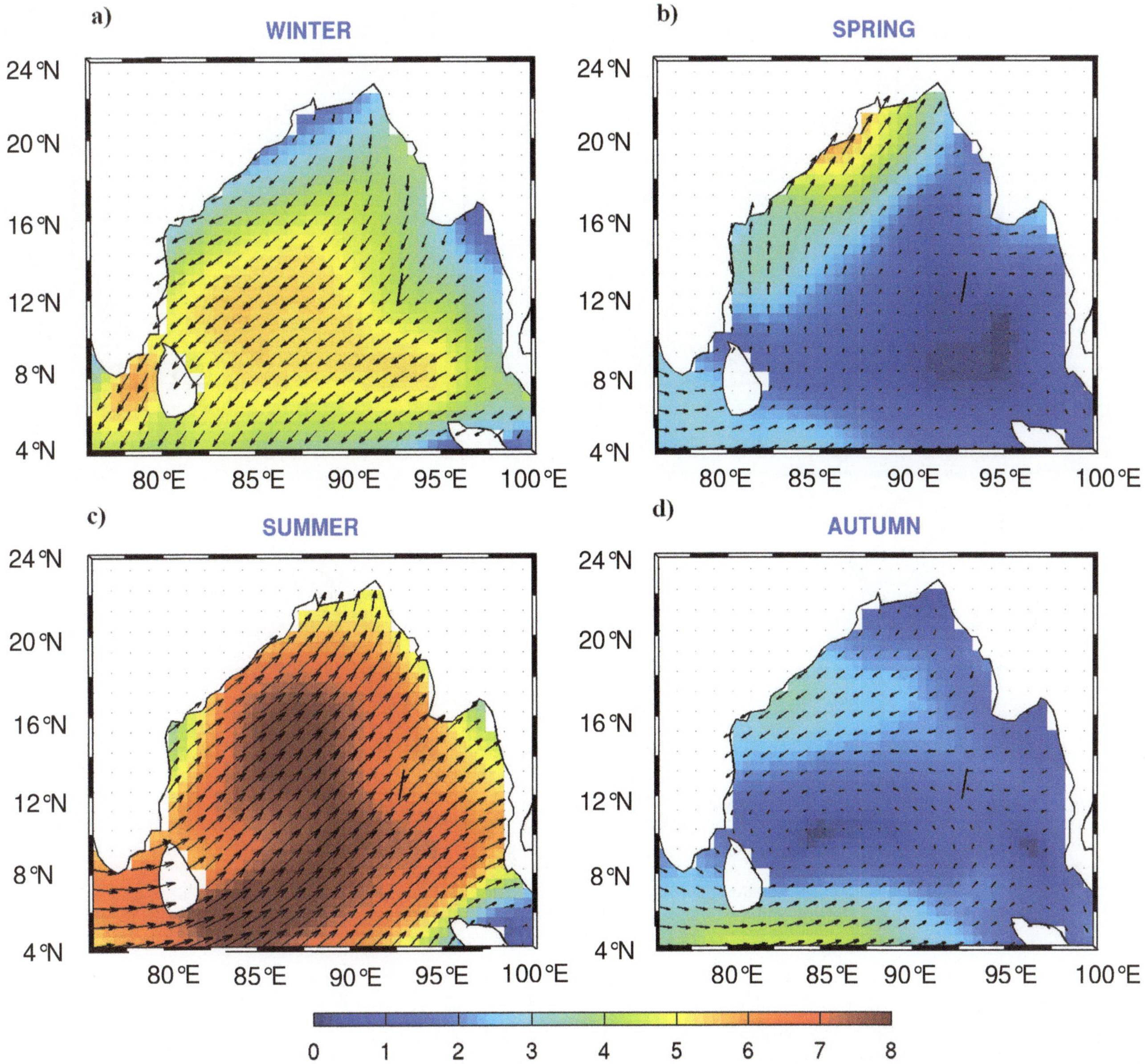

Figure 2. Seasonal wind pattern with magnitude in shaded background for (a) Winter: northeast monsoon (December, January and February), (b) Spring: transition (March, April and May), (c) Summer: southeast monsoon (June, July, August and September) and (d) Autumn: transition (October and November).

and then flows along the west coast of the basin. The bifurcation of this jet is found along 90°E to form an anticyclonic gyre at around 92°E; 10°N with other part to flow northward that may be due to the ridge along 90°E. The northward component deviates westward around 12°N due to anticyclonic gyre in the head of the BOB. But in this anticyclonic gyre, two cyclonic Eddies with centers at around 88°E; 16.5°N and 92°E; 18°N are observed. A small cyclonic Eddy is found with center at around 90°E; 21°N above the anticyclonic gyre. Two anticyclonic Eddies are found in the AnS with centers at around 96°E; 13°N in the north and 96°E; 9°N in the south.

Figure 3b shows that in the spring northward propagating western boundary current is found to develop due to intensification of consecutive anticyclonic Eddies along the boundary (Legeckis, 1987; Johns and Ali, 1992; Somayajulu, 2003; Sil et al., 2010; Sil and Chakraborty, 2011). The cyclonic and anticyclonic Eddies in the head BOB are shifted southward. The equatorial jet also shifts southward to the south of Sri Lanka. The circulation in the AnS is separated to two anticyclonic Eddies; one in the south and another in the north. Due to these two anticyclonic Eddies southward flowing eastern boundary is observed in the BOB. This time flow through the MS is from the BOB to the South China Sea with very less intensity. In the summer, the western boundary current is found to be disorganized due to formation

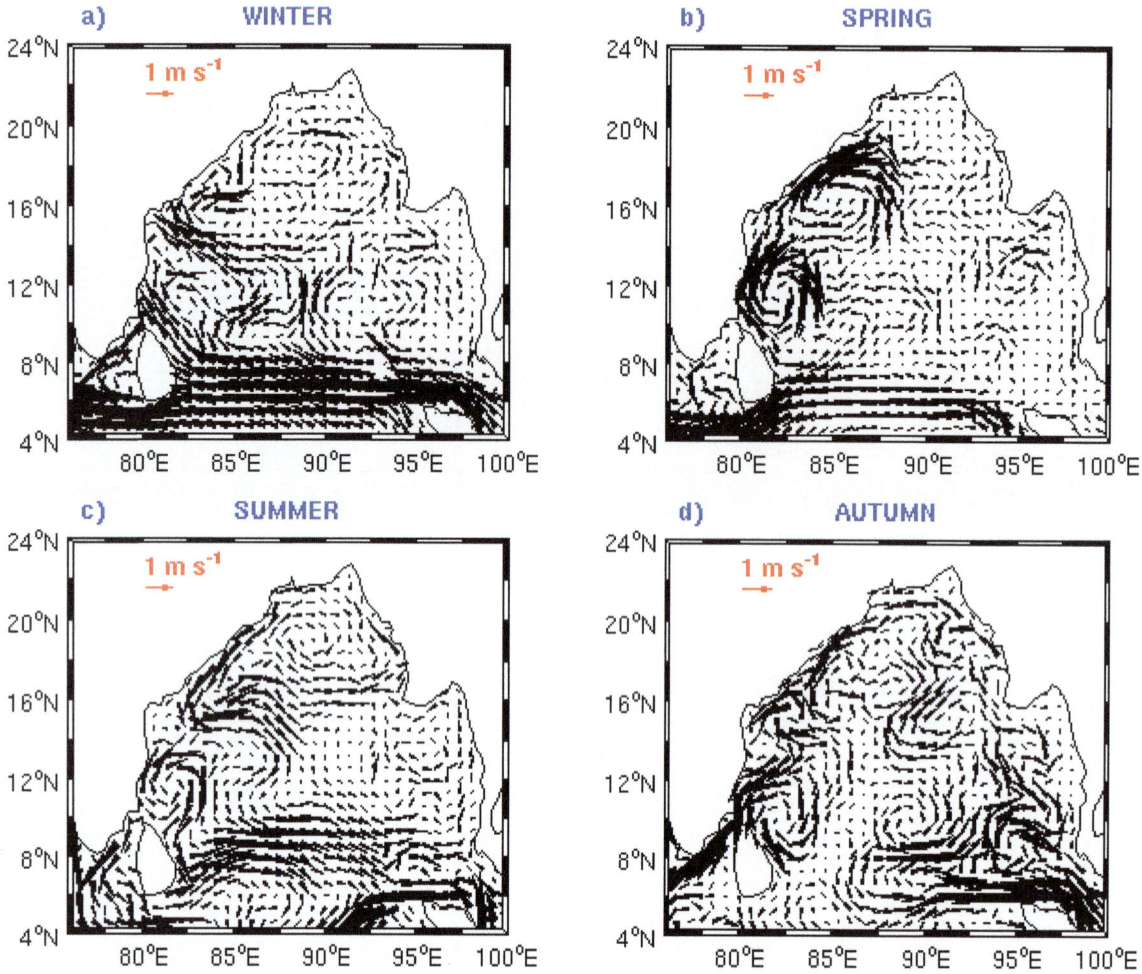

Figure 3. Model simulated current pattern at 10 m depth for (a) Winter, (b) Spring, (c) Summer and (d) Autumn. Western boundary currents during Spring and EICC during Autumn are two important boundary current systems of this basin.

of cyclonic (in the northern coast) and anticyclonic (near the head of Sri Lanka) Eddies along the coast (Figure 3c). This time a cyclonic gyre is found in the head BOB with centers at around 90°E; 19°N. The southwest monsoon current (SMC) is found to flow in the BOB between 80°E and 90°E (Potemra et al., 1991; Suryanarayana et al., 1993; Vinayachandran et al., 1999). This helps to produce anticyclonic Eddies in the head of the Sri Lanka (Vinayachandran and Yamagata, 1998). A cyclonic Eddy is formed at around 95°E; 14°N in the north of AnS.

This time the eastern boundary current is northward being influenced by SMC. It traverses along the coast and then meets with SMC. A flow from the MS does not directly enter the BOB as it is opposed by the SMC to divert around Sumatra. Figure 3d shows that in the autumn the east India coastal current (EICC) flow southward along the coast due to formation of cyclonic Eddies along the coast (Shankar et al., 1996). This time the flow from the MS is found to be strong along the AnS towards head BOB. Due to bathymetry of the AnS, an anticyclonic Eddy is found in the south of the AnS at around 95°E; 8°N. In broad view, a basin wise cyclonic gyre is found in the BOB (Potemra et al., 1991). But the mesoscale features are also obtained from the model simulation. The cyclonic Eddies are found at the central BOB at around 90°E; 9°N; near the Sri Lankan coast, it is found at around 84°E; 10°N; along the coastline at around 85°E; 18°N and 83°E; 17°N.

Circulation at 100 m

In the winter, the current pattern at 100 m depth is similar to the pattern of 10 m in the western pattern of the BOB but it is in opposite sense compared to the 10 m depth current in the AnS (Figure 4a). This time prominent eastern boundary current is found to flow southward. This eastern boundary current meets with the north equatorial jet which attend western boundary current being bifurcated at the Sri Lankan coast. Therefore, a basin wise anticyclonic current pattern is found from the model simulation. At this depth, effect from the MS is not there as depth is less than 100 m. In the spring, the dumbbell shaped current pattern with strong

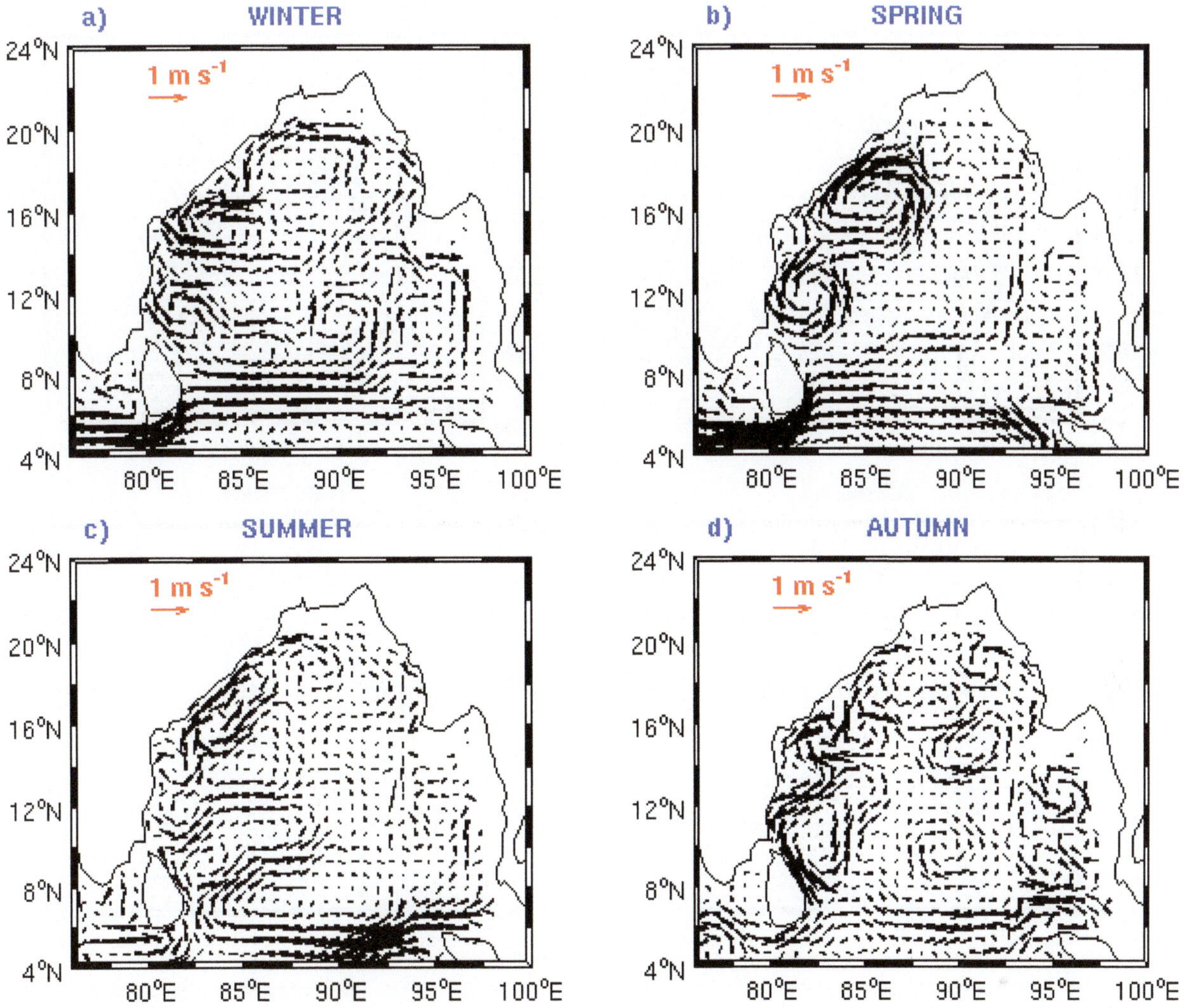

Figure 4. Model simulated current pattern at 100 m depth for (a) Winter, (b) Spring, (c) Summer and (d) Autumn.

boundary current is found in the western boundary similar to the current system at the 10 m depth (Figure 4b). In the AnS, the current pattern is similar to 10 m current. But during this time, a strong current is found to flow southwest from west of AnS which meets the north equatorial current which may be due to the bathymetry of the basin (Figure 1a).

In the summer, the western boundary current is found not well recognized like 10 m current. The eastern boundary current flows southward (Figure 4c). But the SMC is not observed in the south of the basin at this depth like surface. But north equatorial current is found to be reflected at the Sri Lankan coast and enter in the interior BOB. In the head of the BOB, cyclonic Eddies are observed with center at around 91°E; 17°N and 88°E; 18°N. In the autumn, the southward flowing EICC due to formation of cyclonic Eddies along the western coast of the basin is found like 10 m depth current (Figure 4d). The north equatorial current is present at this depth which is not present at the 10 m. This current meets with the EICC and flows towards southwest of the BOB. One cyclonic Eddy is found at around 89°E; 9°N and another two anticyclonic Eddies with centers at around 90°E; 14°N and 91°E; 18°N are observed in the northeast BOB from the model simulation. In the AnS two anticyclonic Eddies are found with centers at around 95°E; 14°N and 84°E; 9°N.

Circulation at 500 m

The current pattern at 500 m in the winter shows WBC is very much deep and strong (Figure 5a). This indicates that strong north equatorial current in the BOB cannot produce WBC in the upper surface due to other surface forcing like river discharge in the head BOB and western coast of the BOB. This WBC is found to form in the upper layer in the spring, but in 500 m depth it is found in the winter. Another thing to observe is that a strong current from west of the AnS to the southwest BOB which was found in the spring at 100 m depth is found at 500 m depth in the winter. Then, this current meets with the counter north equatorial current. The two anticyclonic Eddies in the AnS are also observed at 500

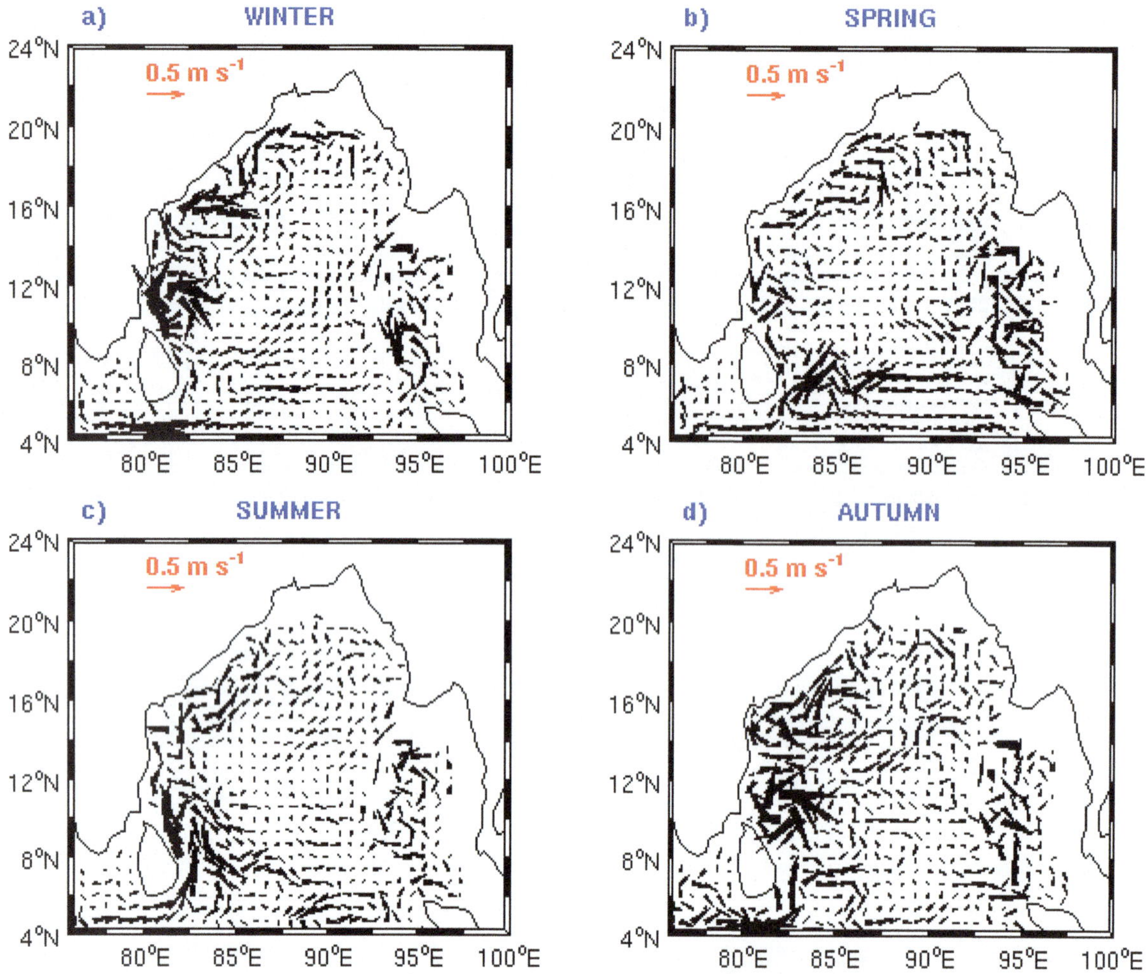

Figure 5. Model simulated current pattern at 500 m depth for (a)Winter, (b) Spring, (c) Summer and (d) Autumn.

m circulation. The current pattern in the spring at 500 m shows that the WBC is not well organized due to formation of cyclonic Eddy at the northern coast and an anticyclonic Eddy at the head of Sri Lanka (Figure 5b). But the dumbbell shaped anticyclonic gyre which is found in the upper 100 m does not appear at this depth; it breaks into two Eddies and they shift towards coast. In the southern BOB, currents from two opposite directions are found at around 6°N to form an anticyclonic Eddy in the southeast of the Sri Lanka. The eastern boundary current is found to flow northward through AnS to the head BOB. The circulation in the AnS is similar to that of winter but the northern Eddy is found to be cyclonic in nature.

In the summer, the current pattern shows the influence of AS to the BOB. The current flows around the Sri Lankan coast and then along the Indian coast (Figure 5c). But it moves back at around 12°N due to the formation of cyclonic Eddies in the northern coast. This time four cyclonic Eddies are formed in the northwest and head of the BOB. In the AnS, two anticyclonic Eddies were found from the model. In the autumn, the EICC is not found to be continuous southward due to the formation of an anticyclonic Eddy at around 82°E; 11°N near the Sri Lankan coast and cyclonic Eddies at the northwestern coast (Figure 5d). This north equatorial current is observed from the model study. The two anticyclonic Eddies which are found in the summer in the AnS are also observed in this season but they close together towards the north.

Circulation at 1000 m

The current pattern shows that in the winter the WBC flows northward. It indicates that the WBC is much deep in this basin (Figure 6a). An anticyclonic gyre is observed in the north of 14°N. Similarly, another anticyclonic gyre was found between 8 and 14°N. In the north of AnS, the anticyclonic Eddy is still there which can be found in the upper depths also. This is due to the bathymetry of this region. It cannot mix with the surroundings once it is formed. But current formed in the southern side of the AnS is observed to meet with the equatorial current. In the spring, the northward propagating WBC is found to flow continuously southward (Figure 6b). This is in opposite direction compared to the

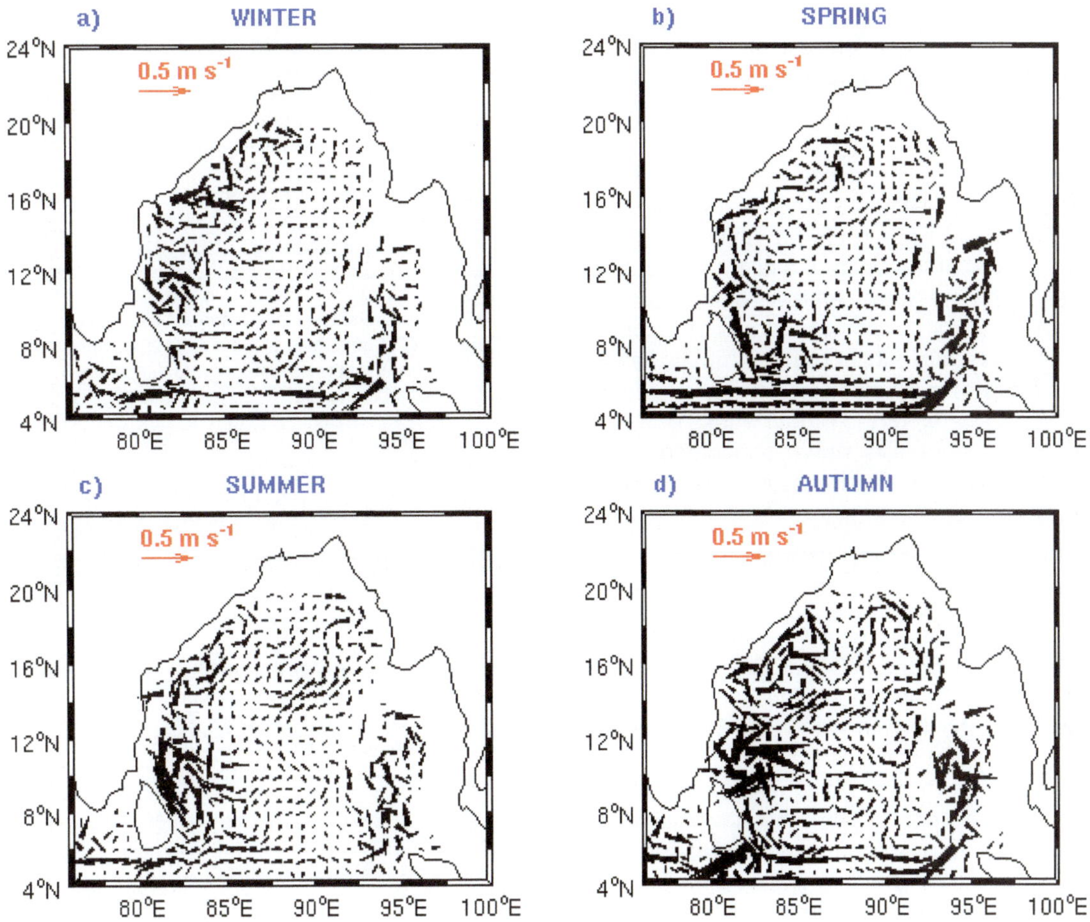

Figure 6. Model simulated current pattern at 1000 m depth for (a) Winter, (b) Spring, (c) Summer and (d) Autumn.

upper depths. The flow from the southern AnS to the southern BOB can also be observed here. The anticyclonic flow also persists at this depth there.

Figure 6c shows that the current along the western boundary is not well organized as the current pattern in summer during the summer. An anticyclonic Eddy is found near the Sri Lankan coast where cyclonic Eddy is found in the northwest coast of the BOB. In the head BOB, an anticyclonic gyre was found from this model simulation. In the AnS, three anticyclonic Eddies were formed in the model simulation. They are due to bathymetry effect of the AnS. In autumn, EICC flows southward in the upper BOB (Figure 6d). But at 1000 m it is not continuous along the western boundary due to formation of consecutive Eddies of opposite sense along the coast line. An anticyclonic Eddy was found at 82°E; 11°N which exists at 500 m also and a cyclonic gyre at 85°E; 14°N.

Conclusions

The seasonal variations in circulation for the BOB has been investigated using high resolution ROMS, driven by the climatological forcing. The outcome of the study are given as follows:

Circulation at 10 m: The formation of WBC at the upper layer and reversal of coastal current are well simulated by the model. The influence of the MS is prominent during the winter season at the 10 m depth. The model simulating north equatorial current and the SMC are quite comparable with previous studies. But the high resolution model is able to detect many more mesoscale features in the BOB basin. Most of the season, the circulation at 10 m depth in the Andaman Sea is driven by two Eddies; one in the north and the other in the south. These two Eddies are very deep.

Circulation at 100 m: The circulations in all seasons at 100 m depth are found to be similar like that at 10 m depth. The winter time influence from the MS is not present here due to its bathymetry which is less than 100 m. But the SMC in the summer and the north equatorial current in the autumn are not found at this depth like the 10 m depth.

Circulation at 500 m: The existence of the WBC at this depth in the winter confirms this current to be deep and strong. But spring time anticyclonic gyre in the western boundary of the basin is not present at this depth. The influence from the AS in the BOB circulation is clear from the model study.

Circulation at 1000 m: The current along the western boundary in the spring in 1000 m current is opposite to that of the upper layer current. A strong anticyclonic Eddy persists near the head of the Sri Lanka at 1000 m depth for almost every season. The overall analysis of the BOB model simulation, using ROMS with climatological southern boundary condition is sufficient enough to simulate seasonal climatological features. In addition, many mesoscale features are also simulated by the model whose existence is also supported by the satellite observations. This model simulation provides the idea about the circulation pattern in deeper oceans of this basin.

ACKNOWLEDGEMENTS

This study is supported by the INDOMOD research project (POM, RP2007) of INCOIS under the Ministry of Earth Sciences, Government of India and OCEANSAT-II utilization project (BBC, RP2008) of SAC, ISRO. The discussion and suggestions provided by Prof. A. Gangopadhyay of UMSSD, USA are greatly acknowledged. We are also grateful to Mr. Sudip Jana for his help. We are also grateful to the anonymous reviewers for their valuable comments and suggestions to ameliorate this revised version of the paper.

REFERENCES

Babu MT, Sarma YVB, Murty VSN, Vethamony P (2003). On the circulation in the Bay of Bengal during northern the spring intermonsoon (March–April 1987). Deep- Sea Res. II. doi: 10.1016/S0967- 0645(02)00609-4.

Chakraborty A, Gangopadhyay A (2011). Development of a high-resolution Multiscale Modeling and Prediction System for Bay of Bengal, Part I: Climatology based Simulations. (in review).

Cutler AN, Swallow JC (1984). Surface currents of the Indian Ocean (to 25S, 100E) compiled from historical data archived by the Meteorological Office. Bracknell, U.K. Inst. Oceanogr. Sci., Wormley, Surrey, England, Rep., 187: 1-8.

Da Silva AM, Young CC, Levitus S (1994). Atlas of surface marine data. Vol. 1, Algorithms and Procedures. NOAA Atlas NESDIS 6, U.S. Department of Commerce; NOAA, NESDIS, Washington, D.C.

Gopalan AKS, Gopala Krishna VV, Ali MM, Sharma R (2000). Detection of Bay of Bengal Eddies from TOPEX and in situ observations. J. Mar. Res., 58: 721-734.

Haidvogel DB, Arango HG, Hedstrom K, Beckmann A, Malanotte-Rizzoli P, Shchepetkin AF (2000). Model evaluation experiments in the North Atlantic Basin: Simulations in nonlinear terrain-following coordinates. Dyn. Atmos. Oceans, 32: 239-281.

Johns B, Ali A (1992). On the formation of a western boundary current in the Bay of Bengal. J. Marine Syst., 3(3): 267-278.

Legeckis R (1987). Satellite observations of a western boundary current in the Bay of Bengal, J. Geophys. Res., 92: 12974–12978.

Lermusiaux PFJ, Robinson AR (2001). Features of dominant mesoscale variability, circulation patterns and dynamics in the Strait of Sicily. Deep Sea Res., 48(9): 1953–1997.

McCreary JP, Kundu PK, Molinari RL (1993). A numerical investigation of dynamics, thermodynamics and mixed-layer processes in the Indian Ocean. Prog. in Oceanog., 31(3): 181-244.

Paul S, Chakraborty A, Pandey PC, Basu S, Satsangi SK, Ravichandran M (2009). Numerical simulation of Bay of Bengal Circulation features from Ocean General Circulation Model. Mar. Geod., 32: 1- 18.

Potemra JT, Luther ME, O'Brien JJ (1991). The seasonal circulation of the upper ocean in the Bay of Bengal. J. Geophys. Res., 96(7) 12 667-12 683.

Sanilkumar KV, Kurvilla TV, Jogendranath D, Rao RR (1997). Observations of the Western Boundary Current of the Bay of Bengal, Deep-Sea Res. I., 44(1)135-145.

Shankar D, McCreary JP, Han W, Shetye SR (1996). Dynamics of the East India Coastal Current. 1. Analytic solutions forced by interior Ekman pumping and local alongshore winds, J. Geophys. Res., 101(6)13 975-13 991.

Shetye SR, Gouveia A, Shankar D, Shenoi S, Vinayachandran PN, Sundar D, Michael GS, Nampoothiri G (1996). Hydrography and circulation in the western Bay of Bengal during the Northeast monsoon. J. Geophys. Res., 101: 114011-14025.

Shetye SR, Gouveia AD, Shenoi SSC, Sundar D, Michael GS, Nampoothiri G (1993). The western boundary current of the seasonal subtropical gyre in the Bay of Bengal. J. Geophys. Res., 98(1) 945-954.

Sil S, Chakraborty A (2011). Simulation of East India Coastal features and validation with satellite altimetry and drifter climatology. Int. J. Ocean and Climate Sys. (under review).

Sil S, Chakraborty A, Ravichandran M (2010). Numerical Simulation of Surface Circulation Features over the Bay of Bengal using Regional Ocean Modeling System. Adv. GeoSci. 24:117-130.

SomayajuluYK, Murty VSN, Sarma YVB (2003). Seasonal and inter-annual Variability of surface circulation in the Bay of Bengal from TOPEX/Poseidon altimetry. Deep-Sea Res., II, 50: 867–880.

Suryanarayana A, Murty VSN, Rao DP (1993). Hydrography and circulation of Bengal during early winter, 1983. Deep Sea Res. I. 40(1)205–217.

Vinayachandran PN, Masumoto Y, Mikawa T, Yamagata T (1999). Instrution of the southwest monsoon current into the Bay of Bengal. J. Geophys. Res., 104: 11,077-11,085.

Vinayachandran PN, Yamagata T (1998). Monsoon response of the sea around Sri Lanka: Generation of thermal domes and anticyclonic vortices. J. Phys. Ocean, 28: 1,946-1,960.

Wilkin JL, Arango HG, Haidvogel DB, Lichtenwalner CS, Durski SM , Hedstrom KS (2005). A regional Ocean Modeling System for the Long-term Ecosystem Observatory. J. Geophys. Res. 110: C06S91, doi:10.1029/2003JC002218.

Yu LL, O'Brien JJ, Yang J (1991). On the remote forcing of circulation in the Bay of Bengal. J. Geophys. Res., 96: 20449–20454.

18

Valuing environmental impacts of coastal development projects: A choice experiment application in Spain

David Hoyos[1]*, Pere Riera[2], Javier Fernández-Macho[3], Carmen Gallastegui[4] and Dolores Garcia[5]

[1]Department of Applied Economics III (Econometrics and Statistics), University of the Basque Country, E48015 Bilbao, Spain.
[2]Institute of Environmental Science and Technology, Autonomous University of Barcelona, Campus UAB, E08193 Bellaterra, Spain.
[3]Department of Applied Economics III (Econometrics and Statistics), Institute for Public Economics, University of the Basque Country E48015 Bilbao, Spain.
[4]Department of Foundations of Economic Analysis I, Institute for Public Economics, University of the Basque Country, E48015 Bilbao, Spain.
[5]Departament d'Economia Aplicada, Universitat de les Illes Balears, E-07122 Palma de Mallorca, Spain.

This paper presents the results of an empirical application aimed at estimating the economic value of the potential environmental impacts to mount Jaizkibel (Spain) resulting from the construction of a new seaport over its hillside. A choice experiment technique is applied for an *ex-ante* natural resource damage assessment in monetary terms. The results revealed that, depending on the extent of the environmental damage, the social welfare loss would be between 172 and 535 million Euros per annum.

Key words: Choice modelling, environmental valuation, social welfare, coastal management.

INTRODUCTION

Sea coast is an appreciated natural resource for human beings because of the ecological, cultural, social and economic values it bears. Human settlements have historically established near the coast. With just 4% of Earth's total land area, coastal areas and small islands house more than one-third of the world's population (Barbier et al., 2008). This is also the case of Spain. With a coastline 8,000 km long (4,000 km of cliffs, 2,000 km of beaches, 1,000 km of low coast, and around 600 km of artificial areas), it is estimated that nearly half of its population live in its coastal zones. Human pressure over the Iberian coast has increased over the last decades. In the Basque Country, a Spanish region in the North of the Iberian Peninsula, land artificialisation has grown 14% between 1987 and 2000, and the surface occupied by seaport areas has grown 72% for the same period, from nearly 366 ha in 1987 to 629 ha in 2000 (OSE, 2007). Developmental monetary benefits are often raised as a justification for diverse coast artificialisation projects, but they are rarely confronted with the environmental benefits that its conservation may entail. The absence of a monetary expression for the goods and services provided by coastal natural ecosystems often implies that they are implicitly equalled to zero. Scarcity of quantitative information for policy decision making has often been a concern of coastal managers (Christie and White, 1997; Chua, 1993).

Economic valuation aims at providing estimates of willingness-to-pay (WTP) of citizens to achieve/avoid certain environmental quality changes. Different economic valuation techniques have appeared to estimate in monetary terms the value of non-market goods. Existing approaches are broadly grouped into revealed preferences methods (like hedonic pricing and travel cost) or stated preferences methods (like contingent valuation and choice experiments). Stated

*Corresponding author. E-mail: david.hoyos@ehu.es.

preference methods have received growing attention mainly due to their flexibility and ability to measure not just use values (as revealed preference methods) but non-use values of natural resources as well (Mitchell and Carson, 1989). The main difference between the Contingent Valuation Method and the Choice Experiment (CE) methodology is that while in the former individuals face the valuation of one good with varying prices, in the latter individuals face the valuation of a bunch of goods (or one good with multiple attributes) and different prices. The underlying idea of CEs is that if human-induced changes in the state of an ecosystem can be coherently represented by a group of attributes, people's choices provide substantial information over their preferences regarding alternative states of the environment (Hoyos, 2010).

This paper examines the use of a CE methodology to assess the economic value of potential environmental impacts that could be associated with the construction of a new seaport over the hillside of mount Jaizkibel (Spain). Jaizkibel is a protected natural resort because of its landscape and geological interest as well as its fauna, flora, and seabed. The values obtained can be employed in an *ex-ante* assessment of the environmental costs associated with coastal development projects useful for decision making.

Case study

Pasaia, a city located on the Spanish Cantabrian coast, near the border with France, has had maritime and commercial activities in its natural port since the 12th century. Even though up to the 19th century it was mainly dedicated to ship building and fishing, its main activity during the 20th century has been the traffic of materials from heavy industry. In recent years, the Port Authority of Pasaia has promoted a project to develop a new seaport outside the bay, in front of mount Jaizkibel. Defenders of this project claim that it will be very profitable to the economic activity of the region while opponents argue that the environmental loss associated with the project advice against its construction. Figure 1 shows the current seaport (A) and three projected seaports (Seaport projects 1, 2 and 3), with different capacity and affections to Jaizkibel natural site.

Mount Jaizkibel is a 2,434 ha natural site located at 1° 50' West longitude and 43° 22' North latitude. It belongs to the European Natura 2000 Network because it is considered a site of Community importance for the Atlantic biogeographical region, thus protected under European environmental legislation (Habitat Directive 92/43/EEC). The landscape of this area is especially interesting because the mountain goes along the coast with abrupt falls in the western part, forming cliffs up to 240 m high. The cliffs are considered of geological interest due to the layout of sandy stratum. They host the *Armeria euskadiensis*, an endemic plant of the Basque coast catalogued as an endangered species. In the Eastern part, the terrain is not so abrupt and there are small beaches and precipices formed by the curse of streams ending in the Cantabrian Sea. In these areas, one can find some interesting species of flora such as tropical ferns (*Woodwardia radicans* and *Trichomanes speciosum*), very rare in the rest of Europe. Certain spaces maintain the original tree cover, oak grove of *Quercus robur* and *Quercus pyrenaica*. The rest of Jaizkibel conforms a non-wooded forest area with some brushes and some pastures associated to local *baserri* (autochthonous farms).

Some colonies of lesser black-backed gull and yellow-legged gull (*Larus fuscus* and *Larus cachinnans*) nest in Jaizkibel's cliffs. Other interesting birds, such as the European storm-petrel (*Hydrobates pelagicus*), Green cormorant (*Phalacrocorax aristotelis*) and Peregrine falcon (*Falco peregrinus*) can be found in this natural area. Over the mainland there are numerous species of amphibious, reptiles and mammals such as Palmite newt (*Triturus helveticus*), Midwife toad (*Alytes obstetricans*), Dark green snake (*Coluber viridiflavus*) and Greater horseshue bat (*Rhinolophus ferrumequinun*). In its seabed, it harbours different types of molluscs, sea urchins and crustaceans. Jaizkibel's seabed also harbours various types of seaweed: green, red and brown. Furthermore, Jaizkibel has one of the most important lands of red seaweed of the Basque coast. In short, Jaizkibel's most outstanding environmental attributes are: landscape, autochthonous fauna and flora, seabed life, and environmental services such as sweet water, clean air and maintaining of current stream, swell and sediment transportation regime.

The construction of a new seaport would involve a number of different stressors that could damage natural site's landscape values, habitats and species. The identification of potential damages was based on published reports, public institutions' and NGOs' reports and assessments made at the site. According to these information sources, anticipated environmental impacts related to the construction and operation of the seaport would include: direct loss of habitat, disturbance to sensitive habitats and species of avifauna, air pollution, increased human access to wildlife, risk of fire and other disturbance to sensitive habitats and species. Detailed information on the environmental values and potential damages to Jaizkibel may be found in Pozueta (2004).

Survey design

A valuation survey was conducted in the Basque Country in order to determine the value of the potential environmental impacts to Jaizkibel area. The questionnaire

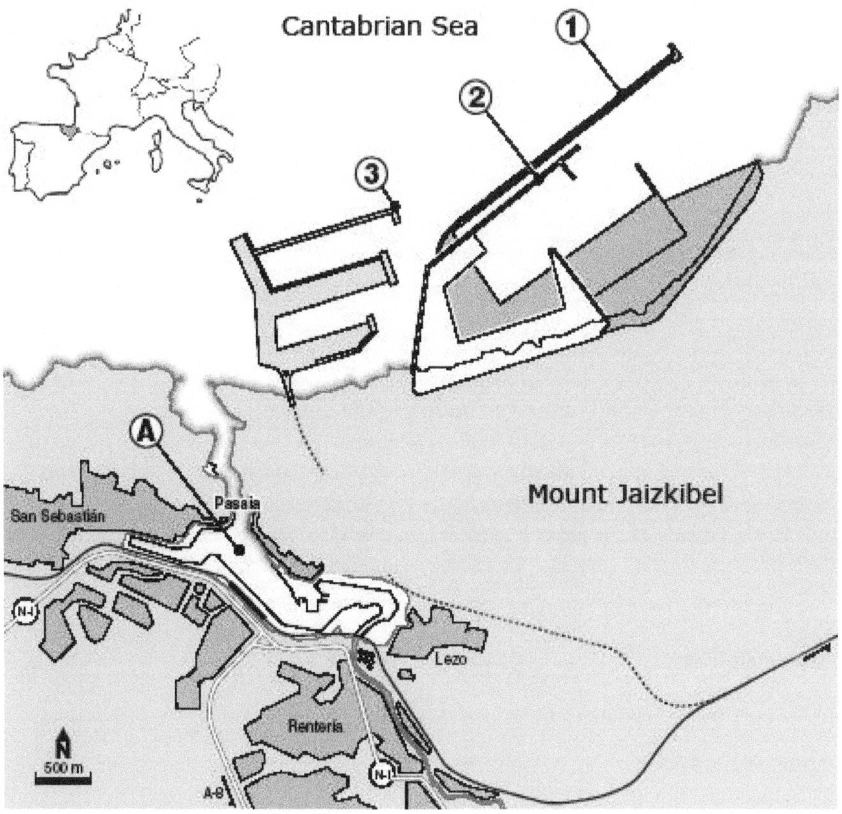

Figure 1. Three scenarios for the future seaport of Pasaia.

started by describing the Jaizkibel natural site. Next, certain changes in the quality of mount Jaizkibel's main attributes were described. It was stated that if the site was not to be protected in the future, it could be affected by human activities. The market institution was detailed next, where compulsory monetary contributions from the Basque residents would contribute to the conservation of the site. The trade-offs between payments and degree of preservation of the different attributes were presented in choice sets of two alternatives and an opt-out situation, as will be detailed shortly. The final part of the questionnaire was devoted to some debriefing and sociodemographic questions. The selection of attributes and levels is an important aspect in CE given that respondents are to make choices characterized by those attributes and levels (Hensher, 2007). Previous investigation on environmental characteristics of Jaizkibel natural site, experts' advice and focus groups facilitated the definition of environmental attributes and levels of provision. Furthermore, a pilot survey using open-ended contingent valuation questions helped in identifying the appropriate levels of the payment attribute.

At the beginning, six non-monetary attributes were identified: landscape, flora, avifauna, seabed, groundwater and air quality. However, the last two attributes (groundwater and air quality) were dropped mainly because of their relative little importance as suggested by experts and focus groups. Table 1 reflects the attributes and levels considered in this study. They were,

(1) Landscape, measured by the percentage surface from which today's landscape could be seen in the future;
(2) Flora, measured by the future level of protection of today's population of *A. euskadiensis* endemism;
(3) Avifauna, measured by the future level of protection of today's population of lesser and peregrine falcons;
(4) Seabed, measured by the future level of protection of today's extension of red algae; and
(5) Annual contribution in euros, varying from 5 to 100 €.

Respondents' understanding of the valuation exercise is a critical aspect of any stated preference survey. Accordingly, the graphic and verbal description of the future environmental changes for the non-monetary attributes was tested in detail in focus groups. Landscape changes seemed to be better understood showing pictures with a different degree of visual affection, as shown in Figure 2. The importance of flora, fauna and seabed attributes seemed to be better captured by its most prominent species, the *A. euskadiensis*, the lesser

Table 1. Attributes and levels considered.

Attribute	Level						
Landscape (%)	40*	60	80	100			
Flora (%)	50*	70	85	10			
Fauna (%)	25*	50	75	100			
Seabed (%)	50*	70	85	100			
Annual payment	0 €*	5 €	10 €	15 €	20 €	30 €	50 €	100 €

*Business-as-usual level.

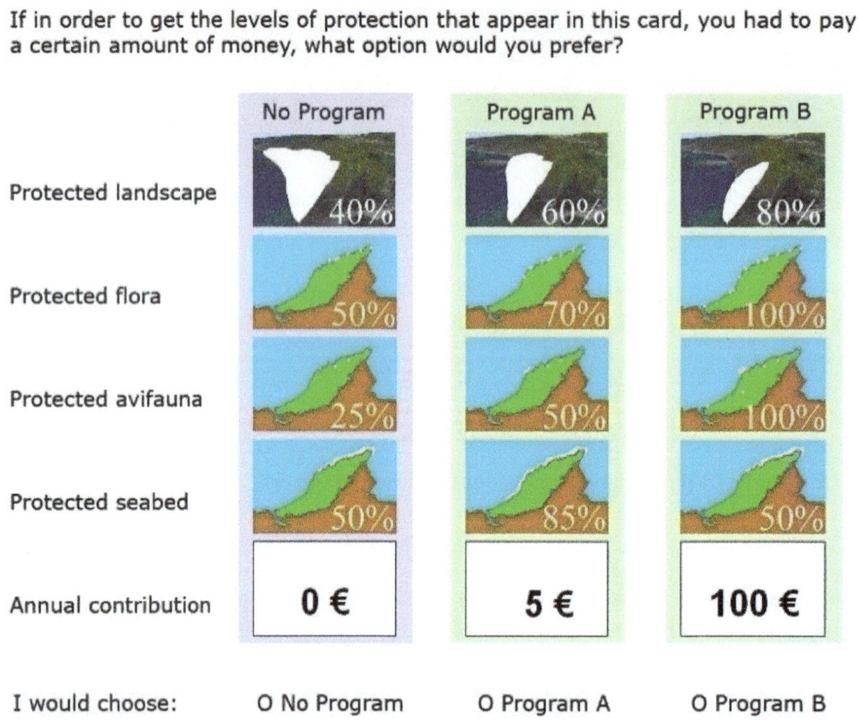

Figure 2. Example of choice set with different protection alternatives used in the valuation exercise.

and peregrine falcon, and the red algae respectively. These attributes' graphical representation corresponded with the actual location of these species' populations according to the experts' advice. When the percentages of protection of these attributes were inferior to 100%, some actual location of its most prominent species' populations were deleted accordingly (Figure 2). None of the levels of protection were set to zero in order to avoid plausibility problems related to the survival of different species.

Combining all these attributes and levels, near two thousand different combinations are obtained ($4^4 x 7^1$). A main effects fractional factorial design with second order interactions was used to reduce the number of alternatives to 96 pairs of protection alternatives (Louviere et al., 2000). Further, the alternatives were grouped in 24 blocks of four choice sets containing two alternative protection programmes plus the business-as-usual (BAU) option. The burden of the choice task was checked in the focus groups. As a result, the final version of the questionnaire had four choice sets, each formed by the BAU option plus two protection alternative programs (Programs A and B) as shown in Figure 2. The proposed payment vehicle was an annual contribution to a Foundation exclusively dedicated to protecting Jaizkibel natural site that all Basque citizens would have to make.

The questionnaire was administered through in-person computer-aided individual home interviews. Respondents could read the questions in the computer's screen and listen to a recorded voice, with three different languages available: Basque, Spanish and French. The relevant population considered was the population from the

Basque Country, accounting for 2.5 million people being at least 18 years old. A pilot survey was conducted in October 2006, while the final survey was undertaken between November and December, 2006. A stratified random sample of 600 individuals was selected from the relevant population. The strata used included age, gender and size of the town of residence, following official statistical information from the Basque Statistics Institute (EUSTAT). In each location, the questionnaires were distributed using random survey routes.

METHODOLOGY

CE belongs to the family of conjoint analysis methods, defined by Green and Srinivasan (1990) as "any decompositional method that estimates the structure of a consumer's preferences given his or her overall evaluations of a set of alternatives that are pre-specified in terms of levels of different attributes." Choice experiments technique is based on Lancaster's characteristics theory of value and random utility theory (Hanley et al., 1998). Following random utility theory, consumers pursue the maximisation of utility in decision-making processes. Thus, the utility an individual *i* obtains from alternative *j* (U_{ij}) can be formalised as;

$$U_{ij} = V_{ij} + \varepsilon_{ij},$$

where V_{ij} is the part of utility that is observable (and thus affected by the attributes of this option), and ε_{ij} is the non-observable part or random component. As a consequence, individual *i* will choose alternative *j* instead of *k* if her utility increases, this is, if:

$U_{ij} > U_{ik}$ for $k \neq j$

However, given the existence of a random component, the choices can be written in probability terms. The probability that individual *i* chooses alternative *j* instead of *k* from a finite set of alternatives *C*, would be:

$$P_{ij} = \text{Prob}\{V_{ij} + \varepsilon_{ij} > V_{ik} + \varepsilon_{ik}; \forall k \in C\}.$$

The stochastic component of utility is usually assumed to be independent and identically distributed (IID) and Gumbel distributed (McFadden, 1974). Thus, the conditional logit model can be written as:

$$P_{ij} = \frac{e^{\omega Vij}}{\sum_{k=1}^{m} e^{\omega Vik}},$$

where ω is a scale parameter, which is inversely proportional to the error term's standard deviation and it is generally assumed to be one so that the variance of the error term is constant.

The equation aforestated can be estimated by means of a multinomial logit (MNL) regression. The MNL model relies on the assumption that choices are consistent with the Independence of Irrelevant Alternatives (IIA). This axiom states that the ratio of the probabilities of choosing one alternative over another is not affected by the presence or absence of other alternatives in the same choice set (Louviere et al., 2000). In case of violation of IIA, the parameters estimation would be biased. The IIA property is usually checked using the test proposed by Hausman and McFadden (1984). The structure of the MNL model depends on the form adopted by the indirect utility function. To estimate the main effects, an additive indirect utility function of the following form may be used:

$$V_{ij} = \beta_0 + \beta_1 Atrib_1 + \beta_2 Atrib_2 + \beta_3 Atrib_3 + ... + \beta_n Atrib_n$$

where β_0 is the constant term and β_1 ... β_n are the coefficients of environmental attributes.

The constant term, β_0 (that can be interpreted as a vector of alternative specific constants, one for each alternative considered in the choice set) reflects the influence on choice of non-observed attributes relative to specific alternatives. Alternative specific parameters, however, may be dropped in dealing with non-labelled experiments (Hensher et al., 2005). Individual marginal WTP estimates represent the average maximum amount of money a person is willing to give in exchange for an additional unit of the environmental good, that is they represent estimations, *ceteris paribus*, of the value of a marginal change in a given attribute. In order to estimate the WTP, the interaction between multiple attributes, if relevant, is to be taken into account as well as the influence of the alternative specific constant. Welfare estimates for MNL models may be obtained from

$$CS = \frac{-1}{\alpha\left[\ln \sum e^{Vi0} - \ln \sum e^{Vi1}\right]},$$

where *CS* represents the compensating surplus, α is the marginal utility of income, and V_{i0} and V_{i1} are the indirect utility functions of alternative *i* in the status quo (0) and in the change considered (1). In our case, the previous expression can be rewritten as:

$$CS = -\frac{1}{\beta_{cost}}\left[\Delta\beta_{landscape} + \Delta\beta_{flora} + \Delta\beta_{avifauna} + \Delta\beta_{seabed}\right]$$

And simplifying the aforestated equation, the marginal value of a change in one attribute with respect to another is measured through the ratio of both coefficients. Therefore, the WTP for each of the environmental attributes is obtained by dividing the coefficient of each attribute by the cost attribute coefficient:

$$WTP = -\frac{\beta_{attribute}}{\beta_{cos\,t}}$$

The cost parameter is interpreted as the marginal utility of income.

RESULTS AND DISCUSSION

A fixed parameter logit model specification was estimated using LIMDEP econometric software (Greene, 2007). The results are shown in Table 2. All the coefficients of the environmental attributes have the expected signs (positive, meaning that conservation is more highly valued than loss) and are significant at 1% level. The negative coefficient of the price attribute is also expected, indicating that the probability of accepting an annual

Table 2. Fixed parameter logit model estimation.

Covariate (attribute)	Coefficient	t-Statistic
Landscape	0.02028	7,38
Flora	0.01272	3,79
Avifauna	0.00998	4,90
Seabed	0.00925	3,90
Cost	-0.01462	-7,17
Log-likelihood		-590.4531
Log-likelihood at 0		-627.1635
Observations		687

Table 3. IIA/IID test statistics for the MNL model.

Alternative dropped	X^2	Df	Probability
Business-as-usual	6.462	5	0.264
Alternative 1	0.926	5	0.968
Alternative 2	6.560	5	0.255

Table 4. Marginal WTP for protecting mount Jaizkibel's environmental attributes, in euros 2006 per person and year.

Attribute	Marginal WTP	95% confidence interval
Landscape	1.39	(0.98,1.86)
Flora	0.87	(0.41,1.31)
Avifauna	0.68	(0.41,0.95)
Seabed	0.63	(0.33,0.96)

95% confidence intervals were calculated using the Krinsky-Robb procedure (Krinsky and Robb, 1986).

contribution for protecting mount Jaizkibel's attributes decreases as the price increases. No relevant second order interactions were found. Regarding the IIA property (Hausman and McFadden, 1984), it was tested whether the full mode, estimated with all three alternative choices, was equivalent to a restricted model where one of the alternatives was eliminated. In every case, the null hypothesis that IIA holds for this data set cannot be rejected, as shown in Table 3. Table 4 shows point estimates and 95% confidence intervals of the marginal WTP estimates for the four attributes. The positive signs of the marginal WTP estimates for the environmental attributes indicate that, everything else being equal, respondents would be better off on average with an increase in the level of those attribute. The individual WTP for a 1% improvement of Jaizkibel's landscape is estimated at 1.39 euros (2006) per person and year.

Similarly, the WTP for a 1% improvement in the quality of the flora, avifauna and seabed is estimated at 0.87, 0.68 and 0.63 euros per person and year respectively. Three damage scenarios to the environmental attributes of mount Jaizkibel can be approximated based on the expected impacts associated with the Seaport projects presented in Figure 1. The scenarios are based on the report by Pozueta (2004) and experts' opinions. They took into account the current geographical location of the *A. euskadiensis*, the lesser black-backed gulls and peregrine falcons, and the red algae. Experts' advice was followed in order to determine the specific impact of each harbour project on the population of these species. Figure 3 summarises the main impacts of each project to the identified environmental attributes. This figure shows, for example, that the largest harbour project (seaport 1) would imply that 40% of current landscape, 50% of current flora, 25% of current fauna and 50% of current seabed would remain preserved in the future.

Annual welfare loss due to the environmental degradation provoked in the construction of each future seaport was calculated using the compensating surplus equation for a linear additive utility function explained before. Table 5 shows the individual WTP estimates for each scenario. The WTP corresponds to the mean amount of money, in 2006 values, that one individual would be willing to pay at most to avoid an environmental damage as described for each scenario. Thus, on average an individual would be willing to pay annually an average of 209 euros to avoid an environmental damage as described in the Seaport 1 degradation scenario, 134 euros for the Seaport 2 degradation scenario and 67 euros for the Seaport 3 degradation scenario. Finally, the annual welfare loss associated with the degradation scenarios described above is calculated by multiplying the mean WTP by the relevant population (2.5 million residents), as shown in Table 6. The annual welfare loss of the deterioration of the environmental quality of Jaizkibel associated to the harbour development may be estimated between 172 and 536 million euros, depending on the project to be undertaken. In the Seaport 1 Scenario, annual welfare loss is estimated at 535 million euros (213 million for landscape, 110 for flora, 130 for avifauna and 81 for seabed). In the Seaport 2 Scenario, annual welfare loss is estimated at 344 million euros (142 million for landscape, 66 for flora, 87 for avifauna and 49 for seabed). And, in the Seaport 3 Scenario, annual welfare loss is estimated at 172 million euros (71 million for landscape, 33 for flora, 43 for avifauna and 24 for seabed).

In sum, this empirical application has attempted to show how economic analysis can be used in coastal management in order to assess *ex-ante* the desirability of a development project according to its environmental effects. The ex ante assessment of potential environmental impacts faces two significant shortcomings: absence of market prices for environmental damages and reliance on hypothetical behaviour that can only be

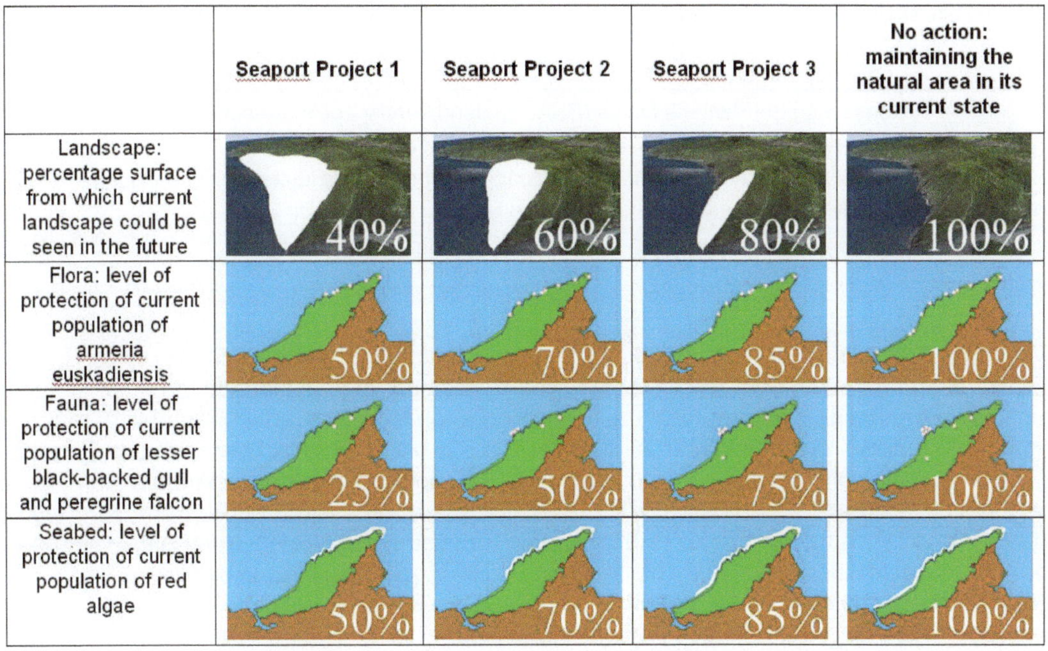

Figure 3. Mount Jaizkibel estimated scenarios according to the impacts of three future seaport projects of Pasaia.

Table 5. Compensating surplus for different degradation scenarios, in euros 2006 per person per year.

Scenario	Level of damage				Individual WTP
	Landscape (%)	Flora (%)	Avifauna (%)	Seabed (%)	
Seaport 1	60	50	75	50	208.74 *(126.26-296.87)*
Seaport 2	40	30	50	30	134.17 *(81.71-190.34)*
Seaport 3	20	15	25	15	67.09 *(40.86-95.17)*

Table 6. Annual welfare loss, in millions of euros 2006.

WTP	Landscape	Flora	Avifauna	Seabed	Aggregate
Seaport 1	212.89 (150.85-287.07)	110.60 (52.57- 168.30)	130.76 (78.32-183.22)	81.16 (42.18-123.01)	535.52 (323.93-761.61)
Seaport 2	141.93 (100.57-191.38)	66.36 (31.54-100.98)	87.17 (52.21-122.15)	48.76 (25.31-73.81)	344.22 (209.63-488.32)
Seaport 3	70.96 (50.28-95.69)	33.18 (15.77-50.49)	43.59 (26.11-61.07)	24.38 (12.65-36.90)	172.11 (104.82-244.16)

captured by SP methods. The use of CE methodology in this context is thus justified by two reasons: first, as a SP non-market valuation technique, it allows to estimate hypothetical environmental quality changes; and secondly, it is able to separately estimate the preference of individuals for different environmental attributes and marginal values. The flexibility of this methodology allows assessing a wide variety of potential damages to a diverse set of environmental attributes.

Some applications of CEs have been reported for the economic valuation of coastal and marine resources (Johnston et al., 2002; Lew and Larson, 2005; Rein, 1999; Samonte-Tan et al., 2007), although, to our knowledge, the use of CE in this context is non existent.

Table 7. Contingent valuation studies of rural landscape changes, with WTP expressed in euros 2006 per person and year.

Study	Landscape change	Population	Individual WTP
Santos (1998)	Conserving the Pennine Dales (ESA, England) landscape's attributes	Visitors	78,84 - 96,17
Willis and Garrod (1991)	Conserving the Yorkshire Dales (UK) today's landscape	Visitors	59,89 - 89,43
Santos (1997)	Conserving today's agricultural landscapes in the Peneda-Geres (NP, Portugal)	Visitors	64,83 - 75,72
Santos (2007)	Multiple study average		42,40 - 64,56
Santos (2007)	Meta-analytical model predictions based on similar studies	Visitors	48,16 - 97,96

Source: Santos (2007).

The WTP results seem to be in line with those from similar studies. Table 7 contains a summary of the mean estimates of different studies and the results from a meta-model built upon contingent valuation studies of environmentally sensitive areas by Santos (2007). In the reported studies, the WTP ranges from 42 to 98 € per visitor and year. These estimates are slightly higher than the WTP for protecting the mount Jaizkibel today's landscape (between 27.80 and 83.40 € per person and year depending on the percentage degradation considered). This difference may be explained, among other things, because the population surveyed in these studies were visitors (normally stating higher WTP than non-visitors) while in Jaizkibel the general population was surveyed.

Conclusions

Coastal managers and policy-makers often need to evaluate policies affecting the welfare of the population. If relevant environmental costs are not incorporated in coastal developmental project assessments, welfare measures will be probably biased. CE methodology is a valuation method that can be used to value impacts related to coastal development projects. This paper has used this methodology to perform an economic assessment of potential non-market environmental damages to Jaizkibel natural site (Spain) associated to the construction of a new seaport in Pasaia. The valuation exercise was based on a questionnaire survey administered in the Basque Country (Spain and France) to a representative sample of the population.

On average a Basque citizen would be willing to pay annually 1.39 euros for avoiding 1% deterioration of mount Jaizkibel today's landscape; 0.87 euros for avoiding 1% deterioration of today's flora; 0.68 euros for avoiding 1% deterioration of today's avifauna; and 0.63 euros for avoiding 1% deterioration of today's seabed. According to this estimates, the economic value of the future environmental damage provoked by a new seaport in Pasaia was estimated between 172 and 535 million euros per annum. This value depends on the future environmental damage that the construction of a new seaport may cause. In case of constructing Seaport 1 (Figure 1), the environmental damage was estimated in 535 million euros per annum. In case of constructing Seaport 2, the environmental damage was estimated in 344 million euros per annum. And last, if Seaport 3 was constructed, the environmental damage was estimated in 172 million euros per annum.

ACKNOWLEDGEMENTS

The authors acknowledge the financial support from the Department of Environment of the Basque Government and IHOBE, S.A., from the Department of Education of the Basque Government through grant IT-334-07 (UPV/EHU Econometrics Research Group), and from CICYT Project No. SEJ2004-00143/ECON and CEDEX project from the Spanish Government.

REFERENCES

Barbier EB, Koch EW, Silliman BR, Hacker SD, Wolanski E, Primavera J, Granek EF, Polasky S, Aswani S, Cramer LA, Stoms DM, Kennedy CJ, Bael D, Kappel CV, Perillo GME, Reed DJ(2008). Coastal Ecosystem-Based Management with Nonlinear Ecological Functions and Values. Science, 319: 321-323.

Christie P, White AT(1997). Trends in development of coastal area management in tropical countries: from central to community orientation. Coastal Manage., 25: 155-181.

Chua TE (1993). Essential elements of integrated coastal zone management. Ocean and Coastal Management, 21:81-108.

Green PE, Srinivasan V (1990). Conjoint analysis in marketing: New developments with implications for research and practice. J. Mark., 54: 3-17.

Greene WH (2007). LIMDEP Version 9.0 User's Manual, Econometric Software, Inc., New York.

Hanley N, Wright RE, Adamowicz V (1998). Using choice experiments to value the environment. Environ. Resour. Econ., 11: 413-428.

Hausman J, McFadden D(1984). Specification tests for the multinomial logit model. Econometrica, 52: 1219-1240.

Hensher DA (2007). Attribute processing in choice experiments and implications on willingness to pay. In: BJ Kanninen (Editor), Valuing environmental amenities using stated choice studies. A common sense approach to theory and practice. Springer, The Netherlands, pp. 135-158.

Hensher DA, Rose JM, Greene WH (2005). Applied choice analysis. A primer, Cambridge University Press, New York.

Hoyos D (2010). The state of the art of environmental valuation with discrete choice experiments, Ecol. Econ., 69(2): 2372-2381.

Johnston RJ, Grigalunas TA, Opaluch JJ, Mazzotta M, Diamantedes J (2002). Valuing estuarine resource services using economic and ecological models: the peconic estuary system study. Coastal Manage., 30:47-65.

Krinsky I, Robb A (1986). On approximating the statistical properties of elasticities. The Review of Economics and Statistics, 68:715-719.

Lew DK, Larson DM (2005). Valuing recreation and amenities at San Diego County Beaches. Coastal Manage., 33: 71-86.

Louviere J, Hensher DA, Swait J (2000). Stated choice methods: analysis and applications, Cambridge University Press, Cambridge.

McFadden D (1974). Conditional logit analysis of qualitative choice behaviour. In: P Zarembka (Editor), Frontiers in Econometrics. Academic Press, New York.

Mitchell RC, Carson RT (1989). Using Surveys to Value Public Goods: The Contingent Valuation Method, RFF Press, Washington, D.C.

OSE (2007). Sostenibilidad en España 2007. Observatorio de la Sostenibilidad en España.

Pozueta J (2004). Estudio comparado de las alternativas de desarrollo del Puerto de Pasajes en relación con su grado de impacto medio-ambiental y sostenibilidad. Departamento para el Desarrollo Sostenible. Diputación Foral de Gipuzkoa.

Rein FA (1999). An economic analysis of vegetative buffer strip implementation case study: Elkhorn Slough, Monterey Bay, California. Coastal Manage., 27: 377-390.

Samonte-Tan GP, White AT, Tercero MA, Diviva J, Tabara E, Caballes C (2007). Economic valuation of coastal and marine resources: Bohol Marine Triangle, Philippines. Coastal Manage., 35: 319-338.

Santos JML (2007). Transferring landscape values: how and how accurately? In: S Navrud and R Ready (Editors), Environmental value transfer: Issues and methods. Springer, The Netherlands, pp. 45-76.

19

An overview of recent technologies on wave and current measurement in coastal and marine applications

P. Kasinatha Pandian[1]*, Osalusi Emmanuel[2], J. P. Ruscoe[2], J. C. Side[2], R. E. Harris[2], S. A. Kerr[2] and C. R. Bullen[2]

[1]Department of Civil Engineering, Tagore Engineering College, Chennai 600048, India.
[2]International Centre for Island Technology, Heriot-Watt University (Orkney Campus), Stromness, Orkney, Scotland, KW16 3AW, United Kingdom.

The coastal ocean is the most important zone for the maritime countries for recreation, mineral and energy exploitation, weather forecasting and national security. Understanding of coastal and oceanic processes is mostly based on field measurements and laboratory experiments. Most coastal processes occur over relatively long time spans and have large spatial extents. They also involve or are impacted by a variety of factors such as waves, wind, tide, storm surges, currents, beach sediment properties, etc. Measurements of waves, both on site and in wave flumes, are carried out using different techniques. This article reviews the different technologies available and presently adopted in wave and current measurements. Different instruments for wave and current measurement have its own advantages and disadvantages depending on the applications and needs. A detailed overview of various studies carried out in worldwide locations on wave and current measurements is presented. Various instruments, their application, advantages, disadvantages and accuracy are discussed in this article. It is essential to choose the type of instrument most appropriate to the application, based on the requirements, necessitating a thorough knowledge of the instruments available, the funding and duration of the project.

Key words: Measurements, waves, currents, buoys, ADCPs, radars, remote sensing.

INTRODUCTION

Most coastal processes occur over relatively long time spans and have large spatial extents and involve or are impacted by a variety of coastal factors such as waves, wind, tide and storm surge, currents and beach sediment properties. Consequently the cost of field measurements can be high. Routine monitoring of waves and currents in the offshore and nearshore regions is of great interest to both coastal and scientific communities. The measurement of waves and in particular their direction, has been one of the most difficult problems in observational coastal engineering and oceanography (Strong et al., 2000). There are many techniques adopted for measuring waves such as wave rider buoys, Acoustic Doppler Current Profilers (ADCP) and remote sensing including High Frequency Radar. In wave measurements, directional wave data are made with a wide variety of instrumenttation techniques. The nature of these instruments dictates that most wave measurements be made in coastal areas. The cost in physical effort and money to maintain an operating wave gauge is relatively high. However, for a variety of reasons, the percentage of time that worthwhile data are obtained is commonly much less than 100% (Sorensen, 1997). Although there are many aspects to the collection of data from the marine measurement, this article only deals with wave and current measurements and their problems, advantages and disadvantages and reviews the different techniques and instruments used by various researchers worldwide.

*Corresponding author. E-mail: pknp2020@gmail.com.

Wave rider buoys

Wave rider buoys have been in use since 1966 and are the principal instrument for national wave measurement programmes for many countries. It has frequently been assumed that the wave rider sensitivity remains within 3% of the specified one volt $m^{-1}$of wave elevation throughout its useful life. During the earlier years of wave rider use, no calibration apparatus was available, although approximate calibrations over a limited range of period of vertical motion could be obtained by suspending the buoy from an elastic cord or spring and maintaining vertical motions manually (Ribe, 1983). In support of their respective wave climate programmes, researchers in the United States, Canada and United Kingdom developed apparatus to calibrate wave riders on a rotating-arm, vertical circle of motion. In Venezuela, a device using pure vertical motion, based on the Scotch-yoke principle has been developed. Such apparatus has enabled improvement in ocean wave measurement accuracy by as much as 5%. Resultant correction factors are a function of period or frequency and, therefore, must be applied to sinusoidal, Fourier components of a wave elevation time series or, as a square of the transfer function, to the coefficients of the variance spectra. The wave rider buoy has a unique accelerometer design consisting of a cantilever-beam sensor mounted in the electrical field of an electrolytic fluid. The fluid conductivity is carefully adjusted to provide the desired wave rider sensitivity and the associated electronic circuitry is designed to match the characteristics of each individual accelerometer. Effects of fluctuations of temperature of the fluid can be considered negligible. However, an increase in the electrical conductivity of the electrolyte of some wave rider buoys has caused a corresponding decrease in sensitivity and has rendered the assumption of insensitivity to temperature fluctuations invalid. This has increased the importance of applying corrections to wave rider measurements.

Magnusson and Donelan (2000) looked at simultaneous measurements taken from both wave rider and vertical lasers at the same location with storms analysed. They discussed the effect of buoy correction and the propagation effect on statistics of highest crests and wave heights from both instrument types. Once the corrections were made, skewness and steepness values from the buoy were seen to converge toward laser measurements. Original records from the wave rider were subsequently corrected for the quasi-Lagrangian behaviour of the buoy and found to similarly converge toward laser measurements. Wikramanayake (2006) investigated the wave climate of Sri Lanka in increasing detail and sophistication over the last two decades. The actual measured wave time series were transformed using a transformation matrix. Comparison of measured and transformed time series showed that there were significant differences between the measurements made using wave rider buoys (used between 1983 and 1997) deployed near the shore and the WAVEC deep water buoy used in the directional measurement programme. Wikramanayake observed that the wave rider buoys are not able to measure the higher waves; particularly the sea waves and, therefore concluded that this under-estimation may be due to the rigidity of the mooring used for the near-shore buoys.

Acoustic doppler current profilers (ADCPs)

Historically the technology for measuring waves and currents has required separate instrumentation for each, but recently it has become possible to measure wave height, wave direction and current from conventional ADCPs. The need to also measure currents frequently confronts the practitioner with the necessity of deploying two instrument systems, such as a data buoy and an ADCP. And because, the ADCPs combine the required functionality to measure both waves and currents in a single compact package, there has been considerable interest in exploring their efficacy as a wave sensor. Pinkel and Smith (1987) and Krogstad et al. (1988) have demonstrated that a Doppler sonar using horizontally-projected beams could provide a high quality measurement of wave direction (Smith, 1989). Most existing ADCPs utilize the Janus beam geometry and do not incorporate a pressure sensor. Terray et al. (1999) proved that it is possible in shallow water to estimate both wave height and direction from a conventional bottom-mounted, upward-looking ADCP. They proved that height and direction spectra compare with a co-located array of pressure gauges. They used the iterative version of Capon's “Maximum Likelihood Method” (Capon, 1969) known as the IMLM (Krogstad et al., 1988). Pawka (1983) estimated the direction of multiple echo arrivals using sparse arrays. Like all high resolution direction-of-arrival estimators, the IMLM is model based and requires a relation between the frequency-direction spectrum and the array covariance. A bottom-mounted, upward-looking ADCP provides a robust means of determining wave height and direction in coastal and deep waters. When equipped with a pressure sensor, the ADCP yields three independent estimates of the non-directional wave height spectrum and hence provides an internal consistency check on the performance of the instrument. Directional spectra obtained from the ADCP tend to be sharper than those from point measurements, such as pressure velocity (PUV) triplets or directional wave buoys and because of the greater number of degrees of freedom in the measurement, the ADCP can resolve complex multi-directional wave distributions. These conclusions were arrived at by Strong et al. (2000) after comparing and analysing of different data sets of ADCPs collected from many parts of the world.

Deepwater current profiling requires a low-frequency

ADCP operating at 38 or 75 kHz. Traditionally these large systems have been hull mounted mostly deployed on research vessels. Utmost care should be taken to ensure the system operates in a low-noise and bubble free environment. Towing an ADCP system offers an attractive alternative to vessel mounted ADCPs. A towed system can be used in ships for different purposes, allows easy maintenance and provides remedies to several problems of vessel-mounted systems. Kaneko et al. (1990) first introduced a towed ADCP. They obtained detailed velocity measurements of the Kuroshio Current with an ADCP mounted inside a sled. The sled was towed behind a ship about 7 m beneath the surface. Internal ADCP sensors measured the heading, pitch and roll of the sled. Vessel-mounted ADCPs have become a standard tool on many research vessels (Joyce, 1989). In addition to the limitations inherent in the ADCP system itself (Chereskin et al., 1989; Chereskin and Harding, 1993), air bubbles near the transducer heads and compass biases (King and Cooper, 1993) often degrade shipboard ADCP observations. New (1992) reported that as the ship heaves, a cloud of air bubbles entrained under the hull of the ship. Monchow et al. (1995) attempted different techniques in which all the measurements utilize the ADCP bottom-tracking capability and an upward-looking self recording 614 kHz ADCP which provides velocity observations of the water column above the towed body. The downward-looking 153 kHz ADCP was connected to a shipboard computer. In this towed ADCP system they found two separate sources of error that is surface gravity waves and compass biases and concluded that this towed ADCP system returns data of at least the same quality as vessel mounted ADCP system.

Towed ADCP systems have been used by many researchers, for example, Anderson and Matthews (2005) developed a towed ADCP system used to support operational monitoring of deepwater currents. The towed body configuration provides a quiet and stable sensor platform that can be readily relocated and deployed from different vessels. High-frequency ADCPs are used in towed bodies for several shallow water applications. The unique Towfish system employed a 75 kHz RDI Long Ranger ADCP packaged in a large Endeco/YSI type 850 V-Fin and was used in the Gulf of Mexico to survey upper ocean currents. The system calibrations were found to be very robust and stable. Towfish was deployed at nominally 20 m depth with tow speeds of 1 to 3 ms^{-1}. One disadvantage of the towed body approach was that the ADCP would be deployed deeper than a hull or side mounted system and as a result, a single down-looking ADCP would not be able to sample the very near surface. Placing a second ADCP configured in an up-looking direction into the V-fin would allow profiling to the surface. ADCPs operating at 300 kHz to 1.5 MHz have been successfully used in towed bodies for some time now and have performed quite well (Anderson and Matthews, 2005).

High frequency radars

HF Radar system is a shore based remote sensing system using the over the horizon radar technology to monitor ocean surface currents, waves and wind direction. Oceanographic High Frequency Radars are simple in concept: electromagnetic waves (EM) sent to the ocean are backscattered on surface waves of exactly half the radio wavelength, just like X-rays are scattered in crystals. Since the ocean is generally covered by waves of many different wavelengths and directions (continuous spectrum), there are always trains of waves propagating toward and away from the radar. The return signal from either train will be Doppler-shifted by the wave velocity, which is known exactly by the gravity wave dispersion relationship. Thus the spectrum of the return echoes consists of two peaks, symmetric with respect to the transmit frequency and in the absence of currents. This long range, high resolution monitoring system operates with radio frequencies 5 - 50 MHz. A vertical polarised electromagnetic wave is coupled to the conductive ocean surface and will follow the curvature of the earth. The rough ocean surface interacts with the radio wave and due to the Bragg Effect back-scattered signals can be detected from ranges of more than 200 km (HELZEL, 2004). This effect was first described by Crombie (1955) and the first radar system using that effect was developed at NOAA in 1977 by Barrick (1977). The WERA HF Radar system was developed at the University of Hamburg by Gurgel et al. (2000) which is operating in a frequency modulated continuous wave mode (FMCW). When a continuously swept rf-signal is transmitted, the reflected signal has a frequency offset compared to the actual transmitted signal, thus the range is frequency coded.

In recent years the evolving capability to use shore based high-frequency (HF) radar systems to continuously monitor vast stretches of coastal ocean surface currents has presented a new possibility for improving our understanding and monitoring capabilities in these marine environments (Paduan and Graber, 1997). HF radar networks exploit radio wave backscatter in the frequency band 3 - 30 MHz to map ocean surface currents. Most systems operate in the range of 12 - 25 MHz and are capable of producing maps up to 60 km from shore with horizontal resolutions of 1 - 3 km (Paduan and Rosenfeld, 1996; Graber et al., 1997). At lower frequencies (around 5 MHz), ranges of 200 km are possible using shore-based systems with horizontal resolution of 5 - 10 km (Barrick, 2003). A major advantage with HF radar systems is that they are unaffected by weather, clouds or changing ocean conditions.

Remote sensing cannot always be counted on to provide

timely observations of the sea surface and is not a direct measurement of ocean currents. Remote sensing of near-surface currents with high frequency (HF) radar was demonstrated more than three decades ago by Stewart and Joy (1974). Subtraction of the theoretical phase velocity of the ocean waves gives radial current velocities (hereafter referred to as radials). Multiple radars are typically deployed so radials have enough angular separation to resolve both the north-south and east-west velocity components referred as totals (Ohlmann et al., 2007). First-order Bragg scattering is due to surface gravity waves of half of the transmitted HF wavelength propagating towards or away from the radar site. HF radars have been operated with frequencies between 7 and 50 MHz. Some of the transmitted HF energy is guided by the sea surface and allows measurements to be made beyond normal radar horizon. Depending on the transmitted frequency, ranges of up to 200 km may be achieved (Gurgel et al., 2000).

Two HF radar technologies are commonly used for oceanographic research. Beam-forming radars electronically point linear arrays of receiving antennas to determine bearing over the sea surface. Examples include the Ocean Surface Current Radar and the Wellen radar (Gurgel et al., 1999). Direction-finding radars rely on directional properties of antenna elements to determine bearing. The most commonly employed direction-finding radar is the Seasonde, which uses two directional antennas and a monopole antenna (Barrick and Lipa, 1997). Spatial coverage of HF radar measurements varies according to transmit frequency. For the ~ 25 and ~ 12 MHz systems, maximum ranges are ~ 42 and ~ 83 km, respectively (Ohlmann et al., 2007). The SeaSonde HF radar system by CODAR Ocean Sensors provides real-time data over large coverage areas, with ranges up to 200 km. The SeaSonde is a compact, non-contact surface current and wave measurement system that can be deployed and maintained easily and can perform even during extreme weather conditions such as hurricanes. CODAR Ocean Sensor's FMCW eliminates the range aliasing and antenna wind-vibration noise inherent to other HF system designs (CODAR, 2008). In recent years, the utility of HF radar-derived surface velocity fields as input to data-assimilating numerical circulation models has been the focus of several studies. The potential benefits of HF radar data are large, particularly in light of the dearth of real-time observations from the marine environment. These data are also potentially important because they can cover significant portions of coastal ocean model domains. They make it possible, for the first time, to track the location and movement of mesoscale oceanic features in a fashion analogous to the superior capabilities provided by data inputs to numerical weather forecast models (Paduan and Shulman, 2004).

The unique advantage of the HF radar is the ability to map the horizontal variability of currents which is needed for several applications (Gurgel et al., 1999). Eddy dynamics, such as propagation and decay, can be studied (Essen et al., 1989), as well as the spatial variability of tidal currents (Prandle, 1987). The capability of measuring surface-wave spectra using WERA has been discussed by Wyatt et al. (1998). Compared to CODAR, the new design of WERA offers increased flexibility in spatial resolution and allows both beam forming and direction-finding techniques, as required by the application. Within the EC project SCAWVEX, the WERA system measured surface currents and wave height directional spectra simultaneously, using the University of Hamburg current algorithm and the University of Sheffield wave algorithm, respectively. This is a further step in research on current-wave interaction. With respect to current measurements the high-resolution mode of WERA (0.3 km) is advantageous for studying near-shore ocean dynamics and for the interpretation of space-borne synthetic aperture radar (SAR) images (Gurgel et al., 1999). Essen et al. (2000) investigated the accuracy of surface current velocities measured by HF radar. Data from the two radar systems of the University of Hamburg CODAR (Coastal Radar) and WERA (Wellen Radar) were compared with *in situ* data. In one experiment, CODAR and a near surface current meter were operated simultaneously over a 19 day period. In addition, WERA was operated for 6 days during that period. In another experiment, WERA and a bottom mounted current meter were operated simultaneously over a 35 day period. Both radars used frequencies of about 30 MHz where backscattering which was due to ocean waves of 5 m wavelength. The influence of the orbital motion of underlying longer waves on radial velocity errors was investigated. In accordance with the theory the measured standard deviations of HF-measured current velocities depend on the sea state. In dependence of the sea state, estimated errors ranged from 3 to 10 cm and explain only part of the RMS difference of 10 to 20 cm found between HF and *in situ* current measurements. The rest was assumed to be due to the differences of the quantities measured, e.g. the spatial averaging.

The wave radar system is a very useful tool for the measurement of waves over a wide area with real-time observation, but it still lacks a method to check its accuracy. The Okinawa Subtropical Remote-sensing Center has developed a new high-frequency ocean radar system called Long Range Ocean Radar (LROR). LROR is a kind of Doppler radar using an HF radio wave; it has the capability to monitor a broad area of the ocean surface, surface currents, waves, etc. LROR has been located in Yonaguni Island and Ishigaki Island; an area vulnerable to serious typhoons every year. It is very difficult to measure waves in the open ocean during a typhoon and as a result, very few accurate measurements are available. More than 20 experimental observations have been made in Japan using HF ocean surface radar system in collaboration with various institutes and universities since 1989 (Koterayama et al., 2003).

Alfonso et al. (2006) compared the CODAR SeaSonde HF Radar data with data collected by wave rider buoys between November, 2005 and February, 2006 from Galician Coast, Spain. They concluded that the statistical results were well correlated with indexes of 0.8 in current U component, 0.7 in current V component and up to 0.9 in wave height. They identified the radar information coverage was rather good in waves (97%) but low in currents (72%) in relation to the buoy. Wave period estimations from radar measurements were very stable but may have been too smooth and seemed incapable of reproducing sudden changes in periods. They also found that some strange oscillations that appeared in radar wave directions might be related to tidal effects. The accuracy of HF radar measurement in the Tsushima Strait located between Japan and Korea has been investigated by Yoshikawa et al. (2006). A comparison between radial velocities measured by HF radar and an ADCP, which provides an upper bound of HF radar measurement error, showed from their study that the root-mean-square (RMS) velocity difference obtained from the principal component analysis was 6.62 ~ 11.3 cms^{-1}. A comparison of velocities measured by two facing HF radars, which provides a lower bound of HF radar measurement error, showed that the variance error of hourly radial velocity was 5.75 ~ 13.3 cms^{-1}. The bias error of HF radar measurement was also found to be reasonably small through a comparison of tidal ellipses estimated from 1 year of HF radar data with those from 5 years of ADCP data. They concluded that the variance error of HF radar measurement was the dominant source of the velocity difference found between ADCP and HF radar.

HF radar derived velocities (radials and totals) compared with velocity estimates from large numbers of simultaneous drifter observations by Ohlmann et al. (2007). Drifter averages were obtained within an area observed by HF radar, thus allowing comparison of velocity estimates on similar time and space scales. The primary goal of their study was to quantify the effects of spatial averaging, over various scales, on measurement differences between HF radar and drifter velocities. They used the HF radar data in the Santa Barbara Channel collected by up to five SeaSondes from 1997 to 2006. Moored current meter and profiler data were also used for validation. Differences between HF radar and current meter-derived velocities near 10 - 15 cms^{-1} have been reported by Holbrook and Frisch (1981), Janopaul et al. (1982) and Schott et al. (1986). Chapman et al. (1997) used shipborne current meter data to suggest that the upper bound of HF radar accuracy is 7 - 8 cms^{-1}. Paduan and Rosenfeld (1996) used both ADCP and drifter data to show that RMS differences with HF radar data are 10 to more than 20 cms^{-1}. The most recent comparisons between HF radar velocities and point measurements show RMS differences between 7 and 19 cms^{-1} (Kohut and Glenn, 2003; Emery et al., 2004; Kaplan et al., 2005; Paduan et al., 2007).

Remote sensing

The high-resolution properties of space borne synthetic aperture radar (SAR) systems, as well as their independence of light and cloud conditions, make SAR imagery a crucial source of information for a number of marine and coastal applications. SAR observations have also demonstrated their ability to routinely monitor different ocean-surface parameters, such as, swell direction and amplitude (Beal et al., 1983; Hasselmann and Hasselmann, 1991). The modification of short surface waves by surface layer winds, air/sea temperatures changes over water masses, the presence of surface currents, surface slicks, bathymetry changes, or coastal plumes have also enabled the identification of atmospheric and oceanic features from SAR images. However, challenges remain in uniquely interpreting such a wealth of high-resolution identified patterns in terms of physical processes in the upper ocean and the atmospheric boundary layer (Kerbaol and Collard, 2005). Hydrodynamic modulation of the surface roughness resulting from wave-current interaction also makes possible the observation of oceanic features, such as current fronts (Marmorino et al., 1994), eddies (Johannessen et al., 1996), internal waves (Alpers, 1985), etc. More recently, ongoing research led to progress in more emerging SAR applications, such as Doppler-based surface current measurements (Chapron et al., 2004) and oceanic and atmospheric features identification, which offers new insights for the observation and modelling of mesoscale meteo-oceanic processes.

SAR is an active microwave instrument producing high-resolution imagery of the Earth's surface, regardless of cloud, dusty and solar illumination. The spatial resolution of space borne SARs typically ranges between a few meters and more than hundred meters, depending on the product type (that is continuous or burst mode). Accordingly, the spatial coverage varies between approximately 100 x 100 km (standard image mode) to 500 x 500 km (wide swath or ScanSAR mode). Presently, three spaceborne SARs namely, ERS-2, Envisat and Radarsat-1 are available. Higher radar frequencies are theoretically expected to be more responsive to wind-speed variations. However, lower radar frequency may be less sensitive to roughness saturation at highest wind speeds. The wave-imaging capability is mostly limited by the so-called SAR azimuth cutoff, which is driven by the root mean square (RMS) of the line-of-sight components of the sea-surface orbital velocities (Kerbaol et al., 1998) weighted by the range-to-velocity ratio. In theory, a lower

1. Various instruments and their functions.

ices	Function		Remarks
e Rider y	To measure near surface waves and currents.	1	Position is not fixed in space in the horizontal direction.
		2	Prone to damage from larger waves, ice, vessel traffic or from vandalism.
		3	Prone to errors in accelerometer due to pitch and roll of the buoys
ustic Doppler ent Profilers	Provides a robust means of determining wave heights and direction in shallow and deep waters.	1	Operating at frequencies between 38 kHz and 1.5 MHz.
		2	Directional spectra sharper than data obtained from point measurements
Frequency ar	To measure ocean surface waves and currents.	1	Radio wave backscatter in the frequency band if 3 - 30 MHz:
	Electromagnetic waves sent to the ocean are backscattered on surface waves of exactly half the radio wavelength	2	Range: 42 - 83 km from shore with Horizontal resolution of 1 - 3 km in 12 - 25 MHz: 200 km with less horizontal resolution of 5 - 10 km in 5 MHz
note Sensing	Synthetic Aperture Radar: a microwave instrument producing high resolution imagery regardless of clouds, dusty conditions and solar illumination.	1	For surface currents, oil slicks, ship detection, bathymetry, wave fields, etc.
		2	Spatial coverage: 100 x 100 km to 500 x 500 km
		3	Spatial Resolution: Between few metres and >100 km

radar frequency should minimize this effect, as a smaller range of wave number components will contribute to the integration of the RMS of the orbital velocities (Kerbaol and Collard, 2005).

DISCUSSION

The various instruments available for the measurement of waves and currents and their functions are presented in Table 1. In addition to traditional sensors, like pressure gauges, wave rider buoys or electric level gauges, radar altimeters are being deployed more frequently in wave monitoring. In comparison to traditional sensors, the radar level gauge, being a remote measuring system, has the benefit of no direct contact with the water (corrosion free and no wave attack on the sensor) (Brumbi, 1995). Nevertheless there are some problems in using standard radar level gauges in wave monitoring such as: (1) Mounting is needed (in contrast to the wave rider buoy); (2) Salinity of the water and sea-ice coverage influences the penetration of the radio-wave pulse into the water; (3) The footprint of the radar has an averaging effect and (4) Reflection of the radio-wave pulse depends on the slope of the water surface.

Wave rider buoy measurements have problems, particularly in breaking seas, where surface-floating instruments are subjected to large accelerations. Under such conditions, wave rider measurements may overestimate or underestimate the actual wave heights. The wave rider buoy measurements are also prone to missing data due to shore-station problems, or damage to the buoy arising from collisions with vessels or sea ice, or due to vandalism. Overall, acoustic-based range measurements offer a promising means of measuring ocean waves from the comparative safety and stability of the ocean floor. Problems with wave rider buoys can be eliminated through use of a better mooring system that minimizes instrument tilt and vertical movements.

It is recognized that wave rider buoys fail to adequately measure many non-linearities and asymmetries in the sea-state (Arhan and Plaisted, 1981; LeBlond, 1982). This is most likely due to the fact that the buoy position is not fixed in space in the horizontal direction, as would be a wave staff device. The slack in the mooring is of the order of the depth and therefore of the wave length or greater in many cases. The buoy probably tends to follow the circular or elliptical path of the water particles under the wave and is thus tending toward some sort of Lagrangian measurement on the scale of a wave length as opposed to the Eulerian measurement made by a wave staff. Care must therefore be taken in attempting to extract information on steepness and wave lengths from the surface elevation signal from the wave rider. Any problems of a decrease in sensitivity of the wave rider buoy due to aging of the accelerometer fluid was found to be either nonexistent in some cases or sufficiently slow that wave programs with a regular calibration schedule should be able to detect and respond to the problem. The wave rider buoy should not be expected to produce the same time history as a wave staff and the data should be integrated accordingly (Wilson and Taylor, 1983).

Extreme wave crests have been of special interest for fixed platforms, since their designs are directly connected to the crest heights and the water velocities. Some activities on the platforms also depend on what crest heights can be expected in a given sea state while in operation. Extreme statistics are usually made from point measurements, but waves often appear in wave groups. Due to the dispersive nature of the waves, these groups can have different maximum crests at different positions along the direction of travel. If the waves ride on an ocean current, the return signal is further Doppler-shifted by the radial component of the current, which can be readily estimated. With two radars some distance apart along the coast, vector currents can be computed. This is a precise concept, which surprisingly has taken more than three decades to be accepted by the oceanographic community (HELZEL, 2004). Yet unlike ADCPs, the scattering targets are well known, and well understood theoretically in HF Radars. ADCPs deployed on ships provide the most viable way to survey ocean current profiler (Anderson and Matthews, 2005).

Different instruments for wave and current measurements have their own advantages and disadvantages depending on the applications and needs. Advantages of wave rider buoys include: (1) Real-time telemetry is facilitated by adding the appropriate wireless antenna and telemetry hardware. In this case there is no need for a cable to shore, or acoustic modems. Depending on the installed telemetry option, the receiving station could even be thousands of kilometers away; (2) Solar power can be employed to maintain power; (3) Attenuation of wave-induced properties (such as pressure, velocity) is not an issue, since the measurements are made at the water surface; (4) Deployment can be relatively simple; e.g. hoist buoy into water and release; hoist anchor into water and release; (5) Inexpensive, easy to handle and accurate for small waves.

Disadvantages include: (1) Buoys are subject to vessel or debris impact; (2) Visibility can lead to vandalism problems; (3) Environmental mooring loads are more severe, since mean flows, wave-induced flows, and winds all act on the buoy; (4) Mooring system requires servicing; (5) Internals (incl. batteries) are under constant acceleration loads, although this has not created problems to date; (6) Fixed, bottom mounted gear (such as up-looking ADCP) provides higher rated accuracy for measurement of mean flows, compared to buoy equipped with down-looking ADCP; (7) Measurement of water depth or tidal range is typically not feasible. Noise in accelerometers may make measurement of low frequency energy problematic; (8) Buoys are not able to measure the higher waves, particularly sea waves; (9) Buoyancy will be inadequate to keep it on the ocean surface during storm in measuring larger waves.

In ADCP, measurement of long term data is a major problem in establishing the effect of wave-current impact, particularly on the marine energy converters. Air frequency measurement is required greater than 1 Hz for a good study. It is difficult to measure data for longer duration which requires larger memory. It is very expensive for long term measurement of air frequency, which requires at least for one year continuous data. It involves continuous deployment deal with harsh turbulent marine environment, make the process difficult. Long term air frequency data is essential as an input for validating theoretical modelling. Therefore, working with short term data will not give good representation of dynamics of marine structures. In ADCPs, aliasing of unwanted high-frequency noise signals such as orbital velocities into the frequency band of interest can be characterised by the variance of an individual recorded ensemble average. In practical situations it is common to deal with noise signals such as waves that are either inherently broadband or must be treated as effectively so due to variability in the peak frequency.

Alternatively, very small gradients and low value currents, which are inferred from sea-surface heights estimated from satellite altimetry, are already routinely used in the global circulation models. However, this approach does not allow direct measurements of sea-surface displacements nor does it achieve the spatial scales that may be obtained from SAR (Kerbaol and Collard, 2005). Sequences of sea-surface temperatures and ocean-colour data can be used either in the Maximum Cross Correlation method (Emery et al., 1986) or the Optical Flow method (Vigan et al., 2000; Yang and Parvin, 2000). Although these methods allow absolute measurements of current vectors, they do not achieve the spatial resolution of SAR-based methods and they are limited by the solar-illumination and cloud-cover conditions. Finally high-frequency (HF) radars offer a very interesting opportunity to retrieve the full vector of sea-surface currents with a temporal sampling of a few minutes. However, the coverage area is more limited than the spaceborne SAR coverage. Furthermore, the deployment and maintenance of such systems often induce additional costs and administrative difficulties. Nevertheless, such a system may be well suited for local measurements and validation purpose (Kerbaol and Collard, 2005).

Many questions still exist, however, about the details and effectiveness of HF radar-derived surface currents as sources for data assimilation. On one level, data descriptors needed to define the requisite error covariance functions are not yet well known. Errors in HF radar-derived currents arise from a variety of sources, including electromagnetic interference, ships, and poorly constrained inversion algorithms. Some insight into these errors can be obtained comparing with data from in situ moored current meters or drifting buoys (Laws et al., 2001). Despite several tasks, it is clear that two-dimensional maps of surface currents from HF radar networks represent a useful and unique resource for the improvement of coastal ocean circulation models, particularly in the critical depth range encompassing the euphotic zone (Paduan and Shulman, 2004).

Differences between surface current velocities from HF radar and other platforms are: (1) measurements from HF radar, drifters, and current meters are all inexact. The frequency resolution of computed radar cross-spectra, which depends on FFT length, limits radial velocity resolution to ~ 5 and 2.5 cms^{-1} for 12- and 25-MHz systems, respectively. Drifters can slip at ~ 1 to 2 cms^{-1} from the ocean water they follow (Ohlmann et al., 2005);

Table 2. Comparison of velocity measurements between various instruments.

Instruments	Parameters	Errors	Authors and Remarks
HF Radar Vs ADCP	RMS velocity difference	6.62 ~ 11.3 cms^{-1}	Yoshikawa et al. (2006): Major error of radial velocity of HF Radar is Variance error rather than bias error
	Variance error of hourly radial velocity	5.75 ~ 13.3 cms^{-1}	
	Bias error	Reasonably small	
HF Radar and Current meter	Velocity	10 - 15 cms^{-1}	Holbrook and Frisch (1981), Janopaul et al. (1982) and Schott et al. (1986)
HF Radar vs. Ship-borne current meter	Velocity	7 - 8 cms^{-1}	Chapman et al. (1997)
HF Radar vs. ADCP and Wave Rider Buoy	RMS velocity difference	10 - 20 cms^{-1}	Paduan and Rosenfield (1996)
		7 - 19 cms^{-1}	Kohut and Glenn (2003), Emery et al. (2004), Kaplan et al. (2005) and Paduan et al. (2007)
HF Radar vs. Single Drifter	RMS velocity difference	10 cms^{-1}	Ohlmann et al. (2007): Minimum of ~ 5 cms^{-1} when 15 – 30 drifter observation averages.

(2) vertical scales of measurement differ. The HF radar gives vertically integrated values from the surface, drifters give integrated values over their drag elements, and current meters give values for specific depths or depth bins; (3) Horizontal scales of measurement differ. Typically, HF radars average over extensive horizontal areas (up to several km^2), while other platforms give point measurements or limited spatial measurements following motion; (4) Measurements are not necessarily coincident in time and Stokes drift may not be reconciled consistently among platforms (Ohlmann et al., 2007).

Velocity measurements using different instruments are compared from different works carried out by various researchers and presented in Table 2. The reduction in RMS velocity differences in the San Diego data with measured antenna patterns supports results of Kohut and Glenn (2003) and Paduan et al. (2007), who attribute errors of more than 10 cm cms^{-1} to poorly known antenna patterns (Ohlmann et al., 2007). Large changes in radar radials, relative to drifter velocities, may not necessarily be evident in RMS difference, for example where the bias (difference) between platforms goes from positive to negative without changing magnitude. The occasional use of drifter comparisons to determine both bias and RMS difference can indicate errors related to antenna patterns, ultimately improving the quality of HF radar data provided to users.

The complicated signal processing for extracting surface currents from backscattered radar signals (de Paolo and Terrill, 2007) yields radial velocities on polar coordinate grid points centered by each antenna location. Radial velocities measured by multiple antenna installations have been combined into current vector field using unweighted least squares fitting (Lipa and Barrick, 1983). The maps of ocean surface currents are incomplete in space and time for the following reasons (Kim et al., 2007); 1. The algorithm on the measured Doppler spectrum does not provide a solution for all bearing angles; 2. Estimate of current vectors along the baseline between two radars where the measurements of radial velocities are nearly aligned suffers from poor geometrical dilution of precision (GDOP), and frequently results in spurious current vectors. The region with radial velocities crossing at angles less than 15 - 20° between two radars is commonly considered to produce unusable current vectors and; 3. Hardware or software problems can lead to the temporary shutdown of individual radar sites.

Relevant applications of surface currents require time and space-continuous data, which requires the interpolation of the incomplete observations to a regularly spaced product. A variety of remote sensing tools have been housed in artificial satellites, which are very useful for monitoring the global environments. Sensors and software for measuring water temperature, wind velocity, current velocity, wave height and chlorophyll on the sea surface have been developed and installed on satellites. Remote sensing using satellites is most effective in its breadth and simultaneity, but it must be calibrated by direct measurements. Ocean engineers have been charged with developing vehicles and buoys for these direct measurements (Koterayama et al., 2003). Surface current measurement by interpretation of radio waves backscattered from surface gravity waves is quickly maturing as an oceanographic observational tool whose data is useful to a broad range of users. The shore-based antenna approach can provide continuous temporal and broad spatial surface current observations, facilitating the delivery of data in near real-time for various applications (Kim et al., 2007).

The SAR is, so far, the only satellite-borne instrument that can measure the directional characteristics of the ocean wave field. In the early 1990s, the derivation of a closed-form expression for the ocean-to-SAR spectral transformation, as well as the formulation of the inversion procedure (Hasselmann and Hasselmann, 1991), significantly improved the understanding about the SAR imaging of ocean waves. A new SAR inversion algorithm for Envisat ASAR Level 2 wave-mode products was developed by Engen et al. (2001). Preliminary results support the fact that the significant wave height estimated from SAR is less than 40 cm in 50% of the cases and the RMS on wave direction is about 30 (Kerbaol et al., 2004). As an example, the detection of current features, including current fronts, internal waves, and eddies, are generally based on the use of the wavelet transform (Wu and Liu, 2003). The future research will focus on understanding the dynamic mechanisms by which surface velocity information impacts subsurface model currents and to explore the sources of error in the HF radar data itself in order to improve error covariance models of data assimilation. Future model improvements should include tidal forcing and special handling of velocity assimilation near the coastal boundary.

SUMMARY

Wave rider buoys have proven to be a cost effective means of collecting ocean current observations. However, buoys report only near surface currents and a Lagrangian technique does not provide a means of activity targeting and monitoring a specific site or area. Acoustic Doppler Current Profilers mounted on moorings and platforms can provide real-time current information but are not always optimally located. HF Radars provide data for a wider area but only of the surface waves/ currents perspective. SAR data provide wider spatial coverage through satellite sensors. It is essential to choose the type of application based on the needs and importance. It is concluded that each of the instrumentation/applications have their own advantages and disadvantages. A thorough knowledge of the particular instrument to be used is essential before it is deployed. Cost may also be an issue, depending on funding availability and duration of project.

ACKNOWLEDGEMENTS

This work is carried out under SRDG/MREDS programme funded by the Scottish Funding Council. The authors are thankful to the Scottish Government for the support and funding provided to the ICIT, Heriot-Watt University, Edinburgh, United Kingdom.

REFERENCES

Alfonso M, Alvarez E, Lopez J.D (2006). Comparison of CODAR SeaSonde HF Radar operational waves and currents measurements with Puertos del Estado buoys. Final Report, March, 32p.

Alpers W (1985). Theory of radar imaging of internal waves. Nature. 314(6008): 245–247.

Anderson SP and Mathews P (2005). A Towed 75 kHz ADCP for Operational Deepwater Current Surveys.

Procs. IEEE/OES Eighth Working Conference on Current Measurement Technology. pp. 46 - 49.

Arhan M, Plaisted RO (1981). Non-Linear deformation of sea-wave profiles in intermediate and shallow water. Oceanologia Acta. 4: 107-115.

Barrick D (2003). Proliferation of SeaSonde coastal current-mapping radars. Hydro Int., 7(1): 30- 33.

Barrick DE (1977). Ocean surface current mapped by radar. Science, 198: 138-144.

Barrick DE, Lipa BJ (1997). Evolution of bearing determination in HF current mapping radars. Oceanography, 10: 72–75.

Beal R, Tilley D and Monaldo F (1983). Large and small-scale spatial evolution of digitally processed ocean wave spectra from Seasat Synthetic Aperture Radar. J. Geophys. Res., 88(C3): 1761–1778.

Brumbi D (1995) Measuring Process and Storage Tank Level with Radar Technology. IEEE Int. Radar Conf., Alexandria, VA, USA.

Capon J (1969). High-resolution frequency - wavenumber spectrum analysis. Proc. IEEE, 57: 1408-1418 also Corrigendum in Proc. IEEE (Lett.), 59, 112 (1971).

Chapman RD, Shay LK, Graber HC, Edson JB, Karachintsev A, Trump CL and Ross DB (1997). On the accuracy of HF radar surface current measurements: Intercomparison with ship-based sensors. J. Geophys. Res. 102 (C8): 18737–18748.

Chapron B, Collard F, Kerbaol V (2004). Satellite synthetic aperture radar sea surface Doppler measurements. Proc. 2nd Workshop Coastal and Marine Applications Synthetic Aperture Radar (SAR), Svalbard, Norway, (ESA SP-565). Noordwijk, Netherlands, pp. 133-140.

Chereskin TK, Firing E, Gast JA (1989). Identifying and screening filter skew and noise bias in acoustic Doppler current profiler measurements. J. Atmos. Oceanic Technol., 6: 1040-1054.

Chereskin TK, Harding J (1993). Modeling the performance of an acoustic Doppler current profiler. J. Atmos. Oceanic Technol. 10: 42-63.

CODAR (2008). Retrieved from the website of CODAR Ocean Sensors, http://www.codaros.com.

Crombie DD (1955). Doppler spectrum of sea echo at 13.56 Mc/s. Nature 175: 681-682.

De Paolo T, Terrill E (2007). Skill assessment of resolving ocean surface current structure using Compact-Antenna style HF RADAR and the MUSIC direction finding algorithm. J. Atmos. Oceanic Technol, in press.

Emery BM, Washburn L, Harlan JA (2004). Evaluating radial current measurements from CODAR high-frequency radars with moored current meters. J. Atmos. Oceanic Technol., 21: 1259-1271.

Emery W, Thomas A, Collins M, Crawford W, Mackas D (1986). An objective method for computing advective surface velocities from sequential infrared satellite images. J. Geophys. Res. 91(C11): 12865-12878.

Engen G, Johnsen H, Høgda K, Chapron B (2001). Envisat ASAR Level 2 Wave Mode Product Algorithm Specification-Software Requirement Document. Technical report. NORUT IT. Doc. 650/1-01.

Essen HH, Gurgel KW, Schirmer F (1989). Surface currents in the Norwegian Channel measured by radar in March 1985. Tellus, 41A: 162-174.

Essen HH, Gurgel KW, Schlick T (2000). On the accuracy of current measurements by means of HF Radar. IEEE J. Oceanic Eng. 25: 4.

Graber HC, Haus BK, Chapman RD, Shay LK (1997). HF radar comparisons with moored estimates of current speed and direction: Expected differences and implications. J. Geophys. Res. 102: 18:749 - 18,766.

Gurgel KW, Antonischski G, Essen HH, Schlick T (1999). Wellen radar (WERA): A new ground wave radar for remote sensing. Coastal Eng. 37: 219–234.

Gurgel KW, Essen HH, Kingsley SP (2000). HF Radars: Physical limitations and recent developments. Elsevier Publications (and in www.helzel.com).

Hasselmann K, Hasselmann S (1991). On the nonlinear mapping of an ocean wave spectrum into a synthetic aperture radar image spectrum and its inversion. J. Geophys. Res., 96(C6): 10713–10729.

HELZEL (2004). Retrieved from the website of Wellen radar. http://www.helzel.com.

Holbrook JR, Frisch AS (1981). A comparison of near surface CODAR and VACM measurements in the Strait of Juan de Fuca, August

1978. J. Geophys. Res. 86: 10 908-10 912.
Johannessen J, Shuchmann R, Digranes G, Lyzenga D, Wackerman, C, Johannessen O, Vachon P (1996). Coastal ocean fronts and eddies images with ERS-1 synthetic aperture Radar. J. Geophys. Res. 101(C3): 6651–6667.
Janopaul MM, Broche P, de Maistre JC, Essen HH, Blanchet C, Grau G, Mittelstaedt E (1982). Comparison of measurements of sea currents by HF radar and by conventional means. Int. J. Remote Sens. 3: 409-422.
Joyce TM (1989). On in situ calibration of ship-board ADCPs. J. Atmos. Oceanic Technol., 6: 169-172.
Kaneko A, Koterayama W, Honji H, Mizuno S, Kawatate K and Gordon RL (1990). Cross-stream survey of the upper 400 m of the Kuroshio by an ADCP on a towed fish. Deep sea Res. 37: 875-889.
Kaplan DM, Largier J, Botsford LW (2005). HF radar observations of surface circulation off Bodega Bay (northern California, USA). J. Geophys. Res, 110, C10020, doi:10.1029/ 2005JC002959.
Kerbaol V,Collard F (2005). SAR-Derived Coastal and Marine Applications: From Research to Operational Products. IEEE J. Ocean. Eng. 30(3): 472-486.
Kerbaol V, Chapron B, Vachon P (1998). Analysis of ERS-1/2 SARwave mode imagettes. J. Geophys. Res. 103(C4): 7833–7846.
Kerbaol V, Johnsen H, Chapron B, Rosich B (2004).Quality assessment of ENVISAT ASAR wave mode products based on regional and seasonal comparisons with WAM model outputs. Proc. ESA ERS/ENVISAT Symp. Salzburg, Austria.
Kim SY, Terrill E, Cornuelle B (2007). Objectively mapping HF radar-derived surface current data using measured and idealized data covariance matrices. J. Geophys. Res., 112(C06021): 1-16.
King BA, Cooper EB (1993). Comparison of ship's heading determined from an array of GPS antennas with heading from conventional gyrocompass measurements. Deep Sea Res. 40: 2207-2216.
Kohut JT, Glenn SM (2003) Improving HF radar surface current measurements with measured antenna beam patterns. J. Atmos. Oceanic Technol. 20: 1303–1316.
Koterayama W, Nakamura M, Sato K, Ikebuchi T (2003). Model and Field Experiments on a Wave Observation Buoy. Int. J. Offshore Polar Eng. 13: 1.
Krogstad HE, Gordon RL, Miller MC (1988). High resolution directional wave spectra from horizontally-mounted Acoustic Doppler Current Meters. J. Atmos. Oceanic Technol., 5: 340-352.
Laws K, Fernandez DM, Paduan JD (2001). Simulation-based evaluations of HF radar ocean current algorithms. J. Oceanic Eng. 25: 481– 491.
LeBlond PH (1982). A preliminary review of non-spectral wave properties: grouping, wave breaking and freak waves. Can. Contract. Rep. Hydrogr. Ocean Sci. 1: 66.
Lipa BJ, Barrick DE (1983). Least squares methods for the extraction of surface currents from CODAR crossed-loop data: Application at ARSLOE. IEEE J. Oceanic Eng. 13(2), pp.507– 513.
Magnusson AK, Donelan MA (2000). Extremes of waves measured by a wave rider buoy and vertical lasers. Proc. Rogue Waves 2000, Brest. 17: 231-245.
Marmorino G, Jansen R, Valenzuela G, Trump C, Lee J, Kaiser J (1994). Gulf stream surface convergence images by synthetic aperture radar. J. Geophys. Res., 99(C9): 18315–18328.
Monchow A, Coughran CS, Hendershott MC, Winant CD (1995). Performance and Calibration of an acoustic Doppler current profiler Towed below the Surface. J. Atmos. Oceanic Technol., 12: 345-444.
New AL (1992). Factors affecting the quality of shipboard acoustic Doppler current profiler data. Deep Sea Res., 39: 1985-1992.
Ohlmann JC, White PF, Sybrandy AL, Niller PP (2005). GPS-cellular drifter technology for coastal ocean observing systems. J. Atmos. Oceanic Technol. 22: 1381–1388.
Ohlmann C, White P, Washburn L, Terrill E, Emery B, Otero M, (2007). Interpretation of coastal HF Radar derived surface currents with High-Resolution Drifter data. J. Atmos. Oceanic Technol., 24: 666-680.
Paduan JD, Graber HC (1997). Introduction to high frequency radar: Reality and myth. Oceanography 10: 36- 39.
Paduan JD, Rosenfeld LK (1996). Remotely sensed surface currents in Monterey Bay from shore-based HF radar (CODAR). J. Geophys. Res. 101: 20,669–20,686.
Paduan JD, Shulman I (2004). HF radar data assimilation in the Monterey Bay area. J. Geophys. Res. 109(C07 S09): 1-17.
Paduan JD, Kim KC, Cook MS, Chavez FP (2007). Calibration and validation of direction-finding high frequency radar ocean surface current observations. IEEE J. Oceanic Eng. in press.
Pawka SS (1983). Island shadows in wave directional spectra. J. Geophys. Res. 88: 2579-2591.
Pinkel R, Smith JA (1987). Open ocean surface wave measurement using Doppler sonar. J. Geophys. Res. 92: 12,967-12,973.
Prandle D (1987). The structure of nearshore tidal and residual circulations revealed by HF radar surface current measurements. J. Phys. Oceanography with High-Resolution Drifter data. J. Atmos. Oceanic Technol. 24: 666-680.
Ribe RL (1983). Accuracy characteristics of the electronics of Wave Rider Buoys used in the ARSLOE. IEEE J. Ocean Eng. OE-8(4) 212-214.
Schott FA, Frisch SA, Larsen JC (1986). Comparison of surface currents measured by HF Doppler Radar in the western Florida straits during November 1983 to January 1984 and Florida current transports. J. Geophys. Res., 91: 8451 – 8460.
Smith JA (1989). Doppler sonar and surface waves: Range and resolution. J. Atmos. and Oceanic Technol., 6: 680-696.
Sorensen RM (1997). Basic Coastal Engineering. Chapman and Hall, New York, 301p.
Stewart RH, Joy JW (1974). HF radio measurements of surface currents. Deep Sea Res. 21: 1039–1049.
Strong B, Brumley B, Terray EA, Stone GW (2000). The Performance of ADCP-Derived Wave Spectra and Comparison with Other Independent Measurements, www.rdinstruments.com/library.html.
Terray EA, Brumley BH, Strong B (1999). Measuring waves and currents with an upward-Looking ADCP. Proc. IEEE 6th Working Conference on Current Measurements, IEEE, pp. 66 - 71.
Vigan X, Provost C, Bleck R, Courtier P (2000). Sea surface velocities from sea surface temperature image sequences, 1. Method and validation using primitive equation model output. J. Geophys. Res. 105(C8) 19499–19514.
Wikramanayake N (2006). The coastal wave-climate of Sri Lanka: Measurements and Modeling. Procs. 7th International Workshop on Indian Ocean Research and Ocean Hazard/ Tsunami, Guangzhou, China.
Wilson JR, Taylor JD (1983) Experiences with wave rider buoys in the Canadian wave climate study. Oceans. 15: 254-257.
Wu S, Liu A (2003). An automated ocean feature detection, extraction and classification algorithm for SAR imagery. Int. J. Remote Sens. 24: 935–951.
Wyatt LC, Thompson SP, Burton RR (1998). Evaluation of HF radar wave measurements. Coastal Engineering and in www.helzel.com .
Yang Q, Parvin B (2000). Feature based visualization of geophysical data. Proc. Computer Vision and Pattern Recognition (CVPR), Hilton Head Island, SC, 2: 276–281.
Yoshikawa Y, Masuda A, Marubayashi K, Ishibashi M, Okuno A (2006). On the accuracy of HF radar measurement in the Tsushima Strait. J. Geophys. Res. 111(C04009): 1-10.

20

Polychlorinated biphenyls and chlorinated pesticides in the mussel *Mytilus galloprovincialis* sampled along the Moroccan Atlantic Coast

Maria Jayed[1]*, Abdelghani Chafik[1], Samir Benbrahim[1], Carlos Vale[2], Salem Bakkas[3], Patrícia Pereira[2] and Ana Maria Ferreira[2]

[1]Laboratory of Chemistry, National Institute of Fischeries, 2 Rue Tiznit , P. O. Box 20000, Casablanca, Morocco.
[2] IPIMAR - National Institute of Biological Resources, Av. Brasília, 1449-006 Lisboa, Portugal.
[3]Laboratory of Organic Chemistry, Faculty of Sciences, El Jadida University Chouaïb Doukkali, 2, Avenue Med Ben Larbi Alaoui, El Jadida, Morocco.

The mussel samples (*Mytilus galloprovincialis*) were collected from 13 sites along 2,000 km of the Moroccan Atlantic coast. Specimens were collected in the wet seasons (winter) of 2003, 2004 and 2007, and in the dry seasons (spring and summer) of 2003, 2005, 2006 and 2007. Polychlorinated biphenyls (PCBs), DDT and its derivatives, aldrin, heptachlor and hexachlorobenzene (HCB), were quantified in the whole mussel. The results of this multi annual survey pointed to low availability of these compounds in the Atlantic coastal environment, decreasing from north to southern areas. Furthermore, the coastal sites appear to be influenced by recent application of p,p'DDT. There is evidence of atmospheric transport of less chlorinated CBs and heptachlor. To our knowledge, this is the first time that organochlorines in mussels from this region are reported.

Key words: *Mytilus galloprovincialis*, pollution, organochlorines, Atlantic Moroccan coasts.

INTRODUCTION

Organochlorines compounds are persistent ubiquitous pollutants in the environment. For several decades, these compounds have been produced and extensively used for various purposes. Most industrial countries have restricted or stopped their use since 1970s, leading to decreasing concentrations in long-term environmental surveillance programmes (Lauenstein and Daskalakis, 1998). However, compounds remained in ecosystems either due to their persistency or transport from developing countries where use in agricultural and industrial purposes is still current. Due to their chemical stability and hydrophobic nature, these compounds are adsorbed onto particles, accumulated in aquatic organisms, and highly biomagnified through the aquatic food webs (Olsen et al., 2000; Tanabe et al., 1993). Many organochlorines have been implicated in a broad range of adverse biological effects, including impaired reproduction and immuno suppression (Manalovani, 2002). Sedentary bivalve species, such as blue mussels, have been extensively used as sentinel organism in surveillance programmes (Farrington et al., 1983; Villeneuve et al., 1999).

This work reports the levels of polychlorinated biphenyls (PCBs), DDT and its derivatives, aldrin, heptachlor and hexhachlorobenzene (HCB), in the mussel *Mytilus galloprovincialis* from 13 sites distributed along 2,000 km of the Moroccan Atlantic coast, between 2003 and 2007. The variation along this coast mirrors the transition between regions with different sources of these compounds.

MATERIALS AND METHODS

Sampling

Wild mussels, *M. galloprovincialis,* were collected from 13 sites between Casablanca and Dakhla (Figure 1). At each site,

*Corresponding author. E-mail: mjayed2003@yahoo.fr.

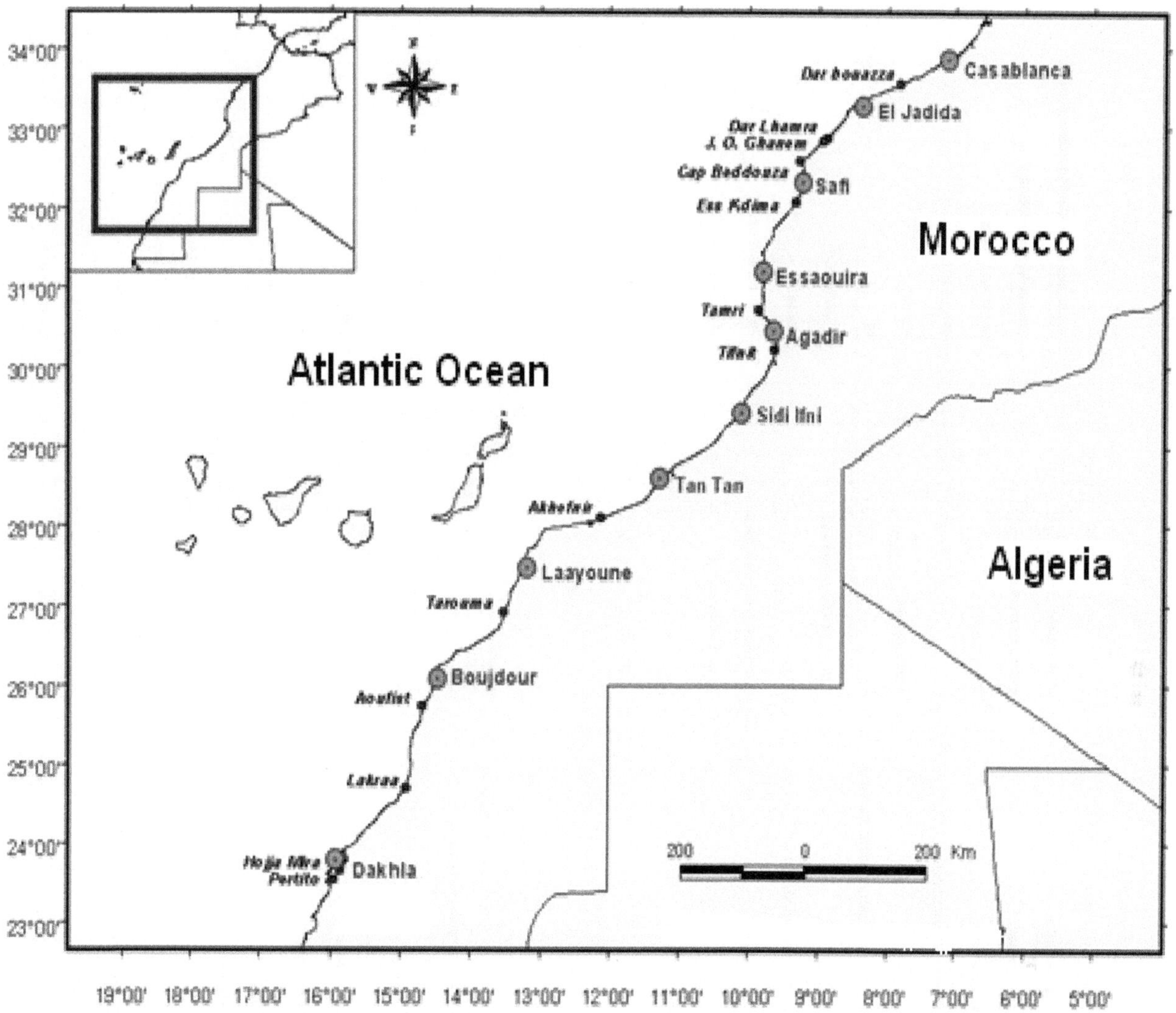

Figure 1. Location of 13 sampling sites along the Atlantic Moroccan Coast. Code of sampling stations from the north to south: DB: Dar Bouazza; DL: Dar Lhamra; OG: Jorf Ouled Ghanem; CB: Cap Beddouza; EK: Essouiria Kdima; TA: Tamri; TI: Tifnit; AK: Akhennir; TAR: Tarouma; Aou: Aoufist; LAK: Lakraa; HM: Hojja Mira; Per: Pertito.

the wet seasons (winter) of 2003, 2004 and 2007 and in the dry seasons (spring and summer) of 2003, 2005, 2006 and 2007. Mussels were depurated in the laboratory, the soft parts removed from their shells, and homogenized in pools of 100 individuals. Composite samples were stored at -20°C before freeze-drying.

Analytical procedures

Each composite sample (5 to 10 g) was spiked with internal standards: 2, 4, 5, trichlorobiphenyl and εHCH. These standards were used to quantify the overall recovery of procedures, the sample were soxhlet extracted for 8 h into 250 ml of hexane, then concentrated down to a few millilitres using a rotary evaporation in a water bath (about 30°C). An aliquot of this extract was taken for treatment with concentrated sulfuric acid to destroy lipids. The aliquot was through the florisil clean up after evaporation to 1 ml. The clean up and fractionation was performed by passing the extract through a florisil (17 g) column which had been activated at 130°C for 12 h and partially deactivated with 0.5% water. From this column, two fractions were extracted, the first fraction with 70 ml of hexane; the second with 50 ml of hexane/dichloromethane. Each fraction was concentrated and injected into a GC (Hewlett Packard HP 6890) equipped with a ^{63}Ni electron capture detector (ECD) and split/splitless injector. The capillary column used was an HP-Pas (phenyl methyl siloxane, 25 m length, 0.32 mm internal diameter, 0.52 µm film thickness). The oven temperature of 60°C (2 min hold) to 160°C at a rate of 20°C min^{-1} and was maintained at 250°C for 2 min. Injector and detector temperature were maintained at 250 and 300°C, respectively. The carrier gas was helium with a flow of 2 ml/min and the make-up was nitrogen with a flow of 60 ml/min. Concentrations of individual organochlorine were quantified from individually resolved peak areas with the corresponding peak areas of standards.

To control the analytical reliability and assure recovery efficiency and accuracy of the results, simultaneously analyse were conducted on PCBs and OC compound reference material freeze dried fish tissue N° 406 provided by IAEA/Monaco. A procedural blank was done to determine any contamination during analyses and the same procedures were applied as to mussels' samples. The quantified compounds were: Six polychlorinated biphenyls (CBs) with several degrees of chlorination (tri CB- 28, tetra CB-52,

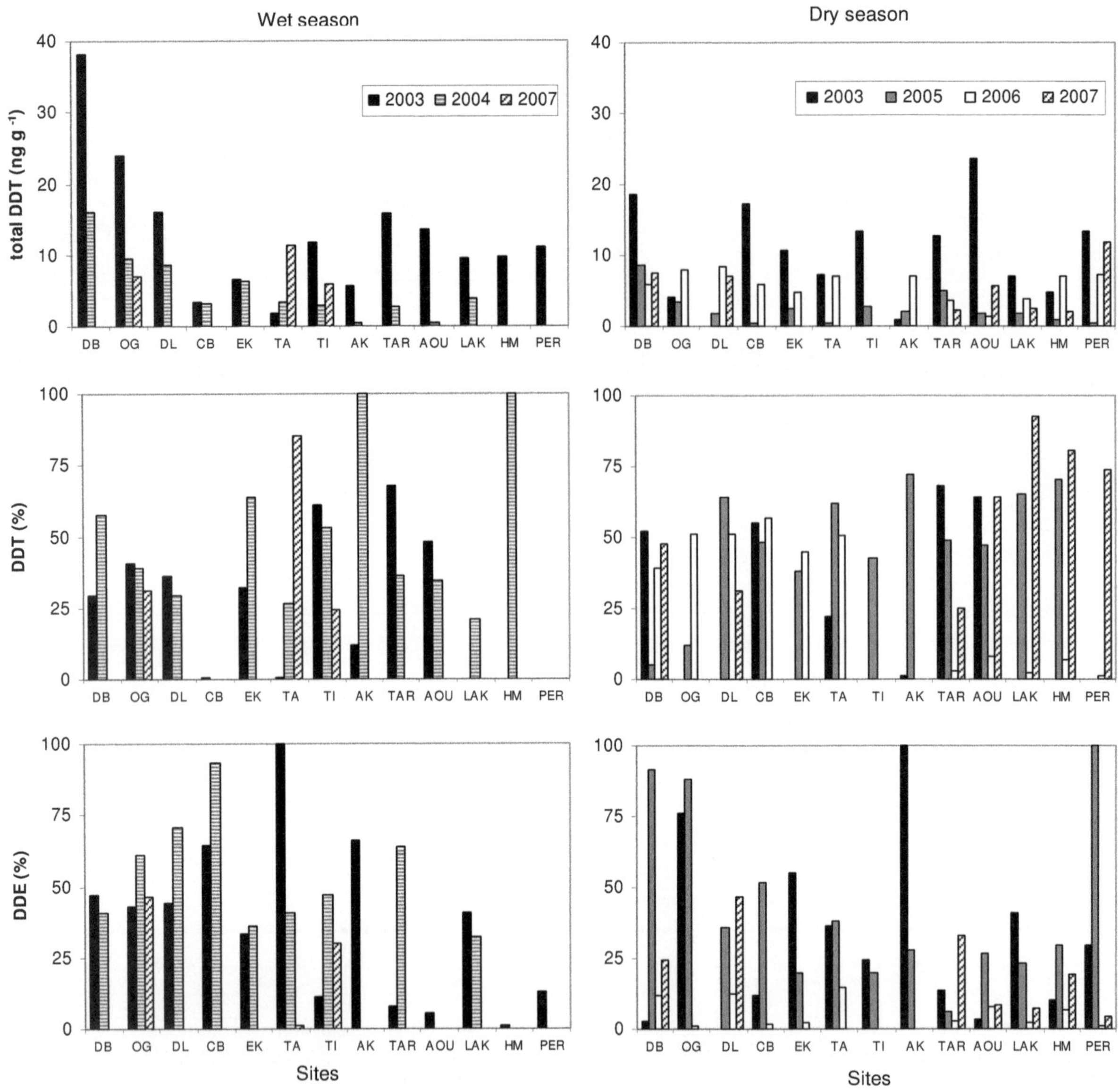

Figure 2. Levels of total DDT and the percentages of p,p, DDT and p,p, DDE in mussels collected in 13 sites along the Atlantic Moroccan coasts.

penta CB-101, penta CB-118, hexa CB-138, hexa CB-153), three DDT compounds (p,p' DDE, p,p' DDD, p,p' DDT), heptachlor, HCB and aldrin. A detailed description of the analytical procedure is in Ferreira and Vale (1997).

Statistical analysis

Statistica software was used for statistical analyses. The non-parametric Mann-Whitney test was used to compare sampling seasons and the difference between seasons were considered significant when $p<0.05$

RESULTS AND DISCUSSION

The total DDT (sum of p,p' DDE; p,p' DDD; p,p' DDT), total PCBs (sum of CBs28, 52, 101, 118, 153 and 138), aldrin, heptachlor and HCB in composite samples of mussels from 13 sites along the Atlantic coast of Morocco, between 2003 and 2007, varied within broad concentration ranges: 0.06 to 38, 0.88 to 91, 0.01 to 8.4, 0.001 to 4.0 and 0.01 to 4.0 ng g^{-1}, respectively. The two to three orders magnitude differences and the marked wet and dry seasons in the western coast of Morocco led to separate samples from all sampling sites into two sets corresponding to the two seasons.

DDT compounds

Figure 2 shows the levels of total DDT and the

contributions of p,p' DDT and p,p' DDE to the total values found in the analysed mussel samples. Considering all data of the 13 sampling sites, no significant ($p>0.05$) differences were observed between wet and dry season for the total DDT. However, it is noticeable elevated levels (max. 38 ng g^{-1}) in mussels collected from the three northern sites (DB, OG and DL) in the wet season of 2003. The increase in comparison to the values registered in the rest of the coast (median = 5.7 ng g^{-1}) points to an addition input of these compounds in the northern sites. Differences between the sampling years were possible to discriminate. The percentage of p,p'DDT was significantly ($p<0.05$) higher in mussels collected in dry season than in wet season (Figure 2). This difference suggests the influence of the atmospheric transport in the distribution of this compound in the Atlantic Moroccan coastal zone.

It should not exclude the possibility of recent application, since p,p'DDT accounted to more than 50% of the total DDT in several samples. The contribution was more pronounced in southern areas (AOU, LAK, HM and PER) either in wet and dry season, reaching 80 to 100% of the total DDT in HM and PER in wet season. The increase of this proportion towards more remote areas in the south of Moroccan coast reinforces the hypothesis of atmospheric transport of recent applications. The proportion of p,p' DDE varied with the site and season, although the percentage in dry season was in general lower than of p,p' DDT (Figure 2).

The registered levels of total DDT in mussels were comparable to the values reported for mussels collected during 1990s on the Mediterranean Moroccan coast (7 to 29 ng g^{-1}) (Albaigés, 2005). However, they were lower than the values reported in the same decade for mussels of Barcelona (up to 189 ng g^{-1}), from the French Mediterranean coast (1.2 to 157 ng g^{-1}) and the Ligurian Sea (28 to 89 ng g^{-1}) (RNO, 2000; UNEP/WHO, 1995; Albaigés, 2005).

Despite the relatively low levels of total DDT in mussels from the Atlantic Moroccan coast in comparison to the northern coast of the Mediterranean Sea, the p,p'DDT emerges as the major metabolite accumulated in mussels in the dry season. This pattern contrasts with the predominance of the p,p'DDE recorded in mussels off Barcelona, suggesting a minor influence of recent applications (UNEP/WHO, 1995).

PCBs

Figure 3 presents the levels of the total quantified PCBs in mussels from the sampling sites in wet and dry seasons. Levels in the two seasons were not significantly ($p>0.05$) different, but values in the northern sites (DB in wet season; DB and OG in dry season) exceeded 2 to 3 times the medians registered for the other sampling sites (11 and 8.2 ng g-1 in wet and dry season, respectively). The contrast of this geographic distribution in the two seasons suggests the influence of localised sources of PCB in the northern Atlantic coast of Morocco. The congeners CB153, CB138 and CB28 were the major compounds accounting to the sum of PCBs. The higher chlorinated congeners (CB153 and CB138) showed no clear trends of variability among sites and surveyed years (Figure 3), and no significant ($p>0.05$) differences between wet and dry seasons. The less chlorinated congener CB28 was significantly ($p<0.05$) higher in dry season than in wet season. The differences were more pronounced in 2003 and in the southern sampling sites. These results clearly point to the importance of atmospheric transport of CB28 in the Moroccan coast. Indeed, this congener is recognised as having high mobility capacity (Shaw and Connely, 1984). The registered total PCB were lower than values reported in mussels collected during1990s on the Spanish coast and Italian Ligurian Sea, 50 to 201 and 120 to 170 ng g^{-1} , (in Aroclor 1254 equiv.), respectively (UNEP/WHO, 1995; Albaigés, 2000).

Aldrin, HCB and heptachlor

The levels of aldrin were significantly ($p<0.05$) higher in mussels collected in the wet season of 2003 (0.01 to 4.0 ng g^{-1}) than of levels registered in the dry season of the four surveyed years (0.01 to 1.8 ng g^{-1}) (Figure 4). Moreover, no geographical pattern was observed, which in conjunction with the seasonal and inter-annual variability may indicate the influence of application in agriculture practices.

The maximum values were higher that levels reported for *M. galloprovincialis* from the coasts of Greece and Turkey (0.1 to 1.0 ng g^{-1}) (Giouranovits-Psylidou et al., 1994; Kucuksezgin et al., 2001). The HCB compound presented enhanced levels in the site AOU in both seasonal periods and elevated values in dry season in the northern site DB. The heptachlor was significantly ($p<0.05$) higher in dry season than in wet season, and in 2003 decreased from northern to southern sampling sites.

Conclusion

Results of this multi-annual survey of organochlorine compounds in wild mussels along 2 000 km of the Atlantic coast of Morocco point to low availability of these compounds in the coastal environment, decreasing from north to southern areas. Furthermore, the milieu is influenced by recent application of p,p'DDT, and there is evidence for relevant atmospheric transport of less chlorinated CBs and heptachlor.

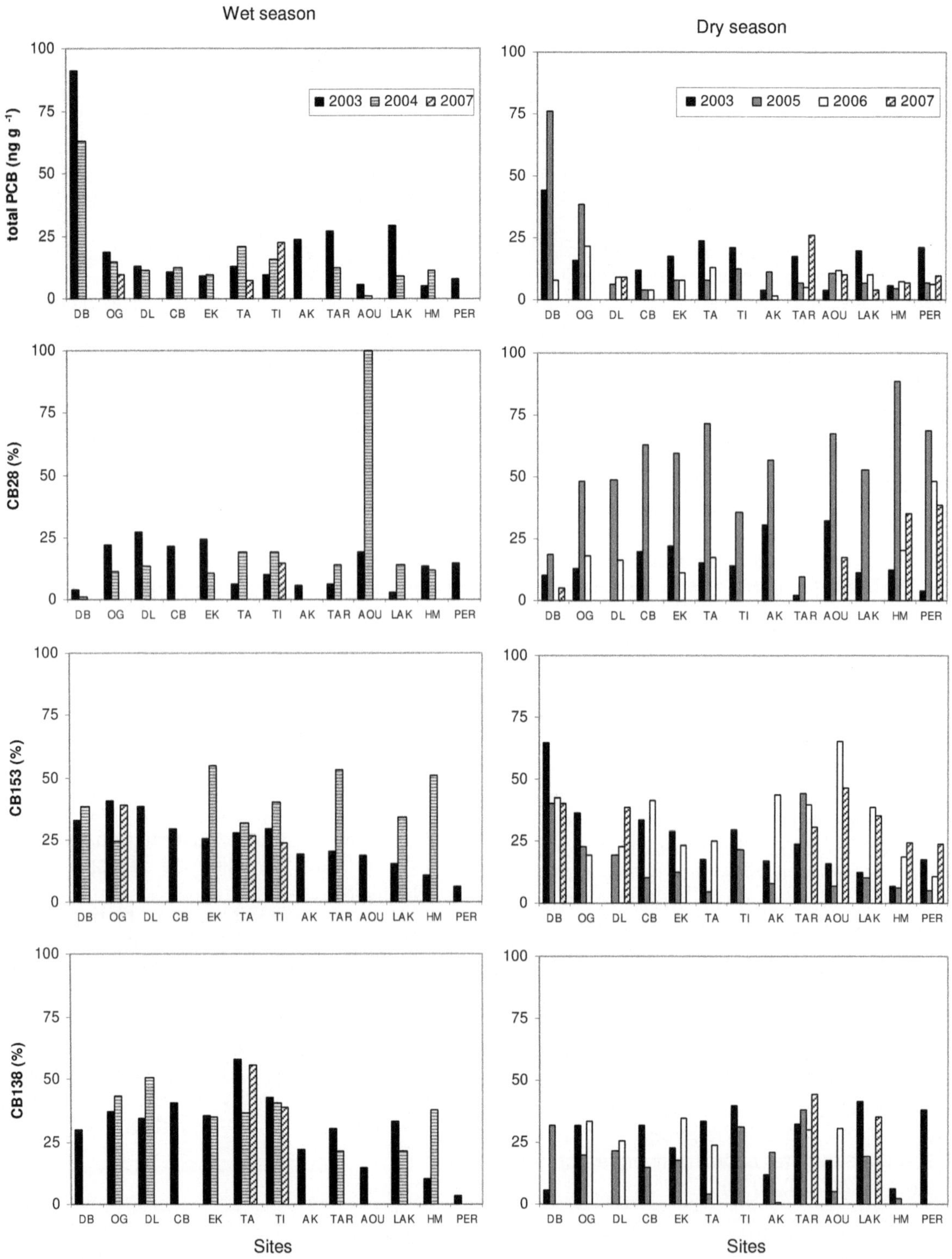

Figure 3. Levels of total PCB and the percentages of the congeners CB28, CB153 and CB138 in mussels collected in 13 sites along the Atlantic Moroccan coasts.

ACKNOWLEDGMENTS

The authors express their thanks to the Institute of Fischeries (INRH) that provided the samples from the sites during the period study, and National Institute of Biological Resources (IPIMAR) for the analysis data.

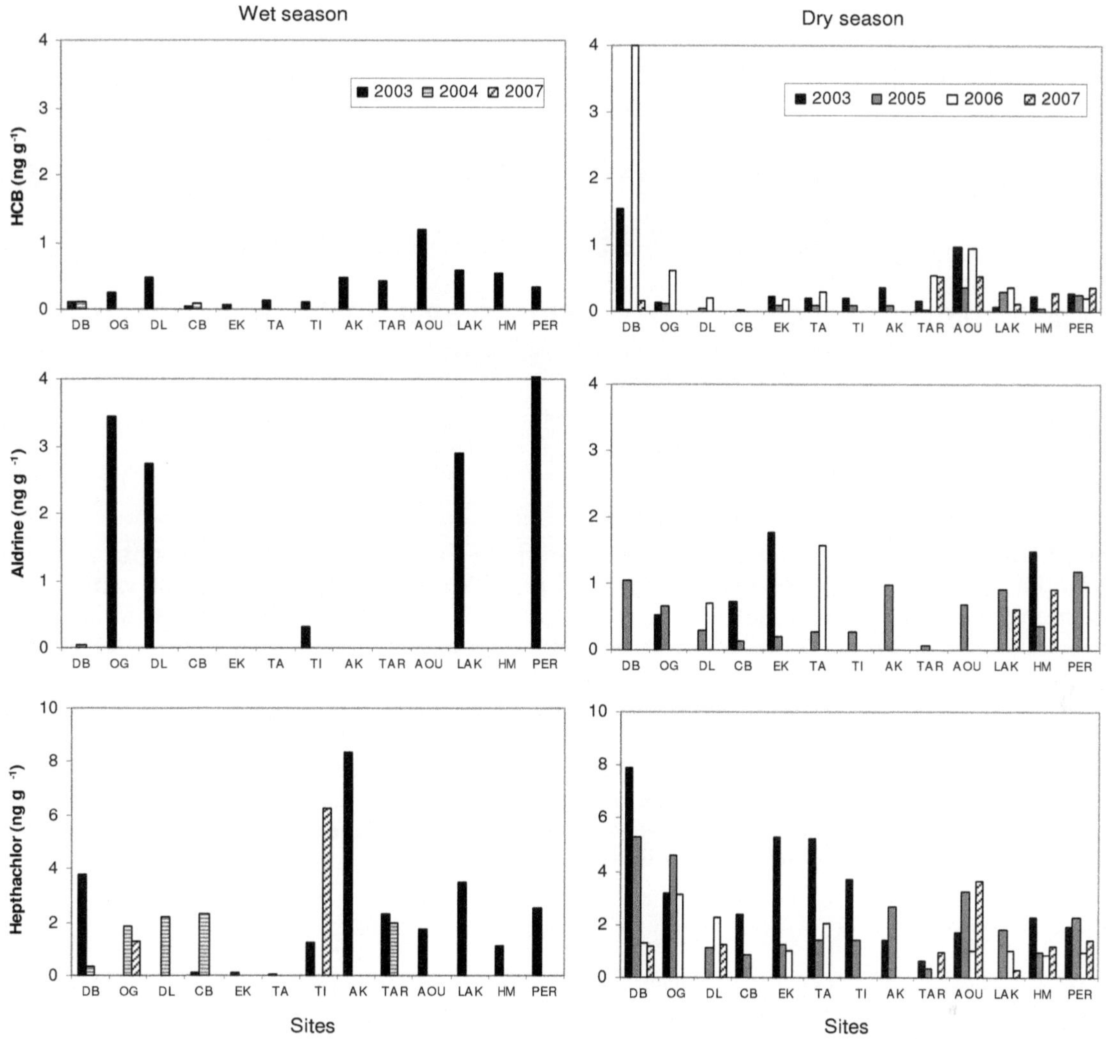

Figure 4. Levels of aldrin, HCB and heptachlor in mussels collected in 13 sites along the Atlantic Moroccan coasts.

REFERENCES

Albaigés J (2005). Persistent Organic Pollutants in the Mediterranean Sea. In: The Handbook of Environmental Chemistry. 2005; vol. 5k. Springer-Verlag, Berlin, pp. 89-149.

Farrington JW, Goldberg ED, Risebrough RW, Martin JH, Bowen VT (1983). US "Mussel Wach" 1976-1978: An overview of the trace metal, DDE, PCB, hydrocarbon, and artificial radionuclide data. Environ. Sci. Technol., 17: 490-496.

Ferreira AM, Vale C (1997). PCB accumulation and alteration of lipids in two length classes of the oyster *Crassostrea angulata* and the clam *Ruditapes decussates*. Mar. Environ. Res. 45: 259-268.

Giouranovits-Psyllidou R, Georgakopoulos-Gregoriades E, Vassilopoulou V (1994). Monitoring of organochlorine residues in red mullet (*Mullus barbatus*) from Greek waters. Mar. Pollut. Bull., 28: 121.

Kucuksezgin F, Altay O, Uluturhan E, Kontas A (2001). Trace Metal and Organochlorine residue Levels in Red Mullet (*Mullus barbatus*) from the Eastern Aegean, Turkey. Water Res., 35: 2327.

Lauenstein CG, Daskalakis KD (1998). US Long-term coastal contaminant temporal trends determined from mollusc monitoring program1965-1993. Mar. Poll. Bull., 37: 6-13.

Manalovani A (2002). Hazard identification and risk assessment of endocrine disrupting chemicals with regard to developmental effects. Toxicology. 181-182: 367-370.

Olsen A, Valters K, Burrean S (2000). Concentrations of organochlorine substances in relation to fish size and trophic position: a study on perch (*Perca fluvialis*). Environ. Sci. Technol., 34: 4878-4886.

RNO (2000). Réseau National de Surveillance du Milieu Marin. IFREMER, France. www.ifremer.fr. p. 29.

Shaw GR, Connely DW (1984). Physicochemical properties controlling polychlorinated biphenyl (PCB) concentrations in aquatic organisms. Environ. Sci. Technol., 18: 18-23.

Tanabe S, Subramanian AN, Ramesh A, Kumaran PL, Miyazakin N, Tatsukawa R (1993). Persistent organochlorine residues in dolphins from Bay of Bengal, South India. Mar. Poll. Bull., 26: 311-316.

UNEP/WHO (1995). Assessment of the state of pollution in the Mediterranean Sea by carcinogenic, mutagenic and teratogenic substances. MAP Technical Reports; Series No. 92, UNEP, Athens, Greece. pp. 18-31

Villeneuve JP, Carvalho FP, Fowler SW, Cattini C (1999). Levels and trend of PCBs, chlorinated pesticides and petroleum hydrocarbons in mussels from the NW Mediterranean Coast: Comparison of concentration in 1973/1974 and 1988/1989. Sci. Total. Environ., 237/238: 57-65.

Benthic biodiversity in the Pichavaram mangroves, Southeast Coast of India

M. Pravinkumar*, P. Murugesan, R. Krishna Prakash, V. Elumalai, C. Viswanathan and S. M. Raffi

Centre of Advanced Study in Marine Biology, Faculty of Marine Sciences, Annamalai University, Parangipettai – 608 502, Tamilnadu, India.

An attempt was made to study the benthic biodiversity in Pichavaram mangrove ecosystem, Southeast coast of India. Monthly (duplicate) sampling was done in 3 stations, (Station I - fresh water zone, Station II - core mangrove zone and Station III - marine zone) for a period of one year from January 2008 to December 2008. As many as 22 species of benthic macro fauna were recorded in the study area (16 species of polychaetes, 4 species of crustaceans and 2 species of molluscs). The population density varied between 394 and 23,888 number/m^2. The species diversity was ranged from 1.52 to 3.87; species richness was from 1.24 to 4.09 and species evenness from 0.71 to 0.96. The maximum diversity was recorded in Station II and minimum in I. As observed in the conventional indices, the values of taxonomic and phylogenetic diversity were also more in Station II compared to Stations I and III. K-Dominance curve drawn paralleled the trend of diversity indices. Cluster and MDS showed the similarity in faunal composition stations within the zones.

Key words: Mangroves, benthos, biodiversity, polychaetes.

INTRODUCTION

Mangrove forests have declined significantly in South Asian countries over the past four decades. The main reasons for mangrove loss and degradation have been population pressure, wood extraction, conversion to agriculture and salt production, tin mining, coastal industrialization, and conversion coastal agriculture (Ong, 1995; Macintosh, 1996). Mangrove is the habitat of very rich faunal, which is comparable to tropical evergreen forest and coral reef ecosystems. Mangroves are extremely important to humans for a variety of reasons including aquaculture, agriculture, forestry, as a source of fire wood and other local use (Kathiresan and Bingham, 2001).

The benthic invertebrates have a profound effect on sediment structure and their biochemical processes by enhancing the porosity and water flow through the sediments. Besides this, their feeding on the sediment surface and plant matter promotes nutrients recycling (Kathiresan, 2000). The distribution, abundance and importance of fauna in mangrove environment were recently realized (Parulekar, 1994). The muddy or sandy sediment of mangrove forest act as a home for a variety of epifaunal and infaunal invertebrates (Kumar, 1995).

Mangroves are inhabited by a variety of benthic invertebrates, such as polychaetes, brachyuran crabs, gastropods, bivalves and sipunculids. Composition of these communities and their importance varies enormously from one habitat to another depending upon the sediment characteristics of the mangroves ecosystem. The saline conditions tolerated by various mangrove species range from brackish water, through pure seawater (30 to 40 ppt), to water concentrated by evaporation to over twice the salinity of ocean seawater (up to 90 ppt).

Similarly, it is understood that faunal assemblages influence the functioning of mangrove ecosystem in various ways. Invertebrates often show marked zonation patterns, and colonize a variety of specific micro-environments. Benthic invertebrates also act as a very good source of food for many bottom feeding fin and shell fishes which enter into the mangroves during high tide for breeding (Sheaves and Molony, 2000).

*Corresponding author. E-mail: sonypravin82@gmail.com.

In spite of rich faunal diversity in mangroves, it has always been given least importance by the scientific community world over. Realizing the importance, a large number of works pertaining to ecology of mangrove fauna of mangroves of east and west coasts of India has been carried out (Patra et al., 1990; Devi and Venugopal, 1989; Chakraborthy and Chowdhury, 1992; Balasubramanyan, 1994; Santhakumaran and Sawant, 1994; Kumar and Anthony, 1994; Kondalrao and Ramanamurthy, 1998; Kumar, 1995, 1997, 2001 and Saravanakumar et al., 2007). The limited number of benthic biodiversity studies was carried out so far in the Pichavaram mangroves. Having the above facts in mind, the objectives of this study was to compare the benthic biodiversity in three zones (FWZ-Fresh water zone; CMZ-Core mangrove zone; MZ-Marine zone) of Pichavaram mangroves. This information will be valuable for monitoring the ecological development of mangroves sites in South Asia.

MATERIALS AND METHODS

Study area and sampling design

Monthly sediment sampling was made from January 2008 to December 2008. Two replicate sediment samples were collected from each of the selected stations, namely fresh water zone (I), core mangrove zone (II) and marine zone (III), (Latitude 11° 27'N, longitude 79° 47'E) - (Figure 1). Peterson grab (0.0251 m^2) was used for collection of sediment samples. Two grab hauls were made at each station, a little quantity of sub-sample (~50 g) was collected for analysis of soil texture and the rest was washed at the sampling site using 0.5 mm sieve. Sediment was then shade dried and the total organic carbon was analysed (El-Wakeel and Riley, 1956).

The percentage composition of sand, silt and clay was calculated, and the sedimentary class was found based on the suggested nomenclature (Shepard, 1962) and soil texture was done by pipette analysis (Kurumbein and Pettijohn, 1938). Grab hauls (sediment samples) were collected according to standard protocol (Mackie, 1994). The environmental parameters such as temperature, salinity, pH and dissolved oxygen (DO) were analyzed by following the standard methods (Strickland and Parsons, 1972).

Benthic fauna

After sieving, all the organisms were carefully fixed in 5 - 7% (neutralized) formalin. Subsequently, the organisms were stained with Rose Bengal solution. Polychaetes (Day, 1967), crustaceans (Barnes, 1980), amphipods (Lyla et al., 1999), gastropods (Rajagopal et al., 1998) and bivalves (Shanmugam et al., 1997) were first segregated into group level and then identified to species level with the help of standard taxonomic references. The organisms were identified and counted under a stereoscopic microscope (Olympus SZ 40X magnification) and density of organisms was expressed as individuals per square meter.

Data analysis

The data were analyzed by various statistical methods namely univariate, graphical/distributional and multivariate methods. Two computer programmes namely SPSS (ver. 11.5.) and PRIMER (ver. 6.) (Clarke and Gorley, 2006) were used.

Univariate methods such as Shannon–Wiener index (H') (Species diversity), Margalef index (d) (Species richness) and Pielou's evenness index (*J'*). Warwick and Clarke (1995) proposed two new diversity indices, capturing the structure not only of the distribution of abundances amongst species but also the taxonomic relatedness of the species in each sample. The first index is taxonomic diversity (Δ) and the second one is total phylogenetic diversity were worked out. *k*-dominance curve were plotted and the results are graphically represented. Multivariate methods such as Bray–Curtis similarity and multi-dimensional scaling (MDS) ordination were carried out and are represented graphically.

BIO-ENV procedure

To ascertain the relationship between biological and environmental variables, the BIO-ENV procedure (Clarke and Ainsworth, 1993) was employed. The basic principle behind this is to measure the agreement between the rank correlations of the biological (Bray-Curtis similarity) and environmental (Euclidean distance) matrices. A weighted Spearman rank correlation coefficient (ρω) was used to determine the harmonic rank correlation between the biological matrix and all possible combinations of the environmental variables.

$$\rho\omega = \frac{6}{N\,(N-1)} \sum_{i=1}^{N} \frac{(ri - si)^2}{ri + sj}$$

The environmental variables such as sand (%), silt (%), clay (%), total organic carbon (mgc/g), temperature (°C), salinity (ppt), dissolved oxygen (mg/L) and pH were used.

RESULTS

Environmental entities

The environmental parameters recorded in three stations are given in Table 1a-c. The water temperature varied from 27 to 31.5°C, with minimum during monsoon and maximum in summer season; salinity ranged between 0 and 34.2 psu with seasonal pattern same as that of temperature with maximum during summer and minimum in monsoon season; pH ranged between 7.2 and 8.4 in Stations I and III with minimum was recorded in monsoon and maximum in summer season. The DO was varied from 3.1 to 5.8 mg/L with minimum value during summer and maximum in monsoon season. Total organic carbon varied from 9.7 to 21.7 mgc/g. As observed in salinity and temperature, the higher value was recorded in summer and minimum in monsoon season. Coming to soil texture, the sand content was found to vary from 20.3 to 72.1%; silt content fluctuated from 0.6 to 7.9% and the clay content ranged from 20.5 to 80.0%.

Among the stations, the maximum sand was recorded in Station III (post monsoon) and minimum was in Station II (pre monsoon), the maximum clay content was in Station II (summer) and minimum was in Station III monsoon) and silt was maximum in Station III (monsoon) and minimum was in Station I (summer). The soils are

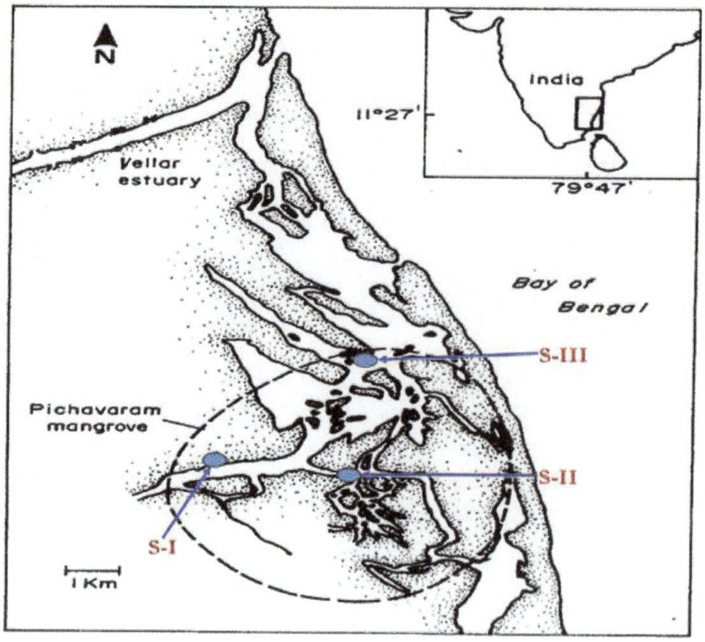

Figure 1. Sampling stations sampled in Pichavaram mangroves (I-III).

Table 1a-c. Seasonal and station wise changes in environmental parameters at Station I, II and III.

S/N	Parameter	Post monsoon	Summer	Pre-monsoon	Monsoon
		(a) Station I			
1	Temperature	28.59±0.52	29.06±0.01	28.33±0.45	28.48±0.51
2	Salinity	11.75±4.49	22.38±4.55	17.98±1.61	6.60±9.01
3	pH	7.70±0.09	8.05±0.08	8.13±0.01	7.82±0.10
4	Dissolved oxygen	3.50±0.15	3.23±0.15	4.23±0.32	5.26±0.15
5	Total organic carbon	11.4±1.44	14.06±0.80	12.36±1.35	12.33±1.76
		(b) Station II			
1	Temperature	29.28±1.14	29.47±0.52	28.61±0.51	28.57±0.485
2	Salinity	17.07±7.50	28.36±4.63	24.3±3.191	8.76±12.39
3	pH	7.90±0.12	8.01±0.109	7.99±0.030	7.79±0.284
4	Dissolved oxygen	3.8±0.2	3.46±0.05	4.66±0.32	5.5±0.1
5	Total organic carbon	15.86±0.51	17.1±0.8	17.1±0.85	19.23±2.45
		(c) Station III			
1	Temperature	29.48±1.59	31.32±1.47	30.08±0.76	28.52±0.50
2	Salinity	22.84±6.67	32.03±2.00	26.34±1.50	23.95±10.59
3	pH	7.9±0.26	8.3±0.1	8.15±0.06	7.97±0.12
4	Dissolved oxygen	4.1±0.36	3.5±0.2	4.46±0.30	5.53±0.251
5	Total organic carbon	15.16±1.68	15.33±0.33	14.5±0.72	15.23±1.05

classified into various soil classes based on the point at which the values meet using the standard diagram (Figure 2a). After calculating the percentage composition, the data were further plotted on soil trigons (Figure 2b-d).

Biological entities

Among 22 species of macrofauna recorded from three stations, polychaetes were found to be the dominant group with 16 species; crustaceans emerged as next dominant group with 4 species and bivalves and gastropods came next in the order with 1 species each. In Station I, a total of 7 species of macrofauna were recorded. Among these, there were 2 species of polychaetes, 3 species of crustaceans, 1 species each of bivalves and gastropods. With respect to Station II, a total of

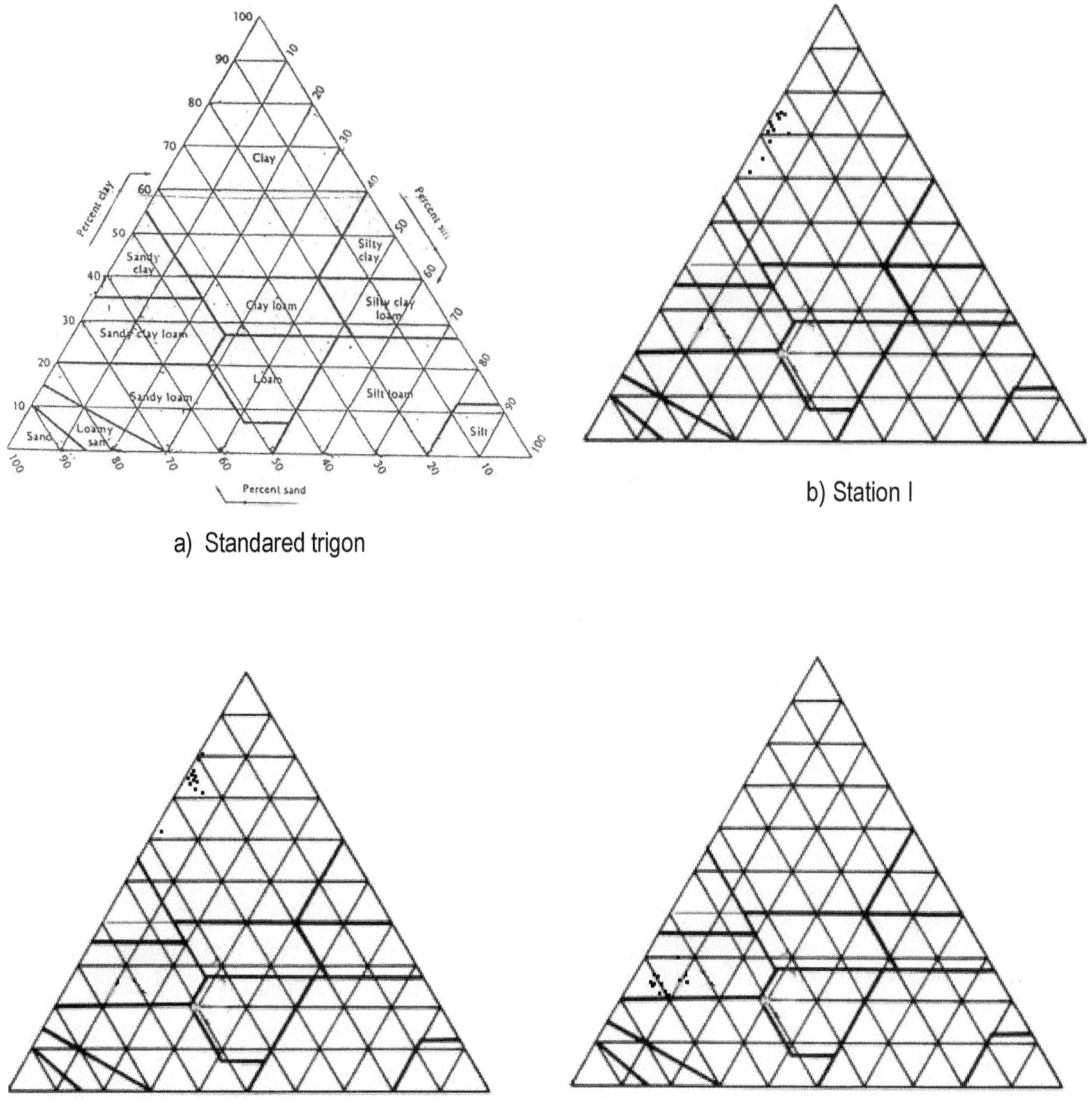

c) Station II d) Station III

Figure 2a-d. Variations in soil texture in the study area.

18 species were recorded. Among these, 12 species belonged to polychaetes, 3 species to crustaceans, 1 each to bivalves and gastropods. At Station III, 12 species of macrofauna were found. Among these, 7 species of polychaetes, 4 species of crustaceans and 1 species of gastropod were recorded.

Among the polychaetes, *Polydora hoplura hoplura*, *P. capensis*, *Exogone clavator*, *Pygospio elegans*, *Euclymene* sp. were found to be common in all the stations during the study period. Concerning the crustaceans, *Apseudus* sp., *Grandidierella* sp. and *Eriopisa chilkensis* were found to occur throughout the study period. In case of bivalves, *Meretrix casta* and gastropod *Cerithidea cingulata* showed consistency in their occurrence in three stations. With respect to population density, in Station I, the density of benthic organisms varied between 394 and 12, 938 NOS/m^2 with maximum during summer and minimum during monsoon season. At Station II, the population density varied from 790 to 23, 888 NOS/m^2. With respect to Station III, the density of organisms varied from 715 to 20, 859 NOS/m^2 (Figure 3). Among the stations, maximum during summer in Station 2 and minimum during monsoon season in Station 1.

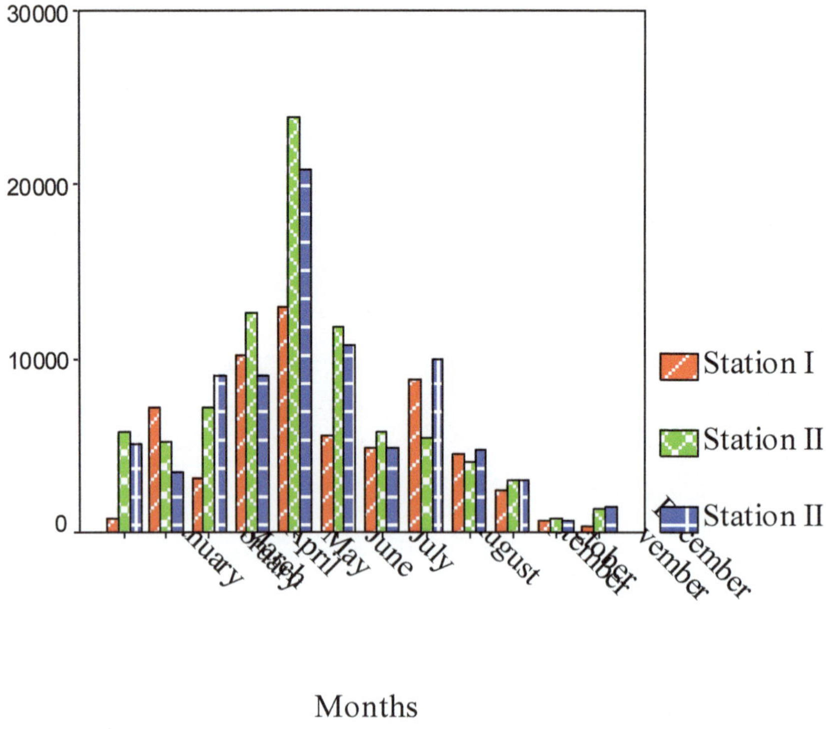

Figure 3. Monthly variations in population density of macrobenthos recorded in Stations I-III during the study.

Regarding the percentage composition of macrofauna, in Station I, polychaetes were found to be the dominant group by constituting 47.2% of the total benthic organisms recorded. Crustaceans formed next dominant group with 42.0%. Gastropods and bivalves came next in the order with a percentage contribution of 6.1 and 4.6% respectively (Figure 4a). As in Station I and II, polychaetes topped the list with 50.6%. Crustaceans ranked second with contribution of 44.6%. Gastropods and bivalves contributed 3.4 and 1.3%, respectively to the total benthic organisms (Figure 4b). With respect to Station III, polychaetes continued to be the dominant group with 47.5%. Crustaceans formed next dominant with a contribution of 41.7%. Gastropods and bivalves contributed 7.3 and 3.5%, respectively to the total benthic organisms collected (Figure 4c).

Diversity indices

The diversity indices calculated for three stations are given in Table 2. In Station I, the species diversity (Shannon-Weiner index) ranged from 1.52 to 3.36; species richness fluctuated between 1.24 and 3.49; species evenness (Pielou's evenness) from 0.74 to 0.93; taxonomic diversity from 32 to 76.93 and the total phylogenetic diversity fluctuated from 180 to 1000. At Station II, the species diversity ranged from 2.24 to 3.87; species richness fluctuated between 1.73 and 4.09; Pielou's evenness varied from 0.71 to 0.96; taxonomic diversity from 55.43 to 76.88 and the total phylogenetic diversity values ranged from 340 to 1120. With respect to Station III, the species diversity (Shannon-Wiener index) varied from 2.70 to 3.61; species richness from 2.04 to 3.59; species evenness from 0.77 to 0.98; taxonomic diversity values were from 65.57 to 79.48 and total phylogenetic diversity varied from 460 to 980. Among three stations, the minimum diversity and richness values were recorded in Station I (fresh water zone) and maximum in Station II (core mangrove zone). As regards species evenness, the minimum value was recorded in Station II and maximum in Station III (marine zone). The taxonomic diversity also showed minimum in Station I and maximum in Station III. Similarly, the total phylogenetic diversity was minimum in Station I while maximum in Station II. When the values were viewed seasonally, summer and post monsoon seasons registered the maximum values and monsoon and premonsoon seasons the minimum values.

The *k*- dominance plot drawn clearly demonstrated the diversity pattern in three stations. When the data of all the stations were plotted together, the curves for Stations II and III (core mangrove zone, marine zone) were found to lie below the curve of I and rose slowly due to the presence of more number of species giving 'ʃ' shape, indicating maximum diversity, whereas the curve for

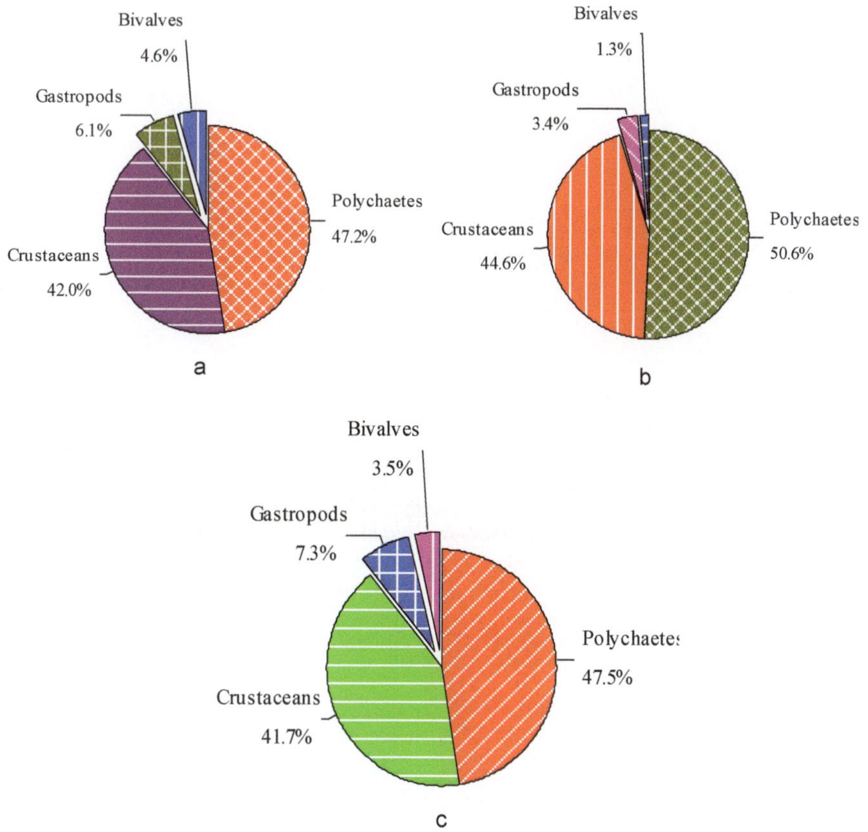

Figure 4. Percentage composition of macrobenthos recorded in (a) Station I, (b) Station II and (c) Station III .

Table 2. Diversity indices in various stations of Pichavaram (I – III) mangroves.

Station	*H'* (log2)	d	J'	Delta	sPhi+
I	1.52-3.36	1.24-3.49	0.74-0.93	32-76.93	180-1000
II	2.24-3.87	1.73-4.09	0.71-0.96	55.43-76.88	340-1120
III	2.70-3.61	2.04-3.59	0.77-0.98	65.57-79.48	460-980

Station I (Fresh water zone) was lying above, showing minimum diversity. This plot also ably proved the rich diversity nature of Stations II and III compared to Station I (Figure 5).

Figures 6 and 7 display results of hierarchical clustering and MDS ordination, respectively. From the resulting dendrogram (Figure 6), among the stations, samples of marine zone got grouped at the highest level of similarity (94%) followed by fresh water zone (93%) and core mangrove zone (90%). Further, cluster of core mangrove and marine zones were formed in a single cluster at the next level of similarity (86%) and cluster of fresh water zone grouped successively to this at the next level (79%).

The non-metric multi dimensional scaling (MDS) plot also revealed the same groupings as recognized in the cluster. The stress value, which overlie on the top right corner of the plot, was also found to be very minimum (0.01) signaling the good ordination pattern of species in stations of various zones (Figure 7).

In the BIO – ENV method, eight variables such as temperature, salinity, pH, dissolved oxygen, total organic carbon, sand, silt and clay were allowed to match the biota. The results showed that, salinity, clay, total organic carbon and temperature were featured as the key variables explaining the best match (0.71) with faunal distributions. Following this, silt, pH and sand were also manifested as important variables influencing the faunal distribution (Table 3).

DISCUSSION

The distribution of mangrove fauna in relation to water quality has been described quantitatively by Deekae and

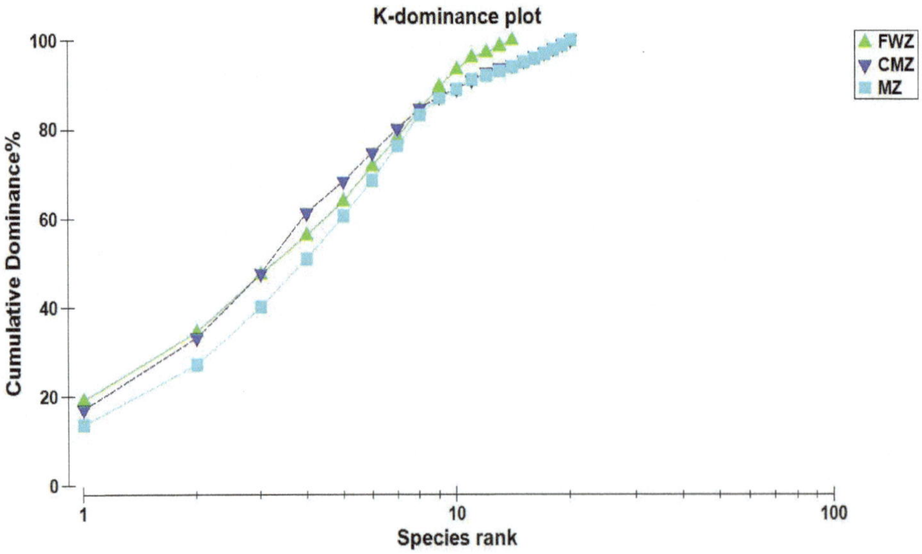

Figure 5. K- dominance curves for the Stations I-III (FWZ; Fresh Water Zone; CMZ; Core Magrove Zone and MZ; Marine Zone).

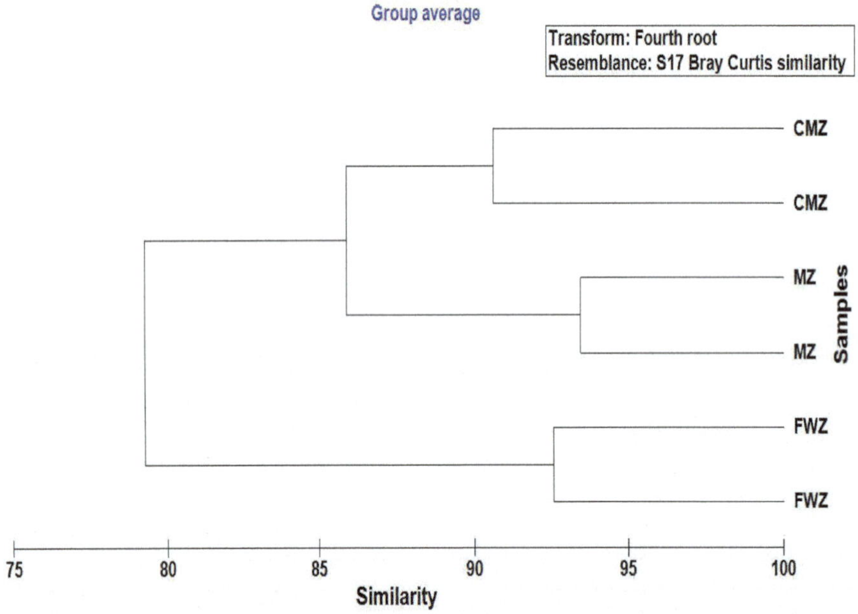

Figure 6. Dendrogram showing grouping of Stations I-III (FWZ; Fresh Water Zone; CMZ; Core Magrove Zone and MZ; Marine Zone) sampled during the study period.

Henrion (1993) and Guerreiro et al. (1996). In the present study, the temperature level varied from 27 to 31°C. All the stations showed a similar trend in seasonal variations with minimum in monsoon and maximum during summer as reported earlier by Kathiresan (2000) and Saravanakumar et al. (2007).

The salinity acts as a limiting factor in the distribution of living organisms, and its variation was due to dilution and evaporation. In the present investigation, dissolved oxygen showed high during monsoon at all stations which might be due to the cumulative effect of higher wind velocity and, it varied from 3.2 to 5.5 mg/L. Similar observation with high value during monsoon and low during summer was reported earlier by Fernando (1987) and Tripathy et al. (2005). The seasonal variations in dissolved oxygen mainly due to the freshwater influx and

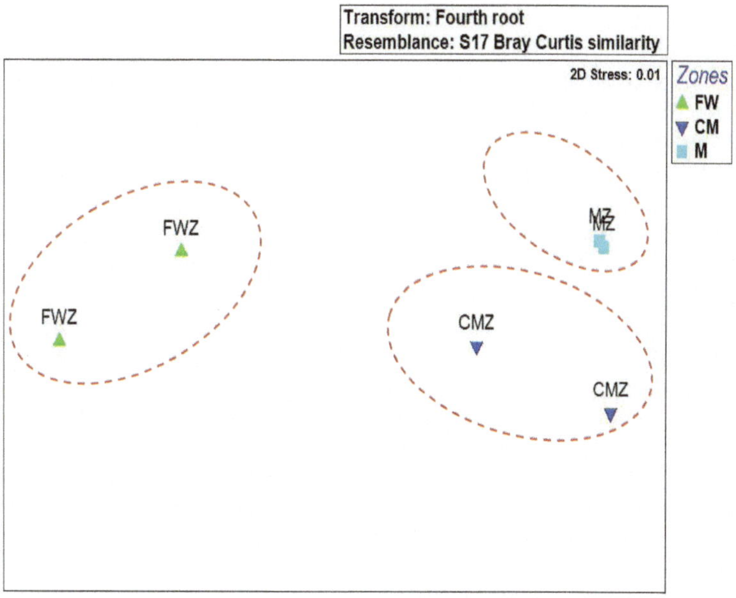

Figure 7. MDS plot for macroenthos in three Stations I-III (FWZ; Fresh Water Zone; CMZ; Core Magrove Zone and MZ; Marine Zone).

Table 3. Harmonic rank correlations (ρω) between faunal and environmental similarity matrices in various stations of mangroves.

No. of variables	Best variable combinations	Correlation (ρω)
4	Salinity-clay-total rganic carbon-temperature	0.71
3	Temperature-Salinity-Silt	0.68
3	Clay-total organic carbon-pH	0.67
2	Sand-salinity	0.65

ferruginous impact of sediments. It is well known that the temperature and salinity affect the dissolution of oxygen (Vijayakumar et al., 2000). The surface water pH remained alkaline at all the stations throughout the study period, with maximum value during summer and minimum during monsoon.

Generally, fluctuations in pH values during seasons of the year are attributed to factors like removal of CO_2 by photosynthesis through bicarbonate degradation, dilution of seawater by freshwater influx, reduction of salinity and temperature, and decomposition of organic matter. The total organic carbon varied from 9.7 to 21.7 mgc/g with maximum value at Station II (summer) and minimum in Station I. The maximum during summer might be due to the thick presence of *Rhizophora* spp. where litter fall was more, which in turn increased the total organic carbon content as reported earlier by Dehairs et al. (2000); Kristensen et al. (2008) and Hasrizal et al. (2009). The muddy or sandy sediments of mangrove forests act as a home of a variety of infaunal and epifaunal invertebrates and their role vary from one habitat to another depending upon the water and sediment characteristics of the mangroves (Kumar, 1995). In the present study, the sediment composition showed striking variations in percentage of sand, silt and clay fractions. The sand, silt and clay composition varied between seasons and stations.

Among the stations, the percentage of sand was high in station III and low in Stations II and I. Similarly the clay and silt were more in Stations I and II and low in station III. The high clay content in Station II might be probable reason for more number of organisms compared to Stations I and III since the most of the benthic organisms are detrital feeders. Ramanathan (1997) reported that in Pichavaram mangroves, the percentages of clay and silt was more, which constituted together 70 - 90% of mangrove sediments followed by sand. Similarly, Reddy and Hariharan (1986) observed that the benthic communities are largely influenced by the texture of the sediment.

Generally, in a benthic sample, polychaetes constitute more than the other faunal groups. The dominance of polychaetes in terms of diversity and species composition in diverse ecological niches is due to their high adaptability to wide range of fluctuation in environmental factors. Their dominance can also be attributed to the high organic carbon content in the sediments.

The present study recorded 16 species of polychaetes out of 22 macro fauna in Pichavaram. In confirmation of this, several benthic researchers studied elsewhere reported that the polychaetes as dominant group followed by other groups. The present observation showed numerical dominance in the order of polychaetes, molluscs (bivalves and gastropods), crustaceans and others has been observed earlier by Ansari et al. (1986); Mohammed (1995); Kumar (2001); Murugesan et al. (2007). The dominance of polychaetes in all the stations might also be due to the conducive environment provided by aerial roots and dense canopy of the mangroves which offered protection against desiccation and this enabled polychaetes to be abundant. In the present study, the macro-benthic faunal density ranged from 394 to 23,888 nos/m^2 in Stations I and II with highest abundance (23, 888 nos/m^2) was recorded at Station II during summer season. This could be due to higher total organic carbon coupled with a stable environment recorded during this season while low population density (394 nos/m^2) recorded during monsoon at station I might be due to the influx of freshwater as it is said to be freshwater zone.

Added to this, Kurian (1984) reports that productivity is also comparatively high in mangroves as the nutrients rich bottom sediments facilitate the luxuriant growth of algae, consequently the abundance of benthic fauna. Similarly, Saravanakumar et al. (2007) and Tolhurst and Chapman (2007) reported the similar range of population density. As regards percentage composition of macrofauna, polychaetes were found to be the dominant group in all the stations followed by crustaceans, gastropods and bivalves. Zhou (2001) and Huang et al. (2002) have reported that the polychaetes, crustaceans and molluscs are the three major groups contributing substantially to the total benthic organisms in mangrove environment.

Species diversity is a simple and useful measure of a biological system. Redding and Cory (1975) found a high level of agreement between species diversity and nature of the environment and hence, regarded the measure of species diversity as an ecologically powerful tool. Moreover, Pearson and Rosenberg (1978) proposed that the use of diversity indices is advantageous for the description of fauna at different stages in the succession. Sanders (1968) also postulated that the species diversity is mainly controlled by the fluctuations in the environment that lead to low diversity. There are reports stating that when the diversity value is more than 3, the system is considered to be pristine and diverse in nature. Species diversity in the present study registered a wide fluctuation between 1.52 in Station I and 3.87 in Station III. Maximum diversity so also the species richness was recorded in Station II during summer and minimum in I during monsoon and higher values recorded in summer in the study area is in conformity with the earlier observations made elsewhere (Chandran, 1987; Murugesan, 2002; Ajmal et al., 2005; Devi, 1994). In the present study, richness values were more during post monsoon season (4.09) in Station II, which is comparable to the studies made by Kumar (1995) and Sunilkumar (1996). The low richness recorded in this study (1.24) in Station I might be due to fresh water inflow that might have induced low saline conditions, which in turn affected the distribution of benthos, particularly the polychaetes.

The above said conventional indices are unduly influenced by the sampling effort and evenness property and these indices can be used only with the quantitative data, unlike taxonomic diversity and phylogenetic diversity which can very well be used with presence / absence data. Not only that, the conventional indices also lack the statistical frame work for comparison of one sample with another. Therefore, the diversity indices based on the taxonomic relatedness of species have been introduced such as taxonomic diversity, and total phylogenetic diversity index by Clark and Warwick (2001). These two indices were used presently with a view to finding out the statistical significance of the difference in diversity between three stations. When the results were viewed, the taxonomic diversity was more in Station III (65.57 - 79.48) than in stations I (32 - 76.93) and II (55.43 - 76.88). But total phylogenetic diversity was more in Station II (340 - 1120) than in Stations I (180 - 1000) and III (460 - 960) indicating the taxonomic breadth of the biota (Clarke and Green, 1988). In addition to the above, k-dominance plot clearly demonstrated the trend as observed in diversity indices. The curves for Stations II and III rose slowly due to the presence of a large number of species. The percentage contribution of each species was added, the curve extended horizontally (species number is evident in the x - axis), before reaching the cumulative 100%.

In the present study, cluster analysis and MDS was used to find out the degree of similarity among the stations of Pichavaram mangroves. Cluster analysis (or classification) is helpful to find out natural groupings of samples, such that samples within a group are more similar to each other than the samples in different groups. In the present study, data of two replicate samples were allowed as input for cluster and MDS analysis. The dendrogram revealed that samples of various zones got grouped individually indicating variation in species composition in particular zone. However, as cluster analysis has inherent disadvantage of linking ultimately with the dissimilar groups, ordination of the samples was also done using MDS. The aim of this method is to represent the samples collected as points in a map (usually low - dimensional space). Samples lying closer have more similarity in species composition and abundance while samples lying far apart have more dissimilarity in species composition and abundance. The

MDS plot also revealed that the stations of respective zones formed groups separately again demonstrating the variation in species composition and abundance in three stations. The stress value, overlying on the right top corner of this map, was also very minimum indicating the good ordination of the samples. Ajmal et al. (2005) also report similar clustering and MDS pattern.

The Pichavaram mangrove forest has been in existence since 1500 years (K. Kathiresan per. com). The distribution of benthos exhibited not only higher density during the post monsoon and summer seasons but also consisted of more diverse fauna at Station II. The productivity and nutrient load was more in Station II and this might be due to the fact that an ample amount of mangrove litter fall occurs which in turn increases the productivity thereby enriching the bottom fauna. Not only that, the age and sprawling area might also be the plausible reason for the higher diversity in Station II than Stations I and III where canopy is less.

The decrease in benthic abundance during monsoon may be attributed to the low temperature and salinity. A medium amount of organic matter and salinity also supports more benthic production. Therefore, it can be deduced from the study that ecological factor like temperature, salinity and total organic carbon have profound influence on the abundance and distribution of benthic organisms in the Pichavaram mangroves. Continue monitoring of benthic diversity may provide the base line data for future monitoring.

ACKNOWLEDGEMENTS

Authors are thankful to Prof. K. Kathiresan, Director and Dean, CAS in Marine Biology, Faculty of Marine Sciences, Parangipettai for providing facilities. We are also grateful to the Ministry of Earth Sciences (OASTC) Government of India for the financial support.

REFERENCES

Ajmal KS, Raffi SM, Lyla PS (2005). Brachyuran crab diversity in natural (Pichavaram) and artificially developed mangroves (Vellar estuary). Curr. Sci. 88(8):1316-1324.

Ansari ZA, Ingole BS, Banerjee G, Parulekar AH (1986). Spatial and temporal changes in benthic macrofauna from Mandovi and Zuari estuaries of Goa, West coast of India. Indian J. Mar. Sci. 15:223-229.

Balasubramanyan K (1994). Micro invertebrate benthic fauna of Pitchavaram mangroves. In: A Training Manual. Sanjay V. Deshmukh and Balagi, V., eds., Conservation of mangroves Forest Genetic Resources: M.S. Swaminathan Res. Foundation, Chennai, India. pp. 257-259.

Barnes RD (1980). Invertebrate Zoology. Saunders College, Philadelphia. P. 1089.

Chakraborthy SK, Chowdhury A (1992). Community structure of macrobenthic polychaetes of intertidal region of Sagar Island, Hoogly estuary, Sunderbans, India. Trop. Ecol. 35(1):97-104.

Chandran R (1987). Hydrobiological studies in the gradient zone of the Vellar estuary. I Physicochemical parameters. Mahasagar – Bull. Natn. Inst. Oceanogr. 17:69-78.

Clarke KR, Gorley RN (2006). Primer v6: User manual/Tutorial. Primer-E, Plymouth. P. 190.

Clarke KR, Ainsworth M (1993). A method linking multivariate community structure to environmental variables. Mar. Ecol. Prog. Ser. 92:205-219.

Clarke KR, Green RH (1988). Statistical design and analysis for a 'biological effects' study. Mar. Ecol. Prog. Ser. 46:231-226.

Clark KR, Warwick RM (2001). Chance marine communities: An approach to statistical analysis and interpretation. Plymouth, PRIMER-E Ltd, U. K. P. 144.

Day JH (1967). A monograph on the polychaeta of Southern Africa. Parts 1 and 2, British Museum Nat. Hist. London. P. 878.

Deekae SN, Henrion R (1993). Multivariate analysis of species distribution: A survey on occurrence of mangrove molluscs in the Bonny and New Calabar rivers of the Niger Delta. Acta Hydrochimicaet, Hydrobiol. 21:273-279.

Dehairs F, Rao RG, Chandra Mohan P, Raman AV, Marguillier S, Hellings L (2000). Tracing mangrove carbon in suspended matter and aquatic fauna of the Gautami-Godavari Delta, Bay of Bengal (India). Hydrobiol. 431:225-241.

Devi KS, Venugopal P (1989). Benthos of Cochin back waters receiving industrial effluents. Indian J. Mar. Sci. 18:165-169.

Devi LP (1994). Ecology of Coleroon estuary: Studies on benthic fauna. J. Mar. Biol. Ass. India, I & II 36:260-266.

El- Wakeel SK, Riley JP (1956). The determination of organic carbon in marine muds. J. Du. Council Permanent International Pourl. Exploration Dela Mer, 22:180-183.

Fernando OJ (1987). Studies on the intertidal fauna of the Vellar estuary. J. Mar. Biol. Ass. India 29:86-103.

Guerreiro J, Freitas S, Pereira P, Paula J, Macia A (1996). Sediment macrobenthos of mangrove flats at Inhaca Island, Mozambique. Cahiers de Biologie Marine 37:309-327.

Hasrizal SB, Kamaruzzaman Y, Sakri I, Ong MC, Noor Azharm MS (2009). Seasonal distribution of organic carbon in the surface sediments of the Terengganu near shore coastal area. Am. J. Environ. Sci. 5:111-115.

Huang B, Zhang B, Lu J, Ou Z, Xing Z (2002). Studies of macrobenthic ecology and beach aquaculture holding capacity in Dongzhai Bay mangrove areas. Mar. Sci. Haiyang-Kexue 26(3):65-68.

Kathiresan K (2000). A review of studies on Pichavaram mangroves, Southeast India. Hydrobiol. 430:185-205.

Kathiresan K Bingham BL (2001). Biology of Mangroves and mangrove ecosystems. Adv. Mar. Biol. 40:81-251.

Kondalrao B, Ramamurthy K (1998). Ecology of inter meiofaunal of the Kakinada Bay (Gautami-Godavari estuarine system), Eastcoast of India. Indian J. Mar. Sci. 17:40-47.

Kristensen E, Bouillon S, Dittmar T, Marchand C (2008). Organic carbon dynamics in mangrove ecosystems: A review. Aquat. Bot. 89: 201-219.

Kurumbein WC, Pettijohn FJ (1938). Manual of sedimentary petrography. Applenton Century- Crofts, New York. P. 549.

Kumar RS (1995). Macrofaunal benthos in the mangrove ecosystem of Cochin backwaters, Kerala (Southwest coast of India). Indian J. Mar. Sci. 24:56-61.

Kumar RS (1997). Vertical distribution and abundance of sediment dwelling macro-invertebrates in an estuarine mangrove biotope, Southwest coast of India. Indian J. Mar. Sci. I 26: 20-25.

Kumar RS, Anthony A (1994). Preliminary studies on the polychaete fauna of the mangrove areas of Cochin. In: Proceedings of the 6th Kerala Science Congress. Kerala, T., and R. Ravikumar, eds., State Committee on Science, Technology and Environment, Thiruvananthapuram, India. pp. 74-77.

Kumar RS (2001). Intertidal zonation and seasonality of benthos in a tropical mangrove. Int. J. Ecol. Environ. Sci. 27:199-208.

Kurian CV (1984). Fauna of the mangrove swamps in Cochin estuary. Proceedings of the Asian Symposium on Mangrove Environment Research and Management. Soepadmo, E. A. N. Rao and D. J. Macintosh, (Eds.), pp. 226-230.

Lyla PS, Velvizhi S, Ajmal Khan S (1999). A monograph on the amphipods of Parangipettai coast. Annamalai University, India. P. 78.

Mackie ASY (1994). Collecting and preserving polychaetes. Polychaete Res. 16:7-9

Macintosh DJ (1996). Mangroves and coastal aquaculture: Doing something positive for the environment. Aquacul. Asi. I 2:3-8.

Mohammed SZ (1995). Observation on the benthic macrofauna of the soft sediment on Western side of the Arabian Gulf (ROPME sea area) with respect to 1991 Gulf war oil spill. Indian J. Mar. Sci. 24:147-152.

Murugesan P (2002). Benthic biodiversity in the marine zone of Vellar estuary. Ph.D., Thesis, Annamalai University, India. P. 330.

Murugesan P, Ajmal Khan S, Ajithkumar TT (2007). Temporal changes in the benthic community structure of the marine zone of Vellar estuary, southeast coast of India. J. Mar. Ass. India 49:154-158.

Ong JE (1995). The ecology of mangrove conservation and management. Hydrobiol. 295:343-341.

Parulekar A (1994). Benthic fauna of mangrove environment In: Conservation of Mangrove Forest Genetic Resources: A training manual, Sanjay V. D. and V. Balaji, eds., CRSARD, Chennai. P. 128.

Patra KC, Bhunia AB, Mitra A (1990). Ecology of macrobenthos in a tidal creek and adjoining mangroves in West Bengal, India. Environ. Ecol. 118:539-547.

Pearson TH, Rosenberg R (1978). Macrobenthos sucession in relation to organic enrichment and pollution of the marine environment. Oceanogr. Mar. Biol. Ann. Rev. 16:229-234.

Rajagopal S, Ajmal Khan S, Srinivasan M, Shanmugam A (1998). A monograph on the gastropods of Parangipettai coast. Annamalai University, India. P. 38.

Ramanathan AL (1997). Sediment characteristics of the Pichavaram mangrove environment, Southeast coast of India. Indian J. Mar. Sci. 26:319-322.

Redding JM, Cory RL (1975). Macroscopic benthic fauna of three tidal creeks adjoining Rhode river, Mary-land. Water Resources Investigation Report, USA. pp. 39-75.

Sanders HL (1968). Marine benthic diversity: A comparative study. Am. Nat. 102:243 -282.

Santhakumaran LN, Sawant SG (1994). Observation on the damage caused by marine fouling organisms to mangrove saplings along Goa coast. J. Tim. Dev. Ass. India II 40:9-19.

Saravanakumar A, Sesh Serebiah J, Thivakaran GA, Rajkumar M (2007). Benthic macrofaunal assemblage in the arid zone mangroves of Gulf of Kachchh, Gujarat. J. Oce. Uni. Chi. IV 6:303-309.

Shanmugam A, Rajagopal S, Nazeer RA (1997). A monograph on the common bivalves of Parangipettai coast. Annamalai University, India. P. 66.

Sheaves M, Molony B (2000). Short-circuit in the mangrove food chain. Mar. Ecol. Prog. Ser. 199:97-109.

Shepard RN (1962). The analysis of proximities: Multidimensional scaling with an unknown distance function. Phychometrika 27:125-140.

Strickland JDH, Parsons TR (1972). A practical handbook of seawater analysis. Bull. Fish. Res. Bd. Canada 167:310.

Sunilkumar R (1996). Distribution of organic carbon in the sediments of Cochin mangroves, Southwest coast of India. Indian J. Mar. Sci. 25: 274-276.

Tolhurst T, Chapman J (2007). Patterns in biogeochemical properties of sediments and benthic animals among different habitats in mangrove forests. Austr. Ecol. 32:775-788.

Tripathy SC, Ray AK, Patra S, Sarma W (2005). Water quality assessment of Gautami-Godavari mangrove estuarine ecosystem of Andhra Pradesh, India during September 2001. J. Earth Syst. Sci. II 114:185-190.

Reddy HR, Hariharan V (1986). Distribution of total nitrogen, total phosphorus and organic carbon in the sediments of Mangalore. Mahasagar-Bull. Natn. Inst. Oceanogr. II 19:119-122.

Vijayakumar S, Rajesh KM, Mirdula RM, Hariharan V (2000). Seasonal distribution and behaviour of nutrients with reference to tidal rhythm in the Mulki estuary, Southwest Coast of India. J. Mar. Biol. Ass. India 42(182):21-31.

Zhou H (2001). Effects of leaf litter addition on meiofaunal colonization of azoic sediments in a subtropical mangrove in Hong Kong. J. Exp. Mar. Biol. Ecol. 256:99-121.

Scenario generation using geographical information system (GIS) based hydrological modelling for a multijurisdictional Indian River basin

Singh A.[1]* and Gosain A. K.[2]

[1]Central Soil and Materials Research Station, Ministry of Water Resources, India.
[2]Department of Civil Engineering, Indian Institute of Technology, Delhi, India.

Lack of information about the futuristic scenarios of possible water allocations acts as a deterrent in resolving the conflicts pertaining to transboundary watercourses. It is increasingly being felt that technology, in the form of simulation modelling, has a very significant role to play in this context. For water resources engineering problem solving, geographical information system (GIS) offers a cognitive spatial representation of complex hydrologic and hydraulic systems. Of specific interest to decision makers is the capability of GIS to visually display information for interpretation of water resource model inputs and outputs, which enables users to take a more dynamic approach with data input, modification, scenario development, and evaluation (Martin et al., 2005). In the present study, GIS based hydrological modelling has been utilized for the purpose of assessment of the total amount of water available in the study area, as well as prediction of the impact of changes in the land management practices on the water availability in the study area, and consequently the amount of water allocated to each of the riparian states. The hydrological model used for the study is SWAT (soil and water assessment tool). The results of the study show that water yield of the basin is inversely proportional to the amount of forest cover. The study demonstrates that simulation modelling can play a very significant role in water resources management by generating a series of scenarios or options for the stakeholders, so as to enable them to take sound rational decisions.

Key words: Geographical information system (GIS), hydrological modelling, soil and water assessment tool (SWAT), scenario generation.

INTRODUCTION

In the present study, geographical information system (GIS) based hydrological modelling has been used to simulate the hydrological regime of Cauvery river basin in India. The hydrological model used for the study is SWAT (soil and water assessment tool) (Arnold et al., 1995). The impact of changes in the land management practices on the water availability in the basin has been modelled. SWAT is the acronym for soil and water assessment tool, a river basin or watershed scale model developed by Dr. Jeff Arnold for the USDA agricultural research services (ARS). SWAT was developed to predict the impact of land management practices on water, sediment and agricultural chemical yields in large complex watershed with varying soils, land use and management conditions over long periods of time. SWAT is a continuous time model, that is, a long-term yield model having the capability of scenario generation, so as to equip the policy makers with a wider range of options, which makes it the ideal tool to be used for such a study (Arnold et al., 1995, 1998; Neitsch et al., 2002).

DESCRIPTION OF THE STUDY AREA

The Cauvery is the fourth largest river in the peninsular of southern India, after Godavari, Krishna, and the Mahanadi. The Cauvery rises at Talakaveri on the

*Corresponding author. E-mail: asingh249@gmail.com or asingh249@rediffmail.com

Brahmagiri Range of hills (12° 25' N, 74° 34'E) in the Western Ghats in the Coorg District of Karnataka, at an elevation of 1,341 m above mean sea-level. In its course of 802 km from the Western Ghats to the Bay of Bengal, the main river flows for 381 km in Karnataka, 357 km in Tamil Nadu, and provides the boundary between the two states for the rest. One of its important tributaries in Karnataka, the Kabini, originates in Kerala; and drainage from Kerala contributes to the Bhavani and the Amaravathy, two of the river's tributaries in Tamil Nadu. Sub-branches of the river irrigate the Karaikal area in Pondicherry before entering the sea.

The Cauvery is thus an interstate river, with all four basin states- Karnataka, Tamil Nadu, Kerala, and Pondicherry- having an interest in the sharing of its waters, although Karnataka in the upper and Tamil Nadu in the lower reaches are by far the principal co-riparians (Guhan, 1993). The climate of Cauvery river basin may be described as essentially a tropical monsoon type, which is a product of the interplay of the two opposing air masses of the South-West and North-East Monsoons. The following four distinct seasons occur in the basin (CFFC, 1972):

1. Cold weather period (January – February);
2. Hot Weather period (March – May);
3. South – West Monsoon period (June – September);
4. North – East Monsoon period (October – December).

The general agricultural practice in the basin broadly has three categories of cultivation, namely (i) wet land irrigated from the rivers and tanks; (ii) garden lands irrigated from pump sets and wells and (iii) rain-fed dry land cultivation. In the wet lands, paddy is the most important crop, whereas ragi, jowar and other millets constitute the important rainfed crops (CFFC, 1972).

MATERIALS AND METHODS

The SWAT model requires data on terrain, land use, soil, weather and man-made structures like reservoirs etc for assessment of water resource availability at the desired locations of the drainage basin. To create a SWAT dataset, the interface needs to access Arc View map themes and database files, which provide topographical, land use/land cover, soil class and climatic information about the watershed. Maps and database files have been prepared prior to making the simulation runs.

Contours

A survey of Indian topographic sheets on a scale of 1:250,000 have been used. Spatial data has been created on the GIS platform using Geomedia Explorer. The polyconic projection system has been used. Two line type feature classes were defined, for digitization of contours and stream network. The contours have been digitized for the purpose of digital elevation model (DEM) generation and subsequently, watershed delineation.

Stream network

Stream network is an optional Arc View map theme which needs to be provided as a shape file. In case it is not provided by the user, the interface generates it using the DEM. In the present case, a polyline shape file of stream network has been obtained from the digitized drainage network. This shape file has been superimposed on the DEM. The stream delineation shape file has been used in the present case because it was felt that in some areas, the relief was so low that the DEM map grid was unable to accurately predict the location of the streams.

Land use data

The input data pertaining to land use/land cover can be input in the shape file as well as grid format. In the present case, the Cauvery basin land use data has been input as a shape file which was obtained after digitizing the land use maps obtained from the Cauvery authorities on a GIS (Geomedia Explorer) platform. This shape file was later converted into the grid format. The categories specified in the land cover/land use map are reclassified into the SWAT land cover/plant types. In case a United States Geological Survey (USGS) land use/land cover map is used, the interface converts it into SWAT land cover/plant types using an inbuilt table that identifies the different SWAT land cover/plant types used to model the various USGS land uses.

Since the USGS maps have not been used in the present case, there were two options available for reclassifying the input map land uses into SWAT land uses. The first option was to type in the 4 letter SWAT land cover/plant type code for each category when the land use map was loaded in the interface, while the second option was to create a look up table identifying the 4 letter SWAT code for different categories. The first option has been used for reclassification and the reclassified land use categories of the Cauvery basin have been shown in Figure 1.

Soil class data

The input data pertaining to land use/land cover can be input in the shape file as well as grid format. In the present case, the Cauvery basin soil data has been input as a shape file which was obtained after digitizing the soil maps obtained from the Cauvery authorities on a GIS (Geomedia Explorer) platform. This shape file was later converted into the grid format. Since the inbuilt AVSWAT database contains information pertaining to the US soils only, the data pertaining to the soil types present in the Cauvery basin has been entered in the user soils database. There are no differences between US and indian soil? The US classification is either well adapted or not (Neitsch et al., 2002). The "Name" specified for each soil in the user soils database is the same as its name in the input soil category map. The soil class categories of the Cauvery basin are shown in Figure 2.

Climatic data

The model requires climatic data pertaining to the precipitation, maximum and minimum temperature, solar radiation, wind speed and relative humidity. Since hydrological processes operate on a daily time step, daily values of these variables are needed to accurately predict the hydrological behaviour of the basin. The model has an inbuilt weather generator which can be used for predicting the daily values of all these variables, provided certain long term weather statistics are available. This feature of the model makes it very useful for ungauged watersheds.

However, Gholami (1999) has shown that the outputs of the

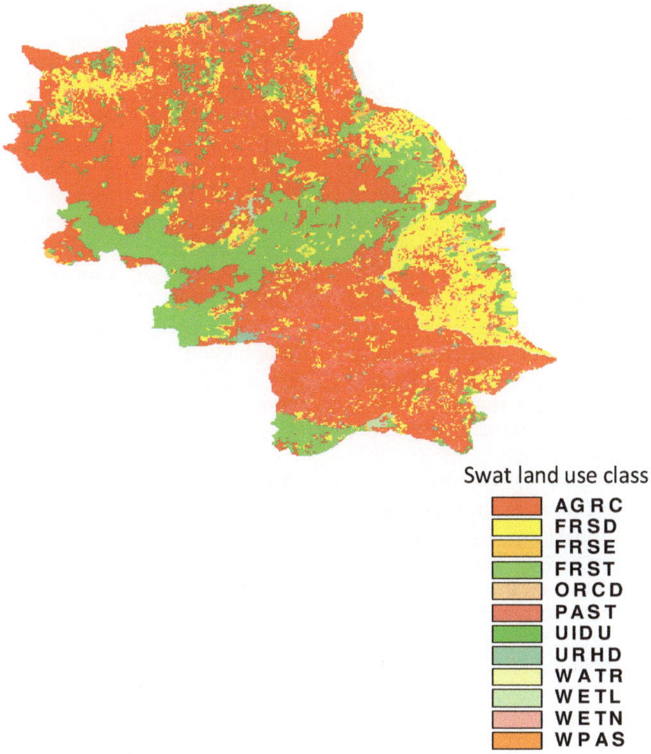

Figure 1. Reclassified land uses for the Cauvery river basin.

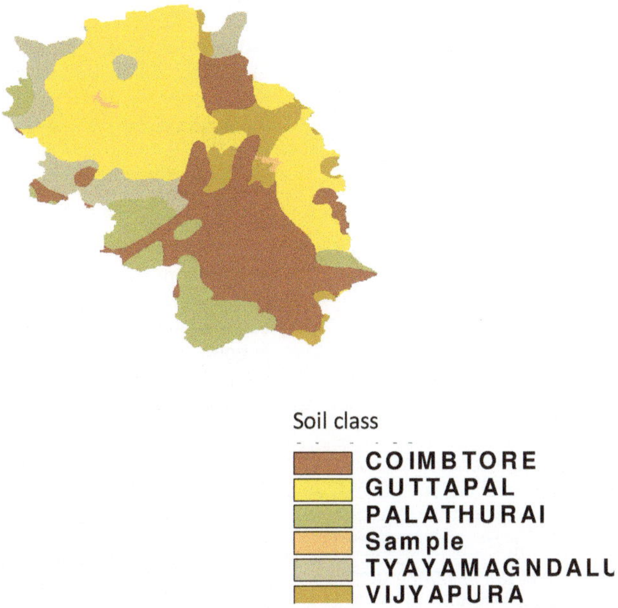

Figure 2. Reclassified soil classes for the Cauvery river basin.

model get closer to the reality if daily observed values are used along with the long term data. The long term (30 year average) data has been provided in the user weather stations database pertaining to input variables like precipitation, maximum temperature, minimum temperature, solar radiation, wind speed and relative humidity. Recently, the Indian Meteorological Department has provided the rainfall data for the whole of India considering a grid with a cell size of 1°×1° (Rajeevan et al., 2006) and the same has been used for the present study. For values of other climatic variables like maximum temperature, minimum temperature and solar radiation, daily values for 3 climatic stations has been used along with the long term averages for the same. The daily values of wind speed and relative humidity have been simulated by the inbuilt weather generator of the model.

Reservoir data

Impoundment structures modify the movement of water in the channel network by lowering the peak flow and volume of flood discharges. SWAT is able to model three types of impoundments. The first type is a small structure with one spillway. Releases occur only when the storage volume of the structure is exceeded and the excess volume is released within one day. The second type of impoundment is a small, uncontrolled reservoir with a principal and emergency spillway. Water is released at a specified rate when the volume of the reservoir exceeds the principal spillway volume. Volume exceeding the emergency spillway storage is released within one day. The third type of impoundment is a managed reservoir. Water may be released from the managed reservoir based on measured outflow or target reservoir volumes. The third type of impoundment has been used in the present study.

A total of nine reservoirs have been simulated in the present study. These are Hemavathy, Harangi, KRS, Kabini, Marconahalli, Suvarnavathy, Mettur, Bhavani and Amravathy. These are the major reservoirs in the basin and the data for these are available, that is the reason for their inclusion in the study. All these have been input as large managed reservoirs with known monthly outflow values. The reservoirs have been input on the main channel network after the watershed delineation (described subsequently) is over. Monthly outflow values for all these reservoirs have been input as separate tables containing monthly outflow data from the time the reservoir became operational.

PREPROCESSING

The following steps have been used in preprocessing:

DEM Generation

Several interpolation methods can be used for generating the unavailable data in GIS which include regression analysis, theissen polygon and inverse distance weighted methods. The method of interpolation used for the present study is the inverse distance weighted (IDW) method. The DEM is shown in Figure 3.

Watershed delineation

Four grid data matrices that are prerequisite to automatic delineation of watershed in the same sequence are:

1. DEM,
2. Flow direction,
3. Flow accumulation, and
4. Stream link.

HRU Generation

Each unique combination of land use and soil class constitutes a "Hydrologic Response Unit" or HRU. Subdividing each sub watershed into unique land use and soil class combination enables the model to reflect the differences in evapotranspiration and other hydrologic conditions for different land covers/crops and soils. Runoff is predicted separately for each HRU and routed to obtain the total runoff for the watershed. This increases the accuracy of the water yield predictions and provides a much better physical description of the water balance. In the present study, the entire basin has been subdivided into 42 sub-basins. There are two options of determining the HRU distribution. In case a single HRU is to be generated for each sub-basin, the HRU is determined by the dominant land use class and soil type within each sub-basin. However, in case multiple HRUs are required within each sub-basin, a two step procedure is followed.

Firstly, land uses are chosen. Once the land uses to be modeled are determined, the different soils for each land use are chosen. One HRU is created for each unique land use/soil class combination. In the present case, since more accuracy is desired, multiple HRUs are chosen within each sub-basin. The threshold levels of land use as well as soil types have been fixed as 5%. The threshold level is used to eliminate minor land uses/soil types in each sub-basin. In the first stage land uses that cover a percentage of the sub-basin area less than the threshold level are eliminated and the area of the remaining land uses is reapportioned, so that 100% of the land area in the sub-basin is modeled. In the second stage minor soil types within each land use area are eliminated and area of the remaining soil types is reapportioned. After fixing the threshold levels, a total of 231 HRUs were obtained.

MODEL CALIBRATION

The daily values of discharges up to Musiri (outlet) have been taken from Cauvery basin authorities for the purpose of model calibration and validation. The model has been validated over a 10 year period from 1980 to 1989 and it was observed that the performance of the model was highly satisfactory. The percentage error in annual average flows over a 10 year period have been estimated as 8.1%, if data from the Central Water Commission (CWC) is used, and 10.8% if data from the Water Resources Development Organisation (WRDO) is used. As a further check, observed and simulated monthly discharges were compared for a two year period and the results have been plotted as shown in Figure 4.

The Nash Sutcliffe coefficient was observed to be 0.934 and the value of coefficient of determination was calculated as 0.9362. This shows that the model has captured the system well and hence there is no need to calibrate the model. The results have been found to be in agreement with Gosain and Rao (2003), wherein it has been stated that "the SWAT model has been used in various Indian catchments of varied sizes and it has been observed that the model performs very well without much calibration."

SCENARIO GENERATION USING HYDROLOGICAL MODELLING

Generation of futuristic scenarios is one of the most important applications of hydrological modelling. Scenario generation helps the decision makers to analyse the potential impacts of their actions. The well known hydrological principle of "Think globally, act locally" can only be applied if sophisticated tools and techniques such as simulation modelling are available. In the present case, a variety of land use changes are simulated and the impact on the hydrological characteristics of the basin are analysed. The aim of the exercise is to generate a series of scenarios or options for the stakeholders, so as to enable them to take sound rational decisions.

The stakeholders, if they so desire, can use simulation modelling for some other sets of land use changes which are more acceptable to them. Hence, the attempt is to empower the stakeholders by providing them with the freedom of choice. Moreover, the impacts of futuristic climate changes on the hydrological regime can be effectively captured through the use of modelling techniques and the remedial actions can be taken well in time. In the present scenario, the (approximate) land uses within the Cauvery basin are:

1. Agriculture = 60%
2. Forests = 35%
3. Others = 5%

Futuristic scenarios have been generated for three categories of

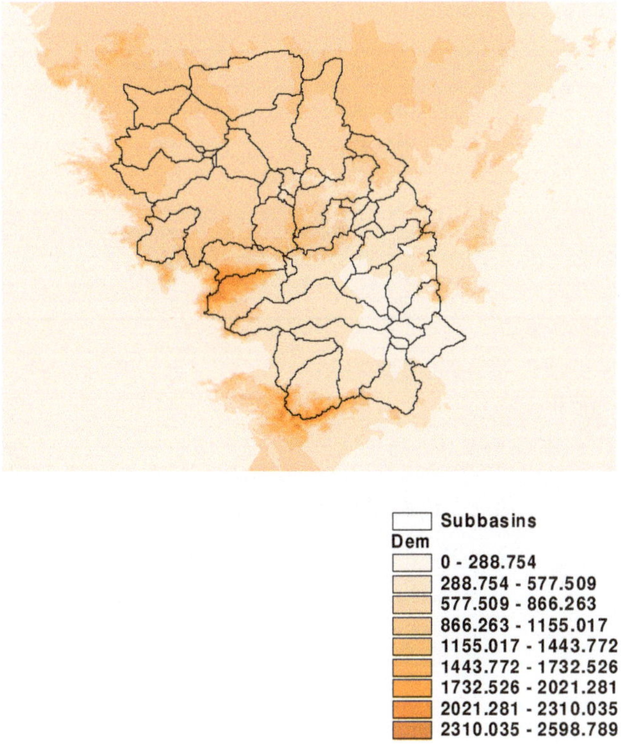

Figure 3. Layout of Cauvery basin and DEM showing elevations in metres.

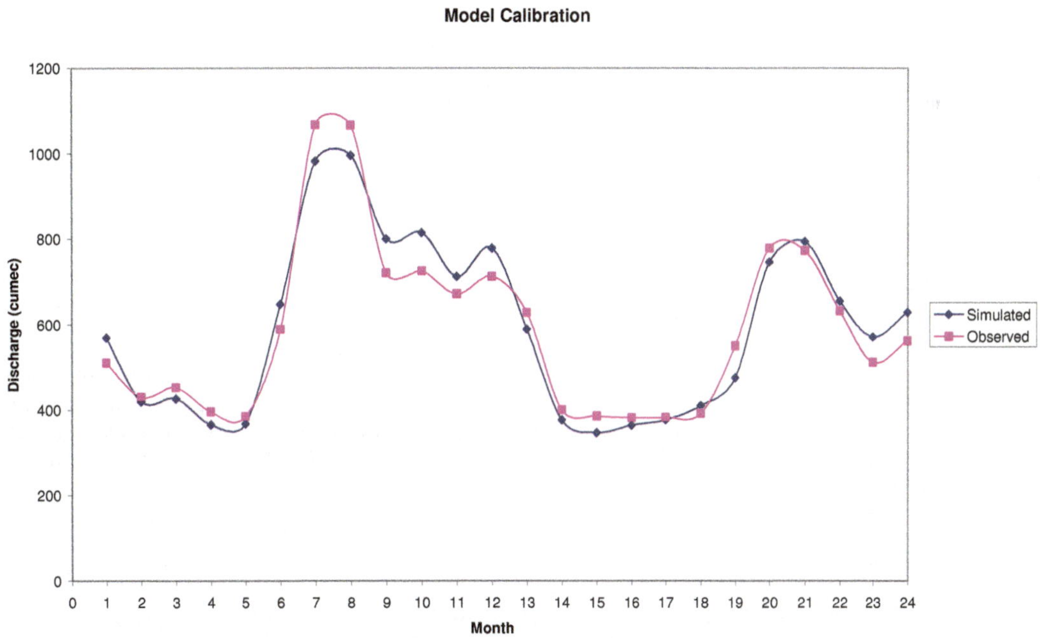

Figure 4. Comparison of average monthly flows over a 2 year period.

land use changes. Table 1 shows four categories of land uses, S_1 to S_4 in which S_4 is the present land use scenario. Hydrological modelling was carried out on GIS (Arc View) platform to analyse the aforementioned categories of land use changes. The resultant variation of water yields of the basin are shown in Table 2. Since it is often argued that water allocations should be on the basis of 75% dependable flows, the same have also been calculated and plotted. Units are in TMC (1 TMC is equal to one thousand million cubic

Table 1. List of land use scenarios.

I	Scenario	%AGRC	%FRST
1	S_1	75	20
2	S_2	70	25
3	S_3	65	30
4	S_4	60	35
5	S_5	55	40
6	S_6	50	45
7	S_7	45	50

Table 2. Variation of average flows.

Agriculture / Forest	45	50	55	60	65	70	75
20							833(S_{11})
25						814(S_{21})	
30					807(S_{31})		
35				795(S_{41})			
40			781(S_{51})				
45		778(S_{61})					
50	757(S_{71})						

Table 3. Variation of 75% dependable flows.

Agriculture / Forest	45	50	55	60	65	70	75
20							597(S_{12})
25						587(S_{22})	
30					584(S_{32})		
35				574(S_{42})			
40			563(S_{52})				
45		562(S_{62})					
50	559(S_{72})						

feet). These values have been plotted in Tables 2 and 3 and Figures 5 and 6.

RESULTS AND DISCUSSION

An analysis of Figures 5 and 6 shows clearly that as the percentage of forests decreases, the water yield increases in the basin. The loss of forest cover reduces interception of raindrops (increasing drop impact energy and soil detachment), reduces evapotranspiration, increasing the amount of water available for infiltration, soil storage, and runoff (Hibbert, 1967; Calder et al., 2009). Therefore, soil moisture capacity is reached with less rainfall, and any excess can produce surface runoff, or Hortonian flow, and increase peakflows and streamflow volumes (Bosch et al., 1982). Hibbert (1967) reported the results for thirty-nine studies of the effect of altering forest cover on water yield. Taken collectively, these studies reveal that forest reduction increases water yield. Studies have also shown that afforestation (re-establishment of forest cover) results in a decrease in water yield, proportionally to the growth rate of the stand (Jason and Matlock, 2009). Note that these findings are for short term period and indeed, precipitation in long term may be negatively affected by loss of forest cover (Hibbert, 1967).

Conclusions

GIS based hydrological modelling has been used to assess the impacts of climate change on the hydrological regime of Cauvery river basin in India. SWAT has been

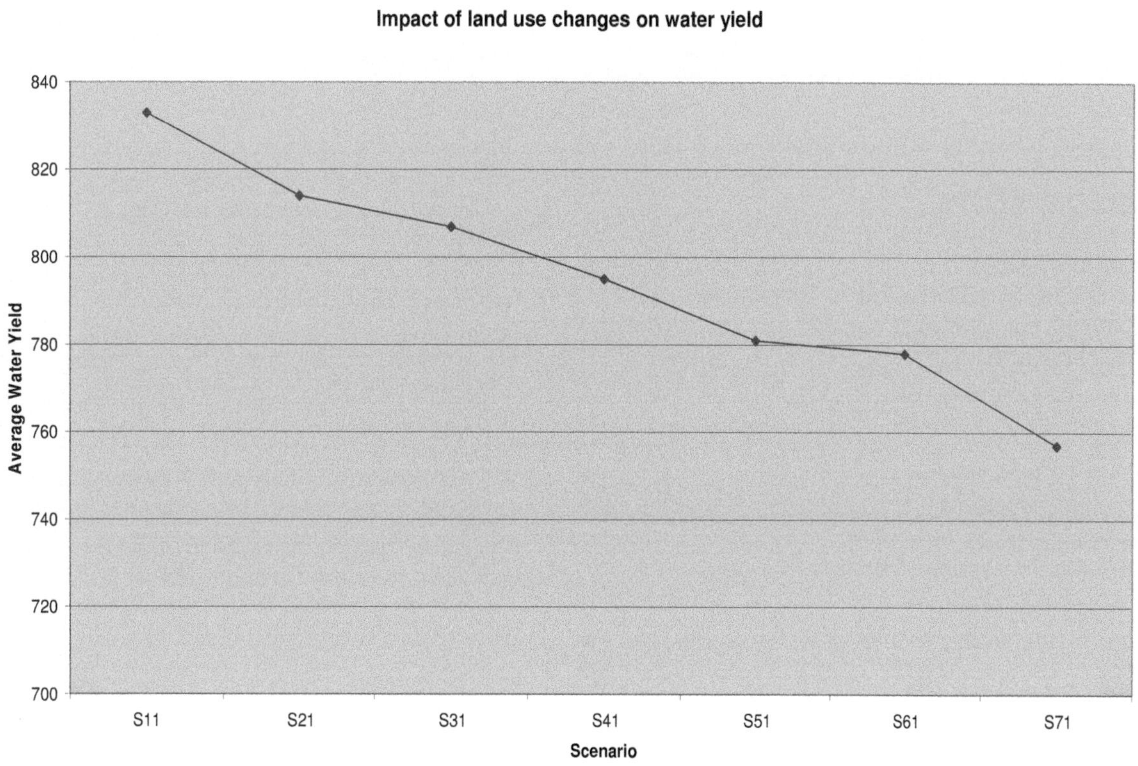

Figure 5. Variation of average water yield in Cauvery under various land use scenarios.

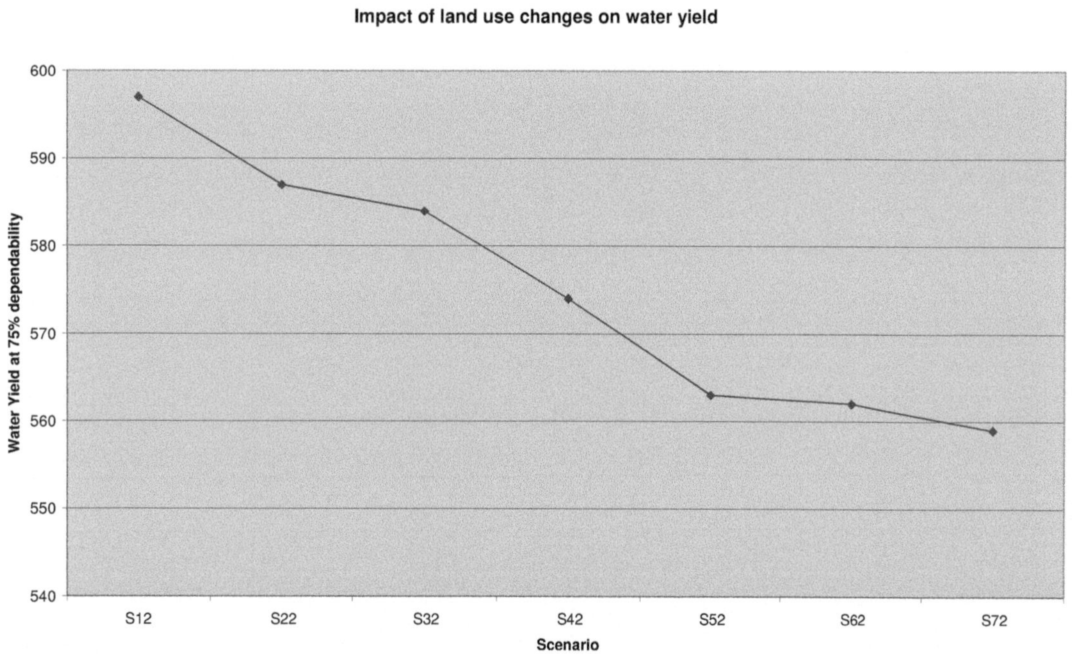

Figure 6. Variation of 75% dependable water yield in Cauvery under various land use scenarios.

utilized for the purpose of assessment of the total amount of water available in the Cauvery basin along with the prediction of the impact of changes in the land management practices on the water availability in the basin. The results indicate that contrary to popular beliefs, as the percentage of forests decreases, water yield increases in the basin. Finally, the study demonstrates that simulation modeling can play a very significant role in water resources management by generating a series of scenarios or options for the stakeholders, so as to enable them to take sound rational decisions.

ACKNOWLEDGEMENTS

The authors are grateful to Shri Manish Gupta, Scientist 'C'; Dr. R. Chitra, Scientist 'D' and Shri Murari Ratnam, Director CSMRS for their kind support in carrying out the aforementioned work.

REFERENCES

Arnold JG, Srinivasan R, Muttiah RS, Williams JR (1998). "Large Area Hydrologic Modelling and Assessment Part I: Model Development." J. Am. Water Resour. Assoc., 34(1): 73-89.

Arnold JG, Williams JR, Maidment DR (1995). "Continuous-time water and sediment-routing model for large basins." J. Hydraulic Eng., 121(2): 171-183.

Bosch JM, Hewlett JD. (1982). "A review of catchment experiments to determine the effect of vegetation change on water yield and transpiration." J. Hydrol., 55: 3-23.

Calder IRT, Nisbet, Harrison JA (2009). "An evaluation of the impacts of energy tree plantations on water resources in the United Kingdom under present and future UKCIP02 climate scenarios." Water Resour. *Res.*, 45, W00A17, doi:10.1029/2007WR006657.

CFFC (1972). "Report of the Cauvery Fact Finding Committee." Ministry of Irrigation and Power, Government of India, New Delhi.

Gholami SA (1999). "Distributed Watershed Modelling of a Mountainous Catchment" PhD dissertation, Department of Civil Engineering, IIT Delhi.

Gosain AK, Rao S (2003). "Impacts of Climate Change on Water Sector." in Shukla, P. R., Sharma, S. K., Ravindranath, N. H., Garg, A., and Bhattacharya, S edited "Climate Change and India: Vulnerability assessment and Adaptation" Universities Press, New Delhi, India.

Hibbert AR (1967). "Forest treatment effects on water yield." In: W. E. Sopper and H. W. Lull (Editors), International Symposium For Hydrology. Pergamon, Oxford, p. 813.

Jason AH, Matlock M (2009). "Forest harvest and water yield". In: Encyclopedia of Earth. Eds. Cutler J. Cleveland (Washington, D.C.: Environmental Information Coalition, National Council for Science and the Environment). [First published in the Encyclopedia of Earth October 6, 2009; Last revised Date October 6, 2009; Retrieved November 16, 2010.

Neitsch SL, Arnold JG, Kiniry JR, Williams JR, King KW (2002). "Soil and Water Assessment Tool Theoretical Documentation." Version 2000.

Rajeevan M, Bhate J, Kale JD, Lal B (2006). "High resolution daily gridded rainfall data for the Indian region: Analysis of break and active monsoon spells." Curr. Sci., 91(3): 296-306.

Impact of environment variables on pelagic fish landings: Special emphasis on Indian oil sardine off Tiruchendur coast, Gulf of Mannar

J. Selvin Pitchaikani[1,2]* and A. P. Lipton[3]

[1]Centre for Marine Science and Technology, Manonmaniam Sundaranar University, Rajakkamangalam, Tamil Nadu, India.
[2]Present address Institute of Environmental Studies and Wetland Management, DD-24, Sector-I, Salt Lake City, Kolkata, West Bengal, India.
[3]Vizhinjam Research Centre, Central Marine Fisheries Research Institute, Vizhinjam – 629 251, Kerala, India.

Results of the present study reported the correlation of various environmental variables such as dissolved oxygen (DO), chlorophyll 'a', primary productivity (PP), salinity, sea surface temperature (SST) and meteorological parameters such as air temperature (AT), air pressure (AP), relative humidity (RH), wind pattern, sunshine hours (SSH) and rainfall (RF) during January, 2008 to December, 2010 with that of the sardine fish landing along the Tiruchendur coast in Gulf of Mannar. The landings of the pelagic fishes like sardine, anchovy and Indian mackerel were high during south-west monsoon months extending from July to September in Tiruchendur coast. Sardines, which contribute 18.24% of the total fish catch, were recorded throughout the study period with a minimum occurrence during January and maximum in August. Large sized fishes were caught during October. It was found that, moderate temperature, high chlorophyll 'a' concentration and phytoplankton abundance are primary factors that enhance sardine fishery. Weather changes, influences fisheries through the modification of habitat characteristics, affecting the organisms to the extent that the physical, chemical and biological conditions that influence their productivity, development, nourishment and reproduction distribution are altered. Subtle changes in key environmental variables can drastically modify the abundance, distribution, and availability of fish populations.

Key words: Environmental parameters, meteorological variability, pelagic fish, salinity, Tiruchendur, Chlorophyll 'a', primary productivity, Gulf of Mannar.

INTRODUCTION

The clupeids comprise a major group among the pelagic resource. Among the clupeids, the Indian oil sardine (*Sardinella longiceps* Val.) (Valenciennes, 1847) is the most predominant species that forms the mainstay of the pelagic fishes of India and its stock is probably the widespread in the Indo-Pacific region. Presently, the west coast of India indicates a declining trend compared to an ascending pattern along the east coast of India. The Indian oil sardine (*S. longiceps* Val.), supports a neritic pelagic fishery contributing 2 to 33% of the annual marine fish production in India. However, due to the seasonal changes and changes in the meteorological variability, sardine-landing pattern has altered. Changes in habitat quality inevitably influence the spatial patterns of abundance of pelagic fish. The distribution of pelagic fish

*Corresponding author. E-mail: selvinocean@gmail.com.

may be related to several interacting environmental factors, which also correlate with each other (Maravelias and Reid, 1997; Maravelias et al., 2000).

Pelagic fishery of the Gulf of Mannar

Several species of fish inhabiting the pelagic zone contribute to the fishery in the Gulf of Mannar (Raju, 2000). Sardines, anchovies, mackerel, carangids, tunas and seer fishes are important pelagic fishery caught in the Gulf of Mannar. Fishing folk of these regions are depending on this artisanal fishery (small scale) for their livelihood. Because artisanal fishers depend on the health of marine ecosystems for their subsistence, the added threat of climate change and sea level rise will not only threaten biodiversity, but human well being as well. Sardines comprising several species of *Sardinella* collectively contribute to a major share of the marine pelagic fisheries resource of India (Rohit and Bhat, 2003). Though several species of sardines supported the pelagic fishery in Tiruchendur coast, the oil sardine, *S. longiceps* alone comprised more than 90% of the catch (Total pelagic fish catch).

Status of Pelagic fishery in other regions

In the Sea of Oman, the sardine population is one of the major consumers of mesozooplankton. Sardine contributes 49% of the total landings in Muscat region and the annual landing of sardines in Oman ranges between 7000 and 18000 metric tonnes (Piontkovski et al., 2012). Due to the rising atmospheric temperature, sea surface temperature (SST), variability of the kinetic energy of eddies, frequency of fish kills and harmful algal bloom incidents, 40% decline in sardine landings were reported by Piontkovski et al. (2012).

The sardine *(Sardinops sugax)* fishery of the Gulf of California is among Mexico's most important fisheries, accounting for the largest catch and providing many productive jobs. During the early 1990s, this fishery collapsed to less than 3% of the production maximum. Surprisingly, after two years of very low catch the fishery recovered quickly. Comparative studies of fish habitat climatology have provided key clues to understanding mechanisms linking physical environment to biological populations.

Fuel consumption per fishing fleet, fuel price and market price are also determining fish catch in Norwegian fisheries rather than environmental parameters (Schau et al., 2009).

Peltonen et al. (2007) studied the pelagic fish abundance in relation to regional environmental variation in the Gulf of Finland, Northern Baltic Sea and documented the fish assemblage consisted of small sprat and herring, which together constituted about 99% of the yield. Sprat made up 83% of the yield, and it was numerically dominant in 19 of the 21 trawl catches. Trawl sampling revealed small differences in fish size distributions in different parts of the basin, but there were large local variations in fish abundance, as well as in the explanatory variables in the region studied by Peltonen et al. (2007).

The results indicate that the pelagic fish preferred regions with deepwater temperature of about 5°C to areas with warmer (especially) and with colder deep water. The avoidance of regions with deepwater temperatures,<4°C is in accordance with the findings of Parmanne et al. (1994), who stated that Baltic Sea sprat prefer temperatures <4°C. Correspondingly, the pelagic fish in the Gulf of Finland in September, 2002, avoided cold temperatures <3°C even though oxygen concentration, for example, was not a limiting factor (Peltonen et al., 2004). Moreover, North Sea herring avoided areas of cold bottom water in summer (Maravelias et al., 2000).

The most important pelagic fish stocks inhabiting Peru's relatively narrow continental shelf are anchoveta, chub mackerel (*Trachurus murphyi*) and Chilean jack mackerel (*Scomber japonicus*), which together represented 95% of the catch volume in 2006. Although historically the standing stocks have been subjected to sharp variations, apparently due more to environmental conditions than to fishing pressure, it is clear that these stocks are potentially large and generally able to recover from periodic declines. Of these, the anchoveta is the main species, comprising 92.5% stocks of the total catch, according to 2005 records.

Although fishing mortality has been a major force driving, the population collapse in Guaymas Basin. Holmgren-Urba and Baumgartner (1993) reported that, analysis of sardine scale deposits in anaerobic sediments off Guaymas Basin have shown large population fluctuations even in the absence of fishing activity because of the natural variability. It is likely that management of the species depends on the ability to differentiate between natural variability and fishing mortality.

Population of fish and other aquatic organisms are reported to be greatly influenced by physical, chemical, biological and meteorological parameters (Biswas et al., 2009). The global climatic changes are expected to affect marine fisheries productivity due to changes in water temperature, ocean currents and other oceanic conditions. This also can be affected by in changes food web structure and species distribution, influenced by the changes in oceanic condition. Among the marine fisheries, small pelagic fishes have undergone considerable variation in both their distribution and abundance over time (Kawasaki, 1984; Lluch-Belda et al., 1989; Navarez-Martinez et al., 2001); influenced by seasonal, inter-annual and decadal climate variations (Lluch-Belda et al., 1989; Bakun and Board, 2003;

de Young et al., 2004). Meteorological fluctuations are expected to affect marine fisheries productivity because of changes in water temperature, ocean currents and other ocean conditions (Pauly et al., 2002). The expected impacts of meteorological change will be seen first on the distribution and abundance of pelagic fishes (Hobday et al., 2006; Lanz et al., 2008). Among the meteorological variability temperature is the key factor, it will alter the SST. Water temperature is a key variable with regard to the distribution, abundance and metabolic activity of fishes. Since, fishes are poikilothermous, changing in water temperature will affect the physiological activity and it leads to decline of biodiversity of fishes. Marine fisheries productivity may be affected by changes in ocean conditions resulting from climate change including changes in food web structure and species distribution. Oceanic conditions were found to affect profoundly the distribution, year-chats production and yields of sardines. Among the pelagic fishery, oil sardine and other or lesser sardine alone contribute about 15% of the total marine fish production in India (Pillai, 1990). In this background, in order to understand the meteorological variability such as air temperature (AT), air pressure (AP), wind speed and wind direction, relative humidity (RH), rainfall (RF) and sunshine hours (SSH) and its effect on sardine fishery was studied for a period of 3 years from January, 2008 to December, 2010. Meteorological parameters and water quality parameters (variability) were correlated with sardine fish landing. In the present study, it is observed that fish landing (oil sardine and other sardine) were highest during South West monsoon seasons.

The metabolism activities and physiological behavior of fish is directly related to the SST and the fluctuation of SST is depending upon the AT, cloud cover and SSH. Changes in AT, cloud cover and SSH will lead to increase or decrease of SST and light penetration. Metabolic activity in fish increases by 10% for every 1°C rise in temperature of aquatic environment that is, fish need 10% of more oxygen for 1°C rise in temperature (Rahman, 1992). In all aquatic environments, phytoplankton acts as primary producer and their growth is an important process, which depends on the concentration of nutrients in water and in external factors such as temperature and light illuminations (Scavia and Park, 1976; Groden, 1977; Desormeau, 1978; Park et al., 1979; Jorgensen, 1980; Biswas et al., 2009; Ramos et al., 2011). The pelagic fishes are not particularly diverse; there are approximately 260 pelagic species, out of 12,000 marine species worldwide. ATs are expected to increase ocean warming, most significantly in the upper layer 500 to 800 m (Bernal, 1993). It causes a rise in sea level, higher SST and a weakened thermocline, which is associated with reduced primary productivity (PP) (Miller and Fluharty, 1992). Extremes in environmental factors such as elevated water temperature, low dissolved oxygen (DO) or salinity and pH can have deleterious effects on fishes (Moyle and Ceeh, 2004).

In this paper, attempt were made to correlate the meteorological parameters such as wind pattern, AP, AT, RH, SSH and precipitation with oil sardine landing at Tiruchendur.

MATERIALS AND METHODS

Tiruchendur is one of the main fish-landing centers town (Latitude: 8.29.19.1 and Longitude: 7.26.62) in Tuticorin District of Tamil Nadu, India. It is located in between Thoothukudi and Kanniyakumari and situated in the bank of Gulf of Mannar, Southeast Coast of India. Traditional sardine fishing grounds (SFGs) were chosen for the present study.

Here, we analyse temporal or seasonal patterns in pelagic fish landings in relation to regional environmental variation in the Gulf of Mannar. We focus mainly on meteorological variation and environmental variability, which can provide new insights into the relationships between sardine fish catch and the environment characteristics by decomposing the variation in response variable(s) into independent components that reflect the relative importance of the groups of predictors and their joint effects.

Meteorological parameters of AT, AP, RH, wind speed, wind direction and SSH were recorded from January, 2008 to December, 2010 in the SFG area. AT and AP were observed by standard mercury thermometer and handheld Aneroid barometer, respectively. RH was observed by whirling hygrometer, wind speed and direction were recorded by using Anemometer and wind vane, respectively. A small meteorological observatory was set up in the fishing boat especially for this study. Day light duration (SSH) and RF data were collected from the Meteorological Observatory of Kanchi School, Tiruchendur. Data on oil sardine and other sardine landings at Amali Nagar center in Tiruchendur in different seasons were collected on every day and pooled into season wise and SST was measured by standard mercury thermometer in the SFG (Traditional fishing ground) off Tiruchendur itself (Figure 1). Salinity was measured by Salinometer (Model E-2), and Chlorophyll "a" was estimated by spectrophotometry following the method of Parsons et al. (1984). Gross PP (GPP) was measured by adopting light and dark bottle method (Strickland and Parsons, 1972).

A simple spearman's correlation was done by using past statistical package. Meteorological parameters and oceanographic variables were correlated with sardine fish landing.

Fish landing data of oil sardine (*S. longiceps*) and other sardines were collected at fish auction center on weekly basis for the period of 3 years from January, 2008 to December, 2010.

RESULTS AND DISCUSSION

Atmospheric temperature at the study area was given (Figure 2) and it varied from 24.0 to 30.5°C in the 3 years. Maximum temperature at the study was uniformly recorded in the month of June during the study period. However, minimum temperatures were consistently recorded only in January, 2009 and 2010, and in 2008, minimum temperature was recorded in the month of December. Northeast monsoon and RF effect causes low temperature and direct solar heating and hot summer effect causes maximum temperature during the study period. Too little is known to even speculate on changes in physiology or morphology of the pelagic fishery that might be driven by elevated atmospheric temperature

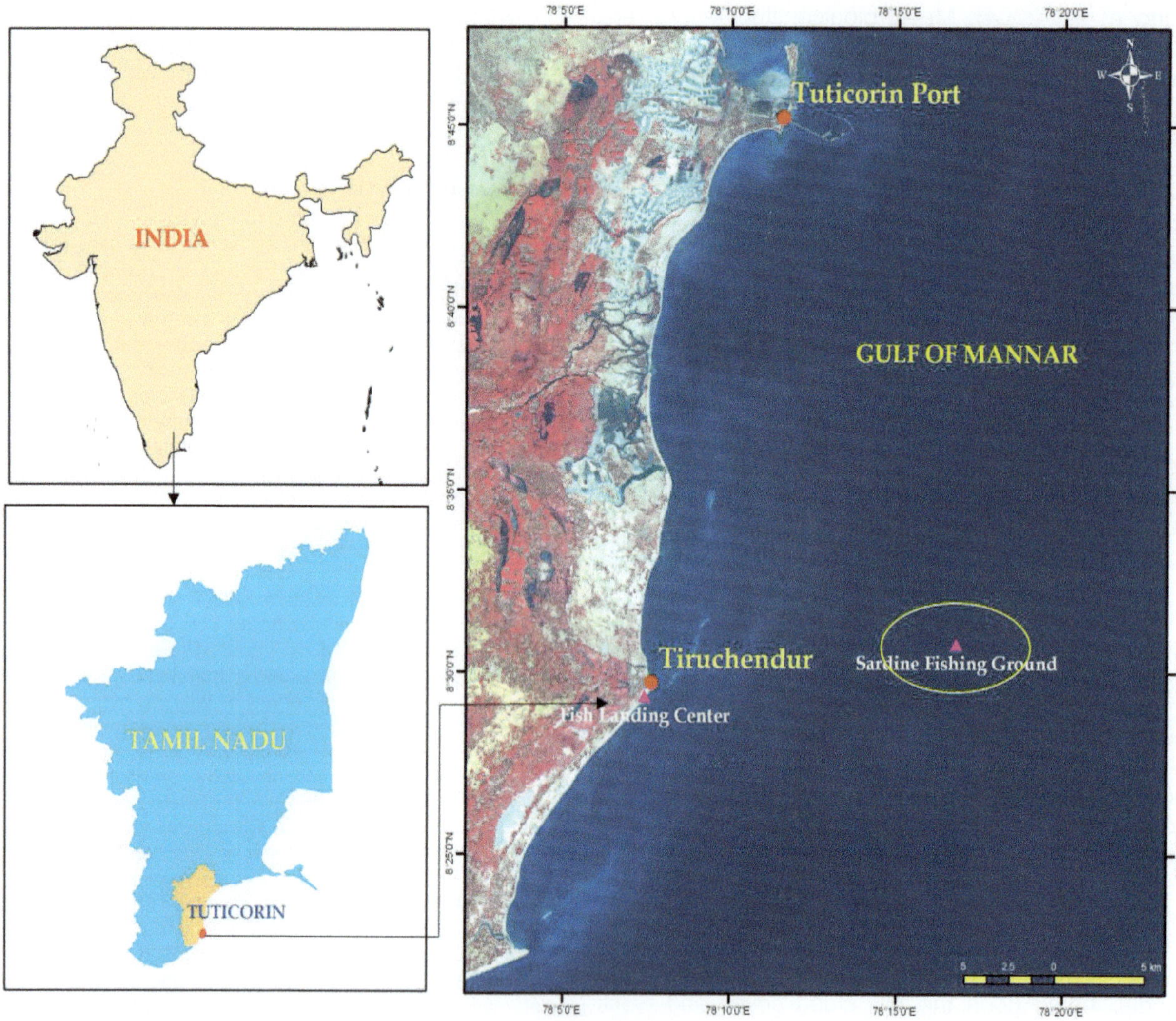

Figure 1. Study area.

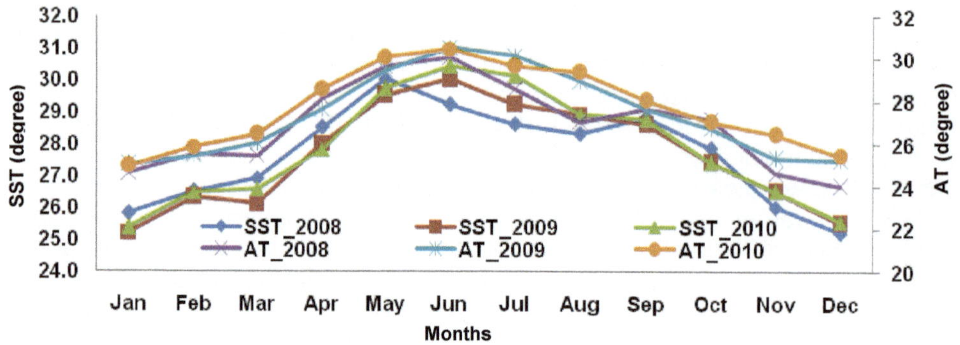

Figure 2. Monthly variation of SST and AT.

(Hobday et al., 2006). AT is directly influencing the SST of the oceanic water body and changes in the ocean temperature especially in the surface layers will have an impact on the distribution of pelagic species. In the present study, temperature is gradually rising from January to June and attaining peak value at June month,

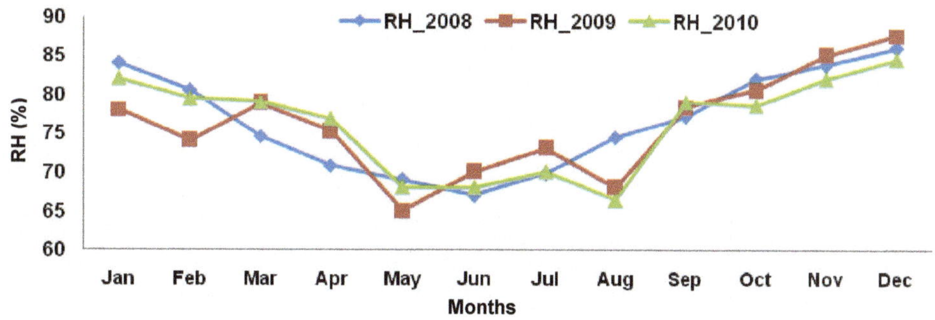

Figure 3. Monthly variation of RH.

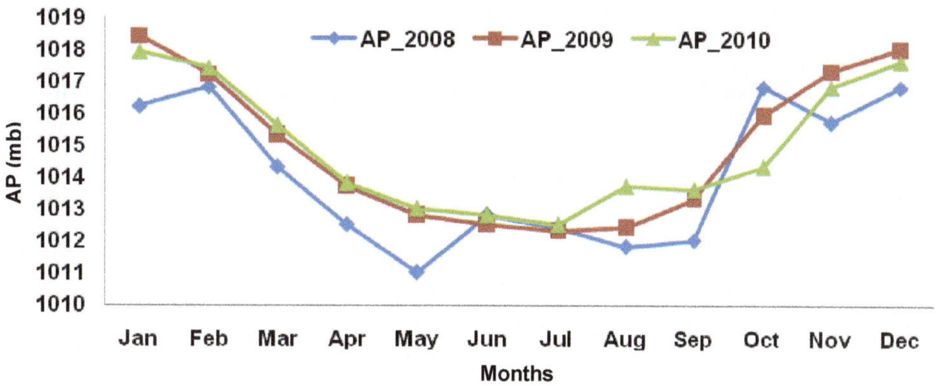

Figure 4. Monthly variation of AP.

thereafter decreasing till December. The same trend has been recorded throughout in the study.

RH was varied from 65 to 87.6% (Figure 3) during the study period and maximum and minimum were recorded in December and May, respectively during 2009 due to the summer effect. Maximum humidity was uniformly observed in the month of December in the 3 years due to the northeast monsoon effect.

Changes in the amount or timing of RF and the associated river runoff affect the salinity of the coastal waters; in the present study, period annual RF was varied from 681 to 1119.9 mm (Table 7). Maximum was observed in 2008 and minimum was observed in 2010. During the study period, maximum RF (786.2 mm) was recorded during monsoon season in 2009 and minimum (16 mm) was recorded during post monsoon in 2010; however, no RF was recorded during pre-monsoon in 2008. APs were varied from 1011 to 1018.4 mb (milli bar) during the study (Figure 4). Minimum value was recorded during summer season throughout the study period owing to the summer effect and high temperature. Maximum value was recorded during northeast monsoon due to the RF and cooling effect was found that pressure decreases during northeast monsoon (Selvin and Selvaraj, 2011). During the present study, AP shows the decreasing trend during summer season and increasing trend persists during northeast monsoon. In general, low pressure was recorded during summer months due to warm temperature, high pressure is always recorded during December due to winter effect, and it was found that a close relationship exists between AT and AP.

Wind indirectly influences pelagic species through mixing of the surface waters (Cury and Roy, 1989). Pelagic and coastal ecosystems are comparatively productive zone depending upon the availability of nutrients, sunlight and stability for phytoplankton production (Cury and Roy, 1989; Bakun and Weeks, 2004). Wind forcing could cause the upwelling in the coastal zone and enhancing the productivity of the coastal and pelagic fishery. In the present observation, wind speed was recorded to range from 3.2 to 15.8 km/h (Figure 5). High wind speed was recorded during June and July months owing to the southwest monsoon effect and during the southwest monsoon period, wind was flowing from southwest with high velocity and low speed could be recorded during March and April months. Northeast wind flow was observed during monsoon and post monsoon (Figure 6). During summer of June and July months, wind flow pattern changed and wind flow coming from Southwest direction due to the Southwest

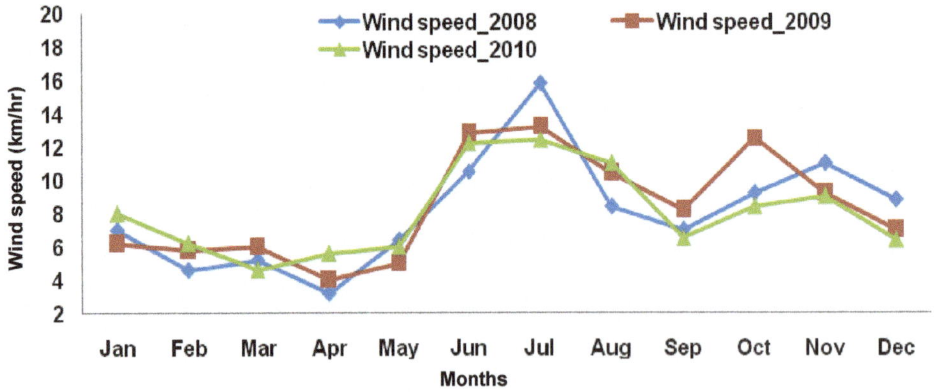

Figure 5. Monthly variation of wind speed.

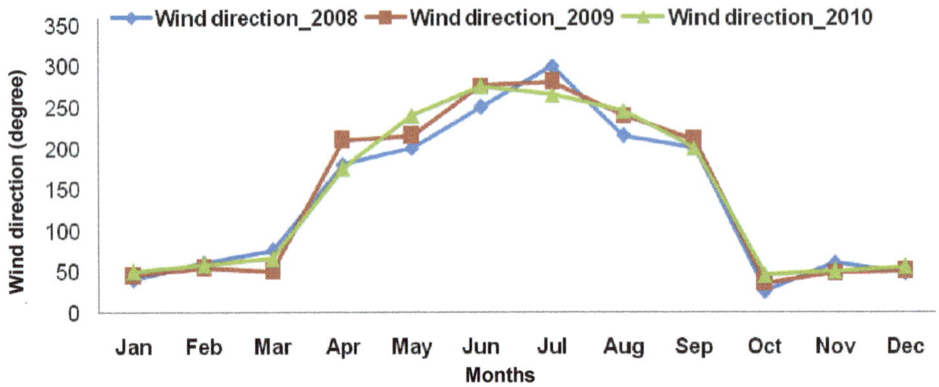

Figure 6. Monthly variation of wind direction.

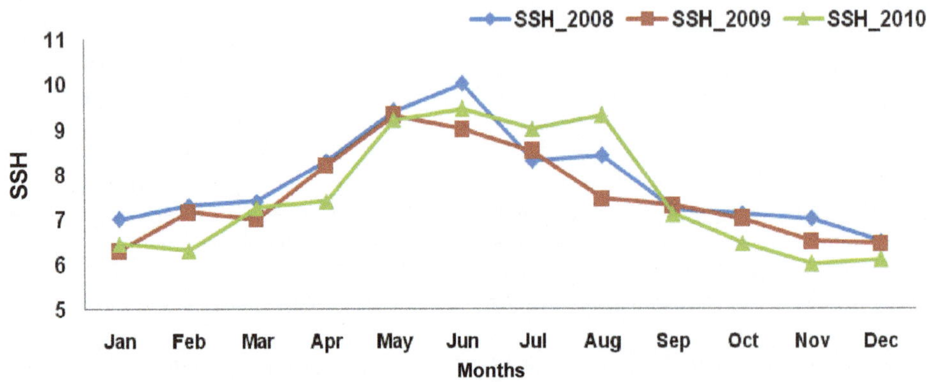

Figure 7. Monthly variation of SSH.

monsoon effect.

SSH could directly influence the salinity, SST and phytoplankton productivity. The intensity of the solar radiation and SSH are determining the euphotic zone of the oceanic ecosystem. SSH was recorded between 6 and 10 h/day (Figure 7) during the study; minimum was recorded in the month of November, 2010 and maximum recorded in June, 2008. SSH or day length is an important factor, which could affect the SST and sea surface salinity. In the present study, it was found that the annual SSH has been decreased from 2008 to 2010; it could be due to the stronger winds (Pant, 2003). Though

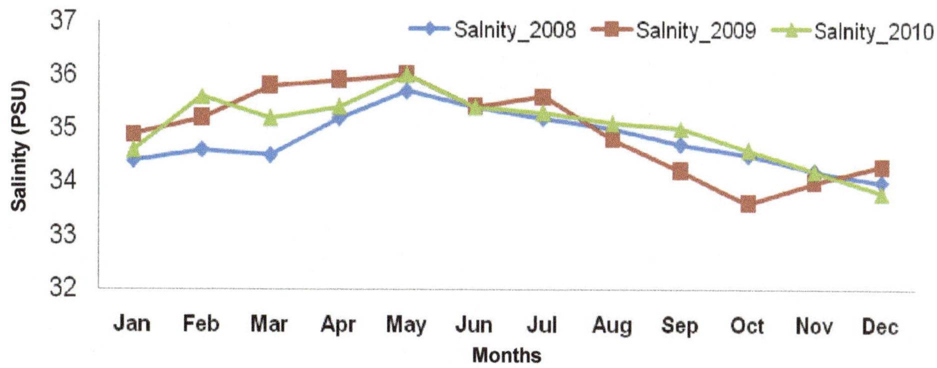

Figure 8. Monthly variation of salinity.

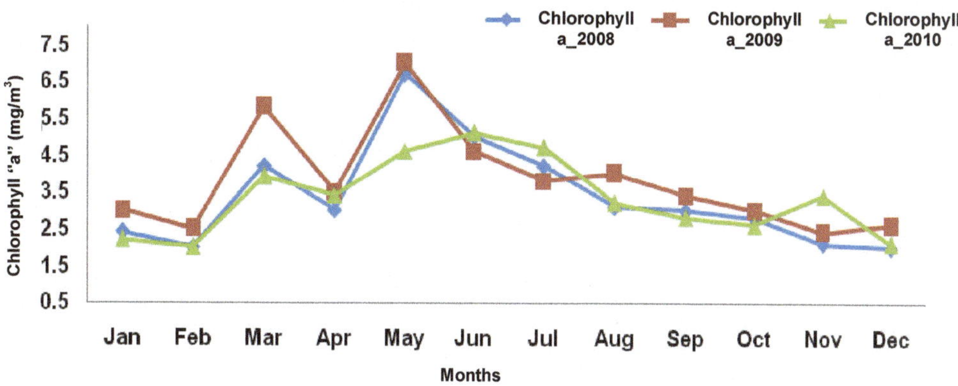

Figure 9. Monthly variation of Chlorphyll "a".

these weather parameters have been influencing the pelagic fishery, other factors like export price of the fish, fuel price, market value of the fishmeal and fish oil are playing important role in the oil sardine fishery. Therefore, these factors also influence the amount of fish caught in this fishery. The oil sardine fishery of the Gulf of Mannar also show declining trend as like other Coast and Gulf regions of the World (like Sea of Oman, Gulf of California and Finland). SST values rises from 25.2 to 30.4°C (Figure 2), maximum was recorded in the month of June, 2010 and minimum was recorded during December, 2008 and January, 2009. SST was greatly influenced by the atmospheric temperature (Asha and Diwakar, 2007) and it is one of the important physical factors, which affects the chemical and biological reactions in water. It will directly affect the pelagic fishery than other layers of the marine fishery and regulates the rate of photosynthesis in aquatic ecosystem. In the present study, it has been observed that high temperature is noticed in the months of May and June due to the longer SSH, bright sunshine and hot air. Low temperature in the months of October to December was due to cloudy sky northeast monsoon effect, resulting RF brought down in the temperature to the minimum range.

Salinity was varied from 33.6 to 36 PSU (Figure 8) during the observation and it has been observed that minimum value was recorded from October to December due to the heavy RF effect and salinity value were gradually increased from January to May and attained high value in the month of May due to the summer effect resulting high evaporation. After that, salinity value decreased up to December raised and reaches high value recorded during May month which might be due to the bright sunshine and hot air. It showed that salinity was negatively correlated with RF. The salinity fluctuated unimodally associated with northeast monsoon in Tuticorin waters. Similar trend in salinity values were also recorded from various parts in Gulf of Mannar (Jayaraman, 1954; Bapat, 1955; Asha and Diwakar, 2007).

Chlorophyll "a" varied between 2 and 7 mg.m^3 (Figure 9). Chlorophyll "a" showed distinct seasonal variation and close direct relationship with the primary production (Gopinathan et al., 1994). Maximum chlorophyll "a" value was recorded in the month of May, 2009 and minimum was recorded in February, 2008 and 2010. Chlorophyll "a" was positively correlated with SST, salinity and PP and negatively correlated with oil sardine. From this

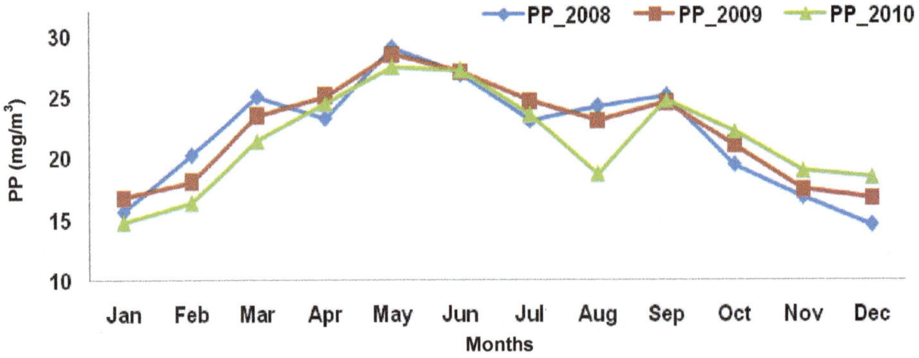

Figure 10. Monhtly variation of PP.

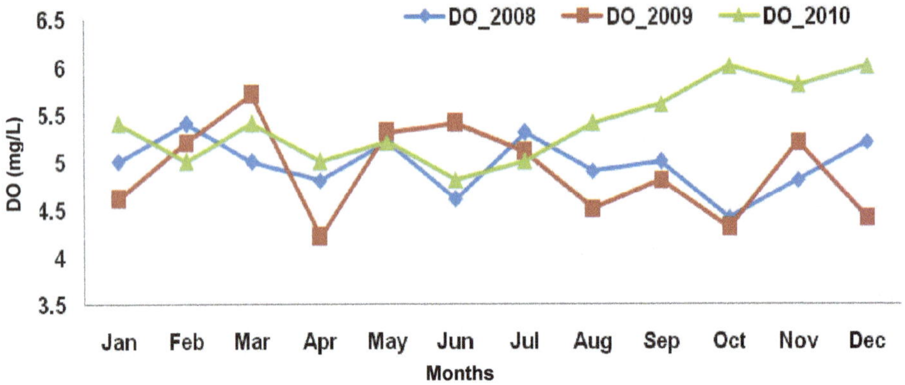

Figure 11. Monthly variation of DO.

study, it is understood that high Chlorophyll "a" was always recorded in summer (in the month of May). Vinayachandran et al. (2004) reported that, in the month of May, there is high Chlorophyll "a" along the Indian coast as well as along the southern coast of Sri Lanka. The Chlorophyll "a" value corresponded to the PP during the observation. The Chlorophyll "a" bloom is fully developed during June to July and there is high Chlorophyll "a" in the Gulf of Mannar, Palk Bay and along the southern coast of Sri Lanka (Vinayachandran et al., 2004), PP in the fishing ground was varied from 14.5 to 29.0 mh/m^3 (Figure 10). The peak value was recorded during summer and low value recorded in the months of December and January. Although photosynthesis is a key component of the global carbon cycle, its spatial and temporal variability is poorly constrained observationally. DO was recorded from 4.2 to 6.0 mg/L (Figure 11). Minimum DO was recorded in the months of April due to the summer effect and maximum was recorded in the months of October and December, 2010. The higher values of DO might be due to the cumulative effect of higher wind velocity coupled with heavy RF and the resultant fresh water mixing (Das et al., 1997; Mitra et al., 1990). The rate of gross primary production shows seasonal variation during the study period. Two distinct peak periods have been noticed, high value of production was observed in the month of May, 2008 and minimum was observed in December, 2008. It was observed that reduced light intensity due to the clouds formation during the study area has affected the PP during December and January in all the 3 years.

During this study period, maximum (total) sardine fish catch (290.87 MT) was recorded in 2008 and minimum (254.05 MT) was recorded in 2010 (Figure 12). Among the total sardine fish catch, oil sardine alone contribute about 200.32 metric tons and it was nearly 18.31% of the total fish catch during the study period. Oil sardine fish catch was high during the southwest monsoon season (mostly in the month of August). High Chlorophyll "a" contents recorded during observation has resulted in good amount of PP attributing to more fish catch. During southwest monsoon, nutrient rich Arabian Sea water brought to Gulf of Mannar by southwest monsoon current caused southwest wind flow. Owing to this process, PP was more during southwest monsoon particularly in the month of August. The probable life span of Indian oil sardine based on the scale studies is 2½ years (Chidambaram, 1950), but difficult to determine because the annual rings in their scales may be formed for several reasons than yearly changes in their environment.

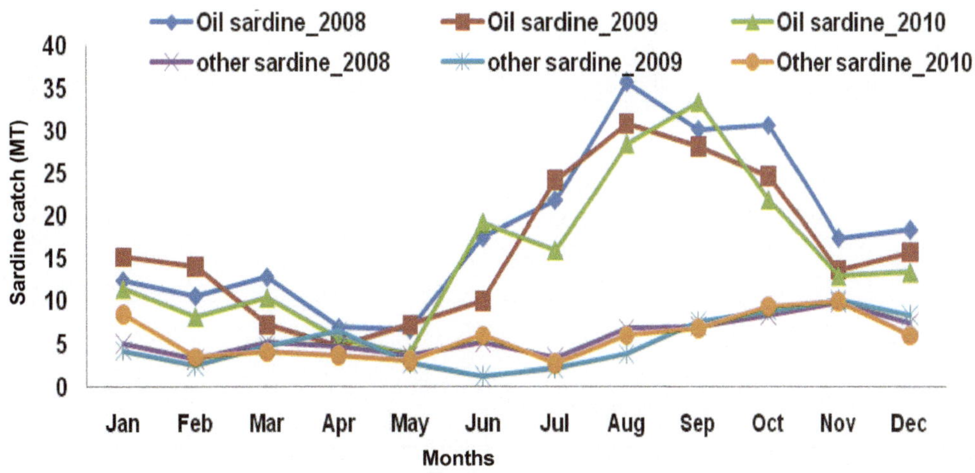

Figure 12. Monthly variation of Oil sardine and other sardine catch.

Table 1. Statistical analyses for weather parameter _2008.

2008	Oil sardine_2008	Other sardine_2008	AT_2008	RH_08	Rainfall_08	Sunshine_2008
Oil sardine_2008	1					
Other_sardine_2008	0.559389318	1				
AT_2008	-0.098398987	-0.459399682	1			
RH_08	0.224349205	0.684617239	-0.96118311	1		
Rainfall_08	-0.132362514	0.643554752	-0.82798808	0.893765099	1	
Sunshine_2008	-0.344211726	-0.605144888	0.967197139	-0.97289210	-0.76964252	1

Table 2. Statistical analyses for weather parameters_2009.

2009	Oil sardine_2009	other sardine_2009	AT_2009	RH_09	Rainfall_09	Sunshine_2009
Oil sardine_2009	1					
Other sardine_2009	0.297418259	1				
AT_2009	0.113677497	-0.496662571	1			
RH_09	0.192629261	0.893349488	-0.82934842	1		
Rainfall_09	-0.040534144	0.93083271	-0.42829032	0.789781192	1	
Sunshine_2009	-0.294754002	-0.606609388	0.915808193	-0.88075589	-0.4055338	1

Statistical analyses

SST was positively and significantly correlated with AT and salinity. Salinity was negatively correlated with fish catch (Oil and other sardine), DO, RF and had positive correlation Chlorophyll "a" and PP (significantly) (Tables 1, 2, and 3). SST had negative correlation with other sardine, DO and positive (insignificant) correlation with oil sardine catch. However, SST had positive and highly significant correlation with salinity, Chlorophyll "a" and PP. Oil sardine has been negatively (insignificant) correlated with RF and other sardine positively correlated with RF.

Fish catch and salinity showed negative correlation. Chlorophyll "a" and PP also showed positive correlation. Atmospheric temperature was negatively correlated with oil sardine and other sardine catch. RH had positive correlation with oil sardine (insignificant) and other sardine (significant). SSH was negatively correlated with fish catches, RH, RF and positive correlation with atmospheric temperature (highly significant). Wind speed was positively correlated with oil sardine and other sardine. Wind direction was positively correlated with oil sardine, Chlorophyll "a", PP, AT, SST, salinity, SSH and negatively correlated with other sardine, AP, RH and RF.

Table 3. Statistical analyses for weather parameters_2010.

2010	Oil sardine_2010	Other sardine_2010	AT_2010	RH_10	Rainfall_10	Sunshine_2010
Oil sardine_2010	1					
Other sardine_2010	0.16943423	1				
AT_2010	0.265871179	-0.62013225	1			
RH_10	-0.259508058	0.772538124	-0.96974815	1		
Rainfall_10	0.079008073	0.95202635	-0.41917228	0.622457829	1	
Sunshine_2010	0.238015242	-0.79872015	0.959667694	-0.99909362	-0.6544085	1

Table 4. Statistical analyses for oceanographic parameters_2008.

2008	Oil sardine_2008	other sardine_2008	DO_2008	SST_2008	Salnity_2008	Chlorophyll a_2008	PP_2008
Oil sardine_2008	1						
Other_sardine_2008	0.559389318	1					
DO_2008	0.043680929	-0.617581943	1				
SST_2008	0.008544102	-0.463567998	-0.082611656	1			
Salnity_2008	-0.247388998	-0.646940341	-0.008826775	0.963252063	1		
Chlorophyll a_2008	-0.416290136	-0.655581606	-0.09819944	0.905630276	0.980370012	1	
PP_2008	-0.181239324	-0.722279044	0.17014525	0.947422433	0.982278393	0.937121493	1

Table 5. Statistical analyses for oceanographic parameters_2009.

2009	Oil sardine_2009	Other sardine_2009	DO_2009	SST_2009	Salnity_2009	Chlorophyll a_2009	PP_2009
Oil sardine_2009	1						
Other_sardine_2009	0.297418259	1					
DO_2009	-0.528888072	-0.842078501	1				
SST_2009	0.13407414	-0.416576943	-0.121837485	1			
Salnity_2009	-0.567207563	-0.94309518	0.81695363	0.435675558	1		
Chlorophyll a_2009	-0.536720312	-0.846452098	0.599765821	0.654710641	0.950881293	1	
PP_2009	-0.147629869	-0.641035235	0.179032947	0.944281793	0.700428956	0.86699938	1

Table 6. Statistical analyses for oceanographic parameters_2010.

	Oil sardine_2010	Other sardine_2010	DO_2010	SST_2010	Salnity_2010	Chlorophyll a_2010	PP_2010
Oil sardine_2010	1						
Other_sardine_2010	0.16943423	1					
DO_2010	0.301388888	0.989944279	1				
SST_2010	0.372202587	-0.624169346	-0.5767589	1			
Salnity_2010	-0.250422816	-0.992601148	-0.9974845	0.633107992	1		
Chlorophyll a_2010	-0.032502423	-0.712032436	-0.7207545	0.914397536	0.760726144	1	
PP_2010	0.009271578	-0.506164645	-0.5224679	0.887669456	0.569962976	0.965888336	1

AP was positively correlated with RH and RF (Tables 4, 5, and 6). Generally, in Tiruchendur landing centre, oil sardine fish catch would be more during June and July. However, during this study period, more amount of sardine fish catch was recorded in the month of August. It could be due to the variability of meteorological characteristics implying climatic changes in coastal ecosystem. Out of the total pelagic fish catch, oil sardine contributed to 75% of fish catch. Low fish catch was recorded during April. Chidambaram (1950) found that

Table 7. Annual rainfall.

	Rainfall (mm)		
Season	**Year_2008**	**Year_2009**	**Year_2010**
Post monsoon	388.4	39.3	16
Summer	104	198	51
Pre- monsoon	0	18	45
Monsoon	627.5	786.2	569
Total	1119.9	1041.5	681

temperature of 26 to 28°C was favorable for inshore migration of oil sardine particularly juveniles and he reported that temperature over 29°C during March to May causes the offshore disappearance of the adult fish. This concept could be confirmed by the present observations.

Fishing season

On the southwest coast, the oil sardine fishery commences soon after the outbreak of monsoon in June and continues until March to April. Along the Kerala coast, catches are high throughout the year except during March to May, while in the Karnataka - Goa belt, the season starts in September/October with peak fishing during October to January period. On the southeast coast, the fishing season is from April to December with peak catches during April to June on the Tamil Nadu coast and July to October along the Andhra coast.

Conclusion

This study is the first to report on *S. longiceps* landings with respect to meteorological changes in Gulf of Mannar. Since no sufficient data on sardine fishery is available in this region, we used only 3 years data collected during the present study. Climate changes are expected to affect the marine fisheries productivity because of changes in water temperature; and other in Tiruchendur coast, *S. longiceps* is available throughout the year with minimum catch during April and maximum in August. Moderate temperature between 24.9 and 28.5°C could be considered as favorable for the sardine fishery. Analyses indicated that oil sardine catch have been decreasing from 2008 to 2010 and about 25 MT of oil sardine fish catch had decreased from 2008 to 2009. About 10.98 MT has decreased from 2009 to 2010. Like other parts of Gulf, oil sardine fishery in Gulf of Mannar also show declining trend. Though, these weather parameters have been influencing the pelagic fishery, other factors like export price of the fish, fuel price, market value of the fish meal and fish oil are playing important role in the oil sardine fishery. Therefore, these factors also influence the amount of fish caught in this fishery. The oil sardine fishery of the Gulf of Mannar also show declining trend like other Gulf regions of the world (like Sea of Oman, Gulf of California and Finland).

From the study, it was understood that July, August, September and October months are favorable for sardine fish catch. High fish catch was recorded during August; however, large size fishes were caught in October. The study result shows the similarity with fishermen opinion. Based on the opinion obtained from the traditional fishermen of the village, it was understood that, sardine fish catch is decreased in the traditional fishing grounds off Tiruchendur and about 20 to 30 years before, they used to get more fish catch than what they are getting now in the same fishing ground. Considering then, continuous monitoring is essential to understand the impact of meteorological variability (changes) on pelagic fishery. The wind pattern in the study area is found to influence the Northeast and Southwest monsoon currents. The currents could be considered as playing the major role in the fish growth and pelagic fish catch in Tiruchendur coast.

REFERENCES

Asha P S, Diwakar K (2007). Hydrobiology of the inshore waters off Tuticorin in the Gulf of Mannar. J. Mar. Biol. Ass. Inida 49(1):07-11.

Bakun A, Broad K (2003). Environmental 'loopholes' and fish population dynamics: comparative pattern recognition with focus on El Nino effects in the Pacific. Fish. Oceanogr. 12:458-473.

Bakun A, Weeks SJ (2004). Greenhouse gas buildup, sardines, submarine eruptions and the possibility of abrupt degradation of intense marine upwelling ecosystem. Ecol. Lett. 7:1015-1023.

Bapat SV (1955). A preliminary study of the pelagic fish eggs and larvae of the Gulf of Mannar and Palk Bay. Indian J. Fish. 2:231-255.

Bernal PA (1993). Global climate change in the oceans: A review. In: Mooney, H. A., Fuentes, E.R., and Kronberg, B.I., (eds.), Earth System Responses to Global change: Contrast between North and South America. Academic Press, San Diego, Ca. pp. 1-15.

Biswas BK, Svirezher YUM, Bala BK, Wahab MA (2009). Climate Change impacts on fish catch in the world fishing grounds. Clim. Change 93:111-136.

Chidambaram K (1950). Studies on the length frequency of the oil sardine *Sardinella longiceps* Cuv. and Val. and on certain factors influencing their appearance on the Calicut coast of Madras Presidency Proc. Indian Acad. Sci. 31(5):252-286.

Cury P, Roy C (1989). Optimal environmental window and pelagic fish recruitment success in upwelling areas. Can. J. Fish. Aquat.

Sci. 46(4):670-680.
Das J, Das SN, Sahoo RK (1997). Semidiurnal variation of some physico-chemical parameters in the Mahanadi estuary, East coast of India. Ind. J. Mar. Sci. 26:323-326.
De Young B, Harris R, Alheit J, Beaugrand G, Mantua N, Shannon L (2004). Detecting regime shifts in the ocean: data considerations. Prog. Oceanogr. 60:143-164.
Desormeau CJ (1978). Mathematical modeling of phytoplankton kinetics with application to two alpine lakes. Report No. 4, Center of Ecological Modeling. Rensselaer Polytechnic Institute, Troy.
Gopinathan CP, Rodrigo JX, Mohamed Kasim H, Rajagopalan MS (1994). Phytoplankton pigments in relation to primary production and nutrients in the inshore waters of Tuticorin, South East coast of India. Indian J. Mar. Sci. 23:209-212.
Groden TW (1977). Modeling temperature and light adaptation ofphytoplankton, Report No. 2, Center of ecological modeling. Rensselaer Polytechnic Institute, Troy.
Hobday AJ, Okey TA, Pdoczanska ES, Kunz TJ, Richardson AJ (2006). Impacts of climatic change on Australian Marine Life. Part – C – Literature Reviews. (Australian Government, Department of the Environment and Heritage. Australian Greenhouse Office). pp. 1-165.
Holmgren-Urba D, Baumgartner T (1993). A 250-year history of pelagic fish abundances from the anaerobic sediments of the Central Gulf of California. Calif. Coop. Oceanic Fish. Investig. Rep. 34:60-68.
Jayaraman R (1954). Seasonal variation in salinity, dissolved oxygen and nutrient salts in the inshore waters of the Gulf of Mannar and Palk Bay near Mandapam (S. India). Indian J. Fish. 1:345-364.
Jorgensen SE (1980). Lake management, Pergamon. Oxford. p. 167.
Kawasaki T (1984). Why do some pelagic fishes have wide fluctuations in their numbers? A biological basis of fluctuations from the viewpoint of evolutionary ecology, p. 1065-1080. In G.D. Sharp and J. Csirke (eds.). Reports of the Expert Consultation to Examine Changes in Abundance and species Composition of Neritic Fish Resources. FAO Fish. Rep. 291(3).
Lanz E, Martinez MON, Martinez JL, Dworak JA (2008). Spatial distribution and species composition of small pelagic fishes in the Gulf of California. Rev. Biol. Trop. (Int. J. Biol. ISSN-0034-7744). 56(2):575-590.
Lluch-Belda D, Crwaford RJM, Kawasaki T, Mac Call AD, Parish RH, Schwartzlose RA, Smith PE (1989). World wide fluctuations of sardine and anchovy stocks: the regime problem. S. Afr. J. Mar. Sci. 8:195-205.
Maravelias CD, Reid DG (1997). Identifying the effects of oceanographic features and zooplankton on prespawning herring abundance using generalized additive models. Mar. Ecol. Prog. Ser. 147:1-9.
Maravelias CD, Reid DG, Swartzman G (2000). Seabed substrate, water depth and zooplankton as determinants of the prespawning spatial aggregation of North Atlantic herring. Marine Ecol. Progress Ser. 195:249-259.
Miller KA, Fluharty DL (1992). El Nino and variability in the northeastern Pacific salmon fishery: implications for coping with climate change. In: Glantz, M. (ed.), Climate Variability, Climate Change and Fisheries. Cambridge University Press, UK. pp. 49-88.
Mitra A, Patra KC, Panigrahy RC (1990). Seasonal variations of some hydrographical parameters in tidal creek opening into the bay of Bangal. Mahasager Bull. Natl. Inst. Oceanogr. 23:55-62.
Moyle PB, Cech JJ (2004). Fishes: An Introduction to Ichthyology, 5thEd. Prentice Hall. Upper Saddle River. NJ. P. 726.
Navarez-Martinez MO, Lluch-Belda D, Cisneros-Mata MA, Santos-Molina JP, Martinez-Zavala MA, Lluch-Coda SE (2001). Distribution and abundance of the Pacific sardine (*Sardinops sagax)* in the Gulf of California and their relations with the environment. Prog. Oceanogr. 49:565:580.
Pant GB (2003). Long-term Climate variability and change over monsoon Asia, J. Ind. Geophys. Union 7(3):125-134.
Park RA, Collins CD, Leung KK, Boylen CW (1979). The aquatic ecosystem model MS CLEANER. In: Jorgensen SE (ed) State – of – the – Art in Ecological Modeling. Pergamon. New York. pp. 579-602.
Parmanne R, Rechlin O, Sjo¨strand B (1994). Status and future of herring and sprat stocks in the Baltic Sea. Dana 10:29-59.
Parsons TR, Maita Y, Lalli CM (1984). A manual of chemical and biological methods for seawater analyses. New York. Pergamon.
Pauly D, Christensen V, Guenette S (2002). Towards sustainability in world fisheries. Nature 4418:689-695.
Peltonen H, Luoto M, Pa¨a¨kko¨nen JP, Karjalainen M, Tuomaala A., Ponni J, Viitasalo M (2007). Pelagic fish abundance in relation to regional environmental variation in the Gulf of Finland, Northern Baltic Sea. ICES J. Marine Sci. 64:487-495.
Peltonen H, Vinni M, Lappalainen A, Ponni J (2004). Spatial feeding patterns of herring (*Clupea harengus* L.), sprat (*Sprattus sprattus* L.), and three-spined stickleback (*Gasterosteus aculeatus* L.) in the Gulf of Finland, Baltic Sea. ICES J. Marine Sci. 61:966-971.
Pillai PKM (1990). The oil sardine Fishery along Northern Tamil Nadu coast with a note on unusually heavy landings at Cuddalore, Pazhayar and Kaveripattinam, CMFRI.
Piontkovski SA, AL-Gheilani HMH, Jupp BP, Al-Azri AR, Al-Hashmi KA (2012). Interannual Changes in the Sea of Oman Ecosystem. Open Marine Biol. J. 6:38-52.
Rahman MS (1992). Water quality management in aquaculture. BRAC Prakashara, 66 Mohakhali, Dhaka, Bangladesh. p. 84.
Raju (2000). Fish and Fisheries of Gulf of Mannar. National Symposium on Eco- Friendly Mariculture Technology Packages-An Update, Souvenir, CMFRI. pp. 44-46.
Ramos S, Vázquez-Rowe I, Artetxe I, Moreira MT, Feijoo G, Zufía J (2011). Environmental assessment of the Atlantic mackerel Scomberscombrus) season in the Basque Country. Increasing the timeline delimitation in fishery LCA studies. Int. J. Life Cycle Assess. 16(7):599-610.
Rohit P, Bhat US (2003). Sardine fishery with notes on the biology and stock assessment of oil sardine off Mangalore-Malpe. J. Mar. Biol. Ass. India 45(1):61-73.
Selvin PJ, Selvaraj P (2011). A study on short term surface weather elements over Kuttapuli coast of Tamil Nadu, Southeast Coast of India. Int. J. Curr. Res. 3(6):029-033.
Scavia D, Park RA (1976). Documentation of selected constructs and parameter values in the aquatic model CLEANER. Ecol. Model. 2:33-58.
Schau EM, Ellingsen H, Endal A, Aanondsen S (2009). Energy consumption in the Norwegian fisheries. J. Clean. Prod. 17:325-334.
Strickland JDH, Parsons TR (1972). A practical handbook of seawater analyses. Bull. Krish. Res. Bd. Canada. 167:310.
Valenciennes M.A (1847). Natural History Fish, Strasbourg, at Ve Levrault, Rue Jews. 20:274.
Vinayachandran PN, Chauhan P, Mohan M, Nayak S (2004). Biological response of the sea around Sri Lanka to summer monsoon. Geophys. Res. Lett. 31:L01302.

Geochemistry and ionic interaction in the Bizerte Lagoon waters (Northern Tunisia)

Amor Ben Garali[1]*, Mohamed Ouakad[1] and Moncef Gueddari[2]

[1]Faculté des Sciences de Bizerte - Laboratoire de sédimentologie et géochimie marine - 7021 Zarzouna - Bizerte. Tunisie.
[2]Faculté des Sciences de Tunis, Université El Manar. 1060 Tunis. Tunisie.

The Bizerte Lagoon is classified in the category of medium salinity environments. The lagoon geochemistry, with regard to the major ions, is essentially controlled by the continental contributions and interactions with the marine environment. The geochemical analyses of the major ions as well as their behaviour in the lagoon water permitted to describe the eventual ability of a solution chemical composition, to precipitate different minerals. Sampling and geochemical analysis of the principal ions (Ca^{++}, K^+, Na^+, Mg^{++}, Cl^-, SO_4^{2-} and HCO_3^-) at the water surface (0 to 2 m) and at the bottom (over 2 m) of the lagoon (9 stations), in the Channel (3 stations) and in the bay of Bizerte (3 stations), was performed monthly during one year. The results for the Bizerte Lagoon waters show that only carbonates and gypsum can precipitate.

Key words: Water, Bizerte lagoon, ionic ratio, precipitation.

INTRODUCTION

The Bizerte Lagoon with an ellipsoidal shape (surface area: ~ 150 km^2) is a littoral lagoon in north-eastern Tunisia, which has two connections with surrounding environment; the first one is continental with the Garaet Ichkeul (the most important fresh water surface in Tunisia (Ouakad, 2007)) by the intermediary of Wadi Tinja, and the second one with the Mediterranean sea by a long channel. The Bizerte region has a sub-humid climate with 600 to 800 mm annual rainfall. Two seasons are distinguished. Winter extends from November to April, with maximum precipitation in February (219.5 mm). Summer and the rest of the year are dry without any freshwater input from Wadis. The drainage network that feeds the lagoon has a catchment of 252 km^2 (Kallel, 1989). The annual balance of water exchanges between the Bizerte Lagoon and the Mediterranean Sea shows that in winter a surplus of fresh water of approximately 534 (±26.03) Mm^3 are lost from the lagoon. However, during the summer, because of the lack of precipitation and strong evaporation, there is an incursion of marine water 84 (±2.45) Mm^3 towards the lagoon (Ben et al., 2009). Generally, water of the Bizerte Lagoon principal affluent shows temperatures oscillating between 11.1 and 20.8°C, with a neutral pH to slightly alkaline (7.13 to 9.22). This water is very rich in dissolved oxygen between 5.8 and 9.5 mg/l. The water salinity varies between 0.5 to 3.8 psu. Only the water of Wadi Tinja during the summer season and Wadi Rharek during all the year, present high contents of dissolved salts.

The mineralization of the streaming water of Bizerte Lagoon drainage basin is rather important. The analysis of the major elements shows for the majority of this water a sulphate-calcareous to chloride calcareous stable facies. Only Wadi Rharek water shows sodium chloride facies (Ben et al., 2010). During the summer season, Wadi Tinja water shows an important concentration in sodium ions and chlorides, with the phenomenon of marine water intrusion from the Bizerte Lagoon towards the Garaet Ichkeul (Ben et al., 2008). Wadi Rharek water presents constantly such high contents of these ions, resulting from scrubbing the Triassic terrains of Djebel Kchabta crossed by this Wadi (Crampon, 1971; Mathlouthi, 1985). The rest of Wadis water presents balanced contents of major ions which could be originated by the hydromorphic soils scrubbing (Ben et

*Corresponding author. E-mail: bgamor@yahoo.fr.

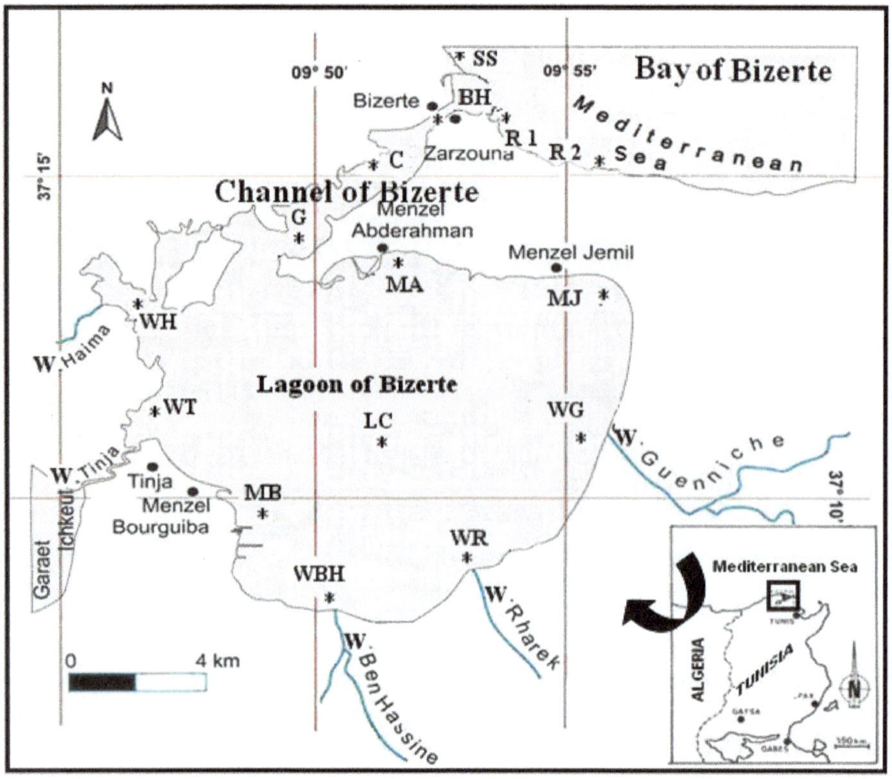

Figure 1. Samplings localisation in the Bay, the channel and the lagoon of Bizerte. SS, Sidi Salem; R1, R'mel 1; R2, R'mel 2; BH, Before Harbour; C, Channel; G, Gully; WH, Wadi Haima; WT, Wadi Tinja; MB, Menzel Bourguiba; WBH, Wadi Ben Hassine; WR, Wadi Rharek; WG, Wadi Guenniche; MJ, Menzel Jemil, MA, Menzel Abderhman and LC, Lagoon Centre.

al., 2010). So, physicochemical parameters, as well as the contents of major ions at Bizerte Lagoon waters are then influenced by those of marine water in summer period and those of the principal affluent in winter period.

The mixture zones of fresh and salty waters constitute an important environment of marine chemical reactions, which are consecutive to the sudden change of environmental conditions (Ouakad, 2007). This type of environment corresponds to estuary, delta or coastal lagoon with continental and marine influences. The biological activity is as important as the geochemical processes. We actually know more than ten enzymes families of carbonic anhydrases, and are universally present in the living organisms. They are used to catalyse in a reversible way the transformation of CO_2 into bicarbonate (Villeneuve, 2002). The Bizerte Lagoon ecosystem is affected by several physicochemical, hydrological and geochemical processes. In order to better understand the processes operating in this disturbed lagoon environment, a geochemical analysis of the major ions was carried out.

The obtained results enabled us to elucidate parameters responsible for precipitation of some minerals.

METHODOLOGY

Monthly water samples were collected from 15 stations in the Bizerte Lagoon between 37° 8' and 37° 14' of northern latitude and 9° 46' and 9° 56' of eastern longitude (Figure 1). Two samples were taken at each station; one at surface and one at bottom (2 m depth, using a hydrological bottle). These samples intended to determine "in situ" physico-chemical parameters. At the laboratory, each water sample undergoes the necessary dilution in order to determine the concentration of major ions (Ca^{++}, K^{+}, Na^{+} and Mg^{++}) using atomic absorption spectrometry of THERMO-ELEMENTAL-SOLAAR-S4 type (Rodier et al., 2005). The alkalinity was determined by titration with the sulphuric acid M/100. The water alkalinity corresponds to the presence of hydroxides, carbonates and bicarbonates. The chlorides were determined by titrimetrically with silver nitrate ($AgNO_3$) (Rodier et al., 2005).

The sulphates were determined with barium chloride ($BaCl_2$. $2H_2O$). The method of sulphates determination is the precipitation of barium sulphate and the determination of the amount of precipitate by turbidimetry (Rodier et al., 2005).

RESULTS AND DISCUSSION

Physico-chemical parameters of water

The spatio-temporal variation of salinity during winter and

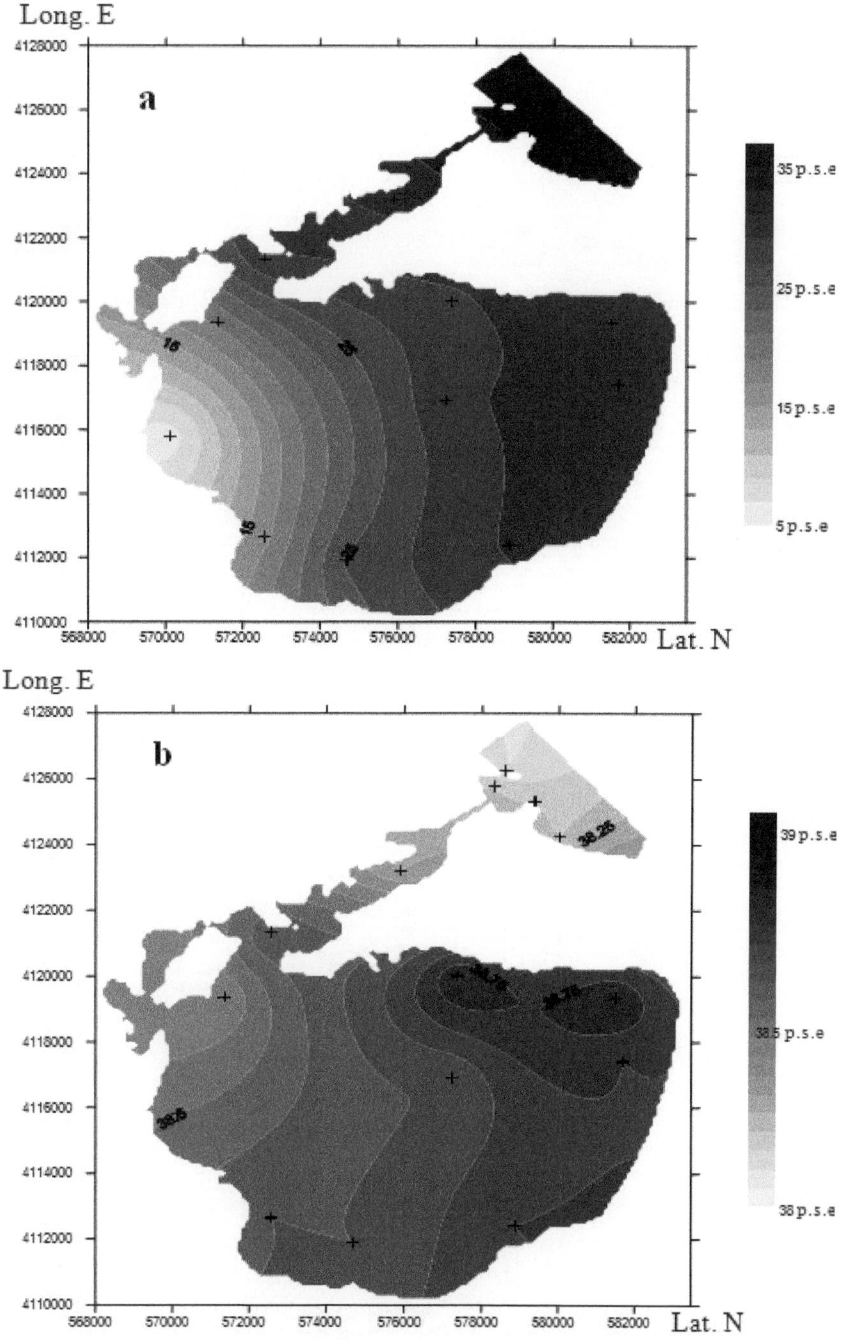

Figure 2. Spatio-temporal variation of Bizerte Lagoon water salinity. a) Surface water (in winter), b) Surface water (in summer).

spring showed that the lagoon water is less saline (from 5.9 to 29.1 psu) than the bay water (always higher than 30 psu) (Figure 2a). This variation is related to the impact of fresh water contributions by Wadi Tinja (Ben et al., 2008). The salinity homogenisation of lagoon waters occurs in summer period (between 36 and 38 psu) (Figure 2b). It should be noted, that the lagoon waters salinity indicated to a small rise compared to bay waters during summer. These results permitted to classify the Bizerte Lagoon, according to the classification of Mars (1966), in the category of the medium salinity environments. The monthly variation of dissolved oxygen (DO) concentrations in Bizerte Lagoon water shows two different periods. These are a winter period, when the concentrations oscillate between 7 and 11 mg/l (Figure 3a) and a summer period, when the concentrations vary

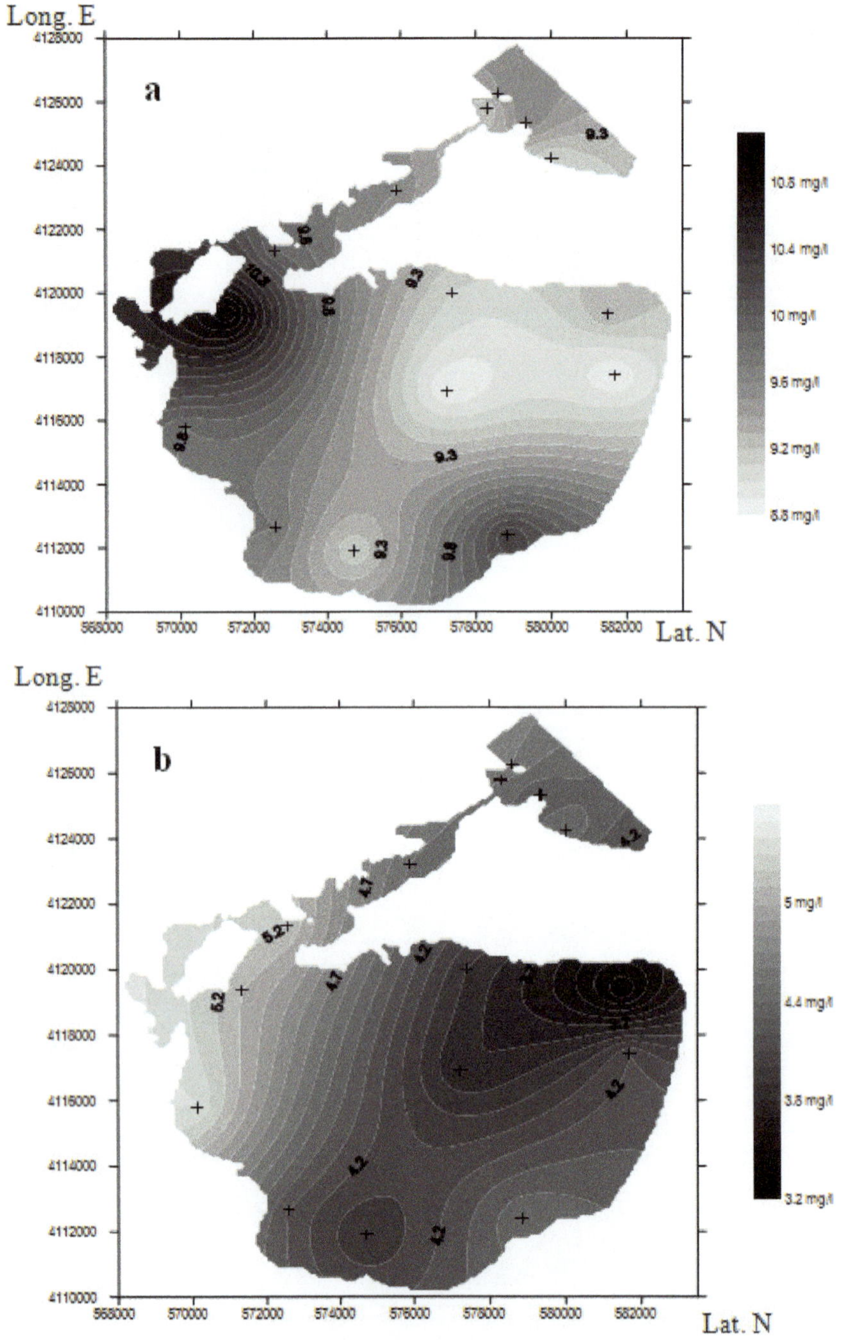

Figure 3. Spatio-temporal variation of the dissolved oxygen concentrations in the Bizerte Lagoon water. a) Surface water (in winter) b) Surface water (in summer).

between 3 and 6 mg/l (Figure 3b).

Geochemistry of water

The bulk geochemical analysis results of the major ions in the lagoon and bay of Bizerte water shows two different periods: The first is a winter period during which the sodium concentrations at station (WT) oscillate between a minimum of 1.82 g/l and a maximum of 9.32 g/l (Figure 4a). The potassium contents do not exceed 0.35 g/l, with a minimum of 0.07 g/l. The magnesium concentrations oscillate around 1g/l with a minimum of 0.22 g/l and calcium concentrations are up to 0.35 g/l, and can have low values up to 170 mg/l on the prodeltaic station of Wadi Tinja. The chlorides concentrations are

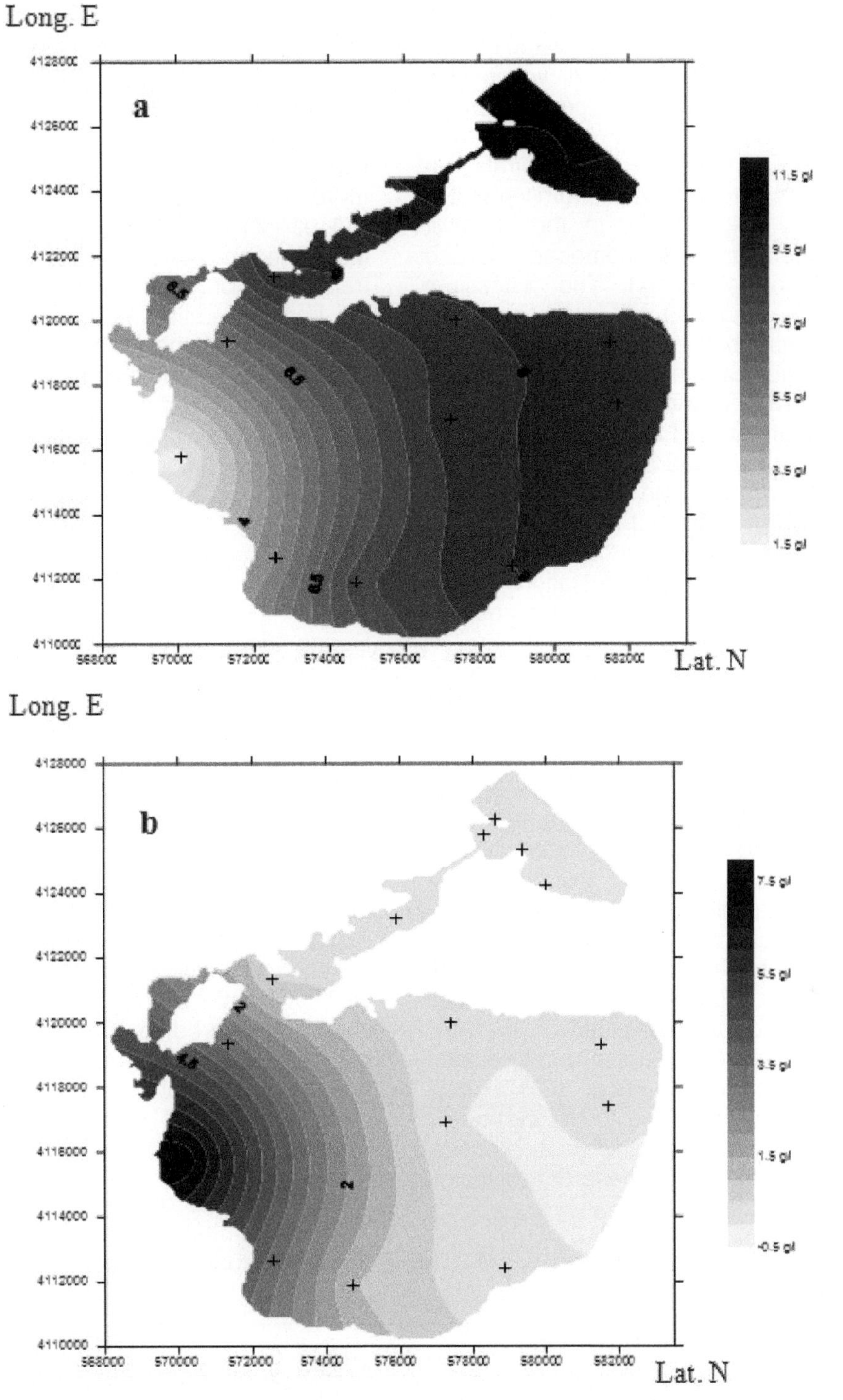

Figure 4. Spatial variation of the sodium content of Bizerte Lagoon water. a) Surface water (in winter), b) Differences in concentrations between surface and bottom water (in winter).

lower than 17g/l with a minimum of 3.28 g/l at the mouth zone of Wadi Tinja. The sulphate concentrations are lower than 2.5 g/l, with a minimum of 0.46 g/l at the surface water station (WT). During the rainy period (in winter), the lagoon water has lower concentrations of major ions than the bay water. It is due to the important continental fresh water contributions, in particular of the Garaet Ichkeul through the Wadi Tinja (Ben et al., 2009). Under these conditions, concentric aureoles around the zones of Wadi mouths are observed (Figure 4a).

It is during this winter period, when the lagoon water stratification develops, that is the surface water is less enriched in major ions than the bottom water. For example, the difference in sodium concentrations between surface and bottom waters becomes very important. At the station WT, the sodium concentration in surface water is 1.82 g/l, whereas at the bottom water it is 9.29 g/l during January (Figure 4b). In the summer period, the sodium concentrations are homogeneous and vary between 10.00 and 11.98 g/l. The potassium concentrations are higher than 0.40 g/l, with a maximum of 0.43 g/l and magnesium concentrations reach 1.43 g/l at several stations. The calcium concentrations are higher than 0.40 g/l, with a maximum of 0.45 g/l. The sulphate concentrations are around 2.5 g/l. Generally, the lagoon and the bay waters showed a homogenisation of bicarbonate concentrations during the year of study, with concentrations from 0.12 to 0.15 g/l. It is only during the winter season and due to the important continental fresh water input rich in bicarbonate, that the stations at the Wadi mouths show surface water richer in bicarbonate than those at the bottom.

In winter, the surface water at the station WT shows high bicarbonate concentration of 0.22 g/l (Figure 5a) compared to 0.15 mg/l for bottom water (Figure 5b). The bicarbonate concentrations in the Bizerte Lagoon are low compared to other environment; this is due to the effectiveness of exchanges between the lagoon and the sea, fresh water inflow, and also to the water dilution. Thus, the lagoon water alkalinity distribution is homogeneous with little variations of pH. The chemical composition of the solution, which is concentrated by evaporation, is variable, but generally, the content of dissolved salts increases. However, the ionic ratios remain constant as far as no salts precipitate.

The obtained results showed a negative correlation between Ca^{++} and HCO_3^- ($R = -0.77$; $p= 0.05$), which indicated that Ca^{++} is not only related to the carbonate rocks dissolution. Indeed, besides this last process, HCO_3^- may be provided by the atmospheric CO_2 diffusion into the lagoon (Villeneuve, 2002).

Behaviour of the major ions

Mg^{++}/Ca^{++} ratio

The temporal distribution of this ratio varies considerably for surface water between summer and winter. Concerning summer values, it was found to be around 3.3 compared to value of 1.4 in winter. The bottom water shows a constant and high ratio. The reduction in the Mg^{++}/Ca^{++} ratio in the surface water of the Bizerte Lagoon during winter period observed especially at the Wadis mouths is due to the important fresh water input. On the other hand, its increase during the summer season could be due to the combined effect of marine water intrusions and the impact of evaporation. This increase is dependent on the precipitation of calcite in the basin, but the involved quantities are low compared to the concentrations of Ca^{++} and Mg^{++}, in water.

During the year of the study, the Bizerte Lagoon bottom waters showed a high Mg^{++}/Ca^{++} ratio. This water is a little richer in Mg^{++} than surface waters, probably because more Ca^{++} was lost by calcite precipitation.

A part of Mg^{++} is lost by the transformation of calcite into magnesium calcite and aragonite. This transformation contributes to the increase of Ca^{++} ions in the basin interstitial waters. The relative increase in this ratio promotes the precipitation of the gypsum, but this increase is possibly limited by the calcic carbonates dolomitization of the lagoon sedimentary filling. Ultimately, we can priori assume that the geochemistry of the Bizerte Lagoon waters, with regard to Ca^{++} and Mg^{++} is primarily controlled by the marine environment.

SO_4^{2-}/Ca^{++}, SO_4^{2-}/Cl^-, Na^+/Cl^- and HCO_3^-/Cl^- ratios

During the summer season, the increase in the SO_4^{2-}/Ca^{++} ratio (6.66 compared to 2.94 in winter) should result in the precipitation of the gypsum partly, but this increase is controlled by the calcite precipitation in the lagoon sedimentary filling. It should be noted that precipitation of gypsum tends to lower the SO_4^{2-}/Cl^- ratio. This ratio increases again, if halite precipitates, with resulting decrease in the Na^+/Cl^- ratio (Medhioub, 1979). However, under these conditions, SO_4^{2-}/Cl^- and Na^+/Cl^- ratios do not show any notable seasonal variations, which may be explained by limited precipitation of gypsum and the absence of halite precipitation. The minimum values of HCO_3^-/Cl^- ratio are reached during the summer season (0.006 compared to 0.01 in winter), which is probably caused by an increase in temperature and precipitation of carbonates. Bottom waters show a low HCO_3^-/Cl^- ratio during all year of the study. This could be related to several phenomena such as:

1. The presence of evolved waters at the surface, which lost a part of their bicarbonate,
2. The photosynthetic activity which can contribute to the calcium carbonate precipitation.

Chemical evolution of waters and saturation test

The possible aptitude of a solution of given chemical composition to precipitate minerals can be checked by the calculation of saturation ratios. The ratio of the ion activities product Q and the theoretical solubility product Ksp is calculated:

$\Omega = Q/Ksp$

Three cases can be distinguished:

$Q < Ksp$: Ω is lower than 1 and the solution is under

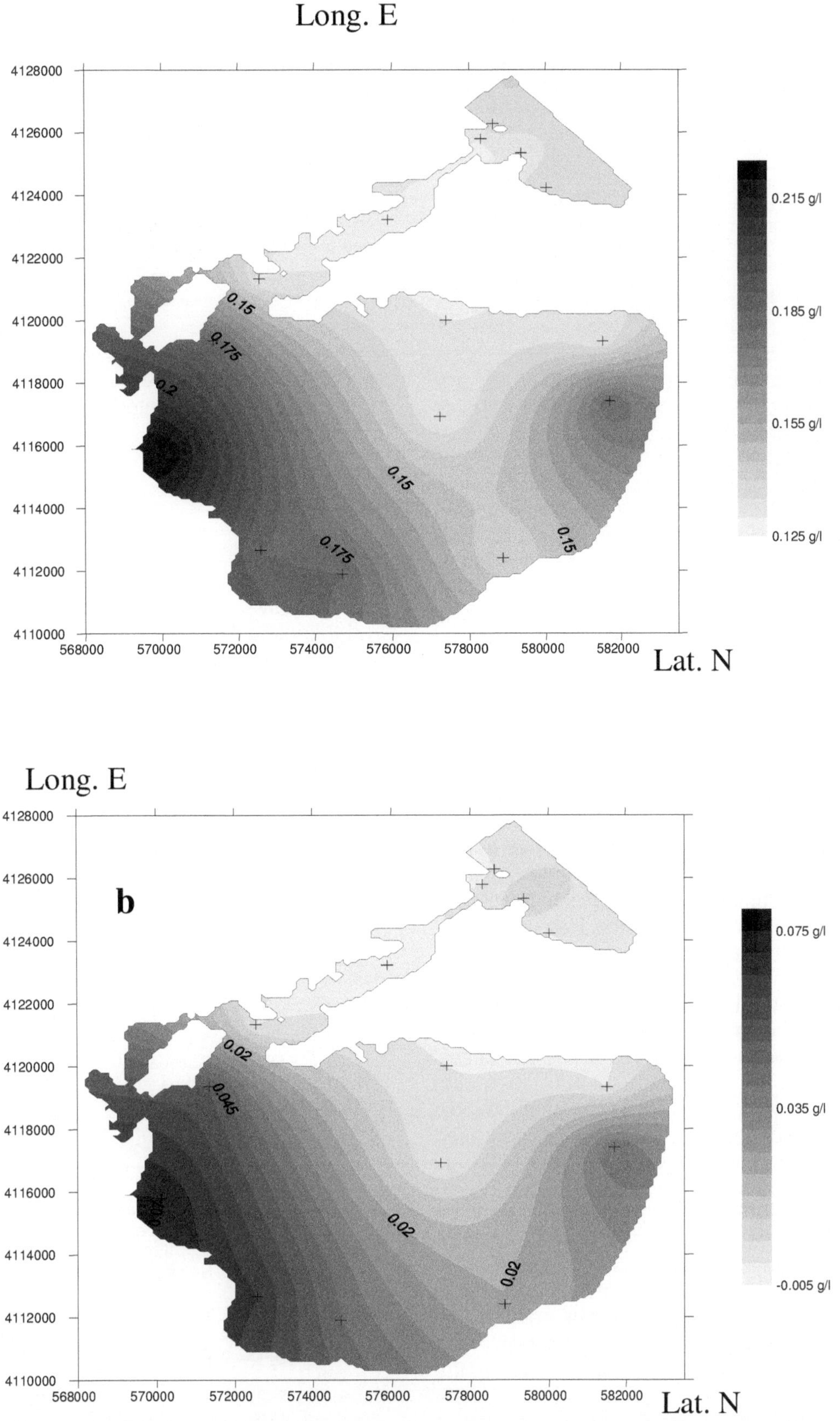

Figure 5. Spatial variations of the bicarbonate concentrations in the Bizerte Lagoon water. a. Water surface concentrations (in winter). b. Variation in concentrations between surface and bottom water (in winter).

Table 1. Saturation ratios calculated for some minerals at 25°C.

Minerals	Log Ksp 299, 15°K	Log Q		Ω = Q / Ksp	
		Winter	Summer	Winter	Summer
Calcite	- 8.36 (a)	- 4.80	- 4.57	1.74	1.83
Aragonite	- 8.22 (b)	- 4.80	- 4.57	1.71	1.79
Dolomite and/or magnesium calcite	- 17.0 (c)	- 8.89	- 7.43	1.91	2.28
Gypsum	- 4.60 (d)	- 4.32	- 3.74	1.06	1.22

(a) Plummer and Bussenberg (1982), Sass and al. (1983), (b) Morse and al. (1980), (c) Truesdell and Jones (1974), (d) Naumov al. (1971).

saturated with respect to a mineral considered.
Q = Ksp: Ω is equal to 1 and the solution is at equilibrium with a mineral considered.
Q > Ksp: Ω is higher than 1 and the solution is oversaturated with respect to a mineral considered (Stumm and Morgan, 1970). The over saturation state is unstable and should theoretically evolve in time towards equilibrium. It could be a slow process due to a precipitation delay related to crystals germination phenomena, to the kinetics reaction etc (Ben, 1988). In particular, the impurities in carbonate minerals (Sr in calcite and aragonite, Mg in magnesium calcite) can not only inhibit (or conversely activate) the precipitation of minerals, but also change the value of a relevant mineral solubility product (Tagorti, 1990).

It is in this context that preliminary evaporation tests of the brine of Sebkha El Adhibate (South-western Tunisian, on the Libyan border) were carried out in order to determine the useful salt sequences deposited (ONM, 1986 in Bouhlila, 1997). The Works of Ben (1988: paralic basin of El Kantara-Gourine, South-eastern of Tunisia), of Tagorti (1990: lagoon complex: Sebkha Halk el Mengel and Assa El jerbia (the Sahel of Tunisia) and of Darragi, (1992: Natron Lake (Tanzania) allowed the simulation of a wider variety of natural brines. From the data on the activity coefficients for ionic species, the stochiometric products are calculated and compared to the theoretical solubility products for minerals. In our case the calculation of Q is based on the concentration of the ionic species and not on activities. The calculations are based on empiric equations proposed by Pitzer and Mayorga (1973). The results are given in Table 1.

Following Table 1, four minerals could precipitate at the water-sediment interface in the Bizerte Lagoon. In spite of obtaining these values from calculations on the basis of concentrations and not activities of the ionic species, they probably are a little bit higher but confirm the possibility of precipitation of these minerals. The calculation of the thermodynamic balance and of the saturation index (SI) by Chaouni Alia and Al., (1997) allowed to specify that the minerals whose potential of dissolution is very high, such as anhydrite, halite, epsomite and brucite, will be dissolved if they are present in the sediments of the Bou-Areg plain (North-Eastern Morocco). Whereas, the solution is in chemical balance or is slightly supersaturated compared to calcite (- 0.12 < SI < +0.93).

Conclusion

The concentrations of main ions in the Bizerte Lagoon waters, that is Na^+, K+, Ca^{++}, Mg^{++}, Cl^-, $HCO_3^-$and SO_4^{2-}, are primarily controlled by the continental contributions and the interactions with the marine environment. During summer (the dry season), lagoon waters are concentrated by strong evaporation and they become increasingly enriched in sulphate, magnesium, and calcium. Based on saturation ratios calculations, carbonate minerals and gypsum can precipitate at the water-sediment interface in the Bizerte Lagoon.

REFERENCES

Ben AL (1988). Geochemistry and Geodynamics water and brine basin paralic El-Kantara Gourine (southeast of Tunisia) seasonal fluctuations in exchange and sedimentation. Thesis 3rd cycle of Geology. Faculty of Sciences of Tunis, p.139

Ben Garali A, Ouakad M, Gueddari M (2008). Hydrology, sedimentology and geochemistry of the wadi Tinja (Northern Tunisia). Bulletin of the National Institute of Science and Technology of the Sea of Salammbo, Tunis. 35,161-168.

Ben GA, Ouakad M, Gueddari M (2009). Hydrological balance of the lagoon of Bizerte (north-east of Tunisia). Journal of Water Science, 22-24 : 523-534.

Ben Garali A, Ouakad M, Gueddari M (2010). Geochemistry of Bizerte Lagoon affluent Water, Northern Tunisia: Principal Component Analysis. Arabian J. Geosci. Published on line on 16 March 2010, pp 15 (3) :273-282

Chaouni Alia A, El Halim N, Walraevens K, Beeuwaert E, De Breuck W (1997). Investigation of the salinity of the Bou-Areg plain (northeastern Morocco). Freshwater Contamination (Proceedings of Rabat Symposium S4, April-May 1997). IAHS Publ. no. 243 : 211-220.

Crampon N (1971). Geological study of the edge of Mogods the "country of Bizerte" and North Hedil (Northern Tunisia). MSc thesis. Natural Science, 3 Vol 1 stratigraphy. University of Nancy I, France. p. 228.

Darragi F (1992). Mineralogical and geochemical study of evaporites and sediments of Lake Natron (Tanzania). Analysis, modeling
Experimental and numerical. Th Univ. L. Pasteur, Strasbourg, p. 152.

Kallel M R (1989). Hydrology of the lagoon of Bizerte. DGRE (Directorate General of Water Resources), Tunis, p. 41

Mars P (1966). Research on some coastal lagoons

French Mediterranean and their molluscan faunas. Life and Milieu, Supp. pp. 20, 359.

Mathlouthi S (1985). Geomorphological study of the surrounding lake system of Bizerte. PhD thesis Specialty. Faculty of Arts and Humanities Tunis, p. 287

Medhioub K (1979). The bahirat el Biban: geochemical and sedimentological study of a lagoon in southeastern Tunisia. Thesis 3rd cycle of Geology. Ecole Normale Superieure,Paris, p. 150

Morse JW, Mucci A, Millero FJ (1980). The solubility of calcite and aragonite in seawater of 35% salinity at 25°C and atmospheric pressure. Geochim. Cosmochim. Acta, 47: 217-233.

Naumov CB, Ryschenoco BN, Khodakovsky TL (1971). Thermodynamic data atlas (In Russion). Atomiz dat, Moscou, p. 239.

ONM (1986). Office National des Mines in Tunisia. Evaporation test brine El Adhibate. Internal Report, p 23.

Ouakad M (2007). Genesis and evolution of lagoonal-lacustrine environments of north-eastern Tunisia (Garaet Ichkeul Reservoirs of Bizerte and Ghar el Melh). MSc thesis. Geological Sciences. University of Tunis el Manar. Tunis, Tunisia p. 453.

Pitzer KS, Mayorga G (1973) .Thermodynamics of electrolytes. Activity and osmotic coefficients for strong electrolytes with one or both univalent. J. Phys. Chem., 77: 2300-2308.

Plummer LN, Busenberg E (1982). The solubility of calcite, aragonite and vaterite in CO_2-H_2O solution between 0 and 90°C an evaluation of the aqueous model for the system $CaCO_3$-CO_2-H_2O. Geochim. Cosmochim. Acta, 46: 1011-1040.

Rodier J, Bazin C, Broutin JP, Chambon P, Champsaur H, Rodi L. (2005). The analysis of water: natural water, wastewater, seawater 8ème édition. DUNOD. Paris, p. 1383.

Sass E,Morse JW, Millero FJ (1983). Dependence of the values of calcite and aragonite thermodynamics solubility products on ionic models. Am. J. Sci., 183: 218-229.

Stumm W, Morgan J (1970). Aquatic Chemistry. An introduction emphasizing chemical equilibria in natural waters. Ed. Wiley Intersciences, p. 583.

Tagorti MA (1990). Dynamic hedging, genesis and evolution of the lagoon complex (Mengel and sabkha Halk el Assa El Jerbia (Sahel of Tunisia). seismograph Analysis, géchimie and seasonal hydrologic exchanges. Thesis 3rd cycle of Geology. Faculty of Sciences Tunis, p. 151.

Truesdell AH, Jones BF (1974). WATEQ, a computer program for calculating chemical equilibria of naturel water. J. Res. U.S. Géol. Suerv., 2: 101-107.

Villeneuve C (2002). Carbon sequestration: Perspectives for Quebec, Report to the National Assembly of Quebec. p. 17

Morphodynamics and features of littoral cell circulation observed from sequential aerial photographs and Davies drifter along a section of the strand coast East of the Niger Delta, Nigeria

Larry Awosika[1]*, Regina Folorunsho[1] and Victor Imovbore[2]

[1]Nigerian Institute for Oceanography and Marine Research, P. M. B. 12729 Victoria Island, Lagos Nigeria.
[2]Environmental Resources Management Limited (ERML) Lagos, Nigeria.

Sequential aerial photographs taken along a section of the Strand coast (longitude 07°35'00" and latitude 8°30' 20"E) East of the Niger Delta Nigeria were anaylsed to understand tidal inundation and resulting morphodynamic features as well as the littoral cell circulation patterns. Records from a Davis drifter deployed in the Qua Iboe River that traveled parallel to the shoreline were also used to confirm the littoral cell patterns. The analysis of the sequential aerial pictures shows that the cell circulation patterns consist of waves with oblique angles which approach in a Southwesterly direction generating an Eastward flowing longshore current. Rip currents are recognised in the surf zone as seaward flowing surfs. Interpretation of a picture of the exposed beach during low tide shows tidally induced morphodynamic features such as rhythmic ripples, perpendicular sand ridges and runnels (hollows) which tend to be welded to the shoreline. Drifter number 26602 deployed inside the Qua Iboe River on 3 March 2001 at 0300 h confirmed the West to East longshore current direction. Though this study employed the use of short term data, it has enhanced the understanding of oceanographic processes, erosional, and depositional processes in the intertidal zone of the stud area.

Key words: Cell circulation, drifter, longshore currents, Niger Delta, rip currents, waves.

INTRODUCTION

Cell circulation in the littoral zone (wave breaking zone) results from combination of wave, tide and wind-induced currents. As waves move into shallow water, the group velocity becomes small, wave energy per square meter of sea surface increases, and non-linear terms in the wave equations become important. These processes cause waves to steepen, with short steep crests and broad shallow trough. When wave slope at the crest becomes sufficiently steep, the wave breaks. When waves break with their crests effectively parallel to the shoreline, a cell circulation is established (Figure 1), which consists of a shoreward mass transport, longshore currents, and narrow seaward-flowing rip currents that extend through the breaker zone and spread out into rip heads (Shepard and Inman, 1951). Tidal fluctuations coupled with cell circulation system and beach profiles result in several types of shore morphodynamic features such as ridge and runnel structures, ripples and sand bars. The morphology and associated dynamics of these features have been well described by several authors like King et al. (1949) and Masselink et al. (2006) amongst others. Previous authors like NEDECO (1961), Burke (1972), Ibe and Awosika (1986), Awosika et al. (1994, 2000) amongst others have described the general dynamics

*Corresponding author. E-mail: larryawosika@yahoo.com.

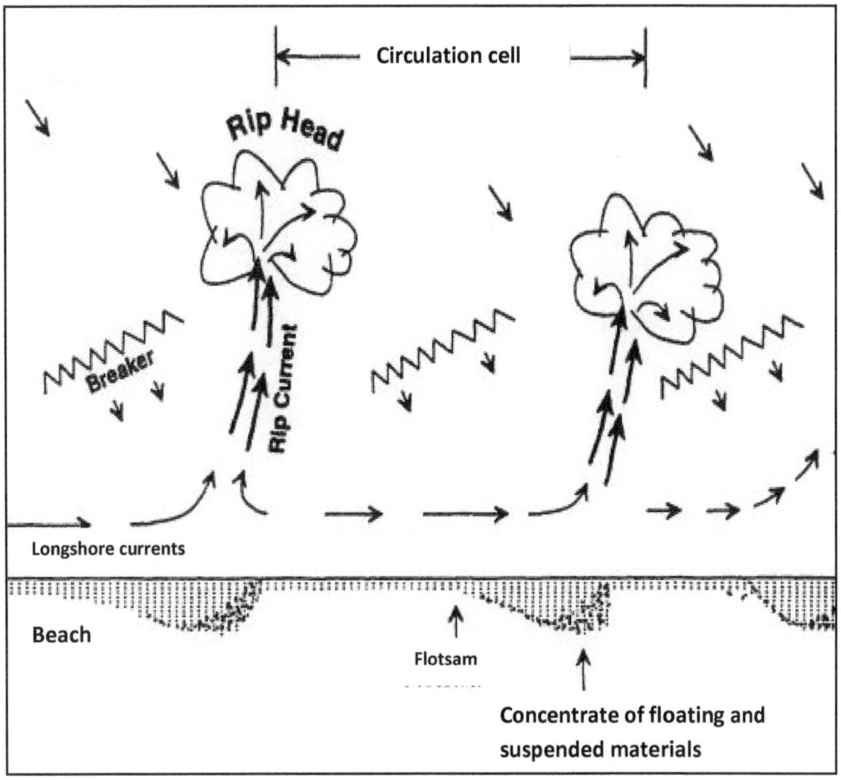

Figure 1. Schematic diagram of nearshore circulation cell consisting of onshore transport by the breaking wave, longshore transport in the surf zone and offshore transport by seaward flowing rip currents (Shepard and Inman, 1951).

along this coast.

Objectives of the study

The primary objectives of this study is to understand the morphodynamics and features of the cell circulation, tidal inundation patterns, and resulting morphological intertidal features along a section of the Strand coast in Nigeria using airborne sequential photographs. Also, trajectory records from a Davies drifter data deployed in the surf zone was used to confirm the littoral cell transport direction.

The study area

The study area lies East of the Qua Iboe River within the Strand coast which is the Easternmost coastal morphological unit in Nigeria (Figure 2). This coastal complex stretches from Imo River Eastwards to the Cross River estuary and within longitude 07°35'00"E and 8°30' 20"E. The coastline is dissected by the Cross, Qua Ibo and Imo Rivers and many small rivulets and tidal creeks. Waves along this coast are wind generated which approach in a South-westerly direction with periods ranging from 8 to 13 s. Tides are generally semi diurnal with two inequalities. Tidal range varies from 1.5 m along the East to 3 m in the Calabar estuary. The study area has a narrow intertidal zone with average of about 100 m. The semi-diurnal nature of the tides affecting this coastline generates tidal currents, are in phase with tidal cycles. The beach is sandy with sediments ranging from fine to coarse sand.

METHODOLOGY

Sequential aerial pictures taken during a helicopter over-flight of the area on 6 November, 2004 were used to identify and verify the dominant circulation patterns along the study area East of the Qua Iboe River. For this analysis, the following oceanographic phenomenon were identified and mapped on the pictures:

i) Wave breaking direction and angle,
ii) Direction of long shore current,
iii) Rip currents.

Rip currents will also be identified by observing the following in the pictures:

a) A channel of churning, choppy water;
b) An area having a notable difference in water color;
c) A line of foam, seaweed, or debris moving steadily seaward and
d) A break in the incoming wave pattern.

The morphodynamics features in the intertidal zone were

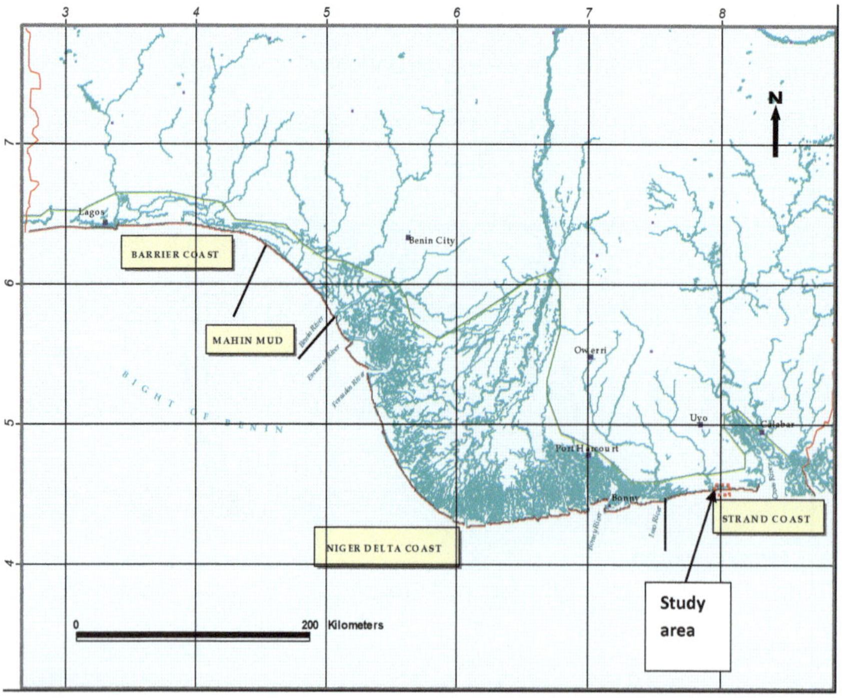

Figure 2. Coastal morphological zones in Nigeria (Awosika et al., 2000) showing the study area.

interpreted from a picture of the shoreline taken during low tide. The direction of the longshore currents within the littoral cell circulation system mapped on the photographs was validated using Davis drifter trajectory records (No. 26602). The drifter (Figure 5) was deployed in the Qua Iboe River on 3 March 2001 at 0300 h. The Davis drifter contained GPS units which transmits coordinates and current direction, and speed to the ARGOS satellites.

RESULTS

Sequential aerial photographs

Analysis of the sequential aerial pictures showed the cell circulation patterns for the study site (Figure 3). The wave trains approaching the shoreline in a South-westerly direction were interpreted in the sequential aerial photographs. The angle of breaking waves also is about 10 to 15°. The resulting direction of the longshore current is West to East. Rip currents between each successive breaking wave train were recognized in the sequential photographs as seaward flowing surf. The rip currents observed from the pictures are narrow but perpendicular to the longshore direction. The rip currents transport material out to the sea but die out within 25 to 50 m seaward of the shoreline.

Using the characteristics (shape, colour tones, and texture) of the morphodynamic ridge and runnel features photographed in the intertidal zone, the dominant tidal direction and patterns have been interpreted. Figure 4 shows the typical beach morphodynamic features (ridge and runnel structures) within the intertidal zone exposed during the low tide. The structures in the picture appear as perpendicular ridges separated by hollows (runnels) which had been shaped by seaward flowing rip currents. The ridges show the toss side facing the ocean and the seaward directional movement of the rip currents. The ridges are rhythmic and tend to weld to the shoreline.

The morphodynamic ridge and runnel structures had been shaped by a combination of flood and ebb tidal currents, longshore, and rip currents. Similar features have also been identified by Short (1999), Short and Aagard (1993) when low waves persist for long periods. Komar (1997) emphasized the importance of low-velocity longshore currents for rhythmic ridge formations with runnels. The hollows (runnels) between the perpendicular sand ridges caused by rip currents shows the competence of the rip current to erode as well as transport materials to sea. Furmanczyk et al. (2002) analysed series of aerial photographs in Pomerainina Bay, Poland and found similar underwater bars and ridges as indicators of rip current activity and circulation system.

Offshore Davis drifter trajectory

Drifter number 26602 was deployed inside the Qua Iboe River on 3 March 2001 at 0300 h. The direction of movement of the Davis drifter is shown in Figure 6. While in the estuary, the drifter was under the ebb currents

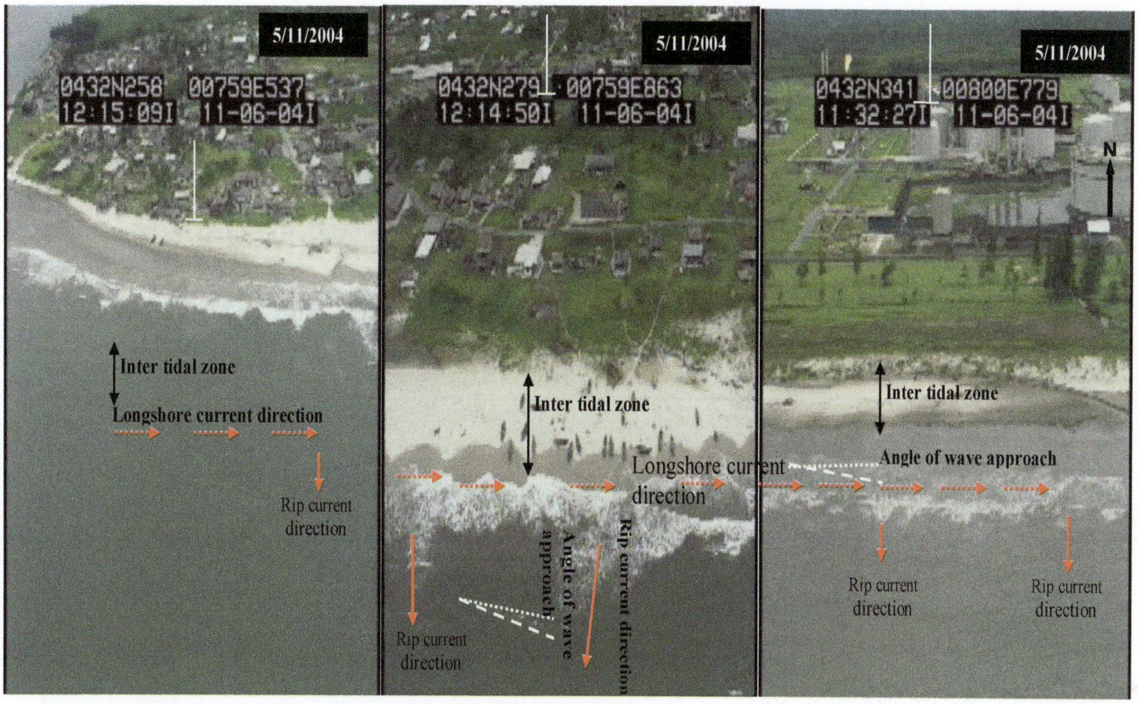

Figure 3. Sequential aerial photograph of a section of the Strand coast east of Qua Iboe River and interpreted cell circulation patterns. (Pictures taken at 4.20 pm on 5 November 2004).

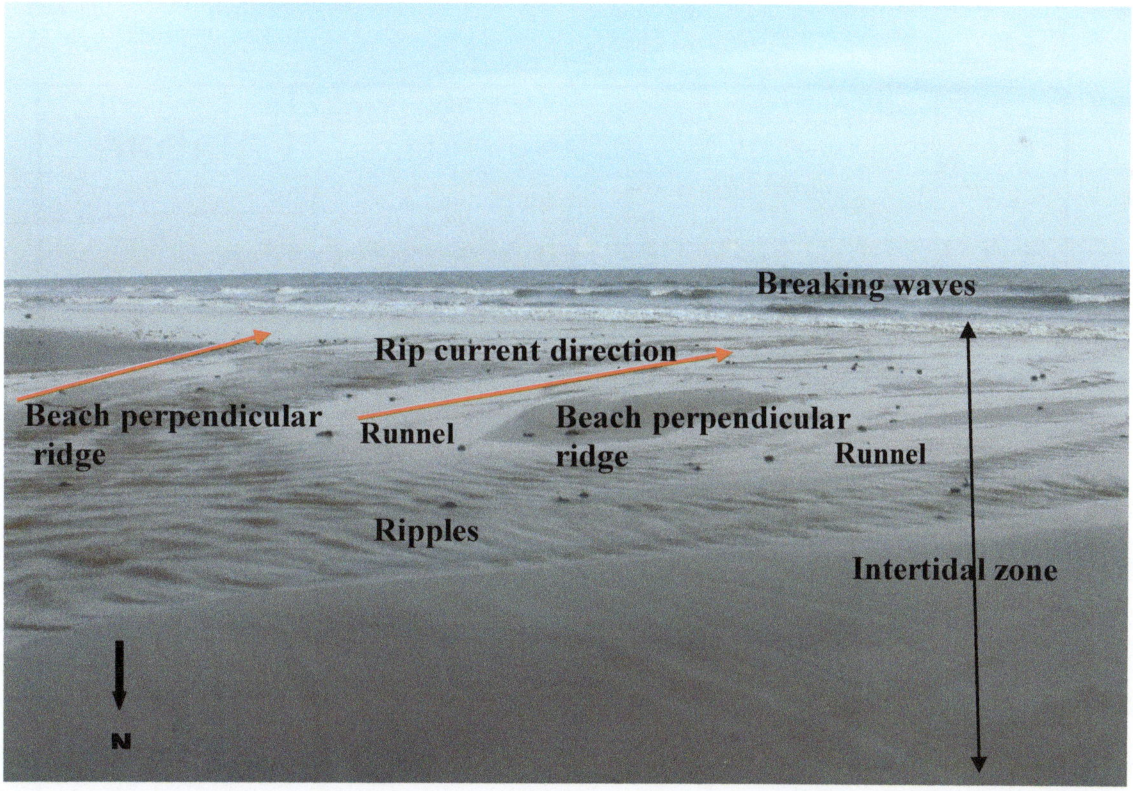

Figure 4. Picture showing morphodynamic features exposed during low tide within the intertidal zone on the shore (Lat. 04°32" 44.3'N; Long. 08°09'59.3"E) East of the Qua Iboe River (Picture taken at 4.25 pm on 5 November 2004).

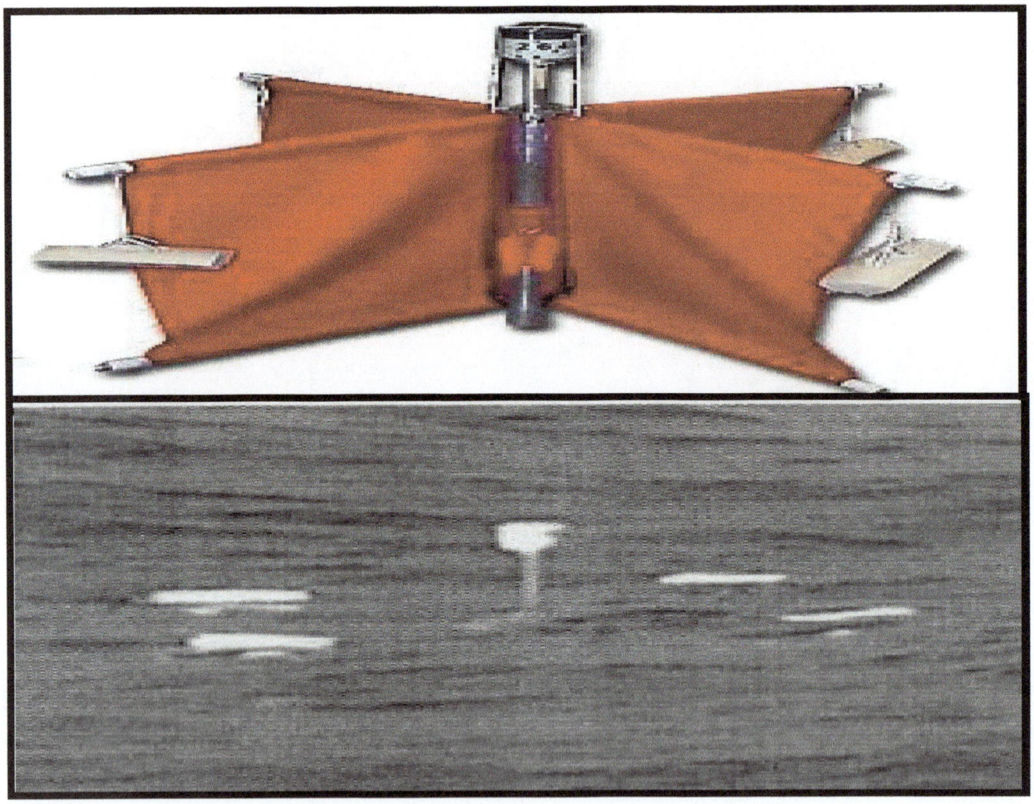

Figure 5. Davis drifter before launch (above) and after launch in water (below).

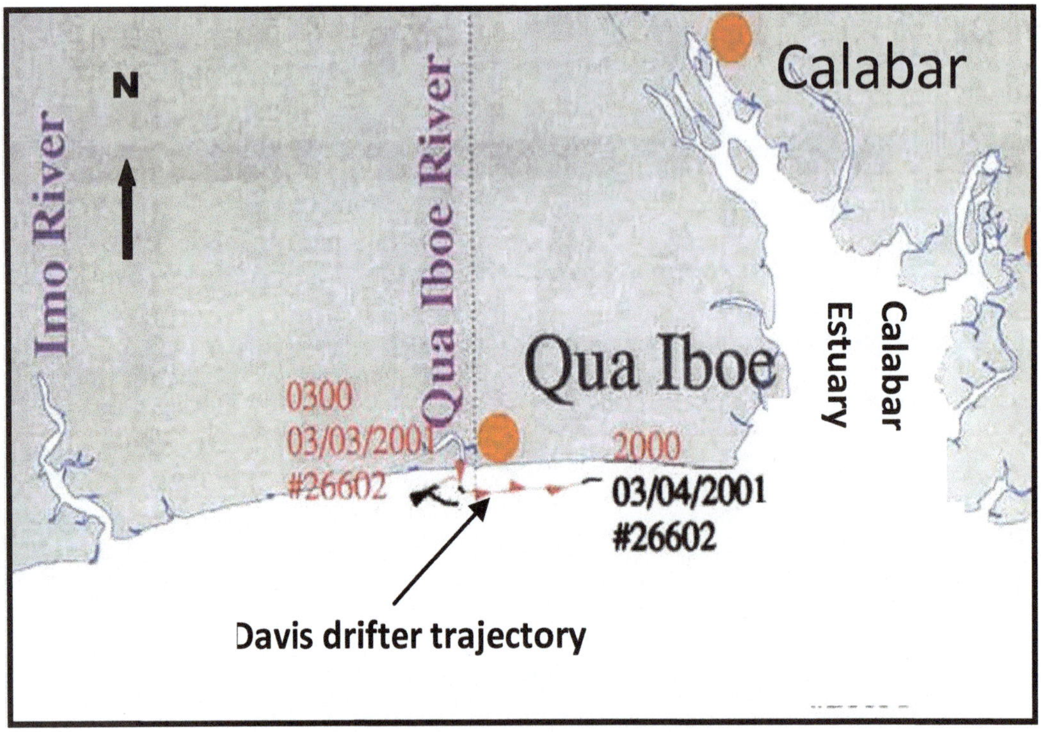

Figure 6. Drifter number 26602 trajectory deployed in the Qua Iboe estuary (Evan Hamilton, Inc., 2001).

which pushed it out of the estuary. The drifter immediately drifted due South out of the Qua Iboe River to sea. Once outside the estuary, the longshore currents were in control which caused it to drift Eastward even during flood tide which would otherwise have reversed the drifter back into the estuary during the next high tide. The drifter maintained this course for the next 24 h and ended its drifting close to the Eastern shore of the Calabar estuary.

DISCUSSION

Waves along the study area are predominantly Southwesterly as a result of the dominant Southwesterly winds. Due to the orientation of the coastline, waves arrive on the coast at oblique angles of between 10 to 15° with the waves opening to the East and hence producing a West-east longshore current. This West to East circulation cell is part of the West-east littoral cell of the eastern Niger Delta (Awosika et al., 1994). Sediment supply to the littoral drift is essentially from the Niger Delta tributaries east of the Akassa point. This coastal circulation cell is important for the transport of sediments eroded from the shoreline or transported downstream by the Qua Iboe River.

As ocean surface waves propagate over the continental shelf and break in shallow nearshore areas, their energy is dissipated in the surf zone. The gradients in the wave heights and momentum give rise to the generation of wave-induced nearshore circulation. Wave-forcing of circulation in the surf zone is a major cause of sediment transport and beach morphology evolution. Lipman and Holman (1991) described the spatial and temporal variability of sand bar (ridge) and runnel morphology. The rip currents as observed in this study have a dominant seaward (Southerly) direction and are strong enough to transport materials out of the surf zone to the inner shelf. Tidal inundation of intertidal zones along the Strand coast also plays important role in the morphology of the beach and beach sand bars. They are also very important for the transport of sediments out to sea and hence of great importance to coastal erosion and its mitigation. Flood current speed in the estuary close to the mouth of the Qua Iboe estuary is estimated at about 0.2 cm/s while ebb tidal currents average 0.6 cm/s (Awosika et al., 2006). Due to the low intertidal, beach topography and tidal range of 1.5 to 2 m, the intertidal zone of the study area as well as a large part of the beach is usually flooded during flood tide. During ebb tide, the waters recede rapidly with currents reach 0.6 m/s. Such tidal patterns are capable of shaping the intertidal zone and producing the ridge and runnel structures observed and described in this study. Similar features have also been observed by Finkl (1994) using satellite imagery.

Conclusion

Surf zones are often neglected in conventional beach surveys due to difficulty of working in this highly dynamic environment. The use of aerial photography to some extent alleviates these constraints by providing remote access to features in a format that is amenable to spatio-temporal analysis.

Relics of the beach morphodynamic structures (ridge and runnel) resulting from the cell circulation patterns and tidal inundation in the study area have also been documented. These morphodynamic structures are controlled by a combination of beach slope, tidal fluctuations, and cell circulations. Though nearshore circulation cell dynamics are normally restricted to the surf zone and the intertidal zone, they play important roles in the shaping of the beach through erosion, transport, and dispersal of sediments to the offshore and general morphology of the beach. Such knowledge is important for coastal erosion mitigation options, transport of pollutants, and oil spill along and especially close to the shoreline. Hence, this pilot study which though utilised one day photos could enhance the understanding of coastal cell circulation pattern along the Strand coast East of the Niger Delta. Future long term series data sets are needed for the holistic understanding of nearshore oceanographic processes and the resulting morphological features in the intertidal Strand coast of Nigeria.

REFERENCES

Awosika LF, Folorunsho R, Dublin-Green CO, Imevbore VO (2000). Review of the coastal erosion at Awoye and Molume areas of Ondo State. A consultancy report for Chevron Niger. Limited P. 75.

Awosika LF (2006). Coastal oceanographic processes and their impacts on the nigerian coastal and marine environment. Sustainable Management of the Nigeria Environment. Ed. MFA Ivbijaro. College Press and Publishers Ltd. Ed. FA Ivijaro et al. pp. 1399-169.

Awosika LF, Ibe AC (1994). Geomorphic features of the Gulf of Guinea shelf and littoral drift dynamics. In Nearshore Dynamics and sedimentology of the Gulf of Guinea. Edited by AC. Ibe, L Awosika, K Aka pp. 21-27.

Burke K (1972). Longshore drift, submarine canyons and submarine fans in the development of the Niger delta. AAPG Bull. 56:1975-1983.

Evan Hamilton, Inc. (2001). Nigeria OPTS Shallow Water Current Measurement Survey; Data Reports 1, 2, 3 and 4.

Finkl CW (1994). Analysis of Nearshore sediment transport patterns using sattellite imagery to delineate turbidity plumes. Proceedings of American Society of Limnology and Oceanography (ASLO) and Phycological Society of America (PSA), (1994 Joint Meeting, Miami, Florida). pp. a-25.

Furmanczyk K Musielak S, Dudzinska J, Szakowski I (2002). Bar Systems as an Indicator of Circulation Patterns in the Pomeranian Bay. In Littoral 2002. The Changing Coast. EUROCOAST / EUCC, Porto – Portugal Ed. EUROCOAST – Portugal, ISBN 972-8558-09-03053.

Ibe AC, Awosika LF (1986). Sedimentology of beaches of barrier bar complexes in Nigeria. NIOMR Tech. Paper No. 28.

Komar PD (1997). Beach Processes and Sedimentation. Englewood Cliffs, New Jersey, Prentice Hall. P. 546.

Lipman TC, Holman RA (1991). The spatial and temporal variability of

sand bar morphology. J. Geophys. Res. 95(14):1575-1590.
Masselink GA, Kroon A, Davidson-Arnott RGD (2006). Morphodynamics of intertidal bars in wave dominated coastal settings: A review. Geomorphology 23:38-49.
NEDECO (1961). The Waters of the Niger Delta. Hague P. 317.
Shepard FP, Inman DL (1951). Sand movement on the shallow inter-canyon shelf at La Jolla, California, US Dept of the Army, Corps of Engineers, Office of the Chief of Engineers, Beach Erosion Board. P. 234.
Short AD, Aagard T (1993). Single and multi bar beach change models. J. Coastal Res. SI 15:141-157.
Short AD (1999). Handbook of Beach and Shoreface Morphodynamics. West Sussex, England: Wiley, P. 379.

The impact of flow regime on the sedimentation pattern of Calabar River, South-East Nigeria

Lynda E. OKON and F. Emile ASUQUO*

Physical Oceanography Unit, Institute of Oceanography, University of Calabar, P. M. B 1115, Calabar, Nigeria.

The sedimentation pattern of the lower reaches of Calabar River, on the south east coast of Nigeria was investigated with a view to evolving criteria for recognition of their ancient counterparts. Using grain size statistical parameters, the sediment-sediment relationships and flow structure of the river were examined from twenty-five georeferenced bottom samples. Results show that the tidal channel is uniformly impacted with eroded terrestrial sediments which have been delineated into A, B, and C distinct lateral facies. Facies A and B portrayed negatively skewed and near symmetric moderately sorted sands with a considerable upstream increase in kurtosis. But, Facie C depicted near symmetric, very leptokurtic well sorted sand, representing more of terrestrial-borne sediments inferring high rate of siltation/sediment accretion. Tidal flow measurements over-spring, mean and neap tidal cycles gave 0.59 m/s highest with maximum current velocity at neap tide flowing southerly. The mean near-bottom tidal velocity indicated a maximum of 0.40 m/s, predicting that bedforms such as ripples are likely to form based on empirical relationship between flow velocity and mean grain size distribution. The observed results are geared towards understanding the impact of flow regime on sediment mobility and the preservation potential of the environment.

Key words: Tidal channel, lateral facies, sediment accretion, tidal current velocities, Calabar River, Nigeria.

INTRODUCTION

Sedimentology is basically aimed at determining and defining the sedimentary materials, sedimentary processes and the products of sedimentation of a sedimentary deposit (Friedman and Sanders, 1978). Coastal fluvial environments constitute an important region with diverse and peculiar sedimentary characteristics. These environments are widespread today as they were in geologic times and may be tidal or non-tidal. Visher (1965) defined a tidal river as an upper estuary or a fluvial estuary characterized by freshwater but subject to daily tidal actions. In a geologic sense however, any fluvial system that is not directly linked to the ocean, but depicts daily tidal action can be considered as a tidal river. The aforementioned is exemplified by the lower reaches of Calabar River. The flow is bi-directional as in an estuary, which is in contrast to a typical fluvial environment. This is similar to reports from Prestrong (1965) who described a tidal river channel within San Francisco bay that showed tidal flow in two opposite directions daily with ebb-dominance. Consequently, sedimentary structures such as asymmetric ripples, channel scours and cross stratification are likely features in such rivers (Selly, 1988).

The grain size characteristics of sediment materials in a tidal river environment are dependent on flow structure, sediment source and distance of transport (Allen, 1965). Materials transported from fluvial portions of the Calabar River therefore dominate the tidal portions of the river. Despite the general view, ebb or flood flow dominance is also dependent on seasonal fluvial discharge, climate, and local energy conditions and to an extent, sea level changes (Yoshida, 2006). Chakrabarti (1971) reported from grain size analysis and flow structure of Riparian tidal river that in the dry season, the flood flow direction dominates, with the reverse being the case in wet season, where finer grained sediments with good sorting

*Corresponding author. E-mail: fe.asuquo@yahoo.com.

were transported downstream towards the estuary during ebb flow. Chakrabarti and Lowe (1981) also showed in a theoretical analysis that flow velocity and flow depth have a strong influence on the rate of deposition of sediments in a river. Sorting of the bed materials particularly appeared to exert a strong control on both grain size and deposition rate.

Accordingly, flow depth and flow velocity were discovered to be related to bedforms, especially small ripples in an experimental study using medium sand (Banks and Collinson, 1975). The details of sedimentary structures such as bed geometry, erosional surface, texture and lateral and vertical sedimentary structural sequences can be used to identify fluid motion and direction in a river channel, and also serve as an evidence for depositional environments (Allen, 1984; Boersma et al., 1981).

Study area

The study area (04°54'15"N, 08°16'08"E) (Figure 1), known as Calabar River, lies geographically between latitude 04°54'15"N and longitude 08°16'08"E; and latitude 04°56'40"N and longitude 08°18'14"E", located in Cross River system, South Eastern Nigeria. The Calabar River takes its rise from the Oban Hills in Akamkpa, Nigeria and flows southwards through the high rain forest of the south east coast of Nigeria before discharging into the Cross River Estuary at Calabar (Akpan, 2002). The river lies in the humid tropical rain forest belt. The zone is characterized by a long wet season between April and November, and a relatively short dry season from December to March. The coastal shoreline is characterized by thick vegetation, changing from freshwater to swamp ecology, to mangrove swamp towards the estuary. On the banks of the Calabar River, nipa palm *(Nypa fruticans)* of the family of *Arecaceae* dominates (Asuquo, 1999).

The geology of Cross River basin consists of crystalline basement complex, overlain by various sedimentary formations. The sediment texture ranged from silty-clay to sandy-clay with the abundance of silt/clay in the lower reaches (estuary) and sandy mud in the upper reaches predominating at regions < 2 m to the shoreline (Asuquo and Ewa-Oboho, 2006). The Calabar River, from its source cuts through black shales, siltstone, clayey and silt deposits, and also alluvial deposits before entering the Cross River at Alligator Island (Asuquo, 1999). The tidal Calabar River is generally devoid of strong ocean waves. However, it is influenced by wind induced waves from the south-west trade winds, and shockwaves produced by moving boats. The Calabar River has a semi-diurnal tidal range of 1.50 to 2.50 m (Asuquo and Tambe, 2010), and experiences both flood and ebb tidal stages, with the flood direction being northwards and southwards in the ebb direction. Activities such as fishing, lumbering, local and large scale agriculture, dredging, craftsmanship, and recreation are among the various occupations of the coastal dwellers around Calabar River.

MATERIALS AND METHODS

Five sets of bed samples were obtained from the river channel (about 6 km stretch) along georeferenced transect spaced at about 1.2 km apart from the river banks (Figure 2). Sediment samples were obtained using a Van veen grab sampler aboard a motor boat and stored in labeled bags. Samples were then oven dried at 75°C and sieved with Ro-tap sieve assemblage (25 mm to 20 µm) to obtain various sand sized fractions. Various size fractions were obtained from sieve analysis. Grain size statistical parameters were obtained using the formulation and verbal description of Folk and Ward (1957) represented as:

$$\text{Mean grain size, } (M_z) = \frac{\phi 16 + \phi 50 + \phi 84}{3}$$

$$\text{Sorting } (\sigma) = \frac{\phi 84 - \phi 16}{4} + \frac{\phi 95 - \phi 5}{6.6}$$

$$\text{Skewness, } (S_{KI}) = \frac{\phi 84 + \phi 16 - 2(\phi 50)}{2(\phi 84 - \phi 16)} + \frac{\phi 95 + \phi 5 - 2(\phi 50)}{2(\phi 95 - \phi 5)}$$

$$\text{Kurtosis, } (K_G) = \frac{\phi 95 - \phi 5}{2.44(\phi 75 - \phi 25)}$$

Where Phi (ϕ) values represent cumulative percentiles.

River depth was also determined at strategic georeferenced locations using a depth sounder (Asuquo and Tambe, 2010). Flow direction and tidal flow velocity were determined using the Langragian method for neap, mean and spring tidal cycles (Asuquo, 1991). Possible bedform types were inferred from the stability diagram by Middleton and Southward (1978). This was achieved by relating tidal current velocity to mean grain size of sediments.

RESULTS

The particle size analyses of the river sediments are presented in Table 1. An analyses of the grain size statistical parameters and surface sediment distribution indicated that the mean grain size for the river channel sediments ranged between 0.85 to 3.48 ϕ, inferring increased deposition of terrestrial materials (Table 2a). Very fine grained sand dominates with patches of fine, medium and coarse sand close to the mouth with fine and medium sized particles further upstream. Sediments generally depict an upward fining pattern towards

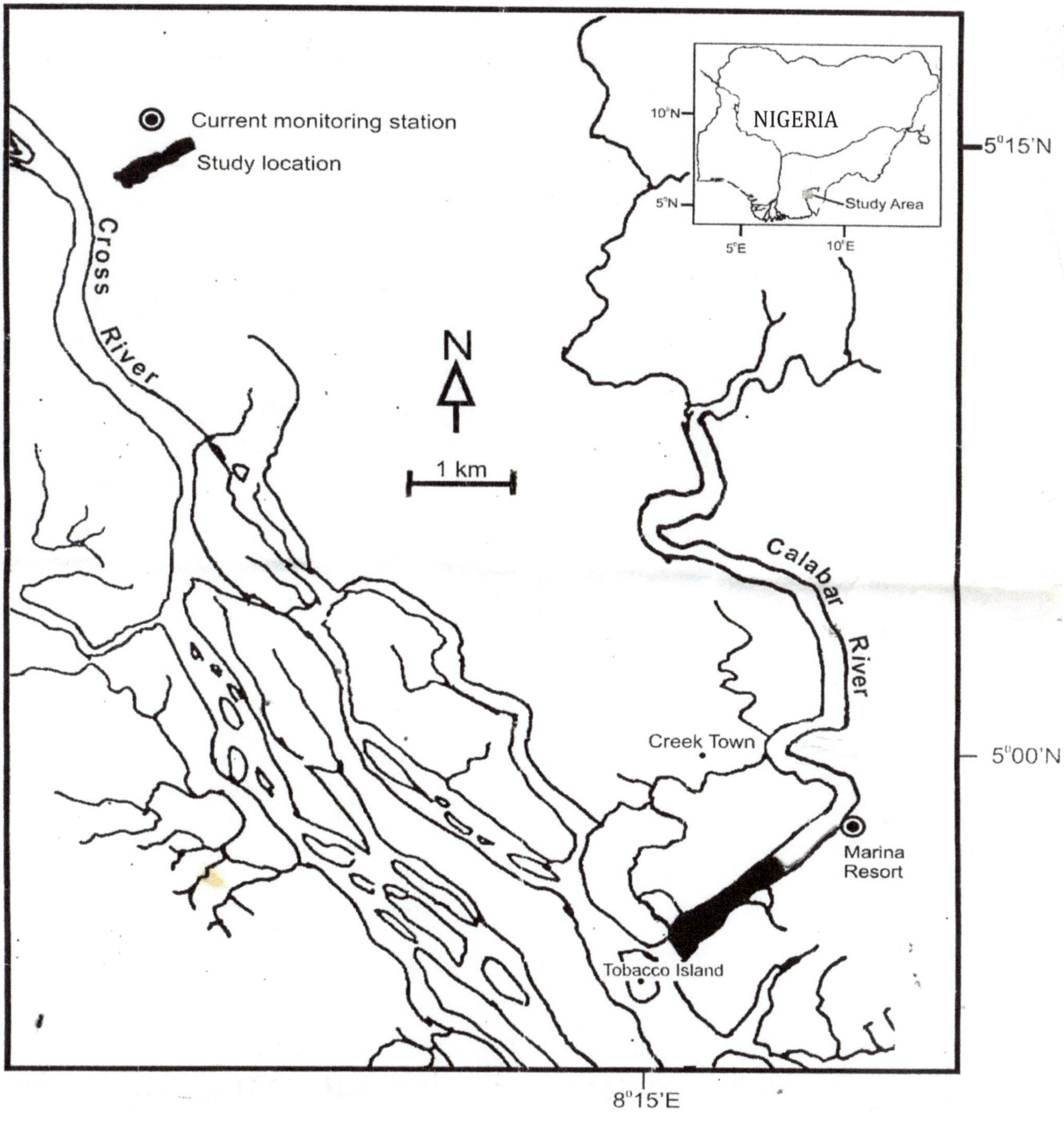

Figure 1. Map of the Cross River system showing Calabar River study location.

the banks from the mid-channel in the downstream direction, except for some portions with minor inclusions of coarse sands at the banks.

Sediments in the river varied from well sorted to poorly sorted sands within the range of 0.41 to 1.42 (Table 2a). The lower reaches of the river, towards the river mouth was generally moderately sorted. However, the upper portion of the studied area was dominated by moderately well sorted sediments. Sorting generally deteriorated away from the mid-channel, and improved downstream.

The river sediments were generally negatively skewed to positively skewed, ranging from -0.4 to 0.31 (Table 2a). Sediments were dominantly negatively skewed within the lower reaches of the river, close to the entrance. The upper portion of the river was dominated by near symmetric sediments. There was a significant decrease

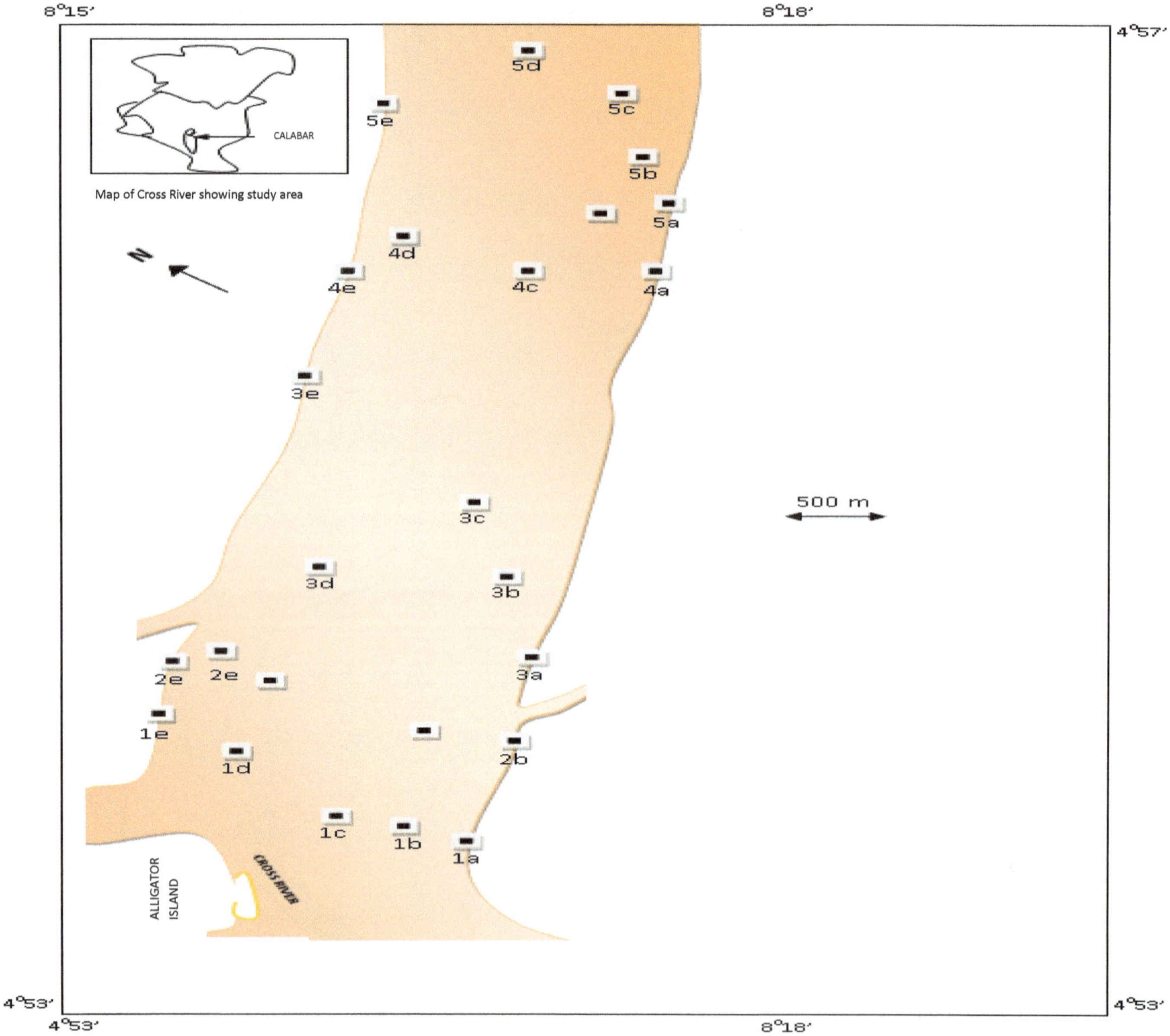

Figure 2. Sediment sampling stations at Calabar River, Nigeria.

in skewness from the east to the west bank, with an increasing trend upstream. Kurtosis within the studied area ranged from platykurtic to extremely leptokurtic (0.78 to 3.09) (Table 2b). Sediments were generally mesokurtic close to the mouth, and dominantly very leptokurtic further upstream. The middle portion of the river was dominated by leptokurtic sediments. The trend revealed a decrease in kurtosis from the east to the west bank. A bathymetric map was produced for the studied segment of the river. This revealed a depth range of about 7 to 11 m in the mid-channel, and about 2 m at the river banks (Figure 3). The mean monthly depth and tidal elevation is presented in Table 3. Study during neap, mean and spring tidal phases revealed maximum surface tidal current velocities of 0.59, 0.40, and 0.37 m/s respectively (Table 4 a, b, and c). Utilizing the stability diagram by Middleton and Southward (1978), (Figure 4) possible bedform types inferred for the studied area were small ripples in fine and medium sands at velocity of 0.40 m/s (maximum surface current velocity of approximately 0.59 m/s).

Three distinct facies delineated in the studied area, based on sorting, skewness and kurtosis were represented as Facies A, B, and C (Figure 5). Facie A, extending about 1.15 km and 11 m thick consists of mainly moderately sorted, near symmetric to dominantly negatively skewed, and mesokurtic sands. Facie B stretching 2.6 km from the distal end of Facie A, and 7 m

Table 1. Results on grain size analyses of Calabar River sediments (%).

Station No.			1					2				3				4				5			
			A	B	C	D	E	A	C	D	E	A	B	C	D	A	B	C	D	A	B	C	D
Weight retained (%)	Grain Size	-1	-	-	0.5	-	-	-	-	-	3.3	-	-	-	-	-	-	-	-	-	-	-	-
		0	-	-	3	-	-	-	-	-	14.7	-	-	-	-	-	-	-	-	-	-	-	1
		1	-	-	7	-	-	-	-	-	31.3	-	-	-	-	-	-	-	-	-	-	3	18.7
		2	-	14	54	16	-	-	9	6	42	-	-	-	6.7	30.7	4	4		34.7	8	68	68
		3	23	24	32	22	25	24	21	42	6	24	20	9.3	15.3	18	17	19	18	24.7	18	21	10.7
		4	57	50	3	44	52	60	54	45.3	2.7	57	53	80	66.3	34.7	67	66	62	23.3	62	7	1.3
		5	20	12	0.5	18	23	16	16	6.7	-	19	48	10.7	14.7	16.6	12	11	20	13.3	12	1	0.3
Cumulative weight (%)	Grain Size	-1	-	-	0.5	-	-	-	-	-	3.3	-	-	-	-	-	-	-	-	-	-	-	-
		0	-	-	3.5	-	-	-	-	-	18	-	-	-	-	-	-	-	-	-	-	-	1
		1	-	-	10.5	-	-	-	-	-	49.3	-	-	-	-	-	-	-	-	-	-	3	19.7
		2	-	14	64.5	16	-	-	9	6	91.3	-	-	-	6.7	30.7	4	4		34.7	8	71	88.7
		3	23	38	96.5	38	25	24	30	48	97.3	24	20	9.3	22	48.7	21	23	18	59.4	26	92	98.4
		4	80	88	99.5	82	77	84	80	93.3	100	81	52	89.3	88.3	83.4	88	89	80	86.7	88	99	99.7
		5	100	100	100	100	100	100	100	100	-	100	100	100	100	100	100	100	100	100	100	100	100

thick consists of moderately sorted, negatively skewed and near symmetric leptokurtic sands. Facie C extending about 2.4 km upstream of Facie B and 5 m thick consists mainly of moderately well sorted to moderately sorted sands, dominantly near symmetric with a leptokurtic to very leptokurtic distribution.

DISCUSSION

Results of the study showed well sorted finely grained sediment in the downstream direction, as well as towards the banks from the mid-channel (Figure 6a). The former conforms to the general rule that mean grain size would decrease in the direction of dominant flow (Shultz, 1941; Plumbey, 1948; Chakrabarti, 1971). Also, the deposition of fines towards the river mouth can be attributed to the effect of flow expansion, downstream at the wider (1.5 km) and deeper (11 m) river mouth. This is because of the dissipation of energy which causes the settling of fine sediments that were in suspension as energy reduces. The fining pattern of sediments towards the banks from the mid-channel is consistent with Semeniuk's (1981) observation, that continuous winnowing in a tidal channel gradually removes finer sediments from the channel and deposits them on the banks, leaving clearly washed coarse fragments within the channel. The slight improvement in the downstream direction indicates a downstream movement of sediments with accompanied accretion/siltation of the channel during ebb tide.

However, in the upstream portion of the studied area, a large patch of moderately sorted sands indicates a deposition of sediments by flood flow in the upstream direction (Figure 6b). These changes in sorting along the channel may be attributed to erosion and deposition caused by the bi- directional flow in the tidal river, which is evident from the flood-ebb time and velocity asymmetry which is almost symmetric in the study area. The observed patterns are consistent with the views of Chakrabarti (1971), and Krumbien and Sloss (1963) who suggested that in the direction of transport, sediments become better sorted. Also, the deterioration in sorting away from the mid-channels is an indication of winnowing of fines to the banks leaving better sorted and coarser sediments within the channel (Seminiuk, 1981; Buller and McManus, 1975; Gibbs, 1973). Skewness in the channel generally depicts a decreasing trend from near symmetric in the downstream direction, to negatively skewed further downstream (Figure 7a). This is consistent with the views of Selley (1988), who suggested

Table 2a. Summary statistics of grain size statistical parameters – mean grain size and sorting.

Transect no.	Sample station	Mean (Mz) (phi)		Sorting (σ)	
		Values	Interpretation	Values	Interpretation
	A	3.45	Very fine sand	0.83	Moderately sorted
	B	3.22	Very fine sand	0.84	Moderately sorted
1	C	1.71	Medium sand	0.71	Moderately sorted
	D	3.10	Very fine sand	1.04	Poorly sorted
	E	3.45	Very fine sand	0.73	Moderately sorted
	A	3.40	Very fine sand	0.61	Moderately well sorted
	B	3.23	Very fine sand	0.79	Moderately sorted
2	C	3.27	Very fine sand	0.82	Moderately sorted
	D	2.97	Fine sand	0.85	Moderately sorted
	E	0.85	Coarse sand	0.95	Moderately sorted
	A	3.45	Very fine sand	0.74	Moderately sorted
3	B	3.82	Very fine sand	0.82	Moderately sorted
	C	3.40	Very fine sand	0.41	Well sorted
	D	3.35	Very fine sand	0.73	Moderately sorted
	A	2.67	Fine sand	1.42	Poorly sorted
	B	3.35	Very fine sand	0.63	Moderately well sorted
4	C	3.30	Very fine sand	0.63	Moderately well sorted
	D	3.48	Very fine sand	0.59	Moderately well sorted
	E	2.53	Fine sand	1.32	Poorly sorted
	A	3.23	Very fine sand	0.75	Moderately sorted
5	C	2.60	Fine sand	0.45	Well sorted
	D	1.73	Medium sand	0.85	Moderately sorted

that finer sediments are weighted to the negative while coarser sediments tend to the positive considering the prevailing energy conditions and the homogeneity of the sediments. Therefore, the decrease in skewness towards the negative in the downstream direction, which corresponds with a downstream decrease in mean grain size of sediments in Calabar River, which is an indication of the dependence of skewness on mean grain size and the flow conditions typical of the studied area. Kurtosis generally decreased from very leptokurtic to mesokurtic downstream (Figure 7b). This can be related to sorting in conformity with the views of Phillip et al. (1968) and Folk and Ward (1957), who associated leptokurtic and very leptokurtic sediments to best sorting (Figures 6b and 7b).

Possible bedform types were inferred for Calabar River based on empirical relationship between mean grain size and mean flow velocity of Middleton and Southard (1978). At maximum near-bottom velocity of 0.42 m/s ripples are formed within very fine sand. The same applies for fine sand and medium sand. In Calabar River, the cross-stratified small ripples are oriented in the ebb-flow direction (downstream). This conforms to the views by Reineck and Singh (1980), who suggested that tidal channels are dominated by asymmetrical ripples with cross stratified internal structure. Chakabarti (1971) also reported a likely maximum sediment transport in the ebb-flow direction during wet season. This proposes a high preservation potential for small ripples as preserved bedforms in Calabar River for rock record reconstruction.

Conclusion

This preliminary report on the lithology of a segment of Calabar River revealed that the channel is characterized by coarse to very fine sand, well to poorly sorted, very negatively to very positively skewed, and platykurtic to extremely leptokurtic sediments. Three distinct facies (A, B and C) delineated based on the interpretation of grain size statistical parameters is highly relevant considering the magnitude and source of the sedimentary materials, which could be utilized as depositional indicators for developing conceptual models for the management of the channel and paleo-environmental reconstruction. The

Table 2b. Summary statistics of grain size statistical parameters – Skewness and kurtosis.

Transect no.	Sample station	Skewness (Ski)		Kurtosis (Kg)	
		Values	Interpretation	Values	Interpretation
1	A	-0.12	Near symmetric	1.36	Leptokurtic
	B	-0.05	Near symmetric	1.03	Mesokurtic
	C	-0.05	Near symmetric	1.01	Mesokurtic
	D	-0.61	Negatively skewed	1.02	Mesokurtic
	E	-0.03	Near symmetric	0.88	Platykurtic
2	A	0.19	Positively skewed	1.18	Leptokurtic
	B	-0.40	Very negatively skewed	1.43	Mesokurtic
	C	-0.21	Negatively skewed	1.31	Mesokurtic
	D	-0.38	Very negatively skewed	3.09	Extremely
	E	0.16	Positively skewed		Leptokurtic
3	A	-0.18	Negatively skewed	1.04	Mesokurtic
	B	-0.24	Negatively skewed	0.83	Platykurtic
	C	0.30	Positively Skewed	1.37	Leptokurtic
	D	0.11	Negatively Skewed	1.42	Leptokurtic
4	A	-0.33	Very negatively skewed	0.78	Platykurtic
	B	-0.13	Negatively skewed	1.51	Very Leptokurtic
	C	-0.03	Near symmetric	1.48	Leptokurtic
	D	0.31	Very positively skewed	1.07	Mesokurtic
	E	-0.06	Near symmetric	0.78	Platykurtic
5	A	-0.22	Near symmetric	1.58	Very Leptokurtic
	C	0.09	Near symmetric	1.73	Very Leptokurtic
	D	0.22	Positively skewed	1.84	Very Leptokurtic

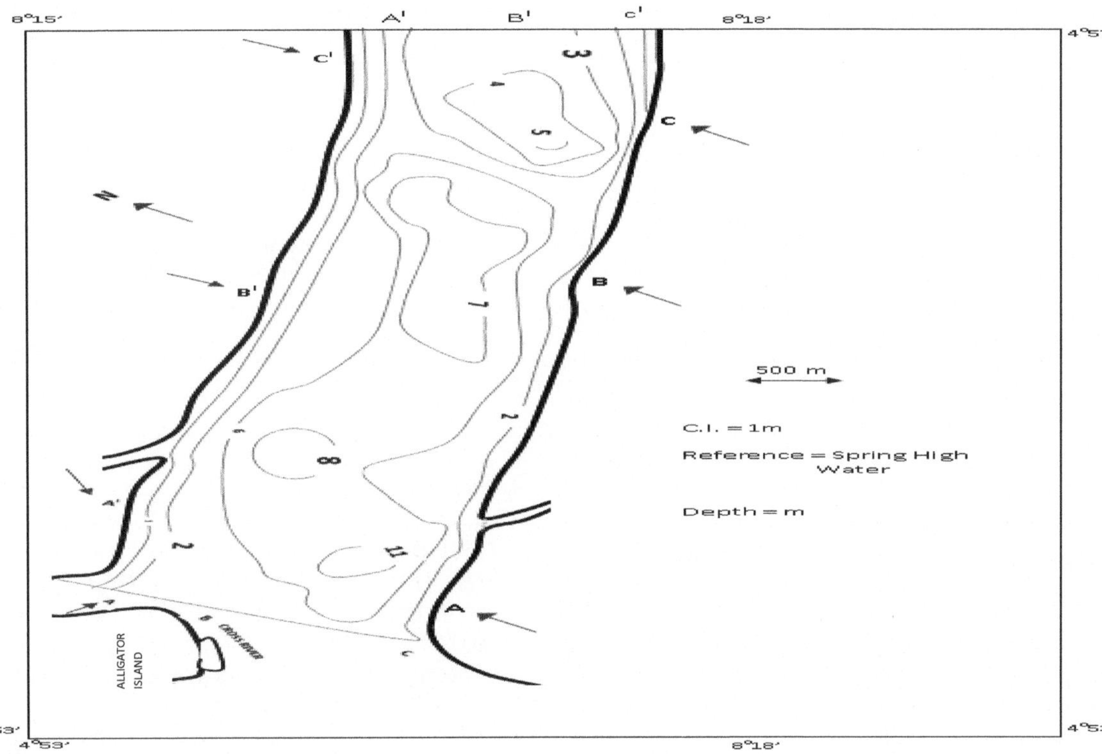

Figure 3. Contour profiles of the bathymetry of lower Calabar River, Nigeria.

Table 3. Values of mean monthly depth and tidal elevations at the sampling locations in Calabar River (April to September 2006).

Transect	Sample stations	Geographical coordinates		Mean monthly Depth (m)	Mean tidal elevation (m)
		Latitude	Longitude		
	A	4°54115"N	08°16'08E	4.25	1.35
	B	4°54^{1}17N	08°16'06E	4.35	1.37
1	C	4°54^{1}32N	08°16'59E	4.40	1.4
	D	4°54^{1}47N	08°16'52E	4.45	1.43
	E	4°54^{1}46N	08°16'38E	4.50	1.45
	A	4°54^{1}39N	08°16'28E	5.05	1.56
	B	4°54^{1}47N	08°16'11E	5.02	1.50
2	C	4°54^{1}54N	08°16'00E	5.00	1.50
	D	4°54^{1}58N	08°16'51E	4.55	1.47
	E	4°54^{1}00N	08°16'48E	4.52	1.45
	A	4°54^{1}52N	08°16'41E	5.13	1.55
	B	4°54^{1}03N	08°16'40E	5.15	1.55
3	C	4°54^{1}18N	08°16'39E	5.20	1.56
	D	4°54^{1}32N	08°16'37E	5.25	1.58
	E	4°54^{1}50N	08°16'41E	5.27	1.58
	A	4°54^{1}00N	08°16'37E	5.45	1.65
	B	4°54^{1}59N	08°16'25E	5.40	1.65
4	C	4°54^{1}58N	08°16'12E	3.35	1.63
	D	4°54^{1}59N	08°16'03E	5.33	1.63
	E	4°54^{1}02N	08°16'55E	5.30	1.60
	A	4°54^{1}12N	08°16'49E	5.47	1.68
	B	4°54^{1}22N	08°16'46E	5.53	1.70
5	C	4°54^{1}26N	08°16'41E	6.00	1.70
	D	4°54^{1}38N	08°16'04E	6.05	1.72
	E	4°54^{1}40N	08°16'14E	6.10	1.75

Table 4a. Data on surface current velocity – Neap tides (April –September 2006).

Time of day	Tidal phase	Range surface current velocity (m/s)	Mean surface current velocity (m/s)	Flow direction
Between 7.00 and 11.30 am	Flood	0.07 – 0.30	0.18(0.01)	North
Slack Tide at 11.45 am	Slack Tide	0.0	0.0	-
Between 12.15 and 18.45 pm	Ebb	0.11- 0.59	0.26(0.02)	South

Values in parentheses are the standard deviation values between 15 to 25 set of readings.

Table 4b. Data on surface current velocities – Mean tides (April to September, 2006).

Time of day	Tidal phase	Range surface current velocity (m/s)	Mean surface current velocity (m/s)	Flow direction
Between 7.00 and 10.00 am	Ebb	0.10 – 0.29	0.19 (0.12)	South
Slack Tide at 10.15 am	Slack tide	0.0	0.0	-
Between 10.30 and 14.15 pm	Flood	0.09 - 0.38	0.19 (0.11)	North
Slack Tide at 14.30 pm	Slack tide	0.0	0.0	-
Between 14.45 and 18.45 pm	Ebb	0.05 - 0.06	0.14 (0.03)	South

Values in parentheses are the standard deviation values between 15 to 25 set of readings.

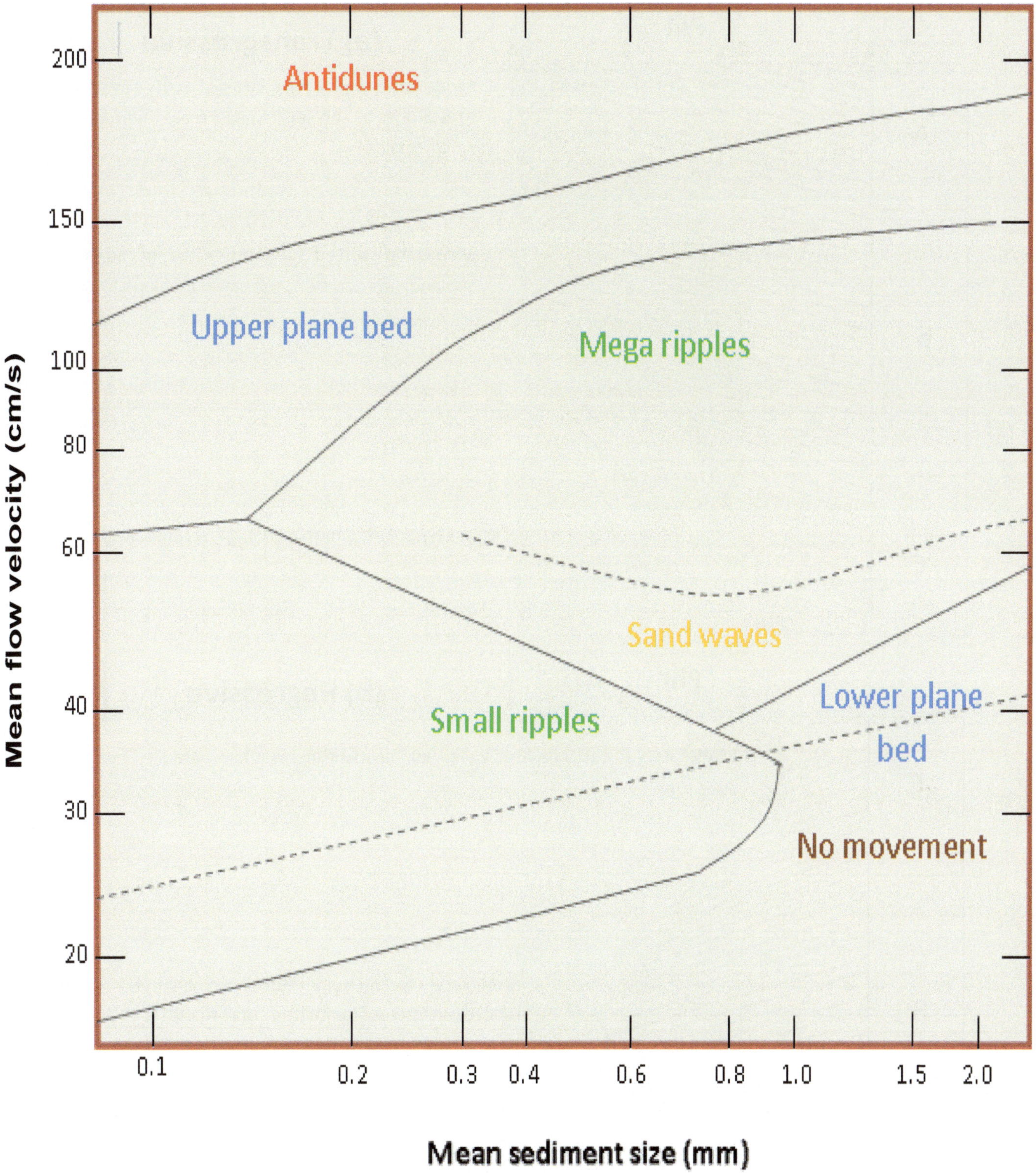

Figure 4. Mean grain size/ flow velocity diagram (Middleton and Southward, 1978).

enhanced mobilization of terrestrial-borne sediments by the inherent flow regime that resulted in frequent siltation of the channel suggest regular opening up /dredging of the navigational channel for accessibility by sea going

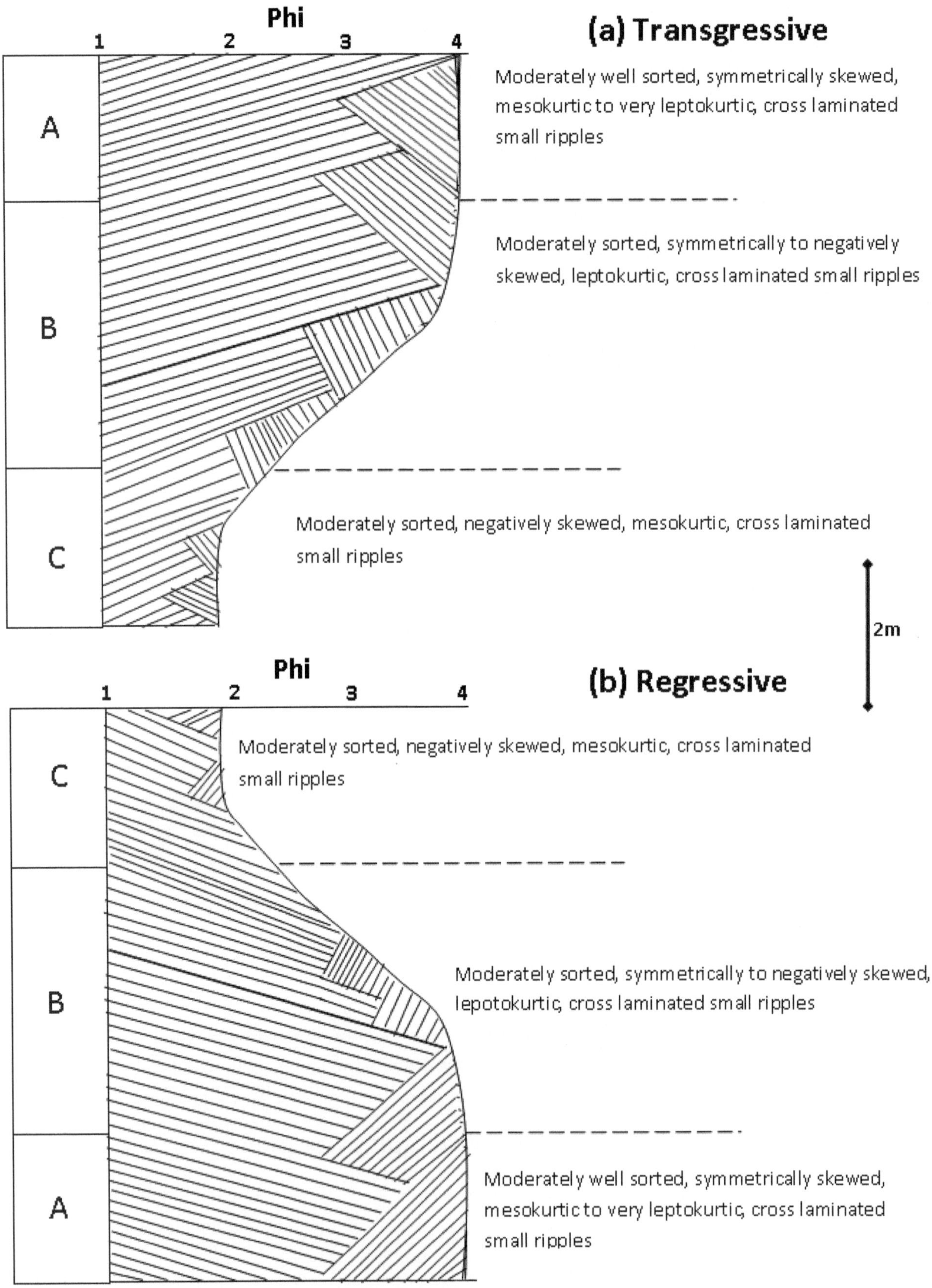

Figure 5. Typical stratigraghpic sequence models of Calabar Tidal River deposits; (a) Transgressive sequence, (b) regresive sequence.

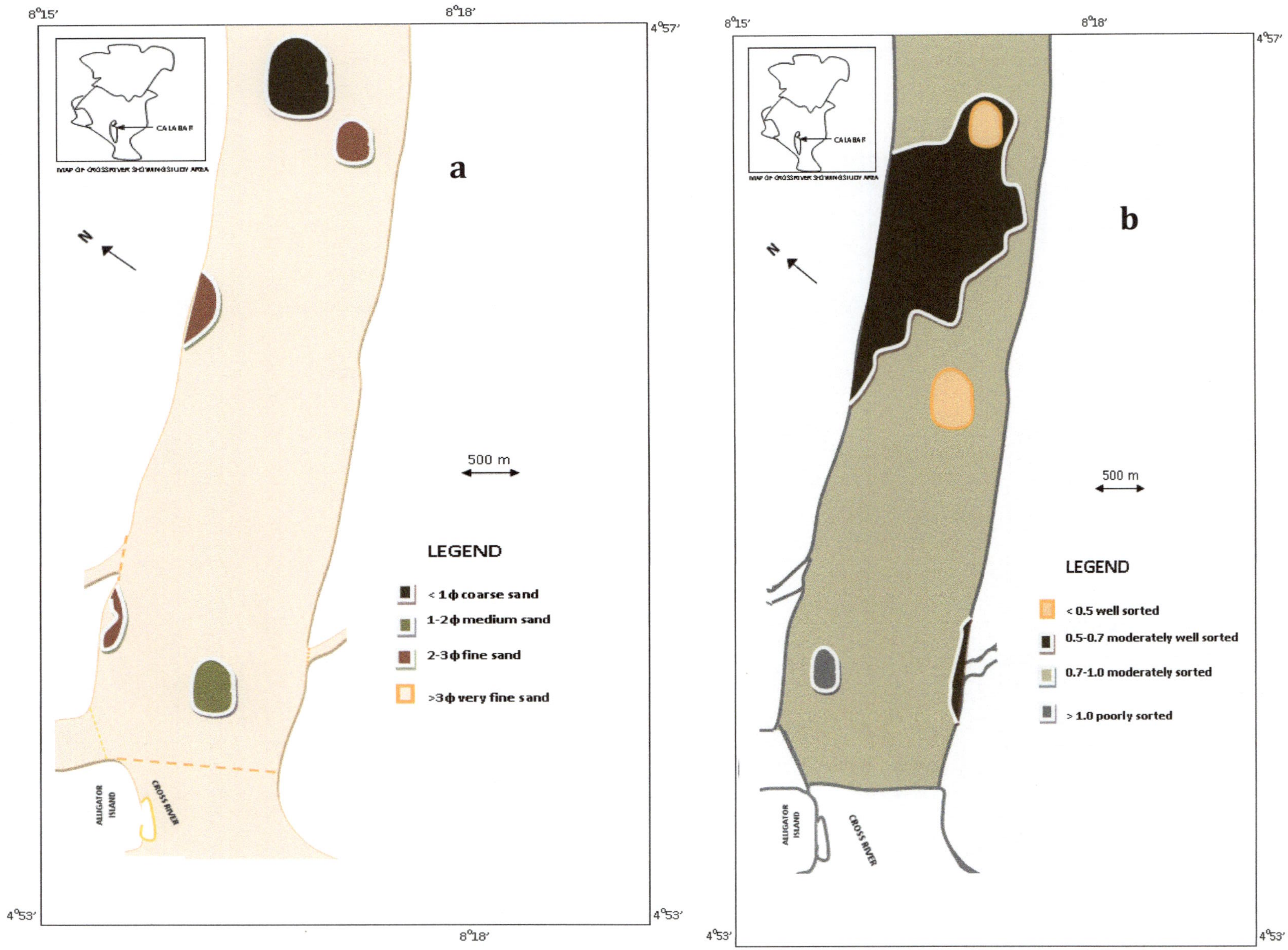

Figure 6. Surface Distribution of Upper Calabar River sediments; (a) Mean (b) Sorting.

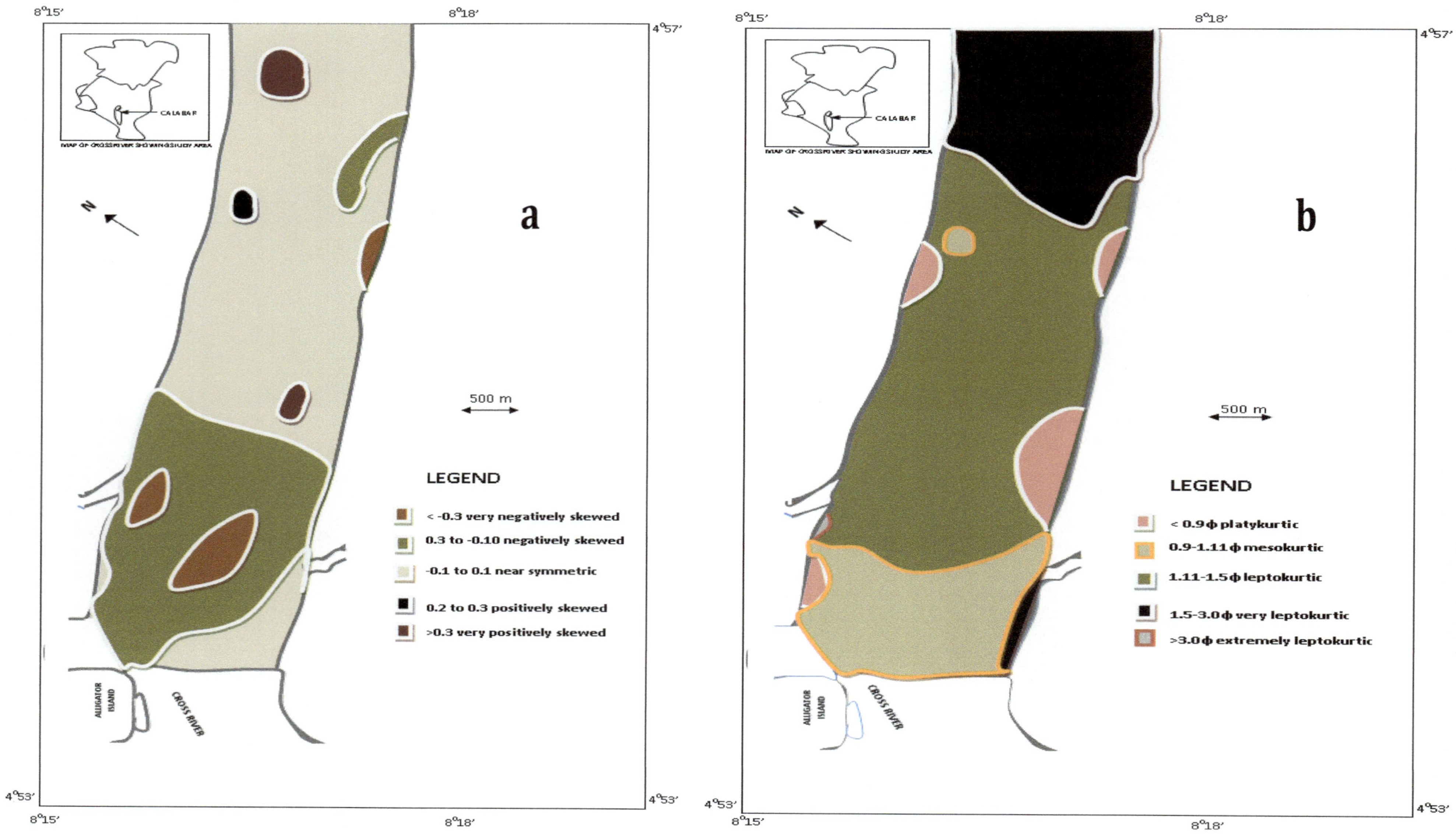

Figure 7. Surface distribution of skewness and kurtosis in lower Calabar River sediments.

vessels.

REFERENCES

Akpan ER (2002). Heavy Metal trends in The Calabar River, Nigeria. Environ. Geol., 42: 47-51.

Allen JRL (1965). Late Quaternary Niger Delta and adjacent areas: sedimentary environments and litho-facies. Bull. Am. Assoc. Pet. Geol., 49: 547-600.

Allen JRL (1984). Sedimentary structures, their characterization and physical basis. Developments in Sedimentology, 30.

Asuquo FE (1991). Tar balls on Ibeno- Okposo beach of South-east Nigeria. Mar. Pollut. Bull., 22: 150-151.

Asuquo FE (1999). Physiochemical characteristics and anthropogenic pollution of Calabar River, Nigeria. Glob. J. Pure Appl. Sci., 5(4): 595-600.

Asuquo FE, Ewa-Oboho I (2006). Chemical Components and Nutrients in Mangal Sediments of Cross and Qua Iboe rivers, South-East Nigeria. J. Environ. Syst., 32: 37-54.

Asuquo FE, Tambe NN (2010). Hydrography and Chemistry of two Nigerian rivers:Calabar and Great Kwa rivers, Southeastern Nigeria. Afr. J. Mar. Sc. (submitted).

Banks NL, Collinson JD (1975). The size and shape of small scale ripples: an experimental study using medium sand. Sedimentology, 22: 583-599.

Boersma JR, Terwindt JHL (1981) Neap-Spring tide sequences of Intertidal shoal deposits in a mesotidal Westerschelede Estuary. Sedimentology, 28:151-170.

Buller AT, Mcmanus J (1975) Sediments of the Tay River Estuary In: Dyer, K. R. (1979) Estuarine hydrography and sedimentation. Cambridge Universality Press, Cambridge, London, pp. 87-130.

Chakrabarti AK (1971). Studies on Sediment movement at the entrance of a tidal river (River Riparian). Sediment. Geol., 6: 111-127.

Chakrabarti C, Lowe DR (1981). Diffusion of sediments on the lee of dune-like bedforms: Theoretical and Numerical analysis. Sedimentology, 28: 531-535.

Folk RL, Ward WC (1957). Brazos River: A Study in the Significance of Grain Size Statistical Parameters. J. Sediment. Petrol., 27: 3-26.

Friedman GM, Sanders JE (1978). Principles of Sedimentology. John Wiley and sons, New York, p. 72.

Gibbs RJ (1973). Amazon River estuarine system In: Hails, J. and Carl, A. (1975) Near shore sediment dynamics and sedimentation. John Wiley and Sons. London, p. 363.

Krumbein WC (1934). Size frequency distribution of Sediments. J. Sed. Petrol., 4: 65-77.

Middleton V, Southard JB (1978) Mechanics of sediment movement. Soc. Econ. Paleotol., Mineralogists short course, 3: 246.

Phillip G, Saadalla, TA (1968). Mechanical analysis and mineral composition of the middle Triassic Ga'ara sandstone (Iraq). Sediment. Geol., 2: 51-76.

Prestrong R (1965). The Development of drainage patterns of tidal marshes In: Prestrong R (1972) Tidal flat Sedimentation at Cooley landing, Southwest San Francisco bay. Sediment. Geol., 8: 257-288.

Plumbey WJ (1948). Black Hills terrace gravels: A study in sediment transport In: Chakrabarti, A. K. (1971) Studies on sediment movement at the entrance of a tidal river (River Riparian). Sediment. Geol., 6: 111-127.

Prestrong R (1965). The Development of drainage patterns of tidal marshes In: Prestrong, R. (1972) Tidal flat Sedimentation at Cooley landing, Southwest San Francisco bay. Sediment. Geol., 8: 257-288.

Reineck HE, Singh IB (1980). Depositional sedimentary environments, 2nd edition. Springer – Verlag, New York, 549 pp.

Selley RC (1988). Applied Sedimentology. Harcourt Brace Jovanovich publishers, Academy press Ltd. London, 346 pp.

Semeniuk V (1981). Sedimentology and stratigraphic sequence of a tropical tidal flat, King Sound, Northwest Australia. Sediment. Geol., 29: 195-221.

Shulits S (1941). Rational equation river bed profiles. In: Chakrabarti (1971) studies on sediment movement at the entrance of a tidal river (Rupnarian). Sediment. Geol., 6: 111-127.

Visher GS (1965). Fluvial processes as Interpreted from ancient and recent fluvial deposits. Soc. Econ. Paleont. Miner. Spec. Publ., 12: 116-132.

Yoshida S (2006). Quantitative 3-D Facies Analysis of a tidal point bar versus shallow marine tidal bars: comparative studies of modern and ancient tidal systems for reservoir characterization. Recent advances for reservoir characterization 1 *SEPM).* AAPG Annual Convention, Technica Program.

The effect of vitamin E and highly unsaturated fatty acid on growth, survival and haematocrit of goldfish (*Carassius auratus gibellio*)

Zeinab Hanaee Kashani*, Mohamad Reza Imanpoor, Ali Shabani and Saeed Gorgin

Department of Fisheries, Gorgan University of Agricultural Sciences and Natural Resources, Gorgan, Iran.

This study was conducted to examine the effects of different content of vitamin E and highly unsaturated fatty acid on growth and survival in goldfish. Goldfish with initial weight 0.69±0.12 g were fed with diets having different levels of vitamin E and highly unsaturated fatty acid (E100 + HUFA, E50 + HUFA, -E + HUFA, -E - HUFA) for 10 weeks. At the end of period, the final weight, specific growth rate, feed conversion ration and condition factor were measured. The result of this study showed that the difference between the final weights, specific growth rate, feed conversion ratio between treatments were not significant (P>0.05) but in control group was less than other groups.

Key words: Fish growth, fish survival, vitamin E, highly unsaturated fatty acid, goldfish.

INTRODUCTION

Goldfish is a relatively small member of the carp subfamily (Cyprininae) that includes the common carp (*Cyprinus carpio*) and Crucian carps (Genus: Carassius) (Zhen, 1988). Vitamin E is a lipid-soluble vitamin that comprises four tocopherols and four tocotrienols in nature. Among them, a-tocopherol has the highest vitamin E activity (NRC, 1993).

One factor that affects the dietary vitamin E requirement is the oxidative stability of the diets (Huang et al., 2004). Addition of vitamin E to rancid diets significantly improved growth performance of the fish (Baker and davis, 1996). Vitamin E is to prevent peroxidation of polyunsaturated fatty acids of phospholipids and cholesterol in cellular and subcellular membranes. Most of the deficiency signs observed in fish, such as nutritional muscular dystrophy, fatty liver degeneration, anemia, erythrocyte hemolysis, hemorrhage, depigmentation, and reduction of fertility, are related to peroxidative damage to cellular membranes (NRC, 1993). As a membrane-bound antioxidant, vitamin E appears to scavenge free radicals at the site of their formation.

High levels of n-3 HUFA, which mainly consist of eicosapentaenoic acid (EPA) and docosahexaenoic acid (DHA), are found in marine organisms. These fatty acids in the phospholipids fraction of the biomembrane in marine species are observed to play a critical role in membrane fluidity (Sargent, 1995; Takeuchi, 1997). The n-3 highly unsaturated fatty acids (HUFA) eicosapentaenoic acid (EPA, 20:5n_3) and docosahexaenoic acid (DHA, 22:6n-3) have been reported to be essential for optimal growth of several species of bivalve larvae (Waldock and Nascimento, 1979; Waldock and Holland, 1984; Delaunay et al., 1993) and juveniles (Langdon and Waldock, 1981; Hurd et al., 1989; Parrish et al., 1995; Knauer and Southgate, 1997). Many authors have studied the effect of vitamin E on growth and immune response in various organisms but studies related to effects of vitamin E on growth, survival and haematocrit in ornamental fishes are scanty. Hence, the present study was undertaken to study the effect of different levels of dietary vitamin E and HUFA on growth, survival and haematocrit in the goldfish, *Carassius auratus gibellio.*

MATERIALS AND METHODS

Experimental diets

Ten experimental diets having different levels of vitamin E and HUFA were formulated and manufactured in the laboratory

*Corresponding author. E-mail: z.h.kashani@gmail.com

Table 1. Ingredient (g /100 g diet) and chemical proximate composition (% Dm[1]) of experimental diets.

Ingredient	E50+ HUFA	-E + HUFA	-E- HUFA	E100+ HUFA
Corn meal	6	6	6	6
Fish meal	20.5	20.5	20.5	20.5
Soybean meal	38.5	38.5	38.5	38.5
Bread flour	10	10	10	10
Rice bran	18.75	18.75	18.75	18.75
Fish oil	0.5	0.5	-	0.5
Soybean oil	-	-	0.5	-
Mineral mixture[2]	2	2	2	2
Vitamin mixture[3]	2	2	2	2
Lysine	0.75	0.75	0.75	0.75
Metionin	0.75	0.75	0.75	0.75
Antifungi	0.25	0.25	0.25	0.25
Proximate composition				
Protein	39	39	39	39
Fat	10.8	10.8	10.8	10.8
Energy	4000	4000	4000	4000

Dm[1]: dry matter; Mineral mixture[2] (g/kg diet) according to delaHiguera et al. (1998): $Ca(PO4H2)2{\bullet}H_2O$(30), $CaCO_3$ (6.5), KCl (2.5), NaCl (4), $MnSO_4{\bullet}H_2O$ (0.2), $FeSO_4{\bullet}7H_2O$ (1.5), $MgSO_4$ (4.6), KI (0.02), $CuSO_4{\bullet}5H_2O$ (0.05), $ZnSO_4{\bullet}7H_2O$ (0.2), $CoSO_4{\bullet}7H_2O$ (0.05), Na_2SeO_3 (0.218•10−2), $Al_2(SO4)_3{\bullet}18H_2O$ (1•10−2); Vitamin mixture (mg/kg diet)[3]: Thiamine (40), riboflavin (60), pyridoxine (30), panthotenic acid (150), niacin (25), folic acid (15), inositol (1000), choline (5000), biotin (3), cyanocobalamin (0.05), vitamin a (1), menadion (25).

according to goldfish nutritional requirements; the test diet were as follow: E100 + HUFA, -E + HUFA, -E - HUFA and E 50 + HUFA. Fish or soybean oil was used as a lipid source for +HUFA and -HUFA diets, respectively (Table 1). The general composition of each experimental diet is given in Table 1.

Experimental design

Goldfish (*C. auratus gibellio*), average 0.69±0.12 g initial weight, were used as an experimental fish and the feeding trial was conducted in the Aquaculture Laboratory, Aquaculture Research Center of Fisheries Department in Gorgan University of Agricultural Sciences and Natural Resource, Iran. Prior to the start of the experiment, the goldfish were reared in 400 L fiberglass tanks for two weeks to acclimatize to the experimental diet and conditions. In the beginning of the experiment, the fish were fasted for 24 h and weighed. Fish of similar sizes were randomly distributed into 30 fiberglass tanks (400 L), and each tank was stocked with 8 fish. Each diet was randomly assigned to triplicate tank. Fish were hand-fed to apparent satiation twice daily for 10 weeks. During the experimental period, the temperature ranged from 21.5 to 22°C, the pH was 7.9 to 8.1.

At the termination of the experiment, the fish were fasted for 24 h before harvest. Blood sampling was carried out from the caudal vasculature with cutting peduncle. The blood was placed in tube containing heparin sodium, then transferred into a 2 ml micro centrifuge tubes and centrifuged for 15 min at 3000 g at 4°C.Haematocrit was read by haematocrit reader.

Total number and mean body weight of fish in each tank were measured. After feeding for 10 weeks, fish fed the diets with different level of vitamin E (50,100 mg kg diet^{-1}) and HUFA (highly unsaturated fatty acid) were selected to investigate the effects of them on growth, survival and haematocrit of goldfish.

Analysis and measurement

Analysis of dietary composition

The samples of diets were dried to a constant weight at 105°C to determine the dry matter content. Protein was determined by measuring nitrogen (N×6.25) using the Kjeldahl method; lipid by ether extraction using Soxhlet (AOAC, 1995).

Calculations

The following variables were calculated:

Specific growth rate (SGR) = (LnWt - LnW0) ×100/ t
Survival rate = Nt * 100/N0
Condition factor (K) = W/L^3*100
FCR) = FED DIET/ W2) feed conversion rate

Where Wt and W0 were final and initial fish weights, respectively; Nt and N0 were final and initial numbers of fish in each replicate, respectively; it is the experimental duration in day.

Statistical analysis

Data were analyzed by one-way analysis of variance (ANOVA). When ANOVA identified differences among groups, multiple comparisons among means were made with Duncan's new multiple range tests. All variances were checked for normality and homogeneity. All percentage values were transformed using arcsine transformation. Data are presented as treatment means ± SD. The values of $P<0.05$ were considered significantly different.

Table 2. Response of Gold fish to the various test diet after 10 weeks of feeding.

Treatment	E100+HUFA	E50+HUFA	-E-HUFA	-E+HUFA
SGR	0.19±0.02	0.21±0.004[a]	0.22±0.02[a]	0.18±0.027[a]
FCR	0.45±0.07[a]	0.37±0.01 [a]	0.39±0.05[a]	0.45±0.09[a]
CF	2.3±0.3[a]	2.02±0.4[a]	1.8±0.02[a]	2.01±0.2[a]
BWI	5.8±1.09[b]	6.5±0.41[b]	12.3±0.5 [a]	10.85±2.1[a]
Survival	100[a]	93.7±8.8[a]	100[a]	93.7±8.8[a]

SGR, Specific growth rate; FCR, feed conversion rate; CF, condition factor; BWI, body weight increase.

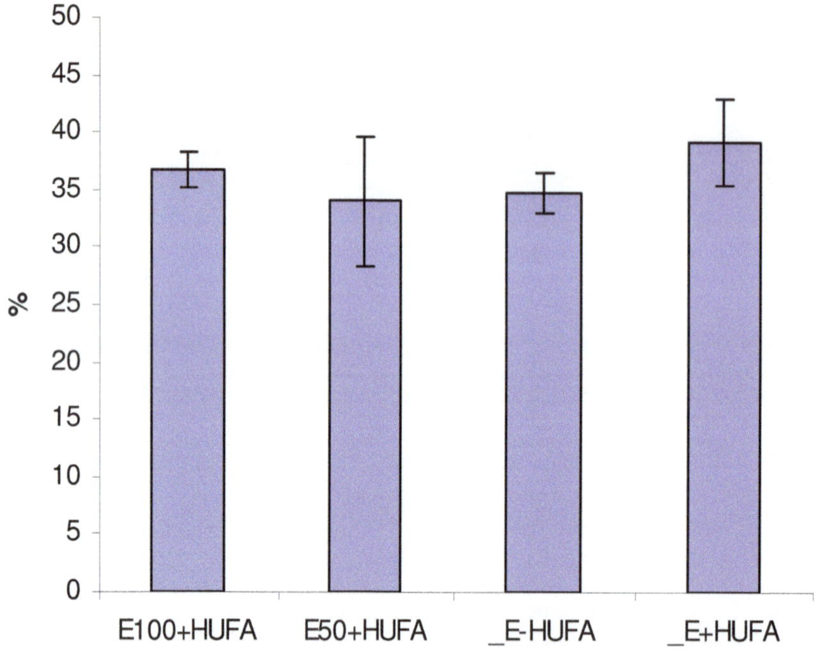

Figure 1. Haematocrit (%) in different treatments of goldfish.

RESULTS AND DISCUSSION

Growth factors are summarized in Table 2. Final weight in fish fed with the diets -E-HUFA, -E + HUFA was significantly higher compared with fish fed with the diet E_{100} + HUFA, E_{50} + HUFA ($P<0.05$). Feed conversion rate (FCR), specific growth rate (SGR) and condition factor (CF) and survival were not significant between groups ($P>0.05$). Haematocrit was not different between groups (Figure 1). Dietary vitamin E requirement of goldfish has been studied with various results. Many factors such as lipid level, source, and other antioxidants or antioxidant enzymes presented in the diet or body system may affect the dietary vitamin E requirement of fish. Satoh et al. (1987) reported that inclusion of 50 to 100 mg α-tocopherol/kg diet was sufficient for tilapia fed diets containing 5% lipid. Our results show that vitamin E addition to diet did not affect the growth of goldfish. Several reports have shown that the supplementation of vitamin E did not improve the growth performance of fish fed oxidized oil. Foster et al. (1988) reported that vitamin E addition to diets containing oxidized herring oil did not affect the growth performance of Coho salmon. Similar results were reported by Mourente et al. (2002) in gilthead sea bream. They found that addition of vitamin E in diets containing oxidized oil did not improve the growth and survival of sea bream although the hepatic antioxidant defense enzyme activities were significantly affected. In contrast, a report from (Baker and Davies, 1996) showed that dietary vitamin E supplementation in rancid diets significantly improved the growth of African catfish. In this study, addition of vitamin E was not significant on growth parameters. Many marine fish larvae are believed to require highly unsaturated fatty acids, especially EPA and DHA (Watanabe, 1982; Sargent et al., 1989, 1995). The nutritional value of high levels of (n-3) HUFA was confirmed by enhanced growth in many marine fish larvae such as turbot *Scophthalmus maximus* (Gatesoupe and Le Milinaire, 1985), Japanese flounder *Paralichthys olivaceus*, Red sea bream, *Pagrus major* (Izquierdo et al., 1989), and gilthead sea bream, *Sparus aurata* (Koven et al., 1993; Rainuzzo et al., 1997). But in present study, HUFA addition did not have effect on growth, survival, SGR, FCR, CF. Addition of vitamin E and HUFA has no effect on growth parameter and haematocrit. In our opinion, addition of vitamin E and highly unsaturated fatty acids in to diets of goldfish is good for experimental fish but their effects were not significant; suggesting that other author use higher doses of vitamin E such as 200-300-500,..., also they use other source of oil in the diet and they can detect effect of dietary vitamin E and HUFA on other blood parameters, reproduction, etc.

REFERENCES

AOAC (1995). 17th Edition, Association of Official Analytical Chemists (AOAC), Washington DC, 21: 447.

Baker RTM, Davies SJ (1996). Oxidative nutritional stress associated with feeding rancid oils to African catfish, *Clarias gariepinus* (Burchell) and the protective role of -tocopherol. Aquac. Res., 27: 795-803.

Delaunay F, Marty Y, Moal J, Samain JF (1993). The effect of monospecific algal diets on growth and fatty acid composition of Pecten maximus (L.) larvae. J. Exp. Mar. Biol. Ecol., 173: 163-179.

Foster I, Higgs D A, Bell GR, Dosanjh B S, March B E (1988). Effect of diets containing herring oil oxidized to different degrees on growth and immuno competence of juvenile coho salmon (*Onchorhynchus kisutch*). Can. J. Fish. Aquat. Sci., 12: 2187-2194.

Gatesoupe FJ, LeMilinaire C (1985). Adaptation de la qualite alimentaire des cltreurs-proies aux besoins nutritifs des larvae de poissons marins. Coll. Fr. Jpn. Oceanography. Marseille, pp. 51-63.

Huang SL, Weng Y M, Huang CH (2004). Lipid peroxidation in sarcoplasmic reticulum and muscle of tilapia is inhibited by dietary vitamin E supplementation. J. Food Biochem., 28: 101-111.

Izquierdo MS, Watanabe T, Takeuchi T, Arakawa T, Kitajima C (1989). Requirement of larval red sea bream Pagrus major for essential fatty acids. Nippon Suisan Gakkaishi, 55: 859-867.

Koven WM, Tandler A, Sklan D, Kissil GW (1993). The association of eicosapentaenoic and docosahexaenoic acids in the main phospholipids of different-age Sparus aurata larvae with growth. Aquaculture, 116: 71- 82.

Mourente G, Dı´az-Salvago E, Bell JG, Tocher DR (2002). Increased activities of hepatic antioxidant defence enzymes in juvenile gilthead sea bream (*Sparus aurata* L.) fed dietary oxidized oil: attenuation by dietary vitamin E. Aquaculture, 214: 343-361.

NRC (National Research Council) (1993). Nutrient Requirement of Fish. National Academy Press, Washington, DC, p. 114.

Rainuzzo JS, Reitan KI, Olsen Y (1997). The significance of lipids at early stages of marine fish: A review. Aquaculture, 155: 103-115.

Sargent JR, Bell JG, Bell MV, Henderson RJ, Tocher DR (1995). Requirement criteria for essential fatty acids. J. Appl. Ichthyol., 11: 183-198.

Sargent JR, Henderson RJ, Tocher DR (1989). The lipids. In: Fish Nutrition, pp. 153-217.

Satoh S, Takeuchi T, Watanabe T (1987). Requirement of tilapia for a-tocopherol. Nippon Suisan Gakkaishi, 53: 119-124.

Takeuchi T (1997). Essential fatty acid requirements of aquatic animals with emphasis on fish larvae and fingerlings Fish. Sci. Rev., 5: 1-25.

Waldock MJ, Nascimento IA (1979). The triacylglycerol composition of *Crassostrea gigas* larvae fed on different diets. Mar. Biol. Lett., 1: 77-86.

Waldock MJ, Nascimento IA (1979). The triacylglycerol composition of *Crassostrea gigas* larvae fed on different diets. Mar. Biol. Lett., 1: 77-86.

Watanabe T (1982). Lipid nutrition in fish. Comparative Biochem. Physiol., 73: 3-15.

Zhen Li (1988). Chinese Goldfish. Foreign Language Press, Beijing, China, pp. 13-26.

Radiological and mineralogical investigation of accretion and erosion coastal sediments in Nile Delta Region, Egypt

Ayman A. El-Gamal[1]* and Ibrahim H. Saleh[2]

[1]Marine Geology Department, Coastal Research Institute, National Water Research Center, 15, St. Elpharanaa, Elshalalat, Postal Code 21514, Alexandria, Egypt.
[2]Environmental Studies Department, Institute of Graduate Studies and Research, Alexandria University, P. O. Box 832, El-Shatby, Alexandria, Egypt.

The Nile Delta coast is a dynamic sedimentary environment experiencing erosion, transport and re-deposition of sand along the coast. The aim of this study was to investigate the effectiveness of using an integrated approach of natural radioactivity of coastal sediments, heavy mineral distribution and grain size information to differentiate between the eroding and accreting areas of the Nile Delta coast. Also, it is to investigate the relationship between grain size and heavy mineral distributions of these sediments on the spatial distribution of the coastal radioactive materials. The relatively higher profile averages of ^{226}Ra, ^{214}Pb, ^{214}Bi and ^{210}Pb (as ^{238}U series) were 104.37±84.66, 45.60±37.83, 38.43±32.49 and 35.69±24.86 Bq/kg, respectively and ^{228}Ra and ^{212}Pb (as ^{232}Th series) were 54.26±56.66 and 42.18±44.66 Bq/kg, respectively. The highest average value of ^{40}K concentration (404.49±125.81 Bq/kg) was detected in the profile located at 2000.1 m west of Rashid estuary. Microscopic study revealed that the more dominant heavy minerals assemblages consisted of Opaques, Hornblende, Augite, Epidote, Biotite and Chlorite. Also, minor amounts of Garnet, Zircon, Rutile, Tourmaline, Kyanite and Monazite have been identified. Relatively high concentrations of ^{238}U and ^{232}Th series members were found in coincident at sites having higher heavy minerals percentages and detected at erosional beach than the accretion one. Coincidence was recognized between the average concentrations of ^{40}K and the percentages of the light minerals collected at 100 m distances of the profiles under investigation.

Key words: Nile Delta, beach erosion, beach accretion, natural radioactivity, heavy minerals.

INTRODUCTION

Coasts are dynamic systems, undergoing adjustments of form and process at different time and space scales in response to geomorphological and oceanographical factors (Cowell et al., 2003). Beach sands are mineral deposits formed through erosion of geological formations which may have been brought to their present location after transport by wind, rivers and glaciers to the coast, and are deposited on the beaches by actions of waves and currents (de Meijera et al., 2001). Nile River is the main source of the Nile Delta beach sands. In the past, Nile River transported black sand from mountain ranges of Sudan and Abyssinia to its delta in the Mediterranean coast of Egypt. This black sand contains anomalies of relatively higher natural radioactive nuclides than the other coastal sands (El-Naggar, 1990; El-Khatib et al., 1993; El-Gamal et al., 2004; Hussein, 2011). This sedimentation was cut off after building of Aswan High Dam.

*Correspondence author. E-mail: ayman_elgamal@yahoo.com.

The principal primordial radionuclides are isotopes of heavy elements (USEPA, 2011). Natural radioactivity in sediments comes from uranium (^{238}U) and thorium (^{232}Th) series and natural potassium (^{40}K) (Veigaa et al., 2006) which are detected in the Egyptian Mediterranean coastal sediments (El-Gamal et al., 2004; Saleh et al., 2004). Traces of artificial radionuclides such as cesium (^{137}Cs) can also be present, as a result of the fallout from weapons testing and from Chernobyl accident (Johanson and Bergström, 1989). The variability in ^{238}U and ^{232}Th series across the shore profiles is mainly explained by variations in heavy and light mineral contents. Typically, higher percentages of heavy minerals will result in higher concentrations of primordial radionuclides (Brai et al., 2006). Such variability can be accounted by redistribution of the sand caused by erosion/deposition processes. Furthermore, parameters such as grain size and heavy minerals content were studied in connection with the distribution of U, Th and ^{40}K by de Meijer (1998), Seddeek et al. (2005) and Vassas et al. (2006). Therefore the enriched areas with U and Th, presumably result from the sorting of sand grains according to their size and density (Vassas et al., 2006).

In the Nile Delta, Egypt, the distribution of heavy mineral assemblages has been recognized as two mineral groups (Frihy et al., 1995; Frihy, 2007). The first group includes heavy minerals of lower density and coarser size (augite, hornblende and epidote). Heavy minerals in this group increase from west to east along the delta as they are easily to entrain and transport the coastal sediments toward the east by wave currents. In contrast, the second group includes heavy minerals of higher-density (opaques, garnet, zircon, rutile, tourmaline and monazite) and these minerals are difficult to entrain and transport by wave-current actions. Hence, minerals in this group form a lag deposit within the delta and beach sand.

De Meijer et al. (1990) and de Meijer and Donoghue (1995) concluded that the origin (provenance) of beach sands can be estimated from the radiometric fingerprints of its light and heavy minerals. The radiometric fingerprint is a set of activity concentrations of ^{40}K, ^{232}Th and ^{238}U (de Meijer, 1998). The heavy minerals such as monazite contain overall an order of magnitude less ^{40}K and two orders of magnitude more of ^{232}Th and ^{238}U, compared to light minerals. Radiometric analysis of various fractions of heavy mineral sands showed that the monazite and zircon sands are highly radioactive contents (U and Th) as compared to other minerals in the heavy mineral suite (Alam et al., 1999; Mohanty et al., 2004; Coelho et al., 2005). The abundance of thorium typically about 10 wt% and that of U about 0.5 wt% are found in monazite crystals. Zircon typically contains 5 to 4000 ppm of U and 2 to 2000 ppm of Th (Deer et al., 1997; Mohanty et al.; 2004). On the other hand, chlorite, biotite, tourmaline, apatite, magnetite and quartz may contain uranium (Dahlkamp, 2009).

Heavy mineral grains are generally more difficult to entrain and transport by currents than other light mineral grains of the same shape and size, and thus, heavy minerals are represented by smaller grains (Kudrass, 1987; Komar, 2007). However, heavy minerals grains with lower mobility values such as zircon, tourmaline, coarse garnet and staurolite tend to remain relatively close to their input sources and thus these minerals are common in the inner-shelf zones near estuaries (Cascalho and Fradique, 2007). The deposition of materials such as heavy minerals requires a high-energy level, which can occur during storm episodes. This may explain why heavy minerals are mainly observed on erosion beaches where the shoreline has retreated, as observed along the Nile Delta (Frihy et al., 1995). High-density materials tend to become concentrated in areas of beach erosion, accounting for the formation of enriched radioactive sand, whereas minerals of lower density and coarse size are selectively entrained by waves and current and preferentially transported to zone of beach accretion where they are deposited (Frihy and Komar, 1993; Frihy et al., 1995). Such a process is widely supported by the inverse correlation observed in sand samples between ^{40}K that is a tracer of light minerals (Seddeek et al., 2005; de Meijer, 1998). These previous studies suggested an inverse correlation between grain size and radioactivity with the finest fraction being enriched by about three orders of magnitude (de Meijer, 1998). Uranium and thorium isotopes have been used commonly as radiotracers for studies of marine sediments, including their sources, transport and sedimentation rates in relation to the sedimentary and geochemical processes in the marine environment (Cochran and Krishnaswami, 1980; Li et al., 1985; Cole et al., 1986; Suman and Bacon, 1989).

The aim of this study was to investigate the effectiveness of using an integrated approach of natural radioactivity of coastal sediments, heavy mineral distribution and grain size information to differentiate between the eroding and accreting areas of the Nile Delta coast. The study of the distribution of primordial radionuclides allows good understanding of the radiological implication of these elements due to coastal processing. The study also aimed to investigate the effects of grain size and heavy mineral distributions in sand on the spatial distribution of coastal radioactivity.

MATERIALS AND METHODS

Study area

During the past century, the Nile Delta was suffering from sever erosion and vast retreating in shoreline especially at Rosetta, El-Burullus and Damietta headlands. So, many protection works have been executed along the Nile Delta coast to protect the eroded areas such as seawalls (for example, El-Burullus headland and the two sides of Rosetta Branch), detached breakwaters (for example, Baltim), groins (for example, eastern side of Rosetta) and jetties like

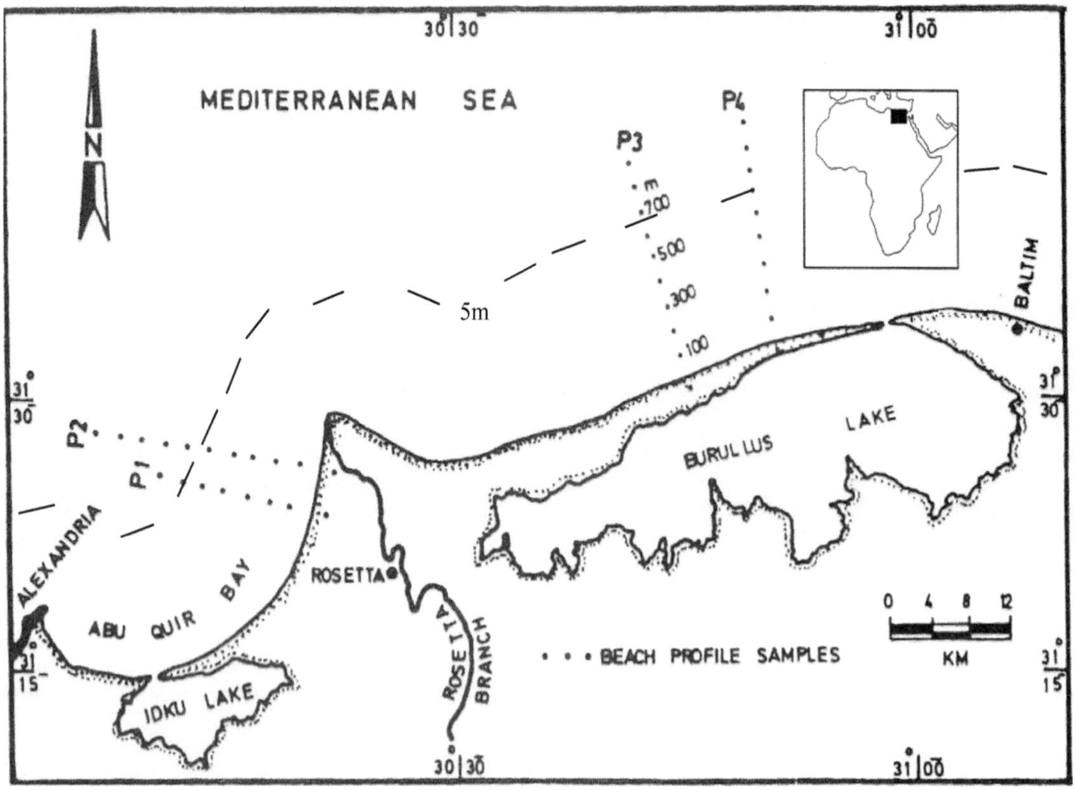

Figure 1. Sampling sites at Nile Delta area.

(for example, El-Burullus outlet) (CoRI, 2002). The hard structures attempt to change the regime of erosion and accretion at the beach. New sites are suffering by erosion while other sites gain more sediments than before the established of these protection tools.

Rashid promontory profiles

Rashid (Rosetta) promontory is located at the western mouth of the Nile River opening into the Mediterranean Sea (Figure 1). The western bank of Rashid was eroded with an average of −4 m/yr between 1978 and 1984 (Smith and Abdel-Kader, 1988). Coastal erosion and land retreat have persisted after the construction of the Rashid seawall (Nafaa, 1995). Downdrift erosion was generated when the seawalls became effective (Kaiser, 2004). Beach profiles have a steep face with no berm and many deep and wide bars and troughs. Many bars and troughs were observed at 25 to 250 m distance offshore at 1 to 5 m water depth. Storm waves destroy the berm as a result of a high intensity wave swash, and sediment was shifted offshore to form one or more bars that extend parallel to the shoreline (Miller, 1976). Beach profile configuration does not change significantly beyond 300 m seaward. The beach face (the seaward section of a beach exposed to and shaped by the action of waves) showed deposited sediment extending 150 m seaward and 2.5 m water depth (Kaiser and Frihy, 2009).

El-Burullus profiles

El-Burullus area is located between Rashid and Damietta promontories (Figure 1). Erosion has caused huge damage to the coastal regions at El-Burullus coastal area (El-Kolfat, 2005). The area west of El-Burullus has experienced continued accretion with rate of 4.9 m/yr due to the high rate of westward drift since the construction of the western jetty in 1972 (Badr and Lotfy, 1999; Frihy and Dewidar, 2003). The measured beach profiles showed a wide berm extending 200 m offshore with wide and shallow troughs and bars at 500 to 700 distance offshore and 5 to 6 m water depth. Beach profile morphology does not change significantly at distances more than 800 m offshore. It moved smoothly with a gentle slope (Kaiser and Frihy, 2009). The high concentrations of the heavy minerals at Rashid and El-Burullus sediments may be related to: 1. Contribution from offshore old sediments of classic Nile branches rather than the present Nile, 2. Contribution from the land itself, where the dunes and backshore contain great amount of these minerals and 3. These minerals are concentrated during severe coastal erosion (El-Fishawi and Badr, 1991).

Sampling

In October 2002 sediment sampling was carried out on west of Rashid (Rosetta) and El-Burullus beaches by Coastal Research Institute (CoRI) team using grab sampler from rubber boat (Zodiac). The marine survey has been carried out using geographical positioning system (GPS) and eco-sounder. A total of 36 surficial (0 to 15 cm depth) sediment samples were collected from 4 profiles (P1 to P4, position in Figure 1). Profiles P1 (31° 26.086' N and 30° 21.219' E) and P2 (31° 27.055' N and 30° 21.572' E) situated at 3892.6 and 2000.1 m distances west of Rashid Estuary, respectively representing the erosion coast. While profiles P3 (31° 33.438' N and 30° 52.946' E) and P4 (31° 34.165' N and 30° 55.524' E) located at 8917.4 and 4621.5 m distances west of El-Burullus Lake outlet, respectively representing the accretion coast. Beach profile lines are perpendicular to the coastline and extend from a fixed baseline in concrete blocks at the beach to 800 m

Table 1. Minimum detectable activities (MDAs) of the measured radionuclides.

Radionuclide	MDA (Bq/kg)
Cs-137	0.004
Pb-210	0.008
K-40	0.259
Pb-214	0.041
Bi-214	0.031
Ra-226	0.224

distances to about 6 m water depth by means of 100 m intervals (CoRI, 2002).

Radioactivity measurements

The dried samples were transferred to polyethylene containers of 100 cm^3. The radioactivity measurements were carried out using well-calibrated gamma spectrometry based on hyper-pure germanium (HPGe) detector, reverse electrode closed- end coaxial and extended range (Canberra Company). The 68 mm diameter and 69 mm length HPGe detector had a relative efficiency of 60%, and its resolution was 2.2 keV at 1.33 MeV, Peak/Compton effect was 60:1. The gamma transmissions used for activity calculations were 46.5 keV (^{210}Pb), 352.9 keV (^{214}Pb), 609.3, 1120.3 and 1764.5 keV (^{214}Bi), and 186 keV (^{226}Ra) for ^{238}U series, 338.4, 911.1 and 968.9 keV (^{228}Ac) for ^{232}Th series, 1460.7 keV for ^{40}K and 661.6 keV for ^{137}Cs. The calibration was carried out using standard sources. The quality control was carried out by parallel measurements of reference samples accumulated in containers with the same geometry and under the same measuring conditions and texture as for the actual samples. The reference materials were supplied by US DOE from mixed-analyte performance evaluation program (MAPEP). The acquisition of the spectrum and its analysis were carried out using Genie 20009 Canberra Software with complementary with packages of Interactive Peak fit, Geometry Composer and Quality Assurance. The minimum detectable activities (MDA) with 95% confidence level of the radioisotopes under investigation are listed in Table 1. The detection of MDA and the detection limit were carried out according to the methods and the instructions of Reguigui (2006). MDA was calculated using the following equations (Reguigui, 2006):

$$MDA = \frac{L_D}{T\ x\ eff\ x\ Br\ x\ m}$$

Where L_D = the detection limit, T = the counting time (sec), eff = efficiency, Br = branching ratio and m = mass.

$$L_D = L_C + K\sigma_D$$

Where Lc = critical level: below which no signal can be detected, σ_D = standard deviation and K = error probability

Heavy mineral detections

Three locations (distances within the profile) were selected from each of the four profiles for heavy mineral investigation: beach samples representing the beach and the swash zone, 100 m distance samples representing the surf zone, and finally, 800 m distance samples that represent the breaker zone. These zones are well defined by Ingle (1966). Samples were sieved with a shaker and the mean grain size values were obtained. 5 g of each dried samples was released into a separating funnel using the dense-liquid technique with bromoform (Specific gravity equal 2.8 g/cm^3). Samples were boiled in 10% hydrochloric acid for two minutes in order to disintegrate the samples. Boiled samples were rinsed with water to remove the hydrochloric acid and dried in an oven in order to separate the fine sand fraction.

The procedure as mentioned by Joshua and Oyebanjo (2009) that the mixture was vigorously stirred and left for about 5 min was observed. The heavy minerals present in the samples with specific gravity > 2.85 g/cm^3 were allowed to settle to the bottom of the separating funnel or after which the filtrate (heavy mineral) were thoroughly washed with acetone to remove any traces of bromoform. The weight percentages of heavy minerals residue (HM%) and the light minerals (LM%) have been calculated. The heavy minerals were mounted on slides with the aid of Canada balsam (Mange et al., 1992). Heavy minerals were identified and counted using polarizing microscope (Carl-Zeiss – West Germany).

Index figure

Ratio of heavy to light minerals calculated as index figure percentage.

Zircon, tourmaline and rutile proportions (ZTR) index

ZTR is the percentages of zircon, tourmaline and rutile in the non-opaque heavy minerals suites. It is to assess the mineralogical maturity of sediments (Garzanti and Andò, 2007). Dill (1995) referred a high ZTR index to a high degree of weathering where the unstable minerals were eliminated by dissolution. The "ZTR" index, which is a quantitative definition of mineral assemblage, was calculated using the following equation:

$$ZTR\%\ Index = \frac{Zircon + Tourmaline + Rutile}{\sum non-opaque}\ x\ 100$$

RESULTS AND DISCUSSION

Depth profile

Normally, it could be found from both status; accretion and erosion at different depths within the same profile (CoRI, 2001) in Nile Delta coast. In order to identify the recent erosion and accretion parts of the four profiles under investigation, matching between data of the survey done during 2001 against the other one on 2002 was carried out. Figure 2 shows the temporal changes of the four profiles; P1 and P2 (west of Rashid) and P3 and P4 (west of El-Burullus). To show the four profiles in one figure, the depth values of each profile subtracted with fixed number per profile (P2 - 2, P3 - 4 and P4 - 6) for the two years data. The first profile (P1) was appeared in Figure 2 as erosion profile except only at 300 m distance at about 3 m depth. While the second profile (P2) was recognized during the sampling time as accretion profile

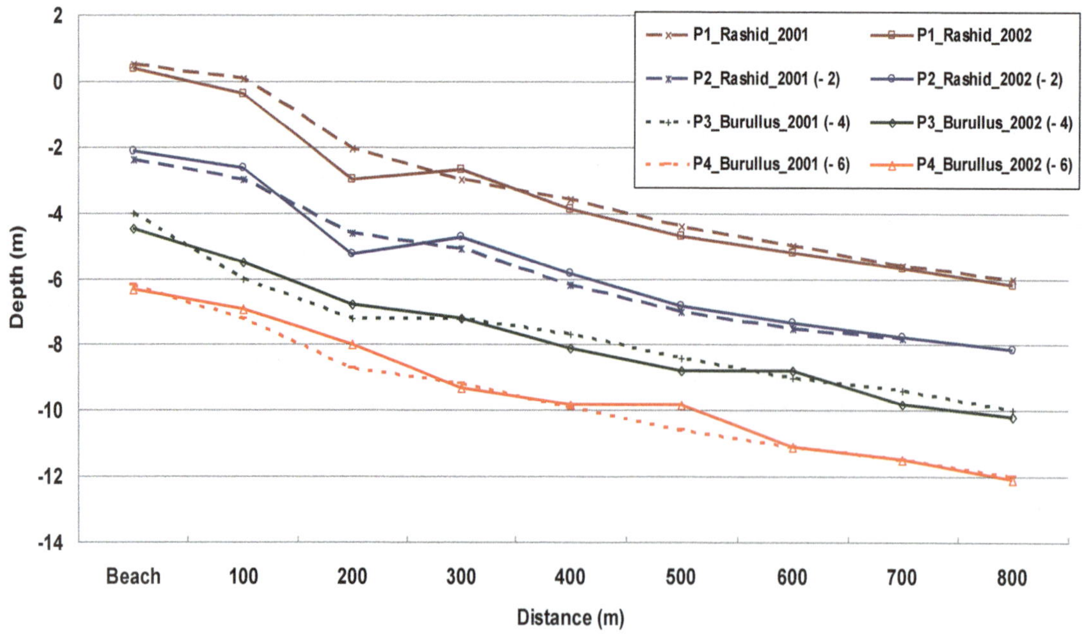

Figure 2. Depth profiles at west of Rashid estuary and El-Burullus outlet during 2001 and 2002.

except at distance 200 m. Longitudinal trough has been observed at Rashid profiles extending parallel to the shore line at 200 m distance from the baseline. Concerning west El-Burullus profiles, fluctuation between erosion and accretion within the same profile has been observed. Measurements indicated that seabed slope was steeper near the shore than at the off shore zones according to the decreasing of the bed grain size. Also, it was noticed that, there are submerged bars parallel to the shoreline. This was confirmed by the finding of El-Kolfat (2005). The beach suffered severe erosion while 100 and 200 m distances sampling point exposed to heavy accretion and it reverse again to erosion and accretion at the rest of the profile of P3. The fourth profile (P4) has heavy accretion situation in many parts, such as at 100, 200 and 500 m distances. The beach of the second profile (P2) was recognized as the only accretion one among the others three erosional beaches. On the other hand, the 100 m distance sampling site at P1 was the only erosional status among the other accretional sites at the same distance.

Mean grain size distribution

The mean grain-size distribution and the sorting of the sediment profiles are presented in Figure 3. The mean grain size of sediments at west El-Burullus beach was found coarser than those in the west of Rashid beach. The detected mean grain sizes of the beach sediments were sorted as 0.239, 0.26, 0.201 and 0.189 mm for P3, P4, P2 and P1, respectively. Reverse distribution was recognized, and sediments grain size at 100 to 400 m distances of P1 were found coarser with lower sorting values than sediments at the same distances of El-Burullus. Moreover, coarser grain size was detected at the second profile (P2) of Rashid, at distances 100 to 200 m less than the distances at El-Burullus. The sediments at distances from 500 to 800 m showed similar trend with a significant decrease of grain size values ranged from 0.069 to 0.097 mm.

Radioactivity distribution

Natural single occurrence (^{40}K) and series (^{238}U and ^{232}Th) radioisotopes were detected in all sediment samples under investigation. The distribution of ^{238}U series members along the four profiles are shown in Figures 4, 5, 6 and 7 and of ^{232}Th series in Figures 8 and 9. The highest averages of ^{226}Ra, ^{214}Pb, ^{214}Bi and ^{210}Pb (as ^{238}U series) were 104.37±84.66, 45.60±37.83, 38.43±32.49 and 35.69±24.86 Bq/kg, respectively and ^{228}Ra and ^{212}Pb (as ^{232}Th series) were 54.26±56.66 and 42.18±44.66 Bq/kg, respectively. These highest averages concentrations have been detected at the third profile (P3) west of El-Burullus outlet. The highest average values of ^{40}K concentration were detected at P2 and P4 (404.49±125.81 and 403.26±20.48 Bq/kg, respectively) ^{137}Cs has been detected at limited number of sites and P4 was found as the relatively higher values compared with the other profiles with average value 0.86 Bq/kg. Generally, in coastal area, the major source of ^{137}Cs is coming from atmospheric deposition (Zhiyong et al., 2010).

The distribution of the detected radionuclides was

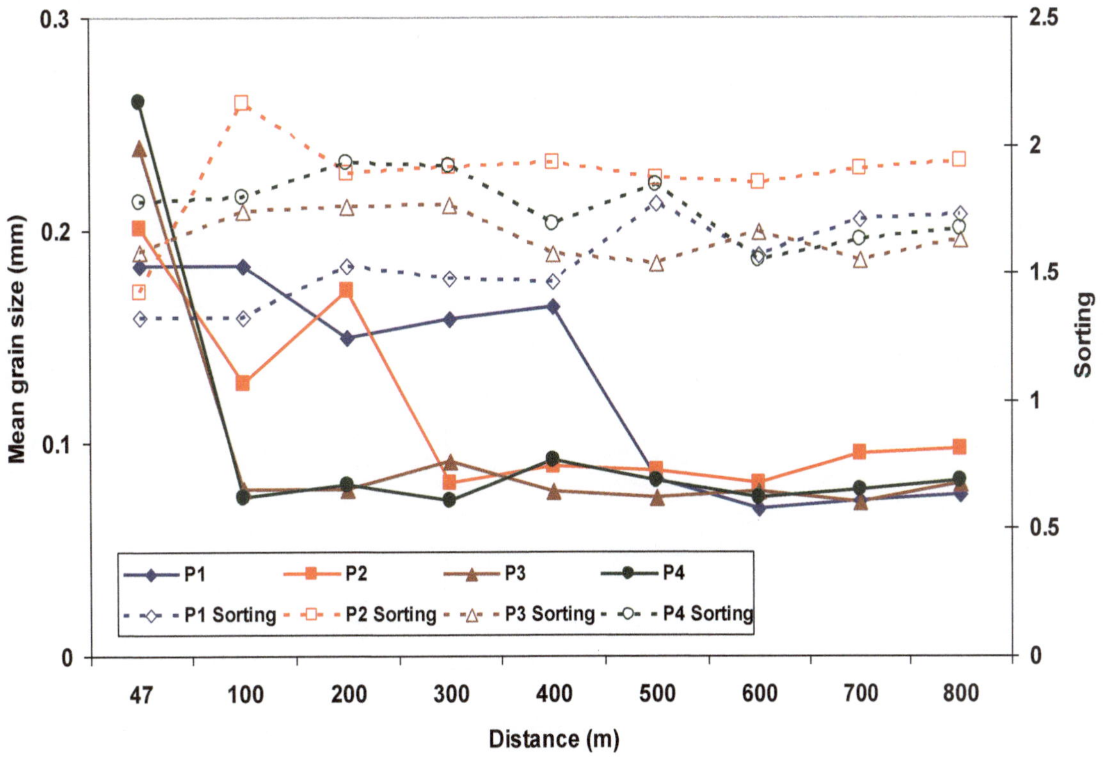

Figure 3. Mean grain size (mm) and sorting (standard deviation) of sediments at P1 to P4 profiles during 2002.

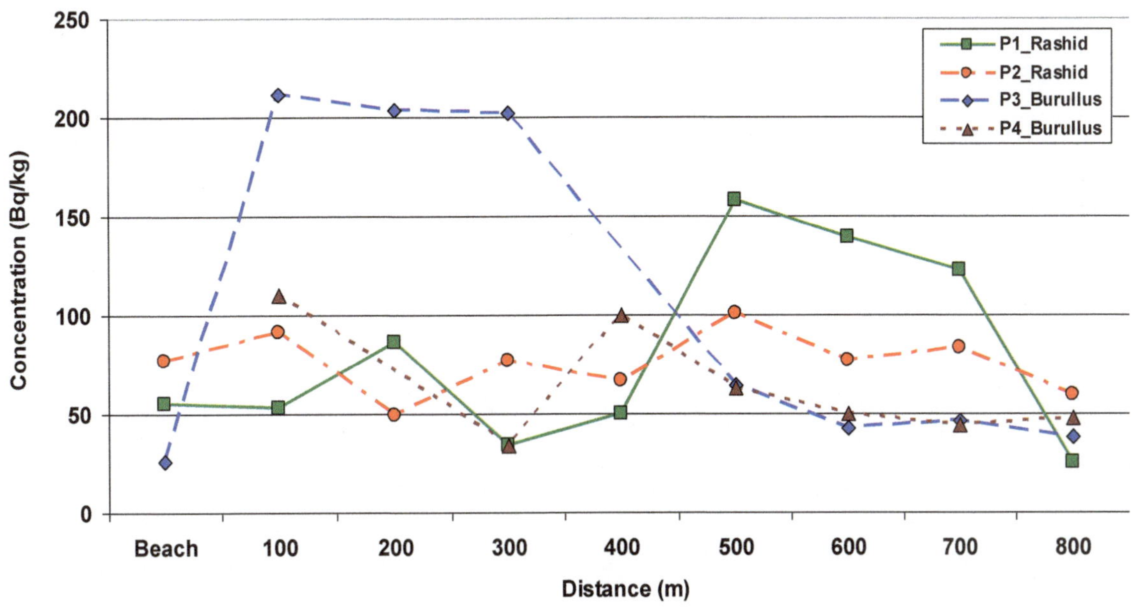

Figure 4. ^{226}Ra concentration (Bq/kg) at Rashid and Burullus coastal sediments profiles during 2002. The highest averages concentrations were detected at the third profile (P3).

varied with depth and distances at each profile as shown in Figures 4, 5, 6, 7, 8, 9 and 10. At the beach, the second profile at Rashid (P2) which is characterized by accretion status, exhibited the highest ^{238}U and ^{232}Th decay series activities, while P3 showed the lowest. Dramatically increase in radioactivity concentration in ^{238}U and ^{232}Th decay series of P3 sediments under its accretion condition was recognized at 100, 200 and 300

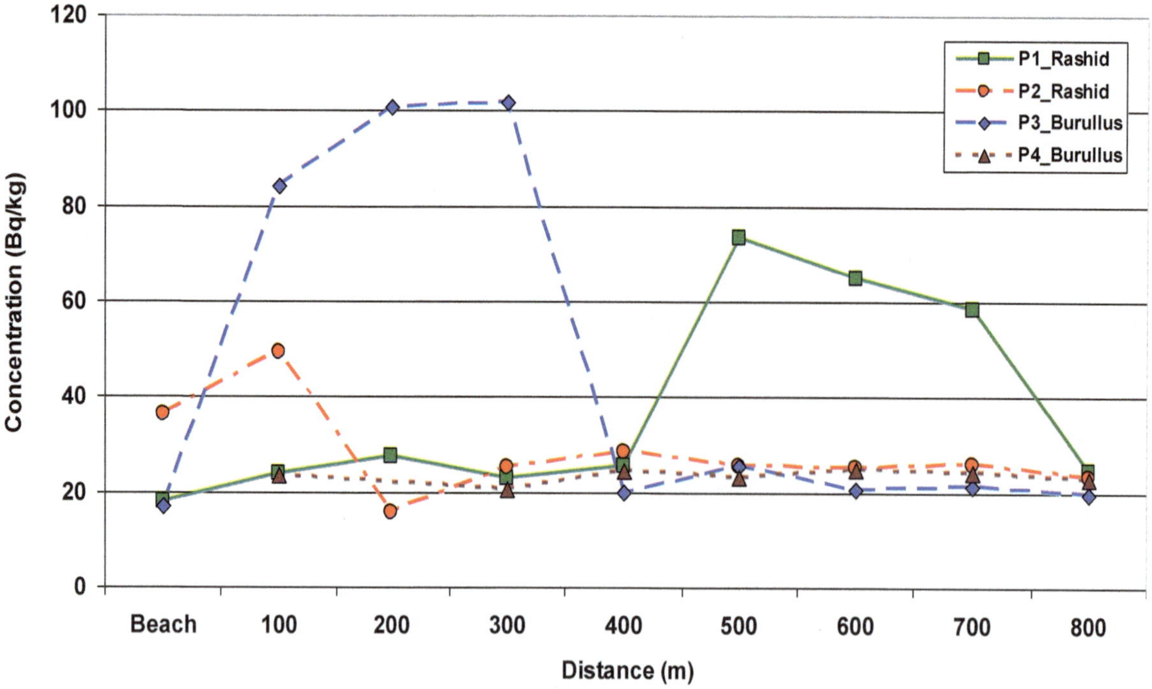

Figure 5. ^{214}Pb concentration (Bq/kg) at Rashid and Burullus coastal sediments profiles during 2002. The highest averages concentrations were detected at the third profile (P3).

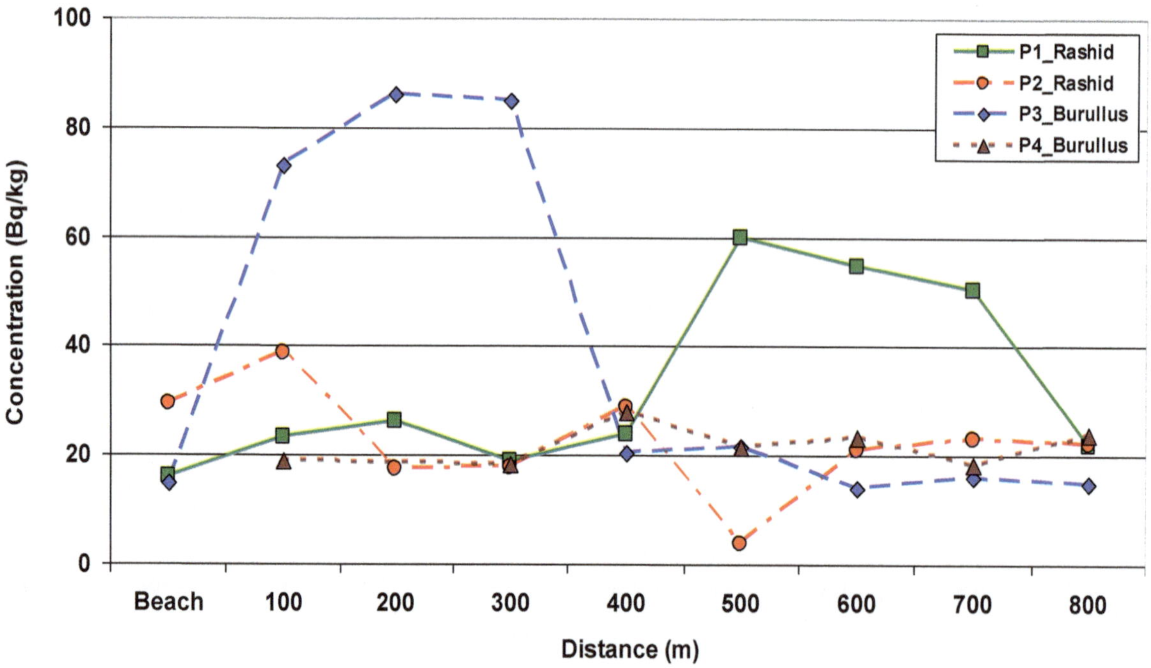

Figure 6. ^{214}Bi concentration (Bq/kg) at Rashid and Burullus coastal sediments profiles during 2002. The highest averages concentrations were detected at the third profile (P3).

m distance samples as shown in Figures 4, 5, 6, 7, 8 and 9. On the other hand, relatively higher concentrations of ^{238}U and ^{232}Th decay series were observed at P1 (Rashid) under its erosion condition at 500 to 700 m distances. This distribution of radioactivity could synchronize with the mean grain size distribution at the same places, which it has finer grains trapped between coarser grains. The relation between radioactivity

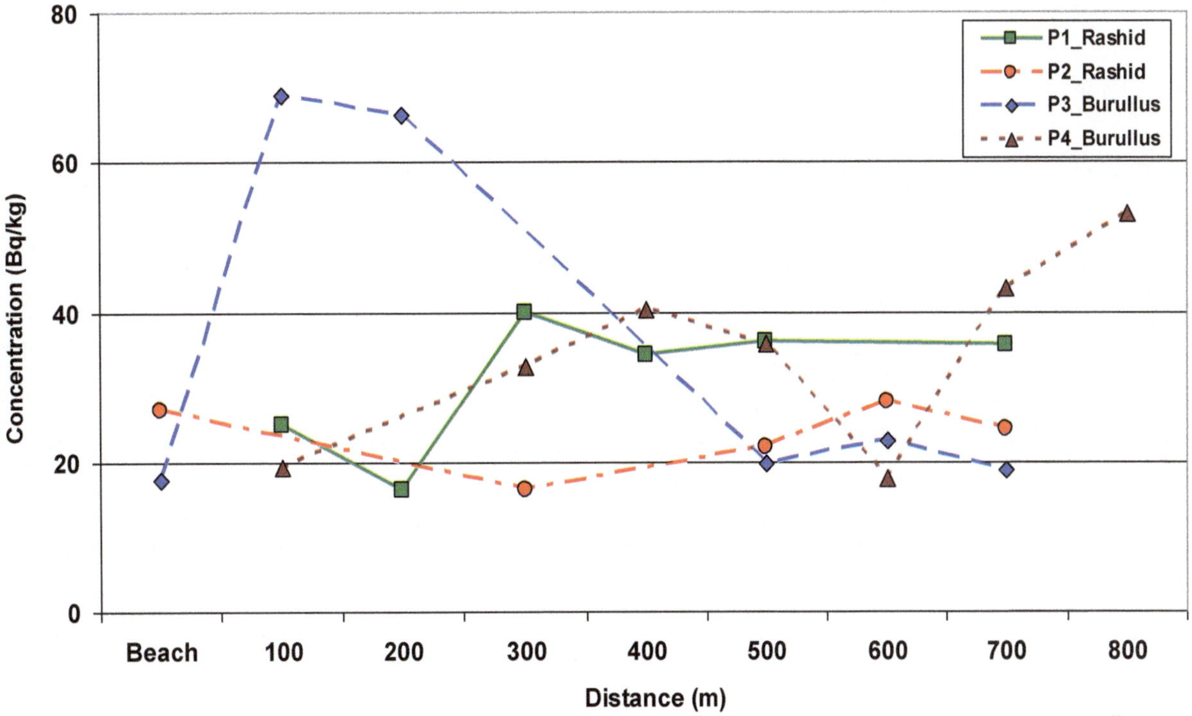

Figure 7. ^{210}Pb concentration (Bq/kg) at Rashid and Burullus coastal sediments profiles during 2002. The highest averages concentrations were detected at the third profile (P3).

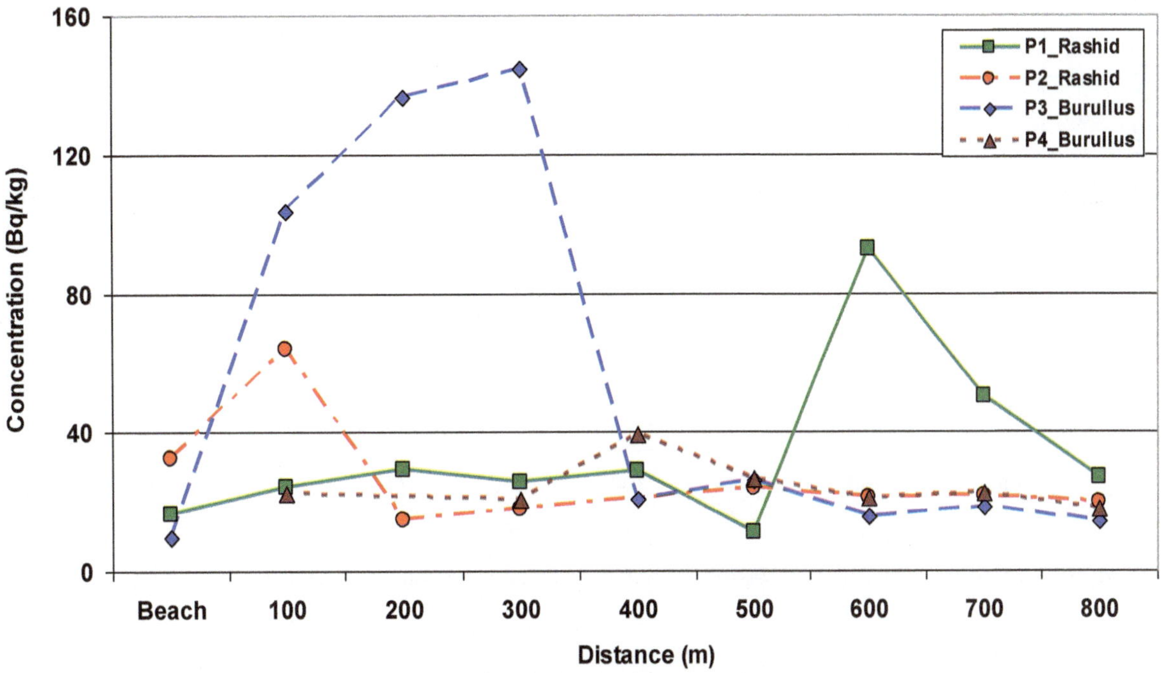

Figure 8. ^{212}Pb concentration (Bq/kg) at Rashid and Burullus coastal sediments profiles during 2002. The highest averages concentrations were detected at the third profile (P3).

distributions with grain size of sediments was confirmed by Vassas et al. (2006). They studied grain size parameters and heavy minerals content in connection with the distribution of U, Th and ^{40}K in a cross-shore

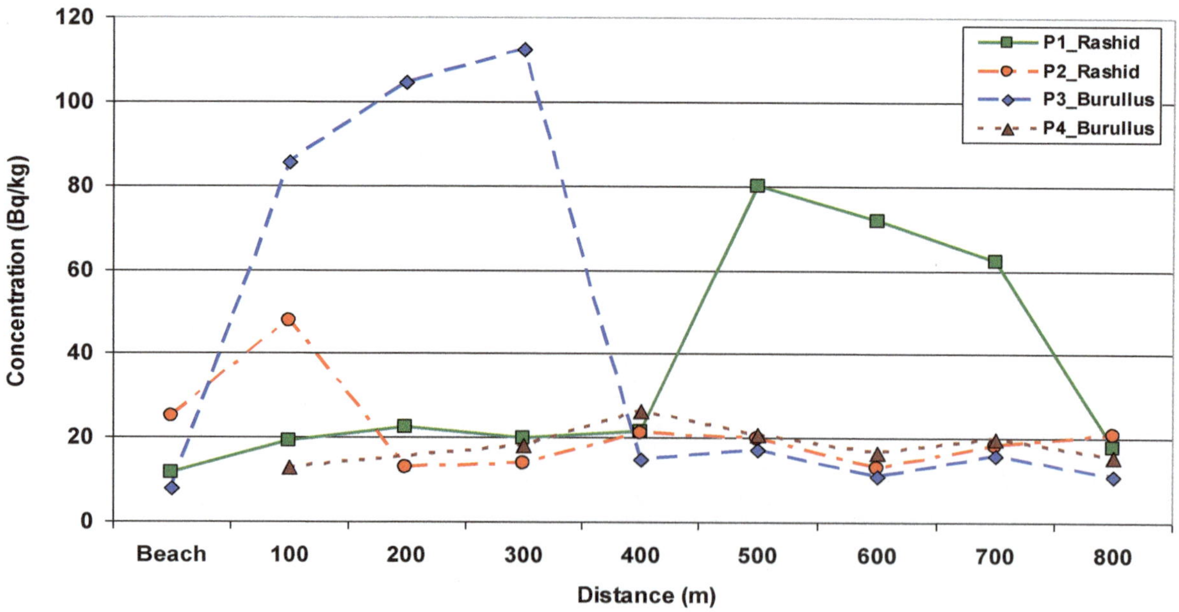

Figure 9. ^{228}Ra concentration (Bq/kg) at Rashid and Burullus coastal sediments profiles during 2002. The highest averages concentrations were detected at the third profile (P3).

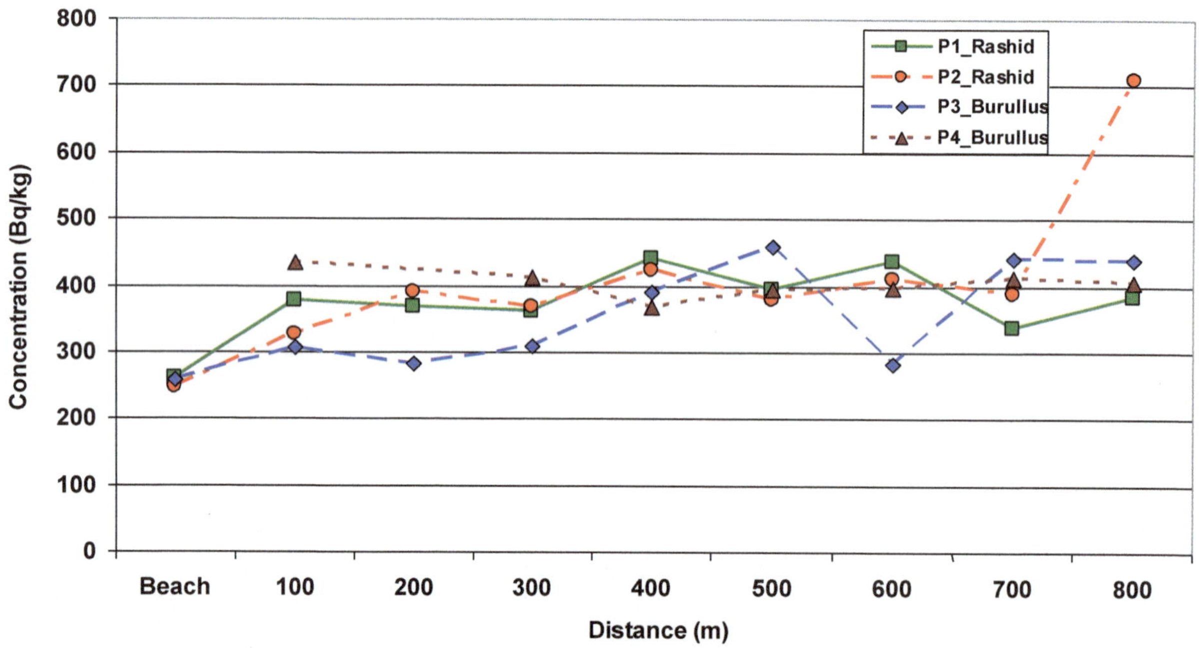

Figure 10. ^{40}K concentration (Bq/kg) at Rashid and Burullus coastal sediments profiles during 2002.

profile and they found that the <0.2 mm fraction contains more than 50% of radioactive elements and heavy minerals (especially zircon). Heavy minerals were the main contributors to high levels of external radiation. Figure 10 illustrate the irregular distribution of ^{40}K at the four profiles. The maximum ^{40}K concentration value (709.10 Bq/kg) was detected at 800 m distance of P2 and minimum value (246.80 Bq/kg) was detected at the beach of P2.

Heavy minerals analysis

Total heavy and light minerals percentages in the profile sediments have been estimated and presented in Table 2

Table 2. Heavy minerals percentages in sediments of the beach, 100 and 800 m distances, along the four profiles situated at the west of Rashid estuary and Burullus outlet.

	Beach				100 (m)				800 (m)			
	P1	P2	P3	P4	P1	P2	P3	P4	P1	P2	P3	P4
HM (%)	4.77	13.79	2.63	0.80	12.76	14.17	38.16	11.75	7.50	22.16	13.72	14.20
LM (%)	95.23	86.21	97.37	99.21	87.24	85.83	61.84	88.25	92.50	77.84	86.28	85.80
IF	0.05	0.16	0.03	0.01	0.15	0.17	0.62	0.13	0.08	0.28	0.16	0.17
ZTR	6.67	17.79	39.32	8.44	1.59	1.87	2.01	5.23	14.67	20.16	0.75	3.35
Heavy minerals												
Opaques	8.00	22.38	22.37	17.20	6.80	23.75	19.07	4.01	23.37	26.70	2.93	18.68
Hornblende	20.00	23.81	14.74	17.74	10.65	15.20	17.44	30.77	15.68	14.49	30.40	31.52
Tremolite + Actinolite	0.00	0.00	0.00	0.00	1.48	0.95	0.00	1.00	0.30	0.00	0.00	0.00
Augite	53.33	33.81	25.26	34.41	31.66	27.79	30.00	32.44	32.84	32.67	23.81	27.63
Epidote	4.00	1.90	0.53	16.13	19.53	18.29	23.26	17.73	3.85	9.09	35.90	12.45
Garnet	0.80	1.43	2.37	0.54	1.48	0.48	0.23	0.00	0.00	0.28	0.73	0.39
Zircon	4.53	11.43	29.47	6.99	0.59	1.19	0.93	4.68	10.06	13.64	0.37	2.33
Rutile	0.27	0.95	0.53	0.00	0.00	0.24	0.70	0.33	0.89	0.85	0.37	0.00
Tourmaline	1.33	1.43	0.53	0.00	0.89	0.00	0.00	0.00	0.30	0.28	0.00	0.39
Apatite	0.00	0.00	0.00	0.54	0.00	0.71	0.00	0.00	0.00	0.28	0.73	0.39
Kyanite	0.80	0.48	0.00	0.00	0.59	0.95	0.00	0.00	1.18	0.28	0.73	0.39
Monazite	0.53	1.43	4.21	1.08	0.30	0.71	0.70	1.00	1.18	0.00	0.37	5.84
Staurolite	0.53	0.95	0.00	1.61	0.00	0.00	0.00	0.00	0.59	0.00	0.00	0.00
Biotite + Chlorite	5.87	0.00	0.00	3.76	26.04	9.74	7.67	8.03	9.76	1.42	3.66	0.00
Total	100	100	100	100	100	100	100	100	100	100	100	100

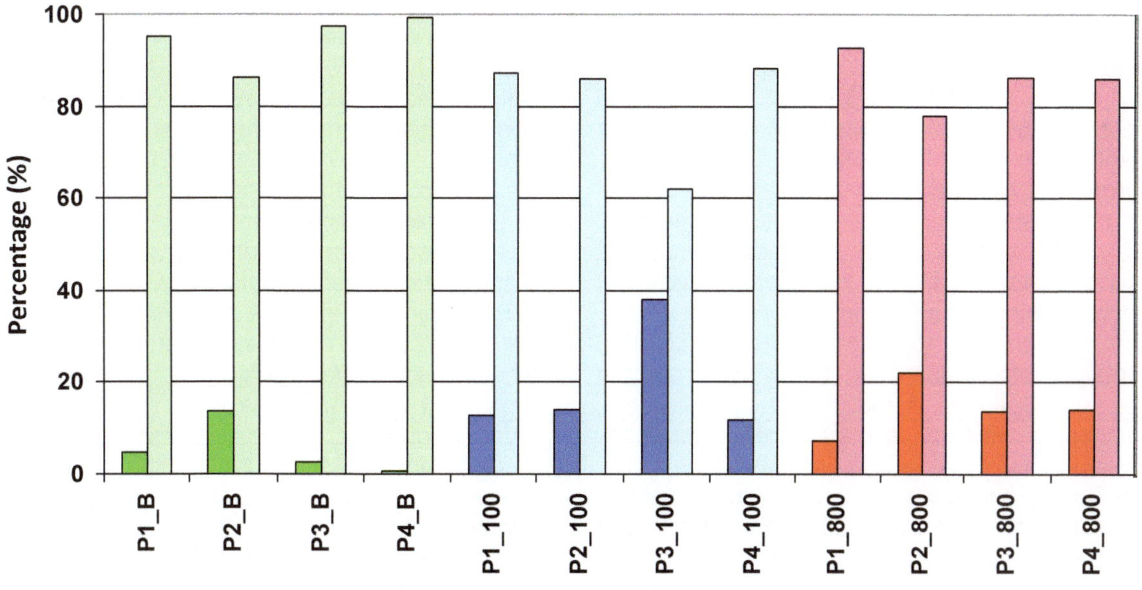

Figure 11. Percentages of heavy minerals (dark color) and light minerals (light color) in coastal sediments at beach, 100 and 800 m distances in four different profiles west of Rashid estuary and Burullus outlet during 2002.

and Figure 11. Index figure (IF) was calculated to recognize the higher percentage of the heavy minerals than the light minerals. Relatively high IF (0.16) was found at beach of P3 compared with the other calculated values of the others beaches under investigation. At 100 m distances, relatively higher IF value (0.62) at P3 which

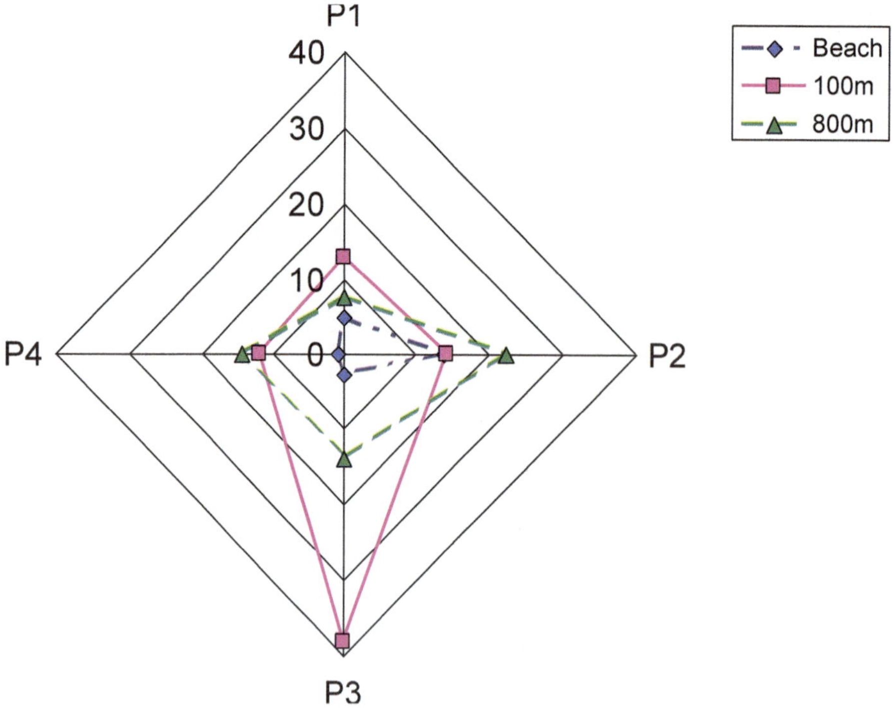

Figure 12. Assembled heavy minerals percents of the four coastal sediment profiles at beach, 100 and 800 m west of Rashid and El-Burullus coastal areas.

indicated the higher percentage of heavy minerals. Both the beach of P2 and at 100 m distance of P3 were characterized by relatively higher ^{238}U and ^{232}Th series compared with the other profiles as shown in Figures 4, 5, 6, 7, 8 and 9. Also, the heavy minerals percentage (38.16%) at 100 m distance of P3 was higher than the beach of P2 (13.79%) and this variation was matched with their radioactive contents variation.

The recorded heavy mineral constituents of the coastal sediments composed of great variety of minerals. Microscopic study revealed that the heavy minerals assemblage consists of opaques, hornblende, tremolite, actinolite, augite, epidote, garnet, zircon, rutile, tourmaline, apatite, kyanite, monazite, staurolite, biotite and chlorite. The percentages of the identified heavy minerals are listed in Table 2. Opaques, hornblende, augite, epidote, biotite and chlorite were the dominant heavy minerals. Also, minor amounts of garnet, zircon, rutile, tourmaline, kyanite and monazite was detected. Tremolite, actinolite, apatite and staurolite were found at scarce amounts at some limited sites along the profiles. The scarcity of these minerals in the sedimentary deposits of the profiles indicated low abundance in the adjacent parent rocks.

Assembled of heavy minerals percents at the beach, 100 and 800 m distances of the four profiles under investigation are plotted together in Figure 12, taking into consideration that three beaches were subject to erosion (P1, P3 and P4) and one (P2) was under accretion condition. The beach site of P3 was subjected to severe erosion where high percentages heavy mineral residues were recorded. Figure 12 illustrated that beaches of the four profiles contain lesser amount of heavy minerals compare to 100 and 800 m distances. Large amount of heavy minerals was found at 100 m distance of the four profiles. This is could be due to the fact that, the 100 m distance acted as the first sink area of the dense and heavy sediment transported from the beach. The highest amount of heavy minerals was observed at 100 m distance of P3 which characterized the highest ^{238}U and ^{232}Th content. The high amount of radioactive materials in this site was mainly attributed to the presence of high amount of radiogenic heavy minerals.

The major contributors to enhance levels of radiation are monazite, and to a lesser extent zircon compared to other minerals (Deer et al., 1997; Mohanty et al., 2004). Significant concentrations of monazite and zircon were found throughout all the profiles. Their distributions at three distances of the four profiles under investigation are presented in Figure 13. Chlorite, biotite, tourmaline and apatite could also play a significant role in uranium and thorium distribution (Dahlkamp, 2009). Rutile showed similar behavior with zircon and tourmaline, for example, strong resistance to weathering and multiple sediment provenances (Mange and Morton, 2007). Rutile displayed only scattered minor concentrations in the study area. Monazite percent at 100 m distance of P3 was recognized as the highest value among the four profiles shown in Figure 13, which demonstrated with the higher values of ^{238}U and ^{232}Th. On the other hand, monazite at the beach sample of P2 showed the highest percentage

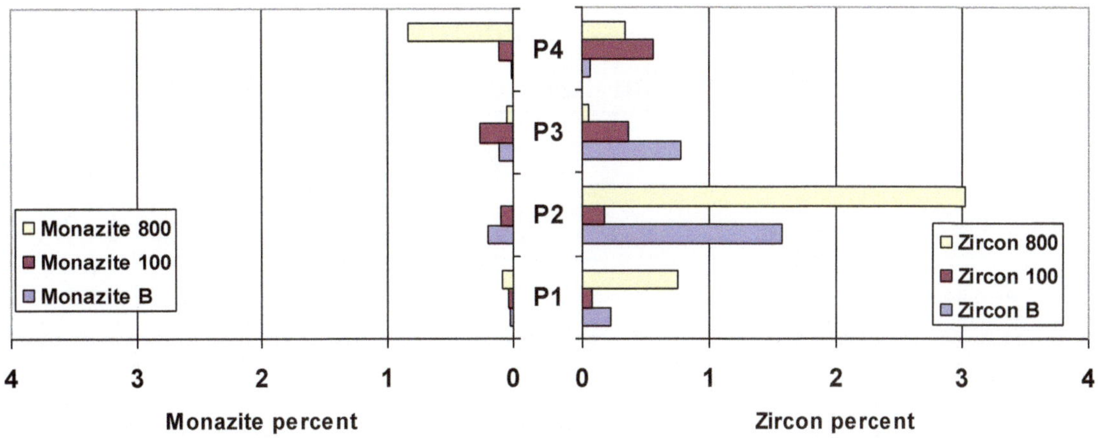

Figure 13. Monazite and zircon percent of heavy minerals detected at beach, 100 and 800 m sediments collected from different profiles at west of Rashid and El-Burullus coasts.

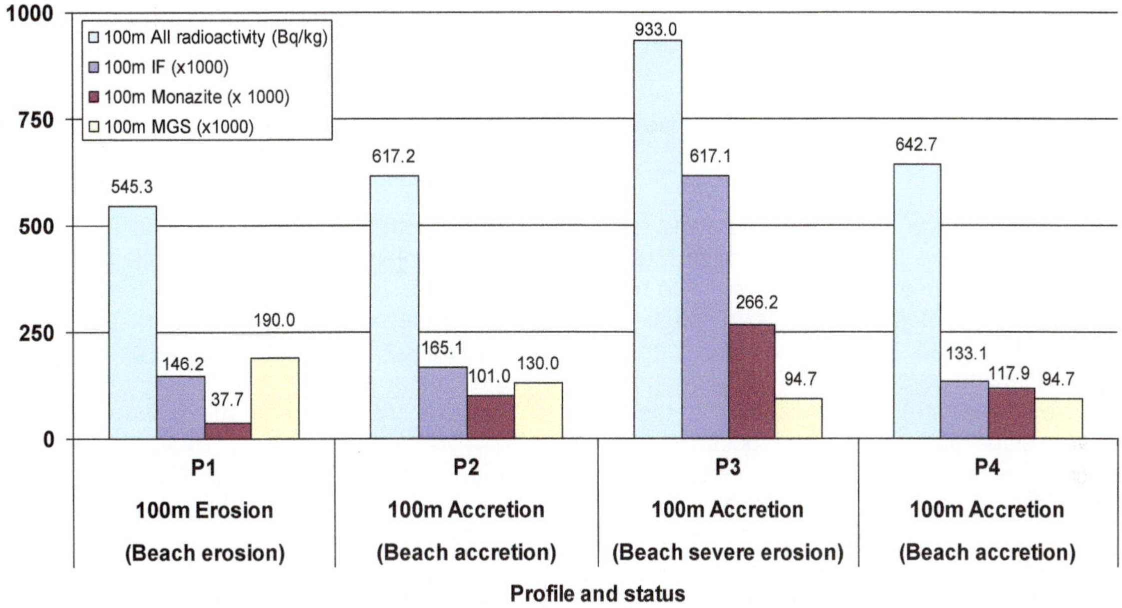

Figure 14. Comparative study of total radioactivity detected (Bq/kg), IF, Monazite percent and mean grain size (mm) in sediments collected at 100 m distance in the four profiles during 2002.

among the beaches and synchronized with the higher values of ^{238}U and ^{232}Th. Conversely, positive correlation (r = 0.7) was observed between the distribution of light minerals percents with the average concentrations of ^{40}K at 100 m distance in the four profiles under investigation. The average concentrations of ^{40}K were 306.5, 326.2, 377.3 and 434.6 Bq/kg, in coincidence with the percentage of the light minerals which are 61.84, 85.83, 87.24, and 88.25% at P3, P2, P1 and P4, respectively.

The higher values of ZTR index, which represented the ultrastable minerals (zircon, rutile and tourmaline) in the superficial sediments, were (39.32, 20.16, 17.79 and 14.67) at the beach of P3, 800 m distance of P2, at the beach of P2 and at 800 m distance of P1, respectively, This relatively high abundance attributed to the provenance. In contrast, the ZTR index values of the other sites, as listed in Table 2, were display relatively low (<10) in these coastal sediments. The highest value of ZTR (39.32) was found at the beach of P3 profile implying severe erosion. The maximum ZTR index of 39.32% obtained shows that the sediments are mineralogical immature.

Figure 14 shows the inter-comparison between the results of activity concentrations of all radioactive nuclides detected at 100 m distance of each profiles, together with heavy minerals percentages demonstrated

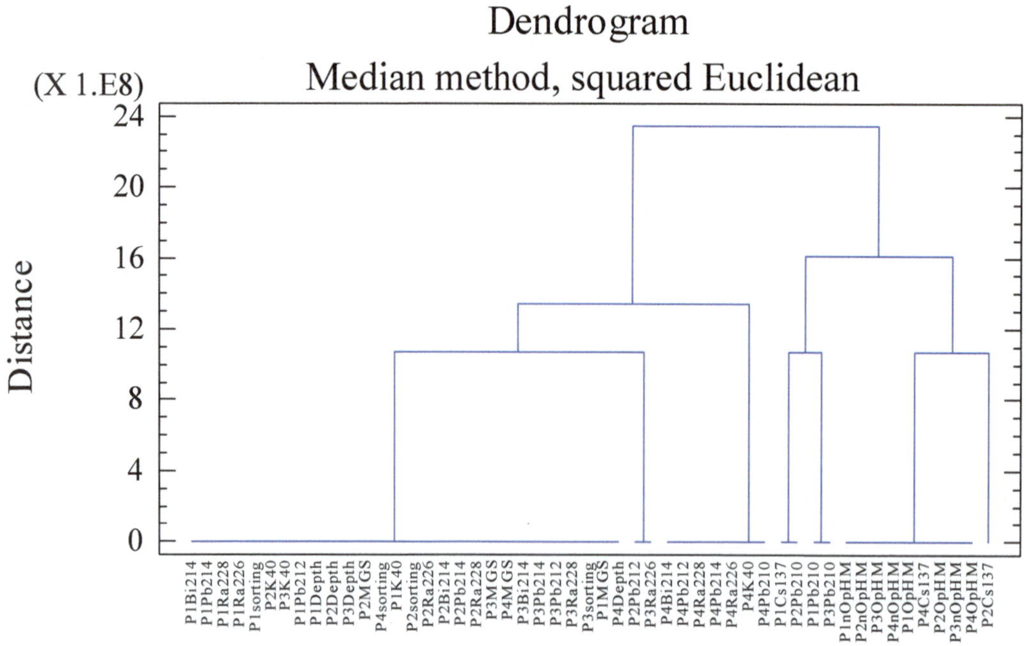

Figure 15. Dendrogram of cluster analysis of all the parameters under investigation. P1 to P4, the four profiles; nOpHM, non-opaque heavy minerals; OpHM, opaque heavy minerals; MGS, mean grain size.

as IF, monazite percentage, mean grain size and erosion and accretion status. The latter three factors multiplied with "1000" in order to show all in one figure. The highest radioactivity values (933 Bq/kg) was determined at 100 m distance of P3 with highest heavy minerals percent (IF = 0.62), highest monazite percent value (0.27%) and lowest mean grain size (0.095 mm). This indicates that there is a strong relationship between the occurrences of radioactive nuclides with the presence of heavy minerals, in particular monazite. Also, this pointed toward importance of the distribution of radiogenic heavy minerals along the coasts. This finding was confirmed by the work of Mohanty et al. (2004) that the highest concentrations of ^{232}Th and ^{238}U were found in monazite compared with the other heavy minerals. Greenfield et al. (1989) confirmed the relation of radiogenic heavy minerals with the transportation of beach sand. Moreover, the grain size distribution plays an important role in this relation. This is in consistent with the finding of Vassas et al. (2006) that across shore profile, the variability in actinides observed at the eastern part of Beauduc spit is mainly explained by variations in heavy and light mineral contents. Such variability can be accounted for by redistribution of the sand caused by erosion/deposition processes occurring in the eastern part of the spit. Severe erosion condition was observed at the beach of P3 and the all accumulated at the next site at 100 m distance which has accretion condition. On the other hand, at P1 the lowest radioactive nuclides (546.3 Bq/kg) are accomplished with low IF, the lowest monazite percent (0.038%) and the highest mean grain size (0.19 mm). Both, the beach and 100 m distance sites at P1 were subjected to erosion condition.

Cluster analysis was performed among all the parameters under investigation using Statgraphics Plus software. Figure 15 shows the dendrogram resulted from the cluster analysis. This procedure has created different clusters from the variables supplied. The clusters are groups of variables with similar characteristics. The first wide cluster is connected with the narrow second at the same distance to gathered all radioactive materials (except ^{210}Pb), mean grain size, depth and sorting values for P1, P2 and P3. Two small clusters are connected at the same distance for the ^{210}Pb of the three profiles P1, P2 and P3. The fourth profile (P4) parameters were clustered as one separate group. Heavy minerals either opaque of non-opaque are gathered together to form one group. ^{137}Cs is disturbed at different clusters. This is because of the large amount of missed data as under the detection limit.

Conclusion

An attempt was made to indicate the relationship between accretion and erosion beaches with the concentrations of different radionuclides and the weight percent of heavy minerals residue of the near shore profile sediments. The distribution of the detected U and Th decay products activities on the beaches of the Nile Delta coast was investigated in order to locate the main areas of enrichment. On the near shore profile, the variability in U and Th observed at the western side of Rashid and El-Burullus outlets can be mainly explained by the heavy and light minerals contents. Such variability is a consequence of the redistribution of the sand due to coastal processes such as erosion and accretion that take place in the study area or at any place in the world with the same situation. The slope of the profile plays an important role in the concentration of heavy minerals. Furthermore, the study of parameters such as ^{40}K activity, grain size, heavy minerals, monazite and zircon content in the near shore profile provides a better understanding of the concentration processes of radioactivity. At this location, the distribution pattern of the radionuclides along the beach profile is dependent mainly on the sorting of sand grains according to size and density and heavy mineral content (especially monazite). Among the beaches under investigation, relatively higher concentrations of ^{238}U and ^{232}Th decay series members were found at sites that have higher heavy minerals contents, and also, it was detected more in erosional beaches than in accretion ones. Finally, the cross analysis of mineralogical and radiometric data provides a useful tool for a better knowledge of the genesis and characteristics of eroding and accreting beaches.

ACKNOWLEDGEMENTS

The authors would like to thank Prof. Abd El-Moneim Badr, the head of Oceanography Department and the director of Abu Qir Research Station of Coastal Research Institute, National Water Research Center for his efforts and help during the determination of heavy minerals of the sediment samples.

REFERENCES

Alam NM, Chowdury MI, Kamal M, Ghose S, Islam NM, Mustafa MN, Miah MMH, Ansary MM (1999). The 226Ra, 232Th and 40K activities in beach sand minerals and beach soils of Cox's Bazar, Bangladesh. J. Environ. Radioact. 46:243.

Badr AA, Lotfy MF (1999). Tracing beach sand movement using fluorescent quartz along the Nile Delta promontories, Egypt. J. Coast. Res. 15:261-265.

Brai M, Bellia S, Hauser S, Puccio P, Rizzo S, Basile S, Marrale M (2006). Correlation of radioactivity measurements, air kerma rates and geological features of Sicily. Rad. Meas. 41:461-470.

Cascalho J, Fradique C (2007). The source and hydraulic sorting of heavy minerals on the northern Portuguese continental margin. In: Mange M.a. and Wright D.T. eds. Heavy minerals in use, Developments in sedimentology 58:75-110, Elsevier, Amsterdam.

Cochran JK, Krishnaswami S (1980). Radium, thorium, uranium, and 'Pb in deep-sea sediments and sediment pore waters from the North Equatorial Pacific. Am. J. Sci. 280:849-889.

Coelho FDS, Couceiro PRDC, Lopes AL, Fabris JD (2005). Iron oxides and monazite from sands of two beaches in Espirito Santo, Brazil. Quim. Nova 28:233.

Cole KH, Guinasso Jr. NL, Richardson MD, Johnson JW, Schink DR (1986). Uranium and thorium series isotopes in recent sediments of the Venezuela basin, Caribbean Sea. Mar. Geol. 68:167-185.

CoRI (2001). Report on analysis of beach profiles from Rosetta to El-Burullus outlet during the period from April/May to Sept/Oct. 1998, Ministry of Water Resources and Irrigation, National Water Research Center, Coastal Research Institute.

CoRI (2002). Technical report on the monitoring of shore changes of Nile Delta along North Egyptian coast, progress report No. 1 for activity during the period from July to December 2002, Ministry of Water Resources and Irrigation, National Water Research Center, Coastal Research Institute.

Cowell PJ, Stive MJF, Niedoroda AW, Vriend De HJ, Swift DJP, Kaminsky GM, Capobianco M (2003). The coastal tract. Part 1: Aconceptual approach to aggregated modelling of low-order coastal change. J. Coastal Res. 19:812-827.

Dahlkamp FJ (2009). Uranium deposits of the world:Asia, Springer reference, Springer-Verlag Berlin Heidelberg.

De Meijer RJ, Lesscher HME, Schuiling RD, Elburg RD (1990). Estimate of the Heavy Mineral Content in Sand and its Provenance by Radiometric Methods. Nucl. Geophys. 4(4):455-460.

De Meijer RJ, Donoghue JF (1995). Radiometric fingerprinting of sediments on the Dutch, German and Danish coasts. Quat. Int. 26:43-47.

De Meijer RJ (1998). Heavy minerals:from Edelstein to Einstein. J. Geochem. Explor. 62:81.

De Meijer RJ, James IR, Jennings PJ, Koeyers JE (2001). Cluster analysis of radionuclide concentrations in beach sand. Appl. Radiat. Isot. 54:535-542.

Deer WA, Howie RA, Zussman J (1997). Rock-forming minerals:Orthosilicates, Geological Society, London. P. 918.

Dill HG (1995). Heavy mineral response to the progradation of an alluvial fan: Implications concerning unroofing of source area, chemical weathering and palaeorelief (Upper Cretaceous Parkstein fan complex, SE Germany). Sediment. Geol. 95:39-56.

El-Fishawi NM, Badr AA (1991). Mineralogical characteristics of the western Nile Delta coast sediments. Act Mineral.-Petrog. Szeged, XXXII:65-75.

El-Gamal A, Saleh I, Nasr S, Naim M (2004). Radiological assessment of the Egyptian Mediterranean coast, International Conference on Isotopes in Environmental Studies- Aquatic Forum 2004, Monte-Carlo, Monaco 25-29 October 2004, IAEA-CN-118/31P, PP. 396-397.

El-Khatib AM, Denton M, El-sabrouti MA (1993). Isotope identification of Egyptian black sands from Rossetta using neutron activation analysis. J. Fiz. Mal. 14:1-10.

El-Kolfat AI (2005). Coastal characteristics and behaviour for eastern Nile Delta region, Egypt (Burullus- Port Said). Alex. Eng. J. 44(6):899-910.

El-Naggar AM (1990). Environmental radioactivity; Sources and effects on man, Regional. Symp. Environ. Stud. (UNARC), Alexandria University, M. El-Raey (edit).

Frihy OE (2007). The Nile delta:Processes of heavy mineral sorting and depositional patterns, In:Mange M.A. & Wright D.T. eds. Heavy Minerals in use, developments in sedimentology 58, 49-74. Elsevier, Amsterdam.

Frihy OE, Dewidar KM (2003). Patterns of erosion/sedimentation, heavy mineral concentration and grain size to interpret boundaries of littoral sub-cells of the Nile Delta, Egypt. Mar. Geol. 199:27–34.

Frihy OE, Komar PD (1993). Long term shoreline changes and the concentration of heavy minerals in beach sand of the Nile Delta. Egypt. Mar. Geol. 115:253.

Frihy OE, Lotfy MF, Komar PD (1995). Spatial variations in heavy minerals and patterns of sediment sorting along the Nile Delta. Egypt. Sediment. Geol. 97:33.

Garzanti E, Andò S (2007). Plate tectonics and heavy minerals suites of modern sands, In: Mange M.A. and Wright D.T. eds. Heavy minerals in use, Developments in sedimentology Elsevier, Amsterdam. 58:741-763.

Greenfield MB, de Meijer RJ, Put LW, Wiersma J, Donoghue JF (1989). Monitoring beach sand transport by use of radiogenic heavy minerals. Nucl. Geophys. 3:231-244.

Hussein AEM (2011). Successive uranium and thorium adsorption from Egyptian monazite by solvent impregnated foam. J. Radioanal. Nucl. Chem. 289:321-329.

Ingle JC (1966). Developments in sedimentology 5, the movement of beach sand, Elsevier Publishing Company, Amsterdam.

Johanson KJ, Bergström R (1989). Radiocaesium from Chernobyl in

Swedish Moose, Environ. Poll. 61: 249-260.
Joshua EO, Oyebanjo OA (2009). Distribution of Heavy Minerals in Sediments of Osun River Basin Southwestern Nigeria. Res. J. Environ. Earth Sci. 1(2):74-80.
Kaiser MFM, Frihy OE (2009). Validity of the equilibrium beach profiles:Nile Delta Coastal Zone, Egypt. Geomorphology 107:25–31.
Kaiser MFM (2004). Monitoring and modelling the impact of engineering structures on coastline change, Nile Delta, Egypt, Ph.D. Thesis, University of Reading, U.K.
Komar PD (2007). The entrainment transport and sorting of heavy minerals by waves and currents, In Mange, M.A. and Wright, D.T. eds. Heavy minerals in use, developments of sedimentology, 58:3-48. Elsevier, Amsterdam.
Kudrass HR (1987). Sedimentary models to estimate the heavy minerals potential of self sediments, In: Teleki P. G. Dobson M.R. Moore J.R. and Stackelberg U.V. eds. Marine Minerals, D. Reidel Publishing Company, Dordrecht pp. 39-56.
Li WQ, Guinasso Jr NL, Cole KH, Richardson MD, Johnson JW, Schink DE (1985). Radionuclides as indicators of sedimentary processes in abyssal Caribbean sediments. Mar. Geol. 68:187-204.
Mange A, Maria A, Heinz-Maurer FW (1992). Heavy minerals in colour. Chapmam and hall, London, p. 230.
Mange MA, Morton AC (2007). Geochemistry of heavy minerals, In:M.a. Mange and D.T. Wright (Eds.), Heavy minerals in use. Developments in Sedimentology, Elsevier; Amsterdam, 58:345-391.
Miller RL (1976). Role of vortices in surface zone prediction: Sedimentation and wave forces, In:Davis, R.A. Ethington, R.L. (Eds.), Beach and Nearshore Sedimentation. Society of Economic Paleontologists and Mineralogists, Tulsa, Oklahoma. Spec. Publ. 24:92-114.
Mohanty AK, Senguptaa D, Dasb SK, Vijayanc V, Sahab SK (2004). Natural radioactivity in the newly discovered high background radiation area on the eastern coast of Orissa, India. Rad. Meas. 38:153-165.
Nafaa MG (1995). Wave climate along the Nile Delta coast. J. Coast. Res. 11:219-229.
Reguigui N (2006). Gamma Ray Spectrometry, Practical Information. http://www.cnstn.rnrt.tn/afra-ict/NAT/gamma/html/Gamma%20Spec%20V1.pdf.
Saleh I, El-Gamal A, Nasr S, Naim M (2004). Spatial and temporal variations of uranium and thorium series along the Egyptian Mediterranean coast, International Conference on Isotopes in Environmental Studies- Aquatic Forum 2004, Monte-Carlo, Monaco 25-29 October 2004, IAEA-CN-118/113P, PP. 550-551.
Seddeek MK, Badran HM, Sharshar T, Elnimr T (2005). Characteristics, spatial distribution and vertical profile of gamma-ray emitting radionuclides in the coastal environment of North Sinai. J. Environ. Radioact. 84:21.
Smith S, Abdel-Kader A (1988). Coastal erosion along the Egyptian Delta. J. Coast. Res. 4:245-255.
Suman DO, Bacon MP (1989). Variations in Holocene sedimentation in the North American basin determined from 23"Th measurements. Deep-Sea Res. 36:869-878.
USEPA (2011). Radiation protection, about TENORM, U.S. Environmental Protection Agency. http://www.epa.gov/radiation/tenorm/about.html.
Vassas C, Pourcelot L, Vella C, Carpe´na J, Pupin JP, Bouisset P, Guillot L (2006). Mechanisms of enrichment of natural radioactivity along the beaches of the Camargue, France. J. Environ. Radioact. 91:146-159.
Veigaa R, Sanchesa N, Anjosa RM, Macarioa K, Bastosa J, Iguatemya M, Aguiarb JG, Santosb AMA, Mosqueraa B, Carvalhoa C, Baptista Filhoa M, Umisedoc NK (2006). Measurement of natural radioactivity in Brazilian beach sands. Rad. Meas. 41:189-196.
Zhiyong L, Shaoming P, Xuying L, Jianhua G (2010). Distribution of ^{137}Cs and ^{210}Pb in sediments of tidal flats in north Jiangsu Province. J. Geogr. Sci. 20(1):91-108.

Permissions

All chapters in this book were first published in JOMS, by Academic Journals; hereby published with permission under the Creative Commons Attribution License or equivalent. Every chapter published in this book has been scrutinized by our experts. Their significance has been extensively debated. The topics covered herein carry significant findings which will fuel the growth of the discipline. They may even be implemented as practical applications or may be referred to as a beginning point for another development.

The contributors of this book come from diverse backgrounds, making this book a truly international effort. This book will bring forth new frontiers with its revolutionizing research information and detailed analysis of the nascent developments around the world.

We would like to thank all the contributing authors for lending their expertise to make the book truly unique. They have played a crucial role in the development of this book. Without their invaluable contributions this book wouldn't have been possible. They have made vital efforts to compile up to date information on the varied aspects of this subject to make this book a valuable addition to the collection of many professionals and students.

This book was conceptualized with the vision of imparting up-to-date information and advanced data in this field. To ensure the same, a matchless editorial board was set up. Every individual on the board went through rigorous rounds of assessment to prove their worth. After which they invested a large part of their time researching and compiling the most relevant data for our readers.

The editorial board has been involved in producing this book since its inception. They have spent rigorous hours researching and exploring the diverse topics which have resulted in the successful publishing of this book. They have passed on their knowledge of decades through this book. To expedite this challenging task, the publisher supported the team at every step. A small team of assistant editors was also appointed to further simplify the editing procedure and attain best results for the readers.

Apart from the editorial board, the designing team has also invested a significant amount of their time in understanding the subject and creating the most relevant covers. They scrutinized every image to scout for the most suitable representation of the subject and create an appropriate cover for the book.

The publishing team has been an ardent support to the editorial, designing and production team. Their endless efforts to recruit the best for this project, has resulted in the accomplishment of this book. They are a veteran in the field of academics and their pool of knowledge is as vast as their experience in printing. Their expertise and guidance has proved useful at every step. Their uncompromising quality standards have made this book an exceptional effort. Their encouragement from time to time has been an inspiration for everyone.

The publisher and the editorial board hope that this book will prove to be a valuable piece of knowledge for researchers, students, practitioners and scholars across the globe.

List of Contributors

V Suneetha
School of Bio Sciences and Technology (SBST), Vellore, Tamil Nadu, India

Karthick Raj
Vellore Institute of Technology University (VIT), Vellore-632014, Tamil Nadu, India

K. Prathusha
Vellore Institute of Technology University (VIT), Vellore-632014, Tamil Nadu, India

Behrouz Behrouzi-Rad
Islamic Azad University, Science and Research Branch, Kuzestan, Iran

Rahman Rasekhi
Shaid Chamran University, Ahwaz, Iran

Sima Momeni
Islamic Azad University, Science and Research Branch, Kuzestan, Iran

Fatemeh Einollahi Peer
Institute Research of Hamoon International Wetland, Zabol University, Iran

Alireza Safahieh
Department of Marine Biology, Faculty of Oceanic and Marine Science, Khoramshar University of Marine Science and Technology, Iran

Hélio Augusto Alves Fracasso
Department of Hydrobiology, Federal University of Sao Carlos, Sao Paulo, Brazil

Joaquim Olinto Branco
Centro de Ensino em Ciências Tecnológicas da Terra e do Mar, Santa Catarina, Brazil

Joanna Burger
Division of Life Sciences, Rutgers University, New Jersey, USA

Luís Fabio Silveira
Museu de Zoologia da USP, CEP 04263-000 - São Paulo - SP - Brazil

José Roberto Verani
Department of Hydrobiology, Federal University of Sao Carlos, Sao Paulo, Brazil

Weiwei Ding
Key Laboratory of Submarine Geoscience of the State Oceanic Administration, Hangzhou 310012, China
Second Institute of Oceanography, State Oceanic Administration, Hangzhou 310012, China

Jiabiao Li
Key Laboratory of Submarine Geoscience of the State Oceanic Administration, Hangzhou 310012, China
Second Institute of Oceanography, State Oceanic Administration, Hangzhou 310012, China

Xiqiu Han
Key Laboratory of Submarine Geoscience of the State Oceanic Administration, Hangzhou 310012, China
Second Institute of Oceanography, State Oceanic Administration, Hangzhou 310012, China

Erwin Suess
Leibniz-Institute for Marine Sciences, 24148 Kiel, Germany

Yongyang Huang
Guangzhou Marine Geological Survey, Guangzhou, 510075, China

Xuelin Qiu
Key Laboratory of Marginal Sea Geology, South China Sea Institute of Oceanology, Chinese Academy of - Sciences, Guangzhou, 510301, China

Mingbi Li
Key Laboratory of Submarine Geoscience of the State Oceanic Administration, Hangzhou 310012, China
Second Institute of Oceanography, State Oceanic Administration, Hangzhou 310012, China

Zengan Deng
Key Lab of Marine Environmental Information Technology, SOA, Tianjin 300171, China
National Marine Data and Information Service, Tianjin 300171, China

Guijun Han
Key Lab of Marine Environmental Information Technology, SOA, Tianjin 300171, China
National Marine Data and Information Service, Tianjin 300171, China

Xuefeng Zhang
Key Lab of Marine Environmental Information Technology, SOA, Tianjin 300171, China
National Marine Data and Information Service, Tianjin 300171, China

Kejian Wu
Ocean University of China, Qingdao 266100, China

Jameel M. Dhabab
Department of Chemistry, College Of Science, Al-Mustunsiyria University, Baghdad, Iraq

Eric O. Odada
Department of Earth Sciences, Marine Geology and Oceanography Unit, University of Nairob, P. O. Box 30197-00100 Nairobi, Kenya

Jesús S. Troncoso
Departamento de Ecología y Biología Animal, Facultad de Ciencias del Mar, Universidad de Vigo, Campus Lagoas Marcosende, 36310 Vigo, Spain

Jeffrey A. Sibaja-Cordero
Departamento de Ecología y Biología Animal, Facultad de Ciencias del Mar, Universidad de Vigo, Campus Lagoas Marcosende, 36310 Vigo, Spain
Centro de Investigación en Ciencias del Mar y Limnología (CIMAR), Universidad de Costa Rica, San Pedro,11501-2060 San José, Costa Rica

G. W Siqueira
Laboratório de Planejamento e Desenvolvimento de Fármacos - LPDF, Campus Universitário do Guamá. Av. Augusto Corrêa n. 1, Universidade Federal do Pará, Belém, PA, # 66075-100, Brasil

F. M. Aprile
Laboratório de Estudos de Ecossistemas Amazônicos - LEEA, Universidade Federal do Oeste do Pará, Campus Rondon. Av. Marechal Rondon s/n Caranazal, Santarém, PA, # 68040-070, Brasil

JianHong Luo
Department of Chemical Engineering, Sichuan University, Chengdu, Sichuan, 610065,People's Republic of China

Jun Li
Department of Chemical Engineering, Sichuan University, Chengdu, Sichuan, 610065,People's Republic of China

Yang Jin
Department of Chemical Engineering, Sichuan University, Chengdu, Sichuan, 610065,People's Republic of China

DongSheng Zheng
Department of Chemical Engineering, Sichuan University, Chengdu, Sichuan, 610065,People's Republic of China

Debasish Mahapatro
Chilika Development Authority, Bhubaneswar, Odisha, India

R. C. Panigrahy
P.G. Department of Marine Sciences, Berhampur University, Berhampur -760007, Odisha, India

S. Naik
P.G. Department of Marine Sciences, Berhampur University, Berhampur -760007, Odisha, India

S. K. Pati
Zoological Survey of India, WRC, Pune, India

R. N. Samal
Chilika Development Authority, Bhubaneswar, Odisha, India

Fabio Aprile
Laboratório de Estudos de Ecossistemas Amazônicos - LEEA, Universidade Federal do Oeste do Pará, Campus Rondon. Av. Marechal Rondon s/n Caranazal, Santarém, PA, # 68040-070, Brasil

Gilmar W. Siqueira
Laboratório de Planejamento e Desenvolvimento de Fármacos - LPDF, Universidade Federal do Pará, Campus Universitário do Guamá. Av. Augusto Corrêa n. 1, Belém, PA, # 66075-100, Brasil

Tamal Mondal
Zoological Survey of India, Andaman and Nicobar Regional Centre, National Coral Reef Research Institute, Haddo,Port Blair-744 102, Andaman and Nicobar Islands, India

C. Raghunathan
Zoological Survey of India, Andaman and Nicobar Regional Centre, National Coral Reef Research Institute, Haddo,Port Blair-744 102, Andaman and Nicobar Islands, India

Ramakrishna
Zoological Survey of India, Prani Vigyan Bhawan, M-Block, New Alipore, Kolkata-700 053, India

T. Panda
Department of Botany, S.N.College, Rajkanika, Kendrapara- 754220, Orissa, India

P. K. Pani
Department of Botany, Agarpara College, Agarpada, Bhadrak, Orissa, India

R. B. Mohanty
Department of Botany, N.C.Autonomous College, Jajpur, Orissa, India

Kaushik Sasmal
Department of Ocean Engineering and Naval Architecture,Indian Institute of Tsechnology Kharagpur, Kharagpur - 721302, India

Hari Warrior
Department of Ocean Engineering and Naval Architecture,Indian Institute of Tsechnology Kharagpur, Kharagpur - 721302, India

Subhendu Maity
Department of Ocean Engineering and Naval Architecture,Indian Institute of Tsechnology Kharagpur, Kharagpur - 721302, India

Sourav Sil
Centre for Oceans, Rivers, Atmosphere and Land Sciences (CORAL), Indian Institute of Technology, Kharagpur, India

Arun Chakraborty
Centre for Oceans, Rivers, Atmosphere and Land Sciences (CORAL), Indian Institute of Technology, Kharagpur, India

David Hoyos
Department of Applied Economics III (Econometrics and Statistics), University of the Basque Country, E48015 Bilbao, Spain

Pere Riera
Institute of Environmental Science and Technology, Autonomous University of Barcelona, Campus UAB, E08193 Bellaterra, Spain

Javier Fernández-Macho
Department of Applied Economics III (Econometrics and Statistics), Institute for Public Economics, University of the Basque Country E48015 Bilbao, Spain

Carmen Gallastegui
Department of Foundations of Economic Analysis I, Institute for Public Economics, University of the Basque Country, E48015 Bilbao, Spain

Dolores Garcia
Departament d'Economia Aplicada, Universitat de les Illes Balears, E-07122 Palma de Mallorca, Spain

P. Kasinatha Pandian
Department of Civil Engineering, Tagore Engineering College, Chennai 600048, India

Osalusi Emmanuel
International Centre for Island Technology, Heriot-Watt University (Orkney Campus), Stromness, Orkney, Scotland,KW16 3AW, United Kingdom

J. P. Ruscoe
International Centre for Island Technology, Heriot-Watt University (Orkney Campus), Stromness, Orkney, Scotland,KW16 3AW, United Kingdom

J. C. Side
International Centre for Island Technology, Heriot-Watt University (Orkney Campus), Stromness, Orkney, Scotland,KW16 3AW, United Kingdom

R. E. Harris
International Centre for Island Technology, Heriot-Watt University (Orkney Campus), Stromness, Orkney, Scotland,KW16 3AW, United Kingdom

S. A. Kerr
International Centre for Island Technology, Heriot-Watt University (Orkney Campus), Stromness, Orkney, Scotland,KW16 3AW, United Kingdom

C. R. Bullen
International Centre for Island Technology, Heriot-Watt University (Orkney Campus), Stromness, Orkney, Scotland,KW16 3AW, United Kingdom

Maria Jayed
Laboratory of Chemistry, National Institute of Fischeries, 2 Rue Tiznit , P. O. Box 20000, Casablanca, Morocco

Abdelghani Chafik
Laboratory of Chemistry, National Institute of Fischeries, 2 Rue Tiznit , P. O. Box 20000, Casablanca, Morocco

Samir Benbrahim
Laboratory of Chemistry, National Institute of Fischeries, 2 Rue Tiznit , P. O. Box 20000, Casablanca, Morocco

Carlos Vale
IPIMAR - National Institute of Biological Resources, Av. Brasília, 1449-006 Lisboa, Portugal

Salem Bakkas
Laboratory of Organic Chemistry, Faculty of Sciences, El Jadida University Chouaïb Doukkali, 2, Avenue Med Ben Larbi Alaoui, El Jadida, Morocco

Patrícia Pereira
IPIMAR - National Institute of Biological Resources, Av. Brasília, 1449-006 Lisboa, Portugal

Ana Maria Ferreira
IPIMAR - National Institute of Biological Resources, Av. Brasília, 1449-006 Lisboa, Portugal

M. Pravinkumar
Centre of Advanced Study in Marine Biology, Faculty of Marine Sciences, Annamalai University, Parangipettai - 608 502, Tamilnadu, India

P. Murugesan
Centre of Advanced Study in Marine Biology, Faculty of Marine Sciences, Annamalai University, Parangipettai - 608 502, Tamilnadu, India

R. Krishna Prakash
Centre of Advanced Study in Marine Biology, Faculty of Marine Sciences, Annamalai University, Parangipettai - 608 502, Tamilnadu, India

V. Elumalai
Centre of Advanced Study in Marine Biology, Faculty of Marine Sciences, Annamalai University, Parangipettai - 608 502, Tamilnadu, India

C. Viswanathan
Centre of Advanced Study in Marine Biology, Faculty of Marine Sciences, Annamalai University, Parangipettai - 608 502, Tamilnadu, India

S. M. Raffi
Centre of Advanced Study in Marine Biology, Faculty of Marine Sciences, Annamalai University, Parangipettai - 608 502, Tamilnadu, India

A Singh
Central Soil and Materials Research Station, Ministry of Water Resources, India

A. K Gosain
Department of Civil Engineering, Indian Institute of Technology, Delhi, India

J. Selvin Pitchaikani
Centre for Marine Science and Technology, Manonmaniam Sundaranar University, Rajakkamangalam, Tamil Nadu, India
Present address Institute of Environmental Studies and Wetland Management, DD-24, Sector-I, Salt Lake City, Kolkata, West Bengal, India

A. P. Lipton
Vizhinjam Research Centre, Central Marine Fisheries Research Institute, Vizhinjam - 629 251, Kerala, India

Amor Ben Garali
Faculté des Sciences de Bizerte - Laboratoire de sédimentologie et géochimie marine - 7021 Zarzouna - Bizerte, Tunisie

Mohamed Ouakad
Faculté des Sciences de Bizerte - Laboratoire de sédimentologie et géochimie marine - 7021 Zarzouna - Bizerte, Tunisie

Moncef Gueddari
Faculté des Sciences de Tunis, Université El Manar. 1060 Tunis. Tunisie

Larry Awosika
Nigerian Institute for Oceanography and Marine Research, P. M. B. 12729 Victoria Island, Lagos Nigeria

Regina Folorunsho
Nigerian Institute for Oceanography and Marine Research, P. M. B. 12729 Victoria Island, Lagos Nigeria

Victor Imovbore
Environmental Resources Management Limited (ERML) Lagos, Nigeria

Lynda E. OKON
Physical Oceanography Unit, Institute of Oceanography, University of Calabar, P. M. B 1115, Calabar, Nigeria

F. Emile ASUQUO
Physical Oceanography Unit, Institute of Oceanography, University of Calabar, P. M. B 1115, Calabar, Nigeria

Zeinab Hanaee Kashani
Department of Fisheries, Gorgan University of Agricultural Sciences and Natural Resources, Gorgan, Iran

Mohamad Reza Imanpoor
Department of Fisheries, Gorgan University of Agricultural Sciences and Natural Resources, Gorgan, Iran

Ali Shabani
Department of Fisheries, Gorgan University of Agricultural Sciences and Natural Resources, Gorgan, Iran

Saeed Gorgin
Department of Fisheries, Gorgan University of Agricultural Sciences and Natural Resources, Gorgan, Iran

Ayman A. El-Gamal
Marine Geology Department, Coastal Research Institute, National Water Research Center, 15, St. Elpharanaa, Elshalalat, Postal Code 21514, Alexandria, Egypt

Ibrahim H. Saleh
Environmental Studies Department, Institute of Graduate Studies and Research, Alexandria University, P. O. Box 832, El-Shatby, Alexandria, Egypt